STRANGENESS NUCLEAR PHYSICS

STRANGENESS NUCLEAR PHYSICS

Proceedings of the APCTP Workshop (SNP'99)

Seoul National University, Seoul, Korea
19–22 February 1999

Edited by

Il-Tong Cheon
Yonsei University, Korea

Seung-Woo Hong
Sungkyunkwan University, Korea

Toshio Motoba
Osaka Electro-Communication University, Japan

World Scientific
Singapore • New Jersey • London • Hong Kong

Published by

World Scientific Publishing Co. Pte. Ltd.

P O Box 128, Farrer Road, Singapore 912805

USA office: Suite 1B, 1060 Main Street, River Edge, NJ 07661

UK office: 57 Shelton Street, Covent Garden, London WC2H 9HE

British Library Cataloguing-in-Publication Data
A catalogue record for this book is available from the British Library.

Proceedings of the APCTP Workshop (SNP '99)
STRANGENESS NUCLEAR PHYSICS

ISBN 981-02-4363-4

Printed in Singapore by Uto-Print

WORKSHOP ORGANIZATION

International Advisory Committee :

T. Bressani (INFN, Torino)
R.E. Chrien (Brookhaven N.L.)
A. Faessler (Univ. Tübingen)
B.F. Gibson (Los Alamos N.L.)
K. Ikeda (Niigata Univ.)
B.T. Kim (Sung Kyun Kwan U.)
D.J. Millener (Brookhaven N.L.)
S. Nagamiya (JHF, KEK)
C. Rim (APCTP, Seoul)
K. Yazaki (Univ. Tokyo)
Y.M. Cho (APCTP, Seoul)
H. Ejiri (RCNP, Osaka Univ.)
A. Gal (Hebrew Univ.,Jerusalem)
E. Hungerford (Univ. Houston)
N. Isgur (T. Jefferson N.A.L.)
T.T.S. Kuo (SUNY and Taiwan)
D.P. Min (Seoul National Univ.)
E. Oset (Univ. Valencia)
H.S. Song (Seoul National Univ.)
Z.Y. Zhang (IHEP, Beijing)

Organizing Committee :

Y. Akaishi (KEK-Tanashi, Tokyo 188-8501)
H.C. Bhang (Seoul National University, Seoul 151-742)
I.T. Cheon* (Yonsei University, Seoul 120-749) **Co-chairman*
K.-I. Imai (Kyoto University, Kyoto 606-8502)
T. Motoba* (Osaka E-C University, Osaka 572-8530) **Co-chairman*
M. Oka (Tokyo Institute of Technology, Tokyo 152-8551)
K.S. Sim (Korea University, Seoul 136-701)

Scientific Secretary :

S.W. Hong (Sung Kyun Kwan University, Suwon 440-746)
Y.D. Kim (Sejong University, Seoul 143-747)

Workshop site :

Sangsan Mathematical Science Building
Seoul National University, Seoul 151-742, Republic of Korea

Host institute :

Asia Pacific Center for Theoretical Physics
Seoul 130-012, Republic of Korea

Partial financial support :

Grant-in-Aid for Scientific Research of the Ministry of Education, Science, Sports, and Culture of Japan

The APCTP workshop on
Strangeness Nuclear Physics (SNP '99)
19–22 February, 1999 Seoul, Korea

PREFACE

This volume contains the papers presented orally in the APCTP International Workshop on Strangeness Nuclear Physics (SNP'99) which was approved and hosted by the Asia Pacific Center for Theoretical Physics in Seoul. The workshop was held at Seoul National University on February 19 - 22, 1999 with about 100 participants from 10 countries.

The unique role of strangeness in nuclear physics has recently attracted much attention from both theoretical and experimental viewpoints. This is not only due to a broad spectrum of possible hadron many-body sytems with strangeness, but also due to the fact that strangeness provides us with an opportunity of studying fundamental baryon-baryon interactions in a new perspective. Our knowledge on this subject has been widened as the scope of hypernuclear experiments is expanded from strangeness-exchange and associated production reactions to hypernuclear weak decays, γ decays, cascade hypernuclei, double-Λ events, electroproduction of strangeness, etc. This trend will be further accelerated by full operations of new laboratories such as TJLab, COSY, DAΦNE, JHF, MAMI, and others. Thus, hypernuclear and kaon physics is surely entering into a new stage. Strangeness is a keyword to a fruitful cross-field where atomic, nuclear and astrophysical concepts develop interdisciplinarily. Accordingly, in addition to 'traditional' nuclear physics approaches, we need various theoretical ingredients such as those taken from meson theories, those based on quark models, those inspired by QCD, etc.

The aim of the workshop was to discuss various aspects of these exciting and important topics and to get a prospect of this developing field of nuclear physics. At the same time, we made an effort to provide young generation with a chance to discuss these challenging subjects in a free and stimulating atmosphere and to join a new scientific collaboration. In this respect we believe that the workshop was successful in encouraging the internationalization and cooperation among the Asia-Pacific physics communities. We hope that these proceedings will help people to understand the present status and prospects of strangeness nuclear physics.

On behalf of the organizers we are very grateful to the APCTP, especially to the President C.N. Yang and Professors Y.M. Cho and C. Rim, for giving us this nice opportunity. The workshop was also financially supported by the Grant-in-Aid for Scientific Research from the Ministry of Education, Science, Sports, and Culture of Japan. We greatly thank the members of the international advisory committee for their valuable advice in the course of our organization. We would like to note that Seoul National University kindly provided us with a nice conference hall and the Ho-Am guest house. We are also indebted to the APCTP staffs, Ms. Ahran Joo and Ms. Eunjin Park, for their assistance in handling the administrative works.

Summer 1999
Editors Il-Tong Cheon
Seung-Woo Hong
Toshio Motoba

CONTENTS

Hyperon-Nucleon Interaction – II

OPENING ADDRESS

Distinguished guests, honourable guests, Ladies and gentlemen!

It is a great honour for me to address this opening message for the APCTP Workshop on Strangeness Nuclear Physics.

The most important element of scientific research activity is that scientists get together in one place to discuss recent development in extensive investigation of current scientific problems.

As you know, the APCTP(Asia Pacific Center for Theoretical Physics) started its scientific activity two and a half years ago. Since then, many activities have been supported in particle physics, nuclear physics and condensed matter physics. So far, we organized one workshop on meson, quark gluon plasma and astro-nuclear physics and one international conference on QCD and QGP. This time, our topics is strangeness in nuclear physics. This is one of the current hot-topics in nuclear physics these days. Strangeness requires roles of hyperons in various processes, i.e., kaon photoproduction, hyperon production, hypernuclei, nucleon spin structure, hyperon-nucleon interaction and so on. Discussion in the SU(2) frame is comparatively easy but strangeness nuclear physics should be described in the SU(3) frame, and it makes our discussions much complicate. Nevertheless, affluent physics might be sprung out of these discussions. It is really exciting for us to learn what is going on in contemparary nuclear physics with strangeness.

Our budget for this workshop is about 14 thousands US dallars, which is supported by APCTP and Japanese Ministry of Education. Although it is not sufficient, we deeply appreciate these financial supports.

On behalf of organizing committee, I would like to thank all of you for participating this workshop. Particularly, I should extend my hearty gratitude to those colleagues from oversea. Your participation is definitely very important to lead this workshop to be successful.

I hope that you will enjoy the workshop and exchange your wonderful ideas and mutual friendship with each other.

Temperature has dropped this morning. However, it will be recovered soon as the weatherman predicts. Take care of yourself during the workshop.

Thank you very much for your attention.

Il-Tong CHEON
Department of Physics
Yonsei University, Seoul

OPENING ADDRESS

T. MOTOBA
Laboratory of Physics, Osaka E-C University, Osaka 572-8530, Japan
E-mail: motoba@isc.osakac.ac.jp

Ladies and gentlemen,

It is my pleasure to give a few words at the opening of the APCTP Workshop on Strangeness Nuclear Physics. I would like to heartily welcome all of you, especially those participants coming to the "Far East" via long way from distant continents. Although Professor Cheon has given the formal opening address, the organizing committee asked me to explain briefly the history and motivation of the Japan-Korea joint organization for this international workshop.

In 1995, major experimentalists and theorists of the hypernuclear physics community in Japan proposed a big collaboration project to the Ministry of Education, Science, Sports, and Culture of Japan. This *project* named "*Strangeness Nuclear Physics*" was approved fortunately with appreciable funds for three fiscal years of 1996–1998. It consists of three experimental sub-projects and one theoretical collaboration project entitled as follows: 1) "Hypernuclear γ-ray spectroscopy with high-performance Ge detectors", 2) "Hyperon-nucleon weak interactions through decay particle spectroscopy", 3) "Study of nuclei with strangeness=-2", and 4) "Theoretical study of strange nuclei". The achievements are expected to play a substantial role in accelerating the realization of the Japan Hadron Facility Project.

In addition to the actual experiments and theoretical studies, the executive committee of the *project* has organized international workshops as well as various kinds of several domestic meetings. In fact, in 1996 we organized a satellite Workshop on Hypernuclear Physics on the occasion of the 25th INS International Symposium in Tokyo, and in 1997 in Beijing we had the Sino-Japan Symposium on Hadrons and Nuclei with Strangeness which was organized jointly with Institute of High Energy Physics (Prof. Zhang Zong-Ye *et al.*). Then, in 1998, we cosponsored the International Workshop on Spectroscopy of Hypernuclei (Sendai, Japan). We think such international activity is very important to encourage and promote the JHF Project, since the proposed 50-GeV proton synchrotron will supply high-intensity Kaon beams and therefore the facility will provide an internationally unique oppotunity for strangeness nuclear physics. We believe that the conferences held in the Asia-Pacific area, especially, should have good influence on the realization of the JHF Project in the full support of international nuclear physics communities. I would emphasize that the *internationalization* is one of the JHF keywords.

These are major motivation and reason that we Japanese and Korean organizers jointly proposed to have this workshop in Seoul. This workshop is the fourth one in the series of the international activities of the *project*. I would like to thank Professors C.N. Yang, Y.M. Cho, and C. Rim of APCTP for their support. I do hope this workshop will be successful both in promoting various stages of international collaborations and in clarifying prospects of the strangeness nuclear physics so that more and more young people will join the research work of this challenging field. Once again I thank all the speakers and participants.

SNP'99

THE NIJMEGEN HYPERON-NUCLEON AND HYPERON-HYPERON INTERACTIONS *

TH. A. RIJKEN

Institute for Particle and Nuclear Study, KEK, Tokyo, Japan and
Institute for Theoretical Physics, University of Nijmegen,
Nijmegen, The Netherlands

V.G.J. STOKS

Physics Division, Argonne National Laboratory, Argonne, Illinois 60439
and Centre for the Subatomic Structure of Matter,
University of Adelaide, SA 5005, Australia

Y. YAMAMOTO

Physics Section, Tsuru University, Tsuru, Yamanashi 402-0054, Japan

In this talk we review results of the Nijmegen soft-core NSC97 OBE-models [1,2,3]. After a general introduction, we first describe the Nijmegen approach to the baryon-baryon problem, and picture the baryon-baryon interactions in the general framework of QCD, flavor $SU(3)$, and chiral $SU(3)\otimes SU(3)$. Stressed is the simultaneous description of all baryon-baryon channels. Then, we discuss the interactions and the treatment of mesons, emphasizing the important role of the scalar-meson nonet. Next, we describe some of the results of the YN-channels ($S = -1$) with the new models NSC97a-f in some detail. Finally, since the nucleon-nucleon and hyperon-nucleon data fitting determined all free parameters in the NSC97 models, using (broken) $SU(3)$-symmetry, predictions for all other baryon-baryon channels can be given. We report briefly on the results for the $S = -2, -3, -4$ $N\Xi$ and YY systems. Here several bound states are predicted.

1 Introduction

The Nijmegen OBE-models comprise two kinds of models. The *hard-core* models, where the short range interactions are described by assuming an infinitely strong repulsion for $r \leq r_c$, with r_c channel dependent. So, the inner region is considered virtually as completely unknown. The YN-version of these models were extensively reviewed by [4,5]. On the other hand, *soft-core* models use gaussian form factors, thus describing both the long/intermediate and short range forces by meson degrees of freedom. The included mesons are nonets of pseudo-scalar, vector, and scalar mesons. The latter are not considered as make ups for other physical effects, but included in their own right of existence. Also, next to the scalar mesons, there are included the diffractive contributions to the potentials. They make strong scalar-exchange compatible with chiral symmetry, see [6]. For a recent review and references, we refer to [1]. Now, for light nuclei and hypernuclei both the long- and the intermediate/short range interaction regions are important. However, for the heavier nuclei especially the intermediate and short range regions are vital, because the 'healing distance' is $\leq$ 1 fm. Therefore, it is important to develop a more detailed physical picture for the short-range forces, in order to make progress in

*TALK GIVEN BY TH.A. RIJKEN, SNP99 WORKSHOP, FEBRUARY 1999

understanding nuclei.

The NN-data have been accumulated and improved since ca 50 years now. Below T_{lab} = 350 MeV the data base contains ca 4300 selected $pp + np$-data on cross sections and a variety of spin correlations. The NN-model is fitted to the Nijmegen phase shift analysis [7]. The experimental data on YN-interactions (ΛN, ΣN, and ΞN) and the YY- YY-interactions on the other hand are rather scarce. In our data base for YN we use ca 35 scattering data [8,9,10,11,12]. These scattering data are at very low energies and provide essentially only information on s-wave interactions. Furthermore, there is rather solid information that there are no YN bound-states. Although, as compared to NN, few in number and most with rather large errors, it is our experience that to fit these YN-data and not allowing YN bound-states *in conjunction* with NN is not at all trivial.

In order to reach a realistic description of the NN- and YN-interactions simultaneously, we followed the following strategy: First, we construct a good NN-interaction by making a fit to the NN-data below T_{lab} = 350 MeV. Then, exploiting broken $SU_f(3)$-symmetry, we extend these interactions to YN, and make a fit to the YN-data. In making this extension, we exploit the fact that YN-systems are more sensitive to some parameters than NN, e.g. the $F/(F+D)$-ratio's and meson mixing angles. Also, we have to make certain 'short range' assumptions. In the HC-models the hard-core's in YN are given freedom w.r.t. the hard-core parameters used in NN. In the SC-models similarly for the form factor cut-off masses. It appears to be possible to reach a very good description for the YN-data using about 5 free parameters. We note that in YN the data are essentially only s-wave data. Therefore, the p-waves are in fact predictions.

The systems with non-zero strangeness, i.e. $S \neq 0$, are potentially very rich in information about the nuclear forces, beyond what could be learned from the nucleon-nucleon interactions. Now, for exploiting both the $S = 0$ and the $S \neq 0$ channels, flavor $SU(3)$-symmetry is very important. From the QCD-viewpoint, $SU_f(3)$-symmetry is an accidental symmetry. It consists of transformations among the u,d, and s quarks. In the standard model the s quark belongs to a different family as the u and d quarks. The gluons are flavor blind, and since the (constituent) masses of the quarks are not very different $SU_f(3)$ is, although not a fundamental, a very useful broken symmetry. The basic difference $m_s \neq m_u \approx m_d$, defines the 'direction' of symmetry breaking, as is shown by the succesful Gell-Mann-Okubo scheme of mass breaking patterns in the meson and baryon $SU_f(3)$-multiplets. Based on this one can also work out breaking of coupling constants. In the Nijmegen models we assume usually that the coupling constants are not broken. In the most recent SC-models [2] we introduced also breaking of the couplings. The kinematical breaking of $SU_f(3)$ is included by using the physical masses of the mesons and baryons. So, the theoretical basis, using $SU_f(3)$-symmetry for the extention of NN-interaction models to YN and YY, is rater solid.

An account of a modern theoretical basis for the soft-core interactions between baryons has been given in a synopsis in [13]. Taking the quark-model picture of the baryons, i.e. QCD, combined with broken chiral symmetry, the effective degrees of freedom at low energies are the so-called constituent quarks and the mesons: pseodo-scalar-, vector-, scalar-mesons etc. The chiral-quark-model (CQM) [14] explains the success of the non-relativistic quark-model (NRQM) and at the same time leads to a description of hadron dynamics through interactions involving exclusively mesons and baryons, using *effective* Lagrangians. The general idea is that the QCD-vacuuum becomes unstable for momemta transfer for which $Q^2 \leq \Lambda^2_{\chi SB} \approx 1$ (GeV)^2. The vacuuum goes through a phase transition, generating the constituent quark masses via $\langle 0|\bar{\psi}\psi|0\rangle \neq 0$, and reducing the gluon coupling α_s. This mechanism thus generates the quarks used in the NRQM. Viewing the pseudo-scalar mesons as Nambu-Goldstone bosons, as a result of the spontaneously broken chiral symmetry, makes it natural to assume quark dressing by pseudo-scalar mesons, and also other mesons in general. In this context, baryon-baryon interactions are naturally described by meson-exchange, using form factors at the meson-baryon vertices. This is the general physical basis for the Nijmegen soft-core models.

The contents of this paper is as follows. In section II the most recent soft-core models NSC97 [2] are discussed. We report on a few results for the $S = -1$-systems. In section III we report on the results obtained in a very recent extension of the NSC97-models to the $S = -2, -3, -4$-systems. Finally, in section IV we offer a summary conclusions.

2 NSC97-models $S = -1$

Recently [2], we constructed SC-models where the form factors depend on the $SU_f(3)$-assignment of the mesons, and not globally depending on the $SU_f(3)$-structure of the BB-channels. The latter was done for the NSC89-model [15,16]. In principle we introduce form factor masses Λ_8 and Λ_1 for each meson-nonet, for respectively the $\{8\}$- and $\{1\}$-members. In the application to YN and YY we allow for $SU_f(3)$-breaking, by using different cut-off's for the strange mesons K, K^*, κ. Also, we used here for the first time $SU_f(3)$-symmetry breaking of the coupling constants, which we realized phenomenologically using the rather succesful $^3P_0$-mechanism of quark-anti-quark pair creation [17], leading to a quark-physics based model for the meson-baryon couplings. The $SU_f(3)$-breaking is generated by assuming that the $Q\bar{Q}$-pair creation constants γ are flavor-dependent, i.e. $\gamma_u = \gamma_d \neq \gamma_s$.

The spin-dependence of the YN-interactions is not yet settled. For example, the ΛN-scattering data can be fitted with a variety of sets of s-wave scattering lengthes (a_s, a_t). See e.g. the review [18]. The first soft-core model [16], called NSC89, provided only one solution. Therefore, we generated the models NSC97a-NSC97f, having different sets (a_s, a_t). This is achieved by taking a series of α_V^m-values, where α_V^m is the $F/(F+D)$-ratio of the tensor-coupling of the vector mesons, and fitting the other free parameters. The latter are still free after the NN-data were fitted. (This is a reasonable procedure, since the NN-data are rather insensitive to α_V^m.) This way models were obtained with an equally good fit to the scattering data, having different spin properties for YN. These models can now be investigated in (precise) hypernuclear calculations. In the next table we give some results for the distinctive models. For more results we refer to the original paper [2].

	α_V^m	a_s	a_t	U_Λ	U_Σ	Γ_Σ	χ^2
(a)	0.4447	-0.71	-2.18	-33.9			15.7
(b)	0.4247	-0.90	-2.13	-34.1			15.8
(c)	0.4047	-1.20	-2.08	-35.3			15.6
(d)	0.3847	-1.71	-1.95	-35.1			15.8
(e)	0.3747	-2.10	-1.86	-34.3	-11.8	14.6	16.1
(f)	0.3647	-2.51	-1.75	-31.1	-11.6	15.5	16.7
NSC89	0.275	-2.73	-1.48	-30.8	-27.1	26.9	15.8
D	0.334	-1.77	-2.06	-40.1	-29.3	10.3	19.4
F	0.588	-2.18	-1.93	-30.8	+5.8	9.6	27.1

NSC97a-f: ΛN-Scattering length's, χ^2 YN-data, etc.

The results in this Table are given as a function of α_V^m. One sees that the singlet-triplet strength of the ΛN-interaction depends smoothly on α_V^m. In the most right-column, it is shown that all NSC97-models have an excellent fit to YN. All results, given for U_Λ, U_Σ, and Γ_Σ were calculated using the Bando-Yamamoto YNG-interactions [19] constructed for the NSC97-models. The Λ well depth $U_\Lambda = -28$ MeV, obtained from analyses of the (π^+, K^+) and (K^-, π^-) cross sections on nuclear targets [20] is reasonably given, in particularly, by NSC97f. Also interesting is that for the NSC97-models Γ_Σ is much smaller than is the case for the NSC89-model. This is more consistent with the recently confirmed ${}^4_\Sigma He$ hypernucleus [21], which has a conversion width $\Gamma \approx 7$ MeV. This rather small Σ-conversion width has been explained by Harrada et al [22], using the SAP-interactions derived using model D [23]. Although the low Γ_Σ suggests that the NSC97 solutions have better ΣN-interactions than NSC89, however the recent work by Dabrowski seems to indicate the opposite [24]. Results for the NSC97 potentials with regard to the Carlson-Gibson [25] computation of the ${}^3_\Lambda He$, ${}^4_\Lambda He$, and ${}^5_\Lambda He$-hypernuclei are not availaible yet. The same is true for the Miyagawa-Gloeckle [26] computation of the hypertriton ${}^3_\Lambda H$, whose binding was reproduced exactly by the NSC89-model.

3 NSC97-models $S = -2, -3, -4$

In the $S = -2$-systems, the experimental information is limited to the ground states of ${}^{6}_{\Lambda\Lambda}He$, ${}^{10}_{\Lambda\Lambda}Be$, and ${}^{13}_{\Lambda\Lambda}B$. From the double-$\Lambda$ states, mentioned above, it is infered that $\Delta B_{\Lambda\Lambda} = 4 - 5$MeV, which indicates a rather strong $\Lambda\Lambda$-interaction. For the $S = -3$- and $S = -4$-systems ther are at present no experimental data.
The estimation for the ${}^1S_0$ $\Lambda\Lambda$-matrix element in ${}^{6}_{\Lambda\Lambda}He$ for model D [23] that $\Delta B_{\Lambda\Lambda} =$ 4 MeV, is in agreement with the experimental observation. Model F [27] gives a repulsive $\Lambda\Lambda$-interaction in this case, in contradiction to the data. For more details we refer to [28]. Yamamoto et al [28] predict $\Lambda\Lambda$-hypernuclei for $A = 7 - 16$ systems, based on a core $+\Lambda + \Lambda$ three-body model with Wood-Saxon interactions. Akaishi and collaborators [29] investigated the ${}^{5}_{\Xi}H$ and ${}^{5}_{\Lambda\Lambda}H$ systems. They predict a bound ${}_{\Lambda\Lambda}H$-state 6.3 MeV below the $t + \Lambda + \Lambda$-threshold. In the Ξ^--channel they have a resonating ${}^{5}_{\Xi}H$-state at 1.7 MeV below the $\alpha + \Xi^-$-threshold, a 'halo' nuclear state The total conversion width is 0.76 MeV, i.e. extremely narrow. Also the analysis by Millener et al [30] gives, using Nijmegen model D, a rather small width ≈ 1.5 MeV.

Since, as noted before, in the NSC97-models all free parameters are determined by fitting the NN- and the YN-scattering data, the results for the $S \leq -2$-systems are really predictions. Below we briefly discuss the most important results of a very recent constructed extension of the NSC97-models [3].

3.1 S-wave Effective-range Parameters

In the SC-models, NSC89 and NSC97, we have nearly ideal-mixing for $Q\bar{Q}$-states, which implies that the scalar-exchange contribution to the potential satisfies

$$|V_{\Lambda\Lambda}(0^+)| < |V_{\Lambda N}(0^+)| < |V_{NN}(0^+)| \ ,$$

which leads to weaker attractive potentials than in the case of hard-core model-D [23], for the $\Lambda\Lambda, \Xi N$-systems. The characteristic feature of model-D is that instead of a scalar nonet, there is only a scalar singlet. This makes the scalar-meson-exchange central attraction BB-channel independent. In this report, we restrict ourselves to the predictions for the ${}^1S_0(\Lambda\Lambda)$ low energy parameters. From Table X of [3] we have that for the models NSC97a-f, $-0.53 \leq a_s(\Lambda\Lambda) \leq -0.27$, while the effective range takes values $7.4 \leq r_s(\Lambda\Lambda) \leq 15.0$. Here, there is the correlation that the smaller a_s-values have the larger r_s-values. In view of the rather large r_s-values it is not so clear whether the $\Lambda\Lambda$-interaction will be too weak to be compatible with the findings of the $\Lambda\Lambda$-hypernuclei.

3.2 Bound-States in S-waves

The S-wave bound-states predicted in the NSC97-models are [3]:

(i) ${}^1S_0(\Sigma\Sigma)$: The binding energies in $\Sigma^+\Sigma^+$ range from –1.53 MeV for NSC97a to –3.07 MeV for NSC97f, while in $\Sigma^-\Sigma^-$ they range from –1.59 MeV for NSC97a to –3.17 MeV for NSC97f.

(ii) ${}^1S_0(\Xi\Sigma)$: The attraction in $\Xi\Sigma$ is even stronger and extends to smaller inter-baryon distances. The binding energies in $\Xi^0\Sigma^+$ range from –3.02 MeV for NSC97a to –16.5 MeV for NSC97f. The presence of the Coulomb interaction in $\Xi^-\Sigma^-$ causes a shift of roughly 1 MeV, resulting in binding energies of –2.30 MeV for NSC97a to –15.6 MeV for NSC97f.

(iii) ${}^1S_0(\Xi\Xi)$: The attraction in $\Xi\Xi$ is also strong enough to support a bound state. The $\Xi^0\Xi^0$ and $\Xi^-\Xi^0$ give almost identical results ranging from –0.10 MeV for NSC97a to –15.8 MeV for NSC97f. Again, the presence of the Coulomb interaction in $\Xi^-\Xi^-$ causes a shift of about 1 MeV, and so the NSC97a model no longer supports a bound state in this channel.

(iv) ${}^3S_1 -{}^3 D_1(\Xi\Sigma)$: The NN and $\Xi\Sigma$ ${}^3S_1$-${}^3D_1$ interactions both belong to the $\{10^*\}$ irrep, and so, in analogy with the deuteron bound state in NN, we also expect bound states in $\Xi\Sigma$. Indeed, in $\Xi^0\Sigma^+$ the binding energies range from –5.64 MeV for NSC97a to –36.1 MeV for NSC97f, whereas in $\Xi^-\Sigma^-$ they range from –4.86 MeV for NSC97a to –35.3 MeV for NSC97f.

The analysis of thee bound states, is facilitated by the fact that all these cases belong to single $SU_f(3)$-irreps. In the first (i)-(iii) the bound-states occur in the $\{27\}$-irrep, the same as ${}^1S_0(NN)$. It is wel known that the latter has almost a bound-state. Going to the heavier systems, one expected bound-states in the past, because of the effect of the larger reduced mass in the Schrödinger equation. For the $S = -1$-systems, this did not occur because of $SU_f(3)$-symmetry breaking. Apparently, in the NSC97-models, the latter is not able to avoid bound states to occur. Viewed upon these predictions this way, it is no really a surprise to find bound states.

4 SUMMARY AND CONCLUSION

As already stated in our NSC97 publication [2], the NSC97 potentials presented here are an important step forward in modeling the baryon-baryon interactions for scattering and hypernuclei in the context of broken $SU_f(3)$ symmetry. The potentials are based on the one-boson-exchange model, where the coupling constants at the baryon-baryon-meson vertices are restricted by broken $SU_f(3)$ symmetry. Each type of meson exchange (pseudoscalar, vector, scalar) contains five free parameters: a singlet coupling constant, an octet coupling constant, the $F/(F+D)$ ratio α, a meson-mixing angle, and a parameter λ which effectively accounts for the fact that the strange quark is much heavier than the up and down quarks. The potentials are regularized with exponential cutoff parameters, which provide a few additional free parameters. Most of these parameters are fixed in the fit to the wealth of accurate NN scattering data, while the remaining ones are fixed in the fit to the (few) YN scattering data. This still leaves enough freedom to construct six different models, NSC97a through NSC97f. They all describe the NN and YN data equally well, but differ on a more detailed level. The assumption of $SU_f(3)$ symmetry then allows us to extend these models to the higher strangeness channels (i.e., YY and all interactions involving cascades), without the need to introduce additional free parameters [3]. The NSC97 models are the first potential models of this kind.

The strangeness-rich systems are very important for understanding e.g. the heavy ion collissions [31], and astrophysical problems [32,33]. Very valuable information on the interactions in such systems can be obtained from possibly existent few-body systems, in particular $\Lambda\Lambda$-hypernuclei like ${}^{6}_{\Lambda\Lambda}He$,${}^{10}_{\Lambda\Lambda}Be$ etc. Strangeness-rich systems could also be exotic multi-quark systems. The so-called H-dibaryon, predicted by Jaffe [34], would be such a system: *uuddss*. But, as shown in [3] they can also be bound states of nucleons (N), hyperons (Y), and cascades (Ξ).

Acknowledgements

Th.A. Rijken and Y. Yamamoto are grateful to Y. Akaishi, T. Harada, T. Motoba, for many stimulating discussions. V.G.J. Stoks likes to thank T.-S.H. Lee and T. Kuo for their continued interest and constructive comments. The work of V.G.J.S was supported in part by the U.S. Department of Energy, Nuclear Physics Division, under Contract No. W-31-109-ENG-38.

References

1. Th.A. Rijken, Nucl. Phys. A**639** 29c-40c (1998).
2. Th.A. Rijken, Y. Yamamoto, V.G.J. Stoks, Phys. Rev. C**59** 21 (1999).
3. V.G.J. Stoks and Th.A. Rijken, *Soft-core Baryon-baryon Potentials for the complete Baryon Octet*, preprint 1999.
4. C.B. Dover and A.Gal, Progr. Part. Nucl. Phys. **12** 171 (1984).
5. Y. Yamamoto and H. Bando, Progr. Theor. Phys. **73** 905 (1985); Progr. Theor. Phys. Suppl. No.81 9 (1985); Progr. Theor. Phys. **83** 254 (1990).
6. J.J. de Swart, Th.A. Rijken, P.M. Maessen, and R. Timmermans, Nuov. Cim. **102**A 203 (1989).
7. V.G.J. Stoks, R.A.M. Klomp, C.P.F. Terheggen, and J.J. de Swart, Phys. Rev. C **49**, 2950 (1994).
8. R. Engelmann, H. Filthuth, V. Hepp, and E. Kluge, Phys. Lett. **21** 587 (1966).
9. G. Alexander, U. Karshon, A. Shapira, G. Yuketieli, R. Engelmann, H. Filthuth, and W. Lughofer, Phys. Rev. **173** 1452 (1968).
10. B. Sechi-Zorn, B. Kehoe, J. Twitty, and R.A. Burnstein, **175** 1735 (1968).
11. V. Hepp and M. Schleich, Z. Physik **214** 71 (1968).
12. F. Eisele, H. Filthuth, W. Fölisch, V. Hepp, E. Leitner, and G. Zech, Nucl. Phys. **B37** 204 (1971).
13. Th.A. Rijken, Proceedings of the XIVth European Conference on Few-Body Problems in Physics, eds. B. Bakker and R. von Dantzig, Amsterdam(1993), few-Body Systems, Suppl. **7** 1 (1994).
14. A. Manohar and H. Georgi, Nucl. Phys. B**234** 189 (1984); S. Weinberg, Physica (Amsterdam) **96**A 327 (1979).
15. M.M. Nagels, Th.A. Rijken, and J.J. de Swart, Phys. Rev. D**17** 768 (1978).
16. P.M.M. Maessen, Th.A. Rijken, and J.J. de Swart, Phys. Rev. C**40** 2226 (1989).
17. A. Le Yaouanc, L. Oliver, O. Péne, and J.-C. Raynal, Phys. Rev. D **8** 2223 (1973).
18. J.J. de Swart, M.M. Nagels, T.A. Rijken, and P.A. Verhoeven, Springer Tracts

Mod. Phys. **60**, 138 (1971).
19. Y. Yamamoto and H. Bandō, Progr. Theor. Phys. **83** 254 (1990).
20. D.J. Millener, A. Gal, and C.B. Dover, Phys. Rev. C **38** 2700 (1988).
21. T. Nagae et al, Phys. Rev. Lett. **80** 1605 (1998).
22. T. Harada, Y. Akaishi, S. Shinmura, and H. Tanaka, see Nuov. Cim. **102**A 473 (1989).
23. M.M. Nagels, Th.A. Rijken, and J.J. de Swart, Phys. Rev. D **15** 2547 (1977).
24. J. Dabrowski, *The isospin dependence of the single-particle potential of the* Σ *hyperon in nuclear matter*, these proceedings.
25. J.A. Carlson, in *Proc. of the LAMPF workshop on* (π, K) *physics, AIP Conf. Proc.* 224, (American Insitute of Physics, New York, 1991), p. 198; J. Carlson and B. Gibson, Particles and Fields **43**, 198 (1990).
26. K. Miyagawa and W. Gloeckle, Phys. Rev. C **48** 2576 (1993).
27. M.M. Nagels, Th.A. Rijken, and J.J. de Swart, Phys. Rev. D **20** 1633 (1979).
28. Y. Yamamoto, T. Motoba, H. Himeno, K. Ikeda, and S. Nagata, Progr. Theor. Phys. Suppl. **117** 361 (1994).
29. K. S. Myint and Y. Akaishi, Progr. Theor. Phys. Suppl. **117** 251 (1994); I. Kumegai-Fuse, T. Koike, and Y. Akaishi, Nucl. Phys. **A585** 367 (1995); I. Kumegai-Fuse and Y. Akaishi, Prog. Theor. Phys. **94** L151 (1995).
30. D.J. Millener, A. Gal, and C.B. Dover, Progr. Theor. Phys. Suppl. **117** 307 (1994).
31. See, e.g., B. Shiva Kumar, Nucl. Phys. **A590**, 29c (1995), and references therein.
32. See, e.g., G. Baym, Nucl. Phys. **A590**, 233c (1995), and references therein.
33. M. Prakash and J.M. Lattimer, Nucl. Phys. **A639**, 433c (1998), and references therein.
34. R.L. Jaffe, Phys. Rev. Lett. **38**, 195 (1977); **38**, 617(E) (1977).

HYPERON-NUCLEON SCATTERING EXPERIMENT (KEK-PS E289)

M.Ieiri, T.Fukuda
Physics Department, IPNS, KEK, High Energy Accelerator Research Organization, Oho 1-1, Tsukuba 305-0801, Japan

J.K.Ahn, H.Akikawa, H.En'yo, H.Funahashi, Y.Goto, T.Haseyama, A.Ichikawa, K.Imai, M.Ishino, Y.Itow, H.Kanda, Y.Kondo, A.Masaike, S.Mihara, Y.Nakada, K.Ozawa, N.Saito, T.Tabaru, H.Torii, K.Yamamoto, S.Yamashita, S.Yokkaichi, M.Yoshida
Department of Physics,Kyoto University, Kyoto 606-8502, Japan

T.Hayakawa, K.Nakai, A.Sato
Department of Physics, Science University of Tokyo, Chiba 278-8510, Japan

A.Higashi
INS, University of Tokyo, Tanashi 188-8501, Japan

M.Hanabata, K.Nakazawa, Y.Ueda
Physics Department, Gifu University, Gifu 501-1193, Japan

F.Takeutchi
Department of Physics, Kyoto Sangyo University, Kyoto 603-8555, Japan

M.S.Chung, K.S.Sim
Department of Physics, Korea University, Seoul 136-701, Korea

Y.D.Kim
Physics Department, Sejong University, Seoul, 143-747, Korea

J.M.Lee
Department of Physics, Yonsei Univ., Seoul 120-749, Korea

Y.M.Park
Department of Physics, Kyungsung University, Pusan 608-736, Korea

Y.M.Shin
Department of Physics, University of Saskatchewan, Canada S7N 0W0

S.V.Golovkin, A.M.Gorin
Institute for High Energy Physics, RU-142284, Protvino, Russia

E.N.Kozarenko, I.E.Kreslo
Joint Institute for Nuclear Research, RU-141980, Dubna, Russia

P.Tlusty
Nuclear Physics Institute, 250 68 Rez, Czech Republic

J.Arvieux
Laboratory National Saturne, F-91191, Gif-sur-Yvette, CEDEX, France

B.Bassalleck
Department of Physics, University of New Mexico, Albuquerque, N.M. 87131, U.S.A.

Hyperon-nucleon scattering experiment(KEK-PS E289)has been carried out at KEK-PS with a Scintillating Fiber detector. Differencial cross sections for Σ^+p, Λp and Σ^-p have been measured. In this paper, we present the status of analysis for Σ^-p elastic channel.

1 Introduction

To study hyperon-nucleon(YN) interaction plays a very important role for understanding strong interaction between baryons. Many theoretical approaches such as One Boson Exchange Potential[1] and quark-cluster model[2] have been attempted to describe the YN interaction. However, as for the YN scattering experiment, which is indispensable to study YN interaction, there are very limited data obtained by bubble chamber experiments in 1960's [3]. They are very scarce in statistics and restricted in low momentum region.

To obtain new data of YN scattering, $\Sigma^+ p$ scattering experiment was performed at KEK-PS(E251)[4], which tried to measure the differential cross section of the $\Sigma^+ p$ scattering with a new technique using a scintillating fiber active target. The Σ^+'s from 300 to 700 MeV/c were produced by the reaction $\pi^+ p \rightarrow K^+\Sigma^+$ with 1.6 GeV/c pion beam at the KEK-PS K2 beam channel. The plastic scintillating fiber block was used as a production target of hyperon, a scattering target of hydrogen and 4π detector for related charged particles around the reaction vertices.

As a next step, the measurement of Σ^-p and Λp scattering as well as $\Sigma^+ p$ scattering (E289) was approved in July, 1995. The scintillating fiber detector was improved so as to get a better spatial resolution and higher light outputs. Our goal is to obtain the YN scattering data which have about two or three times more statistics and cover the higher momentum region(below 1GeV/c) than past experiments.

2 Experimental Procedure

The experiment was carried out at the the K2 beam line of 12GeV/c Proton Synchrotron(PS) at KEK. The experimental setup is shown in Figure 1. Hyperon production reactions(see Table 1) were identified by the spectrometer system. If signals from the spectrometer detectors satisfied the trigger logic, the gates of IIT system were opened and image data were taken.

Beam particles(π^+'s and π^-'s) were identified with the aerogel(BAC) and gas(BGC) Čerenkov counters. Beam intensity was 1.5×10^5/spill(4sec). In identifying (π^-, K^+)

Table 1: Reactions for hyperon production.

Reaction	$\pi^- p \rightarrow K^+\Sigma^-$	$\pi^+ n \rightarrow K^+\Lambda$	$\pi^+ p \rightarrow K^+\Sigma^+$
Beam momentum(GeV/c)	1.33	1.67	
Hyperon momentum(MeV/c)	400~700	250~ 1000	400~800
Hyperon life time $c\tau$(cm)	4.434	7.89	2.396

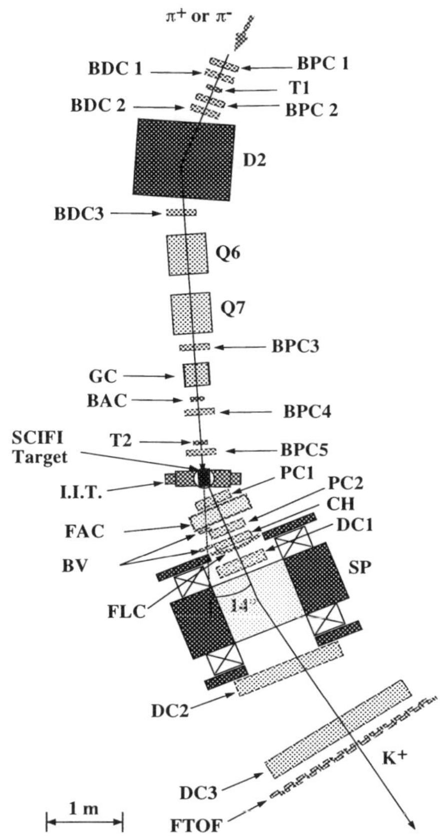

Figure 1: Experimental setup of E289

or (π^+, K^+) reaction, the main backgrounds are pions and protons from the target. Almost all pion background was rejected by the Forward Čereckov counter(FAC). In addition, by using the TOF information obtained by the FTOF counters, mass of outgoing particles were selected by the fast electronics to reduce the proton background at the 2nd level trigger(Mass Trigger).

The scintillating fiber target(SciFi target) consists of 0.3mm $\times$ 0.3mm square plastic scintillating fibers and the effective volume was 200mm $\times$ 100mm $\times$ 100mm. It was made of 0.3mm thick fiber sheets packed in u and v directions alternately. The photon images from both directions were intensified by the chain of Image Intensifier Tubes(IITs) as visual image data and read out by Electron-Bombarded CCD system(EBCCD). The EBCCD itself can amplify signals and the signals from the CCD chips were digitized. The 1st and 2nd IITs were electro-static type and the 3rd IIT was a MCP type. The 1st level trigger made by fast PMT signals opened this MCP gate. More intelligent trigger(Mass Trigger) was applied as the 2nd level trigger to open the EBCCD gate. By the two-dimensional readout, we can reconstruct the three-dimensional event image and identify the hyperons,which would scatter and decay inside the target. The position resolution(track residual) of the SciFi system obtained by the straight track of 1.6GeV/c minimum ionizing π^+ was 160μm.

3 Data Analysis of Σ^-p elastic scattering

3.1 (π^-, K^+) Event Selection

In order to select the (π^-, K^+) reaction, the mass of the scattered particle from the SciFi target has been reconstructed from the measured momentum and velocity. The momentum has been determined by the straight-line fitting to the track both upstream and downstream of the Spectrometer Magnet(SP) separately. The magnetic field of SP was 0.9T for the (π^-, K^+) trigger. The velocity has been obtained by measuring the TOF between T2 and FTOF. The TOF resolution (FTOF-T2) was 140 psec(rms). The K^+ mass resolution was $19MeV/c^2$(rms) at 0.7GeV/c and the momentum resolution ($\Delta P/P$), calculated from the mass resolution and the TOF resolution, was 1.7%(rms) for (π^-, K^+) events. This procedure reduced the number of events from 4.3×10^6 to 3.2×10^5.

3.2 Eye Scanning

The image data of selected (π^-, K^+) event have been analyzed. magnetic field of SP. The selected 3.2×10^5 events were eye-scaned manually and categorization was done. Event topology was recognized as a combination of patterns shown in Figure 2.

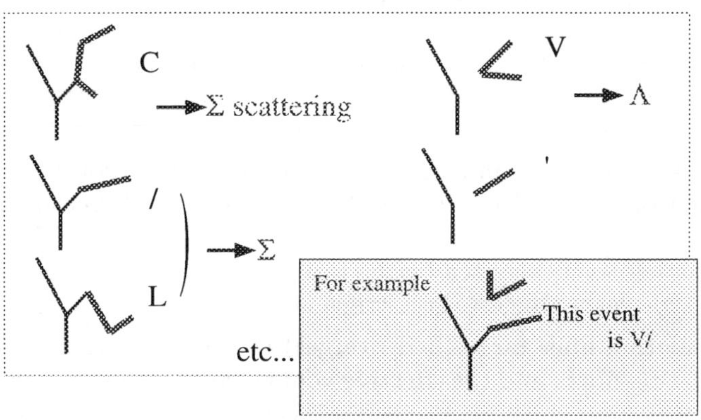

Figure 2: Examples of event topologies.

The event including '/' or 'L' is the Σ^- production candidate, and that including 'C' is the candidate of Σ^-p scattering. We scanned 1/3 of all data and the number of Σ^-p scattering candidats was 81 . For these 81 events, reaction points were marked on the graphic display. We can determine the reaction points 3-dimensionally by these marks. We call this process 'pointing'. As for the Σ^- production candidates, the ratio of the events including '/' or 'L'(the Σ^- production candidates) was 59.6% of all data, and we have carried out pointing for 1500 sample events.

3.3 Σ^- Beam Analysis

In order to reconstruct the kinematics of the Σ^-p elastic scattering, momenta of Σ^- beam should be known. Because protons in carbon nuclei (SciFi target consists of $(CH)_n$) have Fermi momentum, only Σ^-'s produced on hydrogen were used as Σ^- beam. To select hydrogen target events, missing-mass cut and angle-difference cut were adopted. The missing-mass was calculated assuming the p(π^-, K^+)X reaction. The angle-difference was the difference between the predicted momentum vector obteined from π^- and K^+ momentum vector and the momentum vector measured with SciFi detector. The cut condition was 1.16 <missing mass< $1.23(GeV/c^2)$ and |angle difference| $< 0.15(rad)$. Also the track length of Σ^- was required to be more than 5mm to reduce the ambiguity of categolization and pointing. The number of the Σ^- beam is defined as the total flight length of Σ^-, ie. $N_{\Sigma^-} = \sum$(Flight Length-5mm). N_{Σ^-} was $6200 \pm 250mm$ for 1500 event sample. Efficiencies of scanning and hydrogen target selection cut were estimated by the Monte Carlo simulation with dummy image data. Dummy image data were generated to reproduce the response of the SciFi target and the IIT chain. The total efficiency was $89 \pm 3.2\%$ and N_{Σ^-} for 1/3 of all events was estimated to be $(3.4 \pm 0.18) \times 10^5$mm.

3.4 $\Sigma^- p$ Elastic Events

The pointing has been done for 81 Σ^-p scattering candidats, which correspond to 1/3 of all data. Figure 3 shows an example of Σ^-p scattering. We required same conditions as Σ^- beam's case and additionally, recoil Σ^-'s length > 4mm and recoil proton's length > 3mm were also required. To select the elastic scattering, we required following conditions; 1.3 < coplanarity < 1.9(rad), $|\theta_{cal} - \theta_{measure}| <$ 0.3(rad) and $|P_{cal} - P_{measure}| <$ 100(MeV/c). θ_{cal} and P_{cal} denote the calculated scattering angle and momentum of proton assuming the elastic scattering. The angular distribution of survived events is shown in Figure 4. This data is about 1/3 of all data. θ_{cm} represents the scattering angle in CMS. There was no data in the region of $cos\theta_{cm} < -0.4$(backward scattering). This is because, in the backward region the momentum of scattered Σ^- becomes lower and the flight length befor decay becomes shorter, so the efficiency for geometrical cut becomes lower. The vertical axis represents the momentum of the Σ^- 'beam'. The momentum region of this experiment is more than 300MeV/c. This is the first data of the Σ^-p scattering in this momentum region, in which p-wave contribution is exist.

4 Summary

The present status of the analysis of the Σ^-p scattering experiment has been shown. The analysis is, however, still preliminary and further analysis should be done especially in the selection of elastic events and back-ground estimation. Analyses of other channels are also in progress and new data of YN scattering will be available in near future.

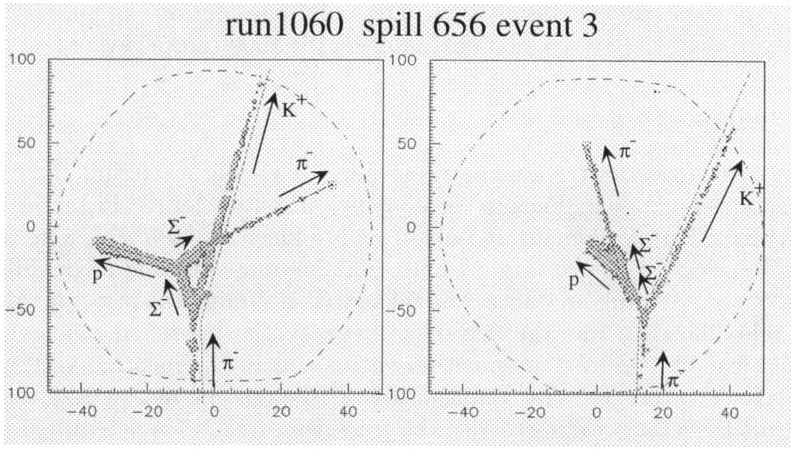

Figure 3: An example of Σ^-p scattering event observed in the SciFi detector.

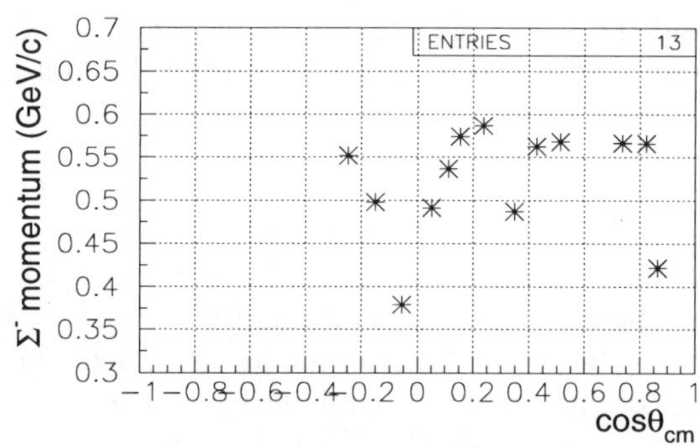

Figure 4: Angular and beam momentum distribution of Σ^-p elasting scattering.

References

1. M.M.Nagels et al., Phys. Rev.D15(1977)2547; M.M.Nagels et al., Phys. Rev.D20(1979)1633; A.Reuber et al., Nucl. Phys. A570(1994)543
2. A.Faessler et al.,Phys. Lett.B183(1987)10; Y.Fujiwara et al., Prog. Theo. Phys.94(1995)215
3. F.Eisele et al., Phys. Lett.37B(1971)204; G.Alexander et al., Phys. Rev.173(1968)1452; B.Sechi-Zorn et al.,Phys. Rev.173(1968)1452; J.A.Kadyk et al.,Nucl. Phys. B27(1971)13; R.Engelmann et al.,Phys. Lett.,21(1966)587;
4. J.K.Ahn et al., Nucl. Phys.A648(1999)263;
   Y.Goto, Ph.D. thesis, Memories of th Faculty of Science, Kyoto University, (Series of Physics, Astrophysics, Geophysics and Cemistry)Vol. 40-1,1(1997)

Ξ^-P SCATTERING AND STOPPED-Ξ^{-12}C REACTION

J.K. AHN[1]*, S. AOKI[4], K.S.CHUNG[11], M.S.CHUNG[10], H. EN'YO[1], T.FUKUDA[5], H. FUNAHASHI[1], Y. GOTO[1], A. HIGASHI[5], M. IEIRI[6], T. IIJIMA[6], M. IINUMA[1], K. IMAI[1], Y. ITOW[1], J.M. LEE[11], S. MAKINO[1], A. MASAIKE[1], Y. MATSUDA[1], Y. MATSUYAMA[5], S. MIHARA[1], C. NAGOSHI[5], I. NOMURA[8], I.S. PARK[10], N.SAITO[1], M. SEKIMOTO[5], Y.M. SHIN[9], K.S. SIM[10], R. SUSUKITA[1] R. TAKASHIMA[2], F. TAKEUTCHI[3], P. TLUSTÝ[5], S. WEIBE[9], S. YOKKAICHI[1], K. YOSHIDA[1], M. YOSHIDA[1], T. YOSHIDA[7], S. YAMASHITA[1]

(KEK-PS E224 COLLABORATION)

[1] *Department of Physics, Kyoto University, Kyoto 606-8502, Japan*
[2] *Kyoto University of Education, Kyoto 612, Japan*
[3] *Kyoto Sangyo University, Kyoto 606, Japan*
[4] *Faculty of Human Development, Kobe University, Kobe 657-8501, Japan*
[5] *Institute for Nuclear Study, University of Tokyo, Tokyo 188-8501, Japan*
[6] *KEK, High Energy Accelerator Research Organization, Tsukuba 305, Japan*
[7] *Department of Physics, Osaka City University, Osaka 558-8585, Japan*
[8] *National Institute for Fusion Science, Nagoya 464, Japan*
[9] *University of Saskatchewan, Saskatoon, Canada*
[10] *Department of Physics, Korea University, Seoul 136-701, Korea*
[11] *Department of Physics, Yonsei University, Seoul 120-749, Korea*

We report upper limits on the cross sections for the Ξ^-p elastic and conversion processes based on the observation of one Ξ^-p elastic scattering events with an invisible Λ decay. The cross section for the Ξ^-p elastic scattering is, for simplicity, assumming an isotropic angular distribution, found to be 40 mb at 90% confidence level, whereas that for the $\Xi^-p \to \Lambda\Lambda$ reaction is 11 mb at 90% confidence level. While the results on the elastic cross section give no stringent constraint on theoretical estimates, the upper limit on the conversion process suggests that the estimate of the RGM-F model prediction could be ruled out. We also report some preliminary results on the obervation of the stopped-Ξ^- hyperon-nucleus interaction with respect to hypernuclear production and existence of doubly-strange H-dibaryon.

1 Introduction

Measurement of cross sections and spin observables for hyperon-nucleon and hyperon-hyperon interactions is of fundamental interest, particularly in the strangeness -2 sector, because it sheds some light on the nature of ΞN and $\Lambda\Lambda$ interactions, thereby extending our knowledge of the SU(3) structure of baryon-baryon forces. Its importance is more highlighted by providing a possible test for predictions of a strangeness -2 dibaryon, the H [1]. However, it is very difficult to achieve enough statistics, due to short lifetime of low-energy Ξ^- particle, in order to quantitatively discuss the nature of ΞN and $\Lambda\Lambda$ interactions using a tagged Ξ^- beam from (K^-, K^+) reaction at present facilities.

Thus far, data exist only for Ξ^-p and Ξ^0p scattering cross sections using a

*present address: Research Center for Nuclear Physics, Osaka University, Osaka, 567-0047, Japan. email: jkahn@rcnp.osaka-u.ac.jp

bubble chamber technique[2] in the range of several GeV/c. Data points for Ξ^- reactions jump then to appear above one hundred GeV/c, which were obtained using a hyperon beam facility at CERN-SPS [3]. Due to higher energy and extremely poor statistics of old bubble chamber experiments, the old data is not as meaningful as we can test theoretical estimates of the cross sections below 1 GeV/c. It is unfortunate and interesting that there has never been reported on the measurement of the $\Xi^- p \to \Lambda\Lambda$ annihilation process, which is of primary interest from the point of view; to provide a quantitative discussion about the formation of the double hypernuclei and most preferably about the fusion into a doubly-strange H dibaryon. Therefore, our understanding of the ΞN and $\Lambda\Lambda$ interactions was totally based on observation of several double-Λ hypernuclear events in emulsion.

Fragmentation of double hypernuclei is of great interest in its own right, since one can unveil the reaction mechanism of Ξ^- atomic capture in nuclei at rest. A quasideuteron model [4] predicts that the reaction $\Xi^- + d \to \Lambda\Lambda + n$ is attributed to double-Λ hypernucleus formation. Yamamoto *et al.*[5] modeled a compound double-Λ hypernucleus, while Yamada *et al.* [6] suggested doorway states as an intermediate state, ${}^{13}_{\Lambda\Lambda}\mathrm{B}^*$, in which $\Xi^- p$ annihilates to $\Lambda\Lambda$ in a nucleus. More recently, Hirata *et al.* calculated formation probabilities of single-, twin- and double-Λ hypernuclei based on a microscopic transport model (Quantal Langevin plus Antisymmetrized Molecular Dynamics). These model predictions could be tested by observing Λ hyperons from the stopped-Ξ^- reaction vertex.

2 $\Xi^- p$ Elastic Scattering and Conversion Processes

Our experiment was carried out using a separated 1.66 GeV/c K^- beam at the KEK proton synchrotron. The heart of the experiment is a novel 4π detector consisting of $0.5 \times 0.5 \mathrm{mm}^2$ plastic scintillating fibers. The scintillating fiber (SCIFI) detector acts as a primary interaction target and visual track detector for recognizing (K^-, K^+) reactions and secondary interactions of hyperons produced in the effective volume of $8 \times 8 \times 10$ cm^3. The momenta of outgoing particles were measured with a K^+ spectrometer of approximately 5 m length, consisting of a dipole analyzing magnet with a field integral of 1.08 T·m, 12 drift-chamber planes, two Cherenkov counters, and three hodoscope counters for timing and triggering purposes. The momentum resolution $(\Delta p/p)$ was 0.5% (rms) at 1.2 GeV/c, and the background ratio of π^+ to K^+ was estimated to be less than 0.2 % [7]. The details of the experimental setup are described elsewhere [7,8].

Our previous analysis of (K^-, K^+) reaction data gave a hint on inelastic cross sections for the $\Xi^- p$ interactions. Measuring escaping probability of the Ξ^- from a nucleus could provide upper limits on the inelastic $\Xi^- N$ cross section. The ratio was found to be $80.1 \pm 2.4^{+0.6}_{-1.3}\%$ [9]. The total cross section for $\Xi^- p$ inelastic scattering was then deduced from the Ξ^- absorption probability for ^{12}C, which is approximately 14 mb assuming an isotropic angular distribution. A recent measurement of quasifree $p(K^-, K^+)\Xi^-$ reaction in emulsion plate yields $12.7^{+3.5}_{-3.1}$ mb in the momentum range 0.4–0.6 GeV/c [10], which is well consistent with our result.

On the other hand, our previous analysis of the $\Lambda\Lambda$ invariant mass spectrum suggests that the $\Xi^- p \to \Lambda\Lambda$ cross section is of order of several mb at $p_{\Xi^-} \sim 0.55$

GeV/c [11]. However, it should be stressed that the model calculation contains uncertainties arising from estimates of the cross sections for the two-step processes, which are not known experimentally. Therefore, it is difficult to resolve $\Xi^- p \to \Lambda\Lambda$ reaction from other $\Lambda\Lambda$ production reactions in the (K^-, K^+) reaction, and also from the contribution of the charge-exchange reaction $\Xi^- p \to \Xi^0 n$, albeit with precise measurement of the ratio.

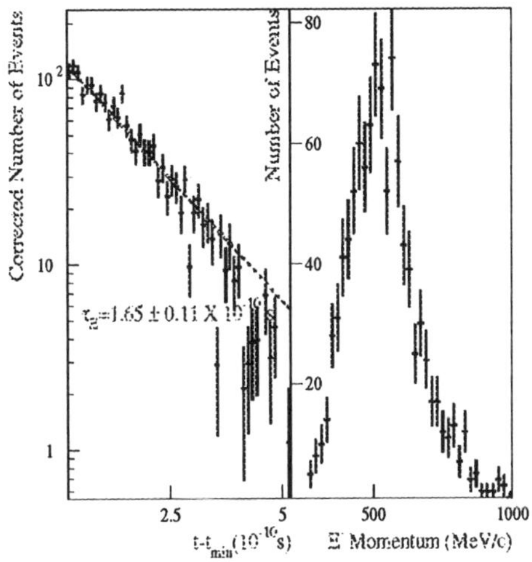

Figure 1. Proper time (left) and momentum (right) distributions of the Ξ^- candidates. The results of the lifetime fit are represented by the overlaid dashed lines. The Ξ^- momentum distribution was measured at the production vertex.

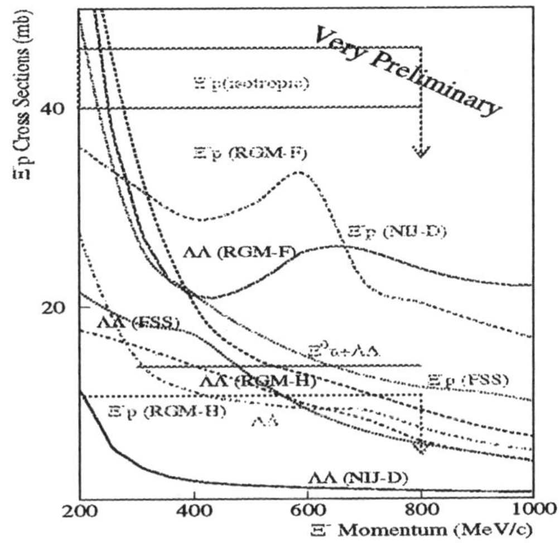

Figure 2. Upper limits on the Ξp cross sections indicated by arrows are compared with theoretical estimates of the RGM-F, the RGM-H, the FSS and the Nijmegen-D model.

We have measured directly Ξ^- interactions with protons and carbon nuclei. Over 5 thousand Ξ^- production events were scanned many times by eye. To make some confidence of the event selection, we calculated a lifetime of the Ξ^- taking the detection efficiency into account for the Ξ^- decay and the subsequent Λ decay simultaneously. The mean lifetimes of the Ξ^- particles (τ_{Ξ^-}=$1.65 \pm 0.11 \times 10^{-10}s$) shown in Fig.1 are consistent with the present world data ($\tau_{\Xi^-}^{PDG}$=$1.639 \pm 0.015 \times 10^{-10}s$).

Out of the Ξ^- event sample, we have found a total of 4 candidate events for $\Xi^- p$ scattering. The $\Xi^- p$ candidate was selected by requiring that the track length of a recoil proton be longer than 3 mm and that of a scattered Ξ^- be longer than 3 mm. None of candidate events has a visible Λ decay following the scattered $\Xi^- \to \Lambda\pi^-$ decay. We have also found 3 candidate events corresponding to $\Xi^- p \to \Lambda\Lambda$ reaction, characterized by two visible Λ decays from a stopping prong at the (K^-, K^+) vertex. Uncertainties in the Ξ^- momentum for the $\Xi^- p$ interaction events result from incomplete kinematics of the $K^- p \to K^+ \Xi^-$ reaction using a composite target; the SCIFI target contains both free proton and carbon nucleus with almost equal amount. We therefore selected only the event correspoding to $\Xi^- p$ scattering off a free proton. A $\Xi^- p$ elastic scattering event was found to be coplanar in the reaction plane. Candidates for the $\Lambda\Lambda$ inelastic channel are all inconsistent with a hypothesis of a two-body scattering off a free proton.

We calculate upper limits on the cross sections for the $\Xi^- p$ elastic and conversion processes based on the observation of one $\Xi^- p$ elastic scattering event with an invisible Λ decay. Since the acceptance is not nearly enough to calculate the cross sections in forward and backward directions, we therefore calculate differential cross sections in the range of the Ξ^- scattering angle $-0.3 < \cos\theta^{cm}_{\Xi^- p} < 0.7$. The cross section for the $\Xi^- p$ elastic scattering is then, for simplicity, assuming an isotropic angular distribution, found to be 40 mb at 90% confidence level, as shown in Fig.2. Regarding the $\Xi^- p \rightarrow \Lambda\Lambda$ reaction, we calculated the cross section with a null result, simply because of kinematical ambiguity in reconstructing the $^{12}\mathrm{C}(\Xi^-, \Lambda\Lambda)X$ reaction. While the results on the elastic cross section give no stringent constraint on theoretical estimates, the upper limit on the conversion process suggests that the estimate of the RGM-F model prediction could be ruled out.

3 Stopped-Ξ^- Reaction

Stopped-Ξ^- reaction candidates were selected by requiring that either neutral(s), a stopping prong, multi-prongs from a stopping point of the Ξ^- track, or a double-kinked track be observed. Due to high momentum transfer ($\sim$ 0.55 GeV/c) in the elementary $K^- p \rightarrow K^+ \Xi^-$ process, it is much less probable to observe a stopped-Ξ^- produced from $p(K^-, K^+)$ reaction, which should leave approximately a 8 cm track. We therefore imposed tight constraints on selection of Ξ^- production from carbon nuclei. Stopped-Ξ^- candidates should be out of coplanar ($|\cos(k_{K^-} \times k_{K^+} \cdot k_{\Xi^-})| > 0.1$) and missing-mass condition ($|m_X - m_{\Xi^-}| > 15$ MeV/c^2).

We then calculated an excitation energy, $E_{^{11}\mathrm{B}}$, relative to mass of a ^{11}B nucleus assuming $K^- {}^{12}\mathrm{C} \rightarrow K^+ \Xi^- {}^{11}\mathrm{B}$ reaction. The excitation enegy spectrum in Fig.3 represents a significant peak at $E_{^{11}\mathrm{B}} = 0$ MeV, and a broad bump above $E_{^{11}\mathrm{B}} = 50$ MeV. The peak can be understood as a sum of the ground state ($3/2^-$), the $1/2^-$ state at 2.21 MeV and the $3/2^-$ state at 5.02 MeV. The small bump at $\sim$ 20 MeV is attributed to the 1s-hole states observed in $^{12}\mathrm{C}(p, 2p)^{11}\mathrm{B}^*$ reaction [12]. The broad bump structure is mainly due to the background from the π^- interaction following the Ξ^- decay. After imposing the requirement that $E_{^{11}\mathrm{B}} < 50$ MeV, 112 events were retained.

The Ξ^--atomic capture reaction at rest could be characterized as either low-energy Λ emission below threshold energies or nearly monoenergetic π^- emission in cascade weak decays. All possible decay channels of the $^{13}_{\Lambda\Lambda}\mathrm{B}^*$ are represented in Ref.[6]. Observation of Λ particles can gives experimental implication on the decay branching ratios of double-Λ plus twin-Λ, single-Λ hypernuclear and no hypernuclear channels. Kinematically, the reaction $\Xi^- {}^{12}\mathrm{C} \rightarrow \Lambda\Lambda + {}^{11}\mathrm{B}$ produces most energetic Λ hyperons at 230 MeV/c, and for single-Λ hypernuclear production, the $\Xi^- {}^{12}\mathrm{C} \rightarrow \Lambda + {}^{12}_{\Lambda}\mathrm{B}$ reaction can produce a Λ below the momentum of 170 MeV/c.

Figure 4 shows a Λ momentum distribution for $\Xi^- {}^{12}\mathrm{C} \rightarrow \Lambda$ X reaction. Hatched area indicates a set of stopped-Ξ^- reaction events, whereas open area shows the data

Table 1. Characteristic π^- kinetic energies for hypernucleus decays

Decay	T_π(MeV)	Decay	T_π(MeV)
${}^{4}_{\Lambda\Lambda}\mathrm{H}\to{}^{4}_{\Lambda}\mathrm{He}+\pi^-$	43.4	${}^{3}_{\Lambda}\mathrm{H}\to{}^{3}\mathrm{He}+\pi^-$	40.7
${}^{5}_{\Lambda\Lambda}\mathrm{H}\to{}^{4}_{\Lambda}\mathrm{H}+p+\pi^-$	53.9	${}^{4}_{\Lambda}\mathrm{H}\to{}^{4}\mathrm{He}+\pi^-$	53.2
${}^{6}_{\Lambda\Lambda}\mathrm{He}\to{}^{5}_{\Lambda}\mathrm{He}+p+\pi^-$	29.5	${}^{5}_{\Lambda}\mathrm{He}\to{}^{4}\mathrm{He}+p+\pi^-$	31.7
${}^{7}_{\Lambda\Lambda}\mathrm{He}\to{}^{7}_{\Lambda}\mathrm{Li}+\pi^-$	34.6	${}^{7}_{\Lambda}\mathrm{He}\to{}^{7}\mathrm{Li}+\pi^-$	41.4
${}^{8}_{\Lambda\Lambda}\mathrm{Li}\to{}^{8}_{\Lambda}\mathrm{Be}+\pi^-$	33.2	${}^{7}_{\Lambda}\mathrm{Li}\to{}^{7}\mathrm{Be}+\pi^-$	36.8

before background subtraction. A strong peak-like structure is observed near 140 MeV/c, where corresponds to the free decay of $\Xi^- \to \Lambda\pi^-$ at rest. However, there was no counterpart π^-. Since we were almost blind for low-energy Λ particles of less than 80 MeV/c, the peak-like structure is accidential as detection efficiency drops significantly. Overlaid lines represent theoretical estimates of the AMD (plus QL method) model for differential yield of Λ emission. Experimental yields integrated over the region of 80–240 MeV/c, $\Delta Y \simeq 0.5$, are smaller than those predicted by the AMD model. This fact implies that there should be larger branching ratio of double-Λ hypernuclear states in ${}^{13}_{\Lambda\Lambda}\mathrm{B}^*$ decay modes than the level predicted by the AMD model.

Kinetic energy spectrum of π^- from the ${}^{13}_{\Lambda\Lambda}\mathrm{B}$ cascade decay is shown in Fig. 5. We assumed all charged particles from the stopped-Ξ^{-12}C reaction vertex to be

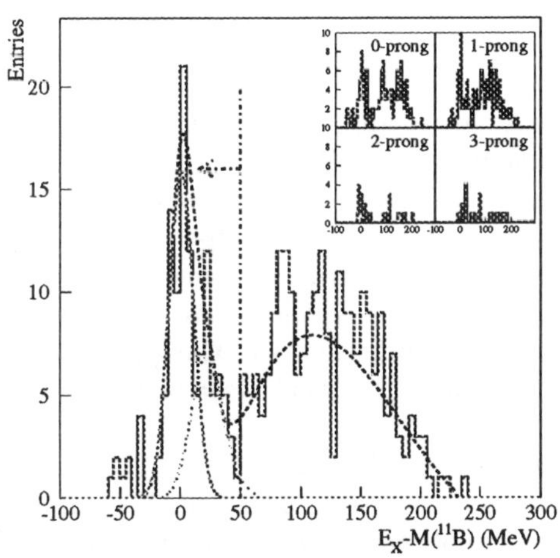

Figure 3. Excitation energy spectrum for $K^{-12}\mathrm{C}\to K^+\Xi^{-11}\mathrm{B}$ reaction relative to the mass of ^{11}B. Insets represent the excitation energy spectra with respect to the number of charged particles at the Ξ^- stopping point.

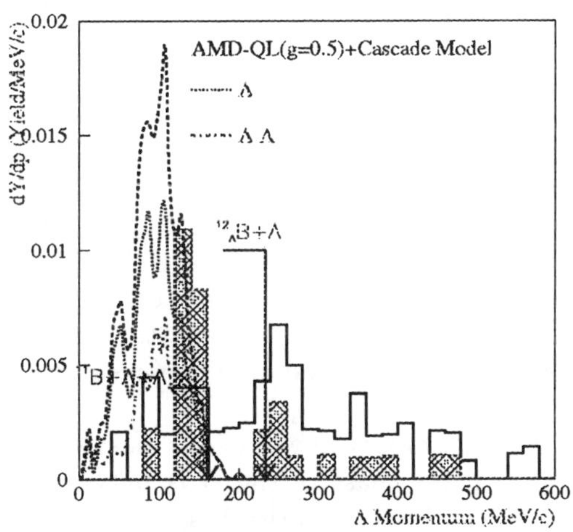

Figure 4. Λ momentum distribution for $\Xi^{-12}\mathrm{C}\to$ Λ X reaction. Hatched area indicates a set of stopped-Ξ^- reaction data, whereas open area shows an event sample before background substraction. Overlaid lines represent theoretical estimates of differential yield for one and two Λ emission probabilities and their sum.

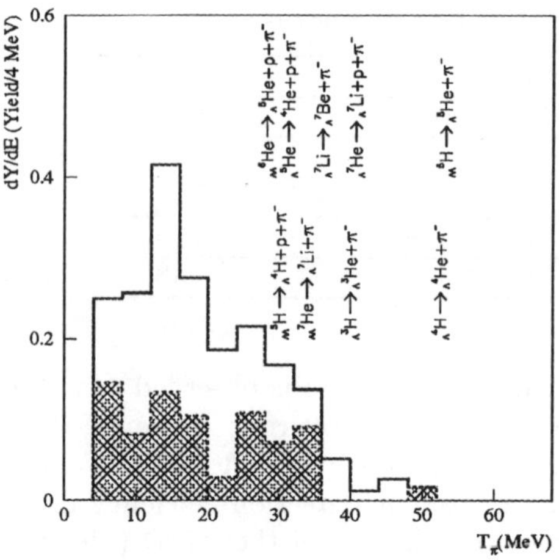

Figure 5. Kinetic energy spectrum of π^- from the ${}^{13}_{\Lambda\Lambda}$B cascade decay. We assumed all charged particles from the stopped-Ξ^{-12}C reaction vertex to be negative pions. Open area represents the spectrum of all events, whereas hatched area does the spectrum of stopped-Ξ^- reaction events. Predicted π^- momenta are shown as texts representing each hypernuclear decay.

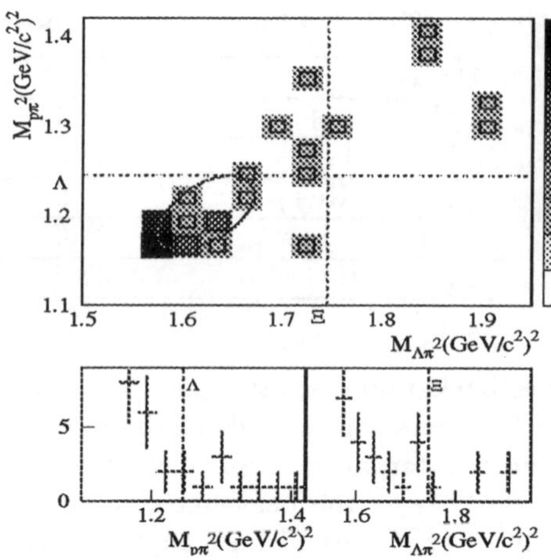

Figure 6. (Top) Dalitz plot for the $H \to \Lambda p \pi^-$ decay is shown as the mass squared of $\Lambda\pi^-$ versus that of $p\pi^-$. Contour shows the maximum phase space for $m_H = 2m_\Lambda$; (Bottom) Projected mass squared spectra for $\Lambda\pi^-$(right) and $p\pi^-$(left) is shown.

negative pions. Open area represents the spectrum of all events, whereas hatched area shows the spectrum of stopped-Ξ^- reaction events. Predicted π^- momenta are shown as texts representing each hypernuclear decay. There is no striking peak resolved corresponding to characteristic π^- weak decay of hypernuclei. The ${}^4_\Lambda$H and ${}_{\Lambda\Lambda}^{\ 6}$He weak decays are characterized by higher energy π^- emission, which could be easily resolved from other decays. Observation of the approximately 50 MeV pions should give information about decay branching ratios of the ${}^{13}_{\Lambda\Lambda}$B* state. We however observed only one event for the ${}^4_\Lambda$H and ${}_{\Lambda\Lambda}^{\ 6}$He weak decays due to the finite size of our target.

The H-dibaryon will decay mostly into $\Lambda p \pi^-$ or $\Lambda n \pi^0$ with the lifetime of a few 10^{-8}s if the mass of the H is between $\Lambda N \pi$ and $\Lambda\Lambda$ threshold. The H$\to$ $\Lambda p \pi^-$ decay can be seen as two V's, one is the Λ and the other is the proton and the pion at the H-decay vertex. Out of two V events, 30 events were selected by requiring that one V be reconstructed as a Λ from the other V's vertex. We required that the reconstructed Λ mass should lie within the interval $\Delta m = 30$ MeV/c^2 of the known mass. After imposing Λ mass cut, 26 events are retained. Calculated proton and pion momenta were checked if those are consistent with the ones measured from the ranges.

A Dalitz plot for the $H \to \Lambda p \pi^-$ decay is shown in Fig. 6. Contour shows the maximum phase space for $m_H = 2m_\Lambda$ and dashed lines indicate reflection of Λ and Ξ^- in the mass squared spectra for $\Lambda\pi^-$ and $p\pi^-$. It is interesting that we observed more events near the Ξ^- mass squared, whereas little background was observed near the Λ mass squared. This fact implies that the proton and pion tracks are not associated with another Λ decay. The peak for the $\Xi^- \to \Lambda\pi^-$ decay can be only understood as the $\Xi^0 n \to \Xi^- p$ reaction. If the Ξ^- track is short enough to reconstruct a Λ mass well whichever we may choose as the vertex, this interpretation makes sense since there should be Ξ^0 production as many as we can observe the secondary interaction.

4 Summary

We report upper limits on the cross sections for the $\Xi^- p$ elastic and conversion processes, 40 mb and 11 mb at 90% confidence level, respectively. Whereas the results on the elastic cross section give no stringent constraint on theoretical estimates, the upper limit on the conversion process suggests that the estimate of the RGM-F model prediction is ruled out. We also report the obervation of the stopped-Ξ^- hyperon-nucleus interaction with respect to hypernuclear production and existence of doubly-strange H-dibaryon. It is necessary to employ two seperate targets for production and interaction of hyperons, in order to suppress unknown background due to hyperon-induced multi-step nuclear processes. This kind of experiments can be carried out at high-intensity machines such like JHP.

Acknowledgments

We would like to thank the staff of the KEK proton synchrotron for their continuous support. We also thank C. Nakamoto for valuable discussions. This work was supported in part by a Grant-in-Aid for Scientific Research (02402008), the Ministry of Science, Culture and Education, Japan.

References

1. R.L. Jaffe, Phys. Rev. Lett. **38**, 195 (1977); **38**, 617(E) (1997).
2. G.R. Charlton, Phys. Lett. **B32**, 720 (1970); R.D.A. Dalmeijer *et al.*, Nouvo Cimento Letters **4**, 373 (1970); R.A. Muller, Phys. Lett. **B38**, 123 (1972); J.M. Hauptman *et al.*, Nucl. Phys. **B125**, 29 (1977).
3. S.F. Biagi, Z. Phys. **C17**, 113 (1983).
4. D. Zhu, C.B. Dover, A. Gal and M. May, Phys. Rev. Lett. **67**, 2268 (1991).
5. Y. Yamamoto, M. Sano and M. Wakai, Prog. Phys. Suppl. **117**, 265 (1994).
6. T. Yamada and K. Ikeda, Phys. Rev. **C56**, 3216 (1997) ; T. Yamada and K. Ikeda, Nucl. Phys. **A585**, 79c (1995).
7. J.K. Ahn *et al.*, Phys. Lett. **B378**, 53 (1996).
8. J.K. Ahn *et al.*, Nucl. Phys. **A625**, 231 (1997).
9. K. Nakazawa, Nucl. Phys. **A639**, 345c (1998).
10. S. Aoki *et al.*, Nucl. Phys. **A644**, 365 (1998).
11. J.K. Ahn,*et al.*, Phys. Lett. **B444**, 267 (1998).
12. H. Toyokawa and M. Yosoi, RCNP-E81.

FEW-BODY SYSTEMS WITH STRANGENESS −1 AND −2

Y. AKAISHI[1], T. HARADA[2] and S. SHINMURA[3]

[1] *Institute of Particle and Nuclear Studies, KEK, Tanashi, Tokyo 188-8501, Japan*
[2] *Sapporo Gakuin University, Ebetsu, Hokkaido 069-8555, Japan*
[3] *Gifu University, Gifu 501-1112, Japan*

It is found that the suppression due to two-body ΛN-ΣN coupling solves the overbinding problem in ${}^{5}_{\Lambda}$He but it, in turn, causes a severe underbinding in the four-body systems. The shortage of this binding is overcome by introducing the Λ-Σ coupling explicitly which is equivalent to the ΛNN three-body force. This three-body force becomes strong in the 0^+ state of ${}^{4}_{\Lambda}$H and ${}^{4}_{\Lambda}$He according to the coherently added enhancement. The 0^+-1^+ splitting in ${}^{4}_{\Lambda}$H and ${}^{4}_{\Lambda}$He is found partly due to the ΛN spin-spin interaction and partly due to the Λ-Σ coupling in the recent Nijmegen soft-core potential. In the strangeness −2 sector there is a possibility that the lightest double-Λ hypernucleus ${}^{4}_{\Lambda\Lambda}$H is abundantly populated with 15% branching by stopping Ξ^- on ^{4}He. A stopped Ξ^- on ^{9}Be also produces efficiently a variety of $\Lambda\Lambda$ hyperfragments, weak-decay pions from which provide a means of mass spectroscopy of double-Λ hypernuclei. It is predicted that the excited state with strangeness halo and the ground state ${}^{5}_{\Xi}$H with 1% Ξ mixture appear in a double-strangeness five-body system.

1 Introduction

There is a long standing problem of fitting the experimental Λ-separation energies of ${}^{3}_{\Lambda}$H, ${}^{4}_{\Lambda}$H, ${}^{4}_{\Lambda}$He and ${}^{5}_{\Lambda}$He consistently. Dalitz *et al.*[1] determined a reasonable ΛN central force that will produce the correct Λ-separation energies in ${}^{3}_{\Lambda}$H and ${}^{4}_{\Lambda}$H. Then this potential is found to be so strong for ${}^{5}_{\Lambda}$He that it gives the binding energy value $B_\Lambda({}^{5}_{\Lambda}\mathrm{He})$ of 5.46 MeV which is much larger than the experimental value of 3.12 ± 0.02 MeV. Shinmura *et al.*[2] gave a solution to this overbinding problem by including a phenomenological ΛN tensor force. On the other hand, the significant effect of coupling between Σ and Λ has also been proposed in theoretical treatments of the hypernuclear problem [3]. In the first part of this paper we investigate the binding mechanism of the *s*-shell hypernuclei and show that some of realistic YN interactions will reproduce all the *s*-shell hypernuclear data by solving the overbinding problem with ΛNN three-body force due to the Λ-Σ coupling.

The formation of double strangeness systems is discussed in the second part of this paper.

2 Hyperon-Nucleon Interactions and Λ-Σ Coupling Model

In order to find some key to solve the overbinding problem we first prepare various types of potentials simplified from realistic hyperon-nucleon (YN) interactions. The potential which is designated as D0 has only central part of ΛN interaction, D1 has central and tensor parts of ΛN channel interaction, and D2 has central parts of both ΛN and ΣN channels. All central and tensor parts of the both channels are included in D3. These D0 $\sim$ 3 potentials give identical phase shifts which are equivalent to the Nijmegen D interaction [4]. The potentials SC89(S) and SC97f(S) are obtained in a same way as D3 from the Nijmegen soft-core SC89 [5] and SC97f [6] interactions.

The Λ separation energies of four-body and five-body hypernuclei are calculated by the Brueckner-Hartree-Fock method on Gaussian basis: Reaction matrices (g_{YN}) from YN interaction are expanded on 20-range Gaussian functions and are used to obtain a hyperon-nucleus Hartree-Fock potential by folding procedure with nucleon wave functions of Gaussian combination. This hyperon-nucleus potential is again expanded on 10-range Gaussian bases, and is used to solve the relative motion between the hyperon and the core nucleus. Matrix elements of potentials with respect to Gaussian functions are calculated without approximation, and the center-of-mass motion of the system is removed by using the hyperon-nucleus reduced mass.

Now let's see some results obtained by restricting the hyperon wave function in the Λ space where Σ-space effects are taken into account via intermediate states of reaction matrices. The D0 potential of only central type ΛN interaction gives a good result $B_\Lambda = 2.44$ MeV for ${}^4_\Lambda$He compared to the experimental one 2.39 ± 0.03 MeV, but causes the overbinding of 6.66 MeV in ${}^5_\Lambda$He as it was calculated by Dalitz *et al.* By employing the D2 potential which includes the ΛN - ΣN coupling of central type, the binding energy in the five-body system is reduced to 3.01 MeV which is well close to the experimental one. This is due to the large suppression of the ΛN - ΣN coupling in the nucleus. While this coupling remedies the overbinding problem in ${}^5_\Lambda$He, the suppression effect in turn causes a serious underbinding problem in the four-body systems, that is, $B_\Lambda = 1.04$ MeV for ${}^4_\Lambda$He which is short by 1.35 MeV.

How to solve this underbinding problem of the four-body system? Gibson *et al.* [7] firstly introduced Σ-space components explicitly into the wave function of ${}^4_\Lambda$H and ${}^4_\Lambda$He as

$$|{}^4_\Lambda\mathrm{He}> = \Phi_\Lambda(\vec{r})|{}^3\mathrm{He}> + \sqrt{\frac{2}{3}}\Phi_{\Sigma^+}(\vec{r})|{}^3\mathrm{H}> - \sqrt{\frac{1}{3}}\Phi_{\Sigma^0}(\vec{r})|{}^3\mathrm{He}> . \qquad (1)$$

We adopt this idea in order to take into account ΛNN three-body force effects due to the Λ - Σ conversion. Then, the hyperon-nucleus potential has a Λ-Σ coupling term, to which the two-body ΛN-ΣN coupling interaction contributes as follows [8];

$$\begin{aligned} &\frac{3}{2}{}^3g_{\Sigma N,\Lambda N} - \frac{1}{2}{}^1g_{\Sigma N,\Lambda N} \quad \text{for } 0^+, \\ &\frac{1}{2}{}^3g_{\Sigma N,\Lambda N} - \frac{1}{2}{}^1g_{\Sigma N,\Lambda N} \quad \text{for } 1^+. \end{aligned} \qquad (2)$$

Since the ΛN-ΣN coupling strength is much stronger in the spin-triplet state than in the spin-singlet state, the Λ-Σ coupling effect on B_Λ in the 0^+ state becomes by about one order of magnitude as large as that in the 1^+ state. This is due to coherent enhancement in the 0^+ state where contributions from each pair of YN coupling are added constructively. On the other hand, contributions from the two pairs are cancelled out in the 1^+ state and the resulting spin-triplet weight is three times smaller than that in the 0^+ state. Therefore, the attractive effect due to the Λ-Σ coupling in the 0^+ state is $9 = (3/2)^2/(1/2)^2$ times larger than that in the 1^+ state. In this respect the 0^+ state is an extraordinary state.

By using the potential D2 the three-body force effect due to the Λ-Σ coupling is estimated to be 1.23 MeV attraction with 1.9% Σ-mixing for the 0^+ state and only 0.01 MeV attraction with 0.01% Σ-mixing for the 1^+ state. Thus, all the existing experimental energies of s-shell Λ hypernuclei are well reproduced by the simple D2 potential which is given as

$$v(r) = V_a \exp\left(-(r/a)^2\right) + V_b \exp\left(-(r/b)^2\right) \tag{3}$$

with $a = 0.5$ fm and $b = 1.2$ fm. The strength parameters are given in Table 1.

Table 1. Parameters of the D2 potential in even states (unit in MeV).

		$T=1/2$		$T=3/2$	
		V_a	V_b	V_a	V_b
^{1}E	$\Lambda N-\Lambda N$	1165.0	−105.12		
^{1}E	$\Sigma N-\Lambda N$	90.547	−45.085		
^{1}E	$\Sigma N-\Sigma N$	927.16	−9.9459	1047.0	−119.24
^{3}E	$\Lambda N-\Lambda N$	763.11	−83.938		
^{3}E	$\Sigma N-\Lambda N$	43.344	55.049		
^{3}E	$\Sigma N-\Sigma N$	539.41	−116.48	1756.1	−35.821

In the D2 case the 0^+-1^+ splitting in $^4_\Lambda$H and $^4_\Lambda$He comes not from the ΛN spin-spin interaction but from the ΛNN three-body force. This confirms Gibson's statement that the 0^+-1^+ energy difference is not a measure of the ΛN spin-spin interaction [9]. However, the D3 potential which simulates most closely the original Nijmegen D potential brings only a small 0^+-1^+ splitting energy of 0.13 MeV in contrast to the D2 potential. This is due to the difference of ΛN-ΣN potential type, which is of central type for D2 but is mainly of tensor type for D3. The success of D2 in solving the four-body underbinding problem does not apply to the Nijmegen D potential.

3 Λ-Σ Coupling of Realistic Hyperon-Nucleon Interaction

Recently the Nijmegen group proposed a new soft-core model for YN interaction [6]. We try to solve the overbinding/underbinding problem by employing the realistic soft-core interaction models [5,6]. Figure 1 summarizes the results calculated for $^4_\Lambda$H together with the experimental data. The left part and the right part for each interaction are the cases without and with the Σ-space component of eq. (1), respectively. The level splitting at the left part is due to the ΛN spin-spin interaction, and the level shifts seen at the right part are the ΛNN three-body force effect due to the Λ-Σ coupling of the hyperon-nucleus potential. The Σ admixture is shown in the figure for the 0^+ state and it is negligibly small for the 1^+ state.

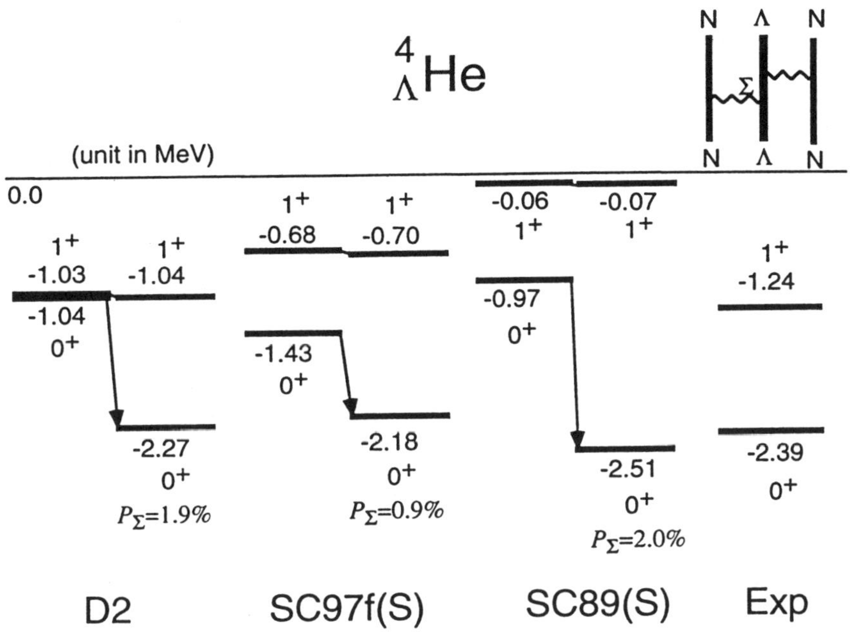

Fig. 1. The Λ energy levels calculated with the D2, SC97f(S) and SC89(S) potentials. The level shift shown by the arrow is mainly due to the ΛNN three-body force.

The SC89(S) potential gives $B_\Lambda(0^+, {}^4_\Lambda\mathrm{He})$ of 2.51 MeV. Carlson obtained 1.6 MeV for it by the VMC method [10] which would be increased a little more by the GFMC method. As seen in the figure the 0^+-1^+ splitting result is too large compared to the experimental one due to too strong ΛN-ΣN coupling of SC89(S). This strong coupling causes too strong suppression in the five-body system giving $B_\Lambda({}^5_\Lambda\mathrm{He})$ of 0.53 MeV which is not inconsistent with Carlson's unbound result [10].

The SC97f(S) potential fits rather well all the experimental Λ binding energies of s-shell hypernuclei. In order to understand the nature of interaction we divide the ΛN-ΣN coupling interaction into its central and tensor parts. The three-body effect on the 0^+-1^+ splitting is due to the ΛN-ΣN central coupling, while the large suppression is mainly due to the ΛN-ΣN tensor coupling. The 0^+-1^+ splitting in ${}^4_\Lambda\mathrm{He}$ is found partly due to the ΛN spin-spin interaction and partly due to the ΛN-ΣN central coupling in the case of SC97f(S). The binding $B_\Lambda({}^5_\Lambda\mathrm{He})$ is calculated to be 2.38 MeV for SC97f(S) and 3.57 MeV for SC97e(S). Since the experimental value 3.12 ± 0.02 MeV is in between them, the result for SC97f(S) is not bad and would be adjusted so as to fit the data without difficulty. A proper ratio of the ΛN-ΣN central coupling and tensor coupling is significant in hyperon-nucleon interactions.

It is found that the Nijmegen SC97f potential meets this requirement.

4 Concluding Remarks to Strangeness −1 Part

The suppression of binding due to the two-body ΛN-ΣN coupling solves the overbinding problem in ${}^{5}_{\Lambda}$He but it, in turn, causes the underbinding problem in the four-body systems. The shortage of this binding is overcome by introducing the three-body force ΛNN due to the Λ-Σ coupling of hyperon-nucleus potential. This three-body force is strong only in the 0^+ states of ${}^{4}_{\Lambda}$H and ${}^{4}_{\Lambda}$He according to the coherently added enhancement. It is found that two different types of YN interactions, D2 and SC97f(S), can solve the overbinding/underbinding problem. The 0^+-1^+ spin doublet splitting in ${}^{4}_{\Lambda}$He is half due to the ΛN spin-spin interaction and half due to the Λ-Σ coupling in SC97f(S). This calculation has justified Gibson's statement that the energy splitting would not be a measure of the ΛN spin-spin interaction [9], first time on the basis of realistic YN interaction.

5 Formation of Light $\Lambda\Lambda$ Hypernuclei

Double-strangeness ($S = -2$) hypernuclei are of particular interest: What nature would be revealed about the $\Lambda\Lambda$ and ΞN interaction? Since two-body scattering data are not available in the $\Lambda\Lambda$ sector, Ξ- and $\Lambda\Lambda$-hypernuclei are unique sources of information concerning the $S = -2$ interaction, which gives a quantitative basis to study the property of multi-strangeness hadronic matter [11]. The improvement of experimental technique in search for double-strangeness systems encourages us to do theoretical investigation of $S = -2$ systems, in which couplings among $\Lambda\Lambda$, ΞN and $\Sigma\Sigma$ channels would play essential roles.

Is there any way to produce abundantly double-Λ hypernuclei? Suppose the case where a Ξ^- hyperon stops on a nucleus and then forms a double-Λ hypernucleus via the elementary reaction process;

$$\Xi^- + p \to \Lambda + \Lambda + 28.33\ \text{MeV}. \tag{4}$$

This reaction Q value of 28.33 MeV is accidentally very close to the binding energy of the alpha particle 28.30 MeV. If an alpha particle is broken by absorbing a Ξ^- hyperon stopped on its atomic orbit, the released Q-value energy is almost exhausted and two produced Λ's cannot get enough energy to escape a hypernucleus. Then, there occurs a large chance for the hyperons to form a double-Λ hypernucleus.

The lightest double-Λ hypernucleus is especially important to be studied, since its simple structure enables us to extract the interaction property. The possible lightest double-Λ hypernucleus is predicted to be ${}^{4}_{\Lambda\Lambda}$H [12]. The formation rate of ${}^{4}_{\Lambda\Lambda}$H via stopped Ξ^- on ^{4}He is estimated by Kumagai-Fuse *et al.* [13]. All the possible reaction processes are the following

$$\begin{aligned}
\Xi^- \cdot {}^4\mathrm{He} &\rightarrow \Lambda + \Lambda + p && +n+n && +0.03\ \mathrm{MeV} \\
&\rightarrow \Lambda + \Lambda + {}^2\mathrm{H} && +n && +2.25\ \mathrm{MeV} \\
&\rightarrow \Lambda + \Lambda + {}^3\mathrm{H} && && +8.51\ \mathrm{MeV} \\
&\rightarrow \Lambda + {}^3_\Lambda\mathrm{H} && +n && +2.38\ \mathrm{MeV} \\
&\rightarrow \Lambda + {}^4_\Lambda\mathrm{H} && && +10.55\ \mathrm{MeV} \\
&\rightarrow \Lambda + {}^4_\Lambda\mathrm{H}^* && && +9.50\ \mathrm{MeV} \\
&\rightarrow {}^4_{\Lambda\Lambda}\mathrm{H} && +n && +2.83\ \mathrm{MeV}.
\end{aligned} \tag{5}$$

The reaction Q values remain small for these processes. The phase volume of final states is proportional to $Q^{1/2}$ for two-body breakup and generally to $Q^{(3n-5)/2}$ for n-body breakup. Thus, the small Q values favor the processes of two-body breakup.

Table 2 shows the formation rate and branching ratio of each process calculated for atomic s-orbit and p-orbit absorptions of Ξ^-. The branching ratios of the four-body and five-body breakup processes are negligibly small due to the small phase volumes. Zhu *et al.* [14] investigated the double-Λ hypernucleus formation via Ξ^- capture in ^{6}Li and showed that the branching ratio of the ${}^6_{\Lambda\Lambda}\mathrm{H}$ formation is 3%. They also argued that in the case of $^4\mathrm{He} \cdot \Xi^-$ the formation of ${}^4_{\Lambda\Lambda}\mathrm{H}$ is forbidden for the p-orbit absorption due to the angular momentum conservation. However, this is not the case in the light nucleus, since the center-of-mass coordinate of the core nucleus differs before and after the reaction.

In the case of the ${}^4_{\Lambda\Lambda}\mathrm{H}$ formation, the rate is suppressed by the final-state interaction for the s-orbit absorption while it is enhanced for the p-orbit absorption. This situation is explained in relation to the existence of a bound state. When the system of ${}^4_{\Lambda\Lambda}\mathrm{H} + n$ has a bound state ${}^5_{\Lambda\Lambda}\mathrm{H}$ which concentrates in it the neutron s-wave strength, the formation rate is largely suppressed in the s-orbit absorption case. On the other hand, when the distortion potential becomes stronger but still has no bound state, the maximum of the formation rate moves toward the smaller-Q side and becomes higher. The ${}^4_{\Lambda\Lambda}\mathrm{H}$ formation rate from the p-orbit absorption is greatly enhanced and amounts to 20%. Now we consider the atomic cascade process of Ξ^- in ^{4}He. The Ξ^- capture fractions from atomic s orbit and from p orbit are estimated to be 25% and 73%, respectively [13]. Thus, a large branching ratio of 15% is obtained for the ${}^4_{\Lambda\Lambda}\mathrm{H}$ formation.

Table 2. The formation rate and branching ratio [%] from Ξ^- stopped on ^{4}He.

Process	s−orbit absorp.(10^{18}/s)	p−orbit absorp.(10^{15}/s)
${}^4_{\Lambda\Lambda}\mathrm{H} + n$	0.04 [1.5%]	0.73 [20.1%]
$\Lambda + {}^4_{\Lambda\Lambda}\mathrm{H}$	0.41 [15.2%]	0.61 [16.9%]
$\Lambda + {}^4_{\Lambda\Lambda}\mathrm{H}*$	1.27 [46.7%]	1.89 [52.2%]
$\Lambda + \Lambda + {}^3\mathrm{H}$	0.94 [34.7%]	0.39 [10.8%]
$\Lambda + {}^3_\Lambda\mathrm{H} + n$	0.05 [1.9%]	0.00 [0.05%]

The breakup of an alpha particle or an alpha cluster benefits the formation of double-Λ hypernuclei. The ^{9}Be nucleus is also a preferable target to populate double-Λ hypernuclei. Suppose that a Ξ^- hyperon stops on ^{9}Be, which has a well-developed cluster structure of $\alpha + \alpha + n$, and reacts with one of the α clusters. Then, a variety of double-Λ hypernuclei such as $^{4}_{\Lambda\Lambda}$H $\sim^{7\sim 9}_{\Lambda\Lambda}$Li can be formed. Now the problem is how to identify the simultaneously produced double-Λ hypernuclei. A useful means is to observe characteristic pion spectra from the weak decay of hypernuclei. For example, $^{4}_{\Lambda\Lambda}$H has a series of two-body weak decays $^{4}_{\Lambda\Lambda}$H $\rightarrow ^{4}_{\Lambda}$H $+\pi^0$, then $^{4}_{\Lambda}$H $\rightarrow \alpha + \pi^-$. These discrete-energy pions are the unique signal of the $^{4}_{\Lambda\Lambda}$H formation. In the case of $^{4}_{\Lambda}$H $\rightarrow \alpha + \pi^-$ and $t+p+\pi^-$, the discrete π^- has a large branching ratio of about 70% [15]. This large ratio is due to the antisymmetrization between a proton born from Λ and a proton included in the triton cluster.

The second lightest hypernucleus is $^{5}_{\Lambda\Lambda}$H. As an impressive example, the double-Λ nucleus formation via a neutron-rich Ξ-state is discussed by Kumagai-Fuse *et al.* [16] The $^{5}_{\Lambda\Lambda}$H nucleus is formed with extremely large branching of 80 % from $^{7}_{\Xi}$H which can be produced by ^{7}Li (K^-, K^+).

6 Double-Strangeness Five-Body System

The $S = -2$ five-body system is composed of $^{5}_{\Xi}$H and $^{5}_{\Lambda\Lambda}$H [17]. An interesting feature of this system is that the $p\Xi^-$ channel comes closer to the $\Lambda\Lambda$ channel due to the large binding energy of the alpha particle. The threshold difference is reduced to 8.51 MeV in this case from the mass difference 28.33 MeV between $p\Xi^-$ and $\Lambda\Lambda$. The mixing of $p\Xi^-$ and $\Lambda\Lambda$ components is interesting since it has some relation to the H-dibaryon in the flavor SU(3) singlet state.

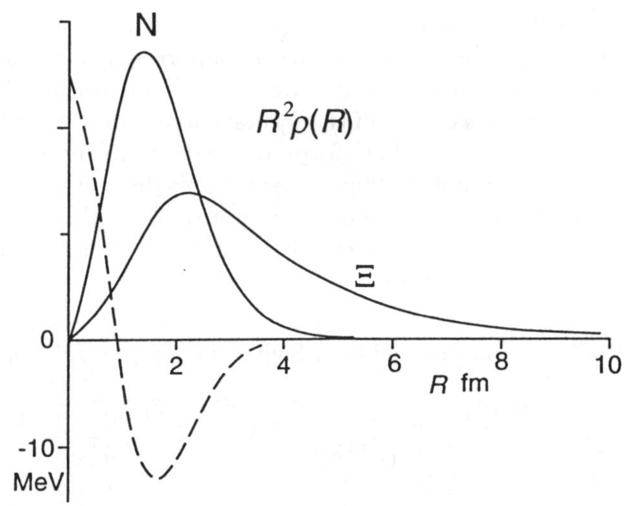

Fig.2. The density distribution of Ξ^- in the $\alpha\Xi^-$ system. The dashed curve denotes the α - Ξ^- potential.

In the system a Ξ-hypernuclear state exists at 6.8 MeV above the $t + \Lambda + \Lambda$ threshold, that is, at -1.7 MeV below the $\alpha + \Xi^-$ threshold. The state is open in the $t + \Lambda + \Lambda$, the ${}^{4}_{\Lambda}\mathrm{H}^* + \Lambda$ and the ${}^{4}_{\Lambda}\mathrm{H}+\Lambda$ channels. From the detailed calculation by Kumagai-Fuse *et al.* [18] it is found that the conversion width of ${}^{5}_{\Xi}\mathrm{H}$ is narrow, which is only $\Gamma = 0.8$ MeV.

The binding mechanism of this system is interesting. In the $\alpha\Xi^-$ system there exists an atomic $1s$ state with 24.0 fm rms distance. The Coulomb interaction gives a binding energy of 0.1 MeV for Ξ^-. The strong interaction between α and Ξ^- brings about a shallow bound state with 0.5 MeV binding energy and 5.3 fm rms distance. When both strong and Coulomb interactions are combined, the binding energy increases to 1.6 MeV but the rms distance remains still large to 4.0 fm due to the long-range nature of the Coulomb interaction. This bound state ${}^{5}_{\Xi}\mathrm{H}$ is a strangeness-halo hybrid state: The Coulomb interaction ensures the existence of this state while the strong interaction maintains the nature of a nuclear state. Both the Coulomb and strong interactions play important roles in forming this exotic state.

There also exists a bound state ${}^{5}_{\Lambda\Lambda}\mathrm{H}$ with $E = -6.3$ MeV below the $t + \Lambda + \Lambda$ threshold. The mixture of Ξ^- component in this state is only 1.0 %. The formation of α from $t + p$ lowers the threshold of the $p\Xi^-$ channel but at the same time weakens the coupling potential with a small spin-isospin weight of the $p\Xi^-$ singlet state which contributes to the conversion process. Thus, the bound state becomes an almost pure ${}^{5}_{\Lambda\Lambda}\mathrm{H}$ state. The two Λ particles in the state can decay into p and Σ^- [19], and if the decay products p and t form an α particle, it would give a larger Q value as ${}^{5}_{\Lambda\Lambda}\mathrm{H} \to \alpha + \Sigma^- + 109.1$ MeV compared to $\Lambda + \Lambda \to \Sigma^- + p + 95.56$ MeV. The mono-energetic pion of 82.5 MeV is a clear signal of the ${}^{5}_{\Lambda\Lambda}\mathrm{H}$ formation. The observation of this discrete energy would provide new information concerning $\Lambda\Lambda$ weak interaction.

7 Concluding Remarks to Strangeness −2 Part

The breakup of an alpha particle/cluster benefits the formation of double-Λ hypernuclei. The conversion reaction to $\Lambda\Lambda$ from $p\Xi^-$ after Ξ^- stopping on ^{4}He produces the lightest double-Λ hypernucleus ${}^{4}_{\Lambda\Lambda}\mathrm{H}$ with a high branching ratio of 15%. The stopped Ξ^- on ^{9}Be produces a variety of double-Λ hypernuclei. Their formation can also be identified with discrete pion spectra.

The ${}^{5}_{\Xi}\mathrm{H}$ state is bound by the cooperation of the Coulomb and the strong interactions. The Coulomb interaction ensures the existence of the bound state, while the strong interaction keeps it as a halo nuclear state. The ${}^{5}_{\Lambda\Lambda}\mathrm{H}$ state is an almost pure $\Lambda\Lambda$ state with 1% admixture of the Ξ^- component. This state can decay through a non-mesonic process $\Lambda + \Lambda \to \Sigma^- + p$ which has never been observed yet.

Through the above glimpse of the "strangeness −2 world", we may have an impression that $S = -2$ nuclear physics is very rich in diversity compared to ordinary one without strangeness. Further experimental investigations of this interesting frontier require high-intensity strangeness beam. The new 50 GeV proton synchrotron being planned in Japan would help to create a unique playground on the strangeness nuclear physics.

The authors would like to thank Prof. Th.A. Rijken for valuable discussion.

References

1. R.H. Dalitz, R.C. Herndon and Y.C. Tang, *Nucl. Phys.* **B47** (1972) 109.
2. S. Shinmura, Y. Akaishi and H. Tanaka, *Prog. Theor. Phys.* **65** (1981) 1290.
3. A.R. Bodmer, *Phys. Rev.* **141** (1966) 1387.
4. M.M. Nagels, T.A. Rijken and J.J. de Swart, *Phys. Rev.* **D12** (1975) 744.
5. P.M.M. Maessen, Th.A. Rijken and J.J. de Swart, *Phys. Rev.* **C40** (1989) 2226.
6. Th.A. Rijken, V.G.J. Stoks and Y. Yamamoto, *Phys. Rev.* **C59** (1999) 21.
7. B.F. Gibson, A. Goldberg and M.S. Weiss, *Phys. Rev.* **C6** (1972) 741.
8. J. Dabrowski, *Phys. Rev.* **C8** (1973) 835.
9. B.F. Gibson, *Proc. U.S.-Japan Seminar, Maui* (1993) 55.
10. J. Carlson, *AIP Conf. Proc.* **224** (1991) 198.
11. S. Balberg, A. Gal and J. Schaffner, *Prog. Theor. Phys. Suppl.* **117** (1994) 325.
12. S. Nakaichi-Maeda and Y. Akaishi, *Prog. Theor. Phys.* **84** (1990) 1025.
13. I. Kumagai-Fuse, T. Koike and Y. Akaishi, *Nucl. Phys.* **A585** (1995) 367c.
14. D. Zhu, C.B. Dover, A. Gal and M. May, *Phys. Rev. Lett.* **67** (1991) 2268.
15. I. Kumagai-Fuse, S. Okabe and Y. Akaishi, *Phys. Lett.* **B345** (1995) 386.
16. I. Kumagai-Fuse and Y. Akaishi, *Phys. Rev.* **C54** (1996) R24.
17. Khin Swe Myint and Y. Akaishi, *Prog. Theor. Phys. Suppl.* **117** (1994) 251.
18. I. Kumagai-Fuse and Y. Akaishi, *Prog. Theor. Phys.* **94** (1995) 151.
19. M. May, *Nuovo Cimento* **A102** (1989) 401.

FOUR-BODY CALCULATION OF ${}^{4}_{\Lambda}$H AND ${}^{4}_{\Lambda}$He WITH REALISTIC YN AND NN INTERACTIONS

E. HIYAMA
Institute for Physical and Chemical Research (RIKEN), Wako, Saitama 351, Japan

M. KAMIMURA
Department of Physics, Kyushu University, Fukuoka 812, Japan

T. MOTOBA
Laboratory of Physics, Osaka Electro-Communication University, Neyagawa 572, Japan

T. YAMADA
Laboratory of Physics, Kanto Gakuin University, Yokohama 236, Japan

Y. YAMAMOTO
Physics Section, Tsuru University, Tsuru, Yamanashi 402, Japan

We carried out four-body calculations of ${}^{4}_{\Lambda}$H and ${}^{4}_{\Lambda}$He taking both the $3N + \Lambda$ and $3N + \Sigma$ channels explicitly with the use of realistic NN and YN interactions. The Σ-channel component plays an important role in binding energies of the $A = 4$ hypernuclei though the adimixture is approximately 1 %. The $\Lambda N - \Sigma N$ coupling is found to be of central-force type in the Nijmegen model D and of tensor-force type in the model D

1 Introduction

One of the main goals in hypernuclear physics is to obtain useful information on the hyperon-nucleon (YN) interaction. Several types of YN interactions have been proposed. However, these interactions have a lot of ambiguity, since the YN scattering data are very limitted. Therefore, as a complementary approach, the detailed study of hypernuclear structure should be helpful in elucidating the YN interaction properties. Structure studies of ${}^{3}_{\Lambda}$H can be used [1] for tests of YN interactions. However, since ${}^{3}_{\Lambda}$H is very weakly bound sytem, the off-shell effect in the YN interactions cannot be well examined. On the other hand, ${}^{4}_{\Lambda}$H and ${}^{4}_{\Lambda}$He are strongly bound systems and both of spin-doublets states have been observed Therefore, four-body calculations of these hypernuclei become important. Especially, it is expected to obtain information on the role of $\Lambda N - \Sigma N$ coupling in the binding mechanisum of the $A = 4$ hypernuclei. In this study, we report a four-body calculation of ${}^{4}_{\Lambda}$H and ${}^{4}_{\Lambda}$He taking the $\Lambda N - \Sigma N$ channel coupling into account explicitly.

2 Model and Method

In order to solve four-body problem accurately, we employ a coupled variational method [2,3] with the use of Jacobian coordinate Gaussian basis functions which span all the rearrangement channels in Fig. 1. This method has been succesfully applied to the three- nucleon bound states [3,4] with realistic two-body and three-body force and to muonic molecules [2], unstable nuclei [5] and hypernuclei [6,7,8].

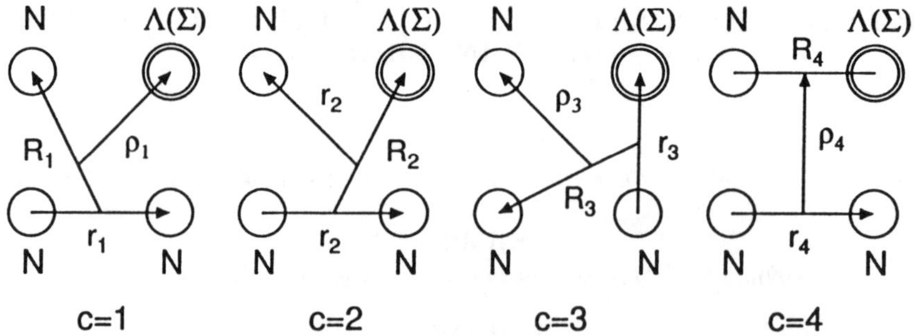

Figure 1: Jacobian coordinates for all rearrangement channels of $(3N+\Lambda)+(3N+\Sigma)$.

The method works excellently in describing both the short-range correlations and the long-range tail behaviour.

We take AV8 NN realistic potential For YN interaction, We take Gaussian soft-core potential with ΛN ΣN coupling terms which simulate the phase shifts obtained from Nijmegen model D (ND) and model F (NF). Theses ND and NF-like soft-core potentials are derived by cone of the present authors (Y. Y) and consists of central term and tensor term. The Coulomb force is included. The four-body wave function is written as

$$\Psi_{JM}({}^4_\Lambda\text{H},{}^4_\Lambda\text{He}) = \sum_{Y=\Lambda,\Sigma}\sum_{c=1}^{4}\sum_{I,j_0}\sum_{l,L,\lambda}\sum_{s_0,s,S,t_0,t_0}\sum_{n,N,\nu} A^{(c)}_{Y,nlNL\nu\lambda Ij_0s_0sSt_0t}$$
$$\times\mathcal{A}_N\Bigg[\Big[\big[\phi^{(c)}_{nl}(\mathbf{r}_c)\psi^{(c)}_{NL}(\mathbf{R}_c)\big]_I\xi^{(c)}_{\nu\lambda}(\rho_c)\Big]_{j_0}$$
$$\times\Big[[[\chi^{(\sigma)}_{1/2}(N_1)\chi^{(\sigma)}_{1/2}(N_2)]_{s_0}\chi^{(\sigma)}_{1/2}(N_3)]_s\chi^{(\sigma)}_{1/2}(Y)\Big]_S\Bigg]_{JM}$$
$$\times\Big[[[\chi^{(\tau)}_{1/2}(N_1)\chi^{(\tau)}_{1/2}(N_2)]_{t_0}\chi^{(\tau)}_{1/2}(N_3)]_t\chi^{(\tau)}_{t_y}(Y)\Big]_{T=1/2}, \quad (1)$$

where n, N and ν are used to denote the size of the Gaussian basis function, and $\mathcal{A}_N$ are the three-nucleon antisymmetrization operator. The eigenenegies of Hamiltonian and coefficients C are determined by Rayleigh-Litz variational method.

3 Results

All the calculations shown below have been done both for ${}^4_\Lambda$H and ${}^4_\Lambda$He, but we present only the case of the former since the result for the latter is very similar to that for the former. Calculated energies of the 0^+ ground state and the 1^+ excited state with the use of ND and NF are illustrated in Fig.2 together with the observed values. The energies are measured from the ^{3}H+Λ threshold. In the case of only the $3N+\Lambda$ channel included in the calculation, the 0^+ state is barely bound and the 1^+ state is unbound. When the $(3N)_{t=1/2}+\Sigma$ channel is further adopted, both of the

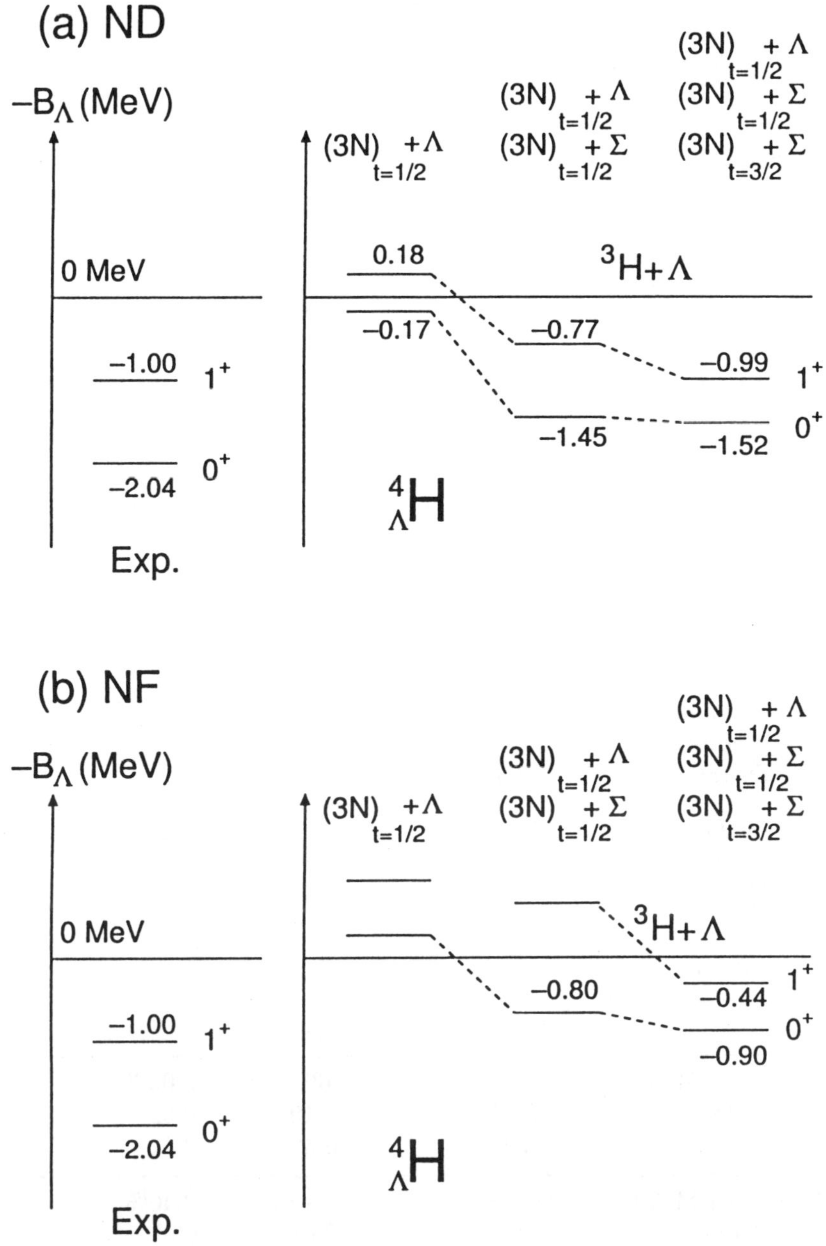

Figure 2: Calculated energy levels of $^4_\Lambda$H with (a) ND and (b) NF and AV8 NN potentials.

two states become bound evidently. Finally, further inclusion of the $(3N)_{t=3/2}+\Sigma$ channel makes the 1^+ state significantly more bound but with the 0^+ state almost unchanged. We understand that the Σ-channel component plays a very important role both in the energy gain and the relative splitting of the 0^+ and 1^+ states. However, the calculated binding energy of the 0^+ state is about 0.5 MeV smaller and hence the $0^+ - 1^+$ splitting is about a half of the observed value. In Fig.2(b), the level structure calculated with NF is similar to the case of ND.

We calculated the probability of the $3N+\Sigma$-channel admixture in ${}^4_\Lambda$H with the use of ND and NF. The tendency of the admixture is different between the two cases. In the case of ND, the probability is $\sim$ 1 % for the 0^+ state and $\sim$ 0.5 % for the 1^+ state. But, in the case of NF, it is $\sim$ 0.7 % for the both states. Precise percentage of the Σ-channel componet is listed in Table 1(a) and (b). In the 0^+ state, the probability of the $(3N)_{t=1/2}+\Sigma$ channel is much larger than that of the $(3N)_{t=3/2}+\Sigma$ channel, but in the 1^+ statc, they are nearly the same. We therefore understand that $(3N)_{t=3/2}+\Sigma$ channel is specially important in the 1^+ state. This tendency holds in the case of NF, too.

Table 1: The S-state and D-state probability of Σ component in ${}^4_\Lambda$H with the use of (a)ND and (b)NF potentials

(a) ND

	$(3N)_{t=1/2}+\Lambda$	$(3N)_{t=1/2}+\Sigma$	$(3N)_{t=3/2}+\Sigma$
0^+ (−1.52 MeV)	98.8 %	1.11 %	0.03 %
S-state	90.9	0.91	0.02
D-state	7.9	0.20	0.01
1^+ (−0.99 MeV)	99.5 %	0.24 %	0.23%
S-state	91.5	0.18	0.17
D-state	8.0	0.06	0.06

(b) NF

	$(3N)_{t=1/2}+\Lambda$	$(3N)_{t=1/2}+\Sigma$	$(3N)_{t=3/2}+\Sigma$
0^+ (−0.99 MeV)	99.3 %	0.69 %	0.03 %
S-state	91.8	0.10	0.02
D-state	7.5	0.59	0.01
1^+. (−0.44 MeV)	99.3 %	0.33 %	0.40%
S-state	91.5	0.01	0.01
D-state	7.8	0.32	0.39

The S-state and D-state probabilities of the channels are also listed in Table 1 (a) and (b). As long as the $3N + \Sigma$ channel in the 0^+ and 1^+ states is concerned, the S-state component is dominant in the case of ND while the D-state is dominant in the case of NF. Therefore, we understand that the $\Lambda N - \Sigma N$ coupling is central-force type in ND, but is tensor-force type in NF. This is verified, as seen in Table 2, from the expectaion value of the $\Lambda N - \Sigma N$ coupling potential. Namely, the central-force contribution is dominant, while the tensor-force type one is dominant in NF.

Table 2: The expectation value of the central and the tensor term of $\Lambda N - \Sigma N$ coupling, 2 $\langle V_{\Lambda N - \Sigma N} \rangle$, in ${}^4_\Lambda$H with the ND and NF

	ND		NF	
	0^+	1^+	0^+	1^+
Central(MeV)	−3.20	−2.19	−0.33	−0.11
Tensor (MeV)	−0.62	−0.59	−5.54	−6.74

4 Summary

We have developed the calculational method of four-body bound-state problems so that it becomes possible to make complicated four-body calculation of ${}^4_\Lambda$H and ${}^4_\Lambda$He taking both the $(3N + \Lambda)$ and $(3N + \Sigma)$ channels explicitly with the use of realistic NN and YN interactions. We are now able to investigate precisely the role of the $\Lambda N - \Sigma N$ coupling in ${}^4_\Lambda$H and ${}^4_\Lambda$He as follows. 1) the probability of Σ-channel mixing in ${}^4_\Lambda$H is about 1 % in the ND and NF cases. This probability play an important role in obtaining large energy gain of ${}^4_\Lambda$H. 2)$(3N)_{T=3/2} + \Sigma$ channel is important for the 1^+ states. 3) The $\Sigma N - \Lambda N$ coupling in ND case is central-force and that in NF case is tensor-type. However, the observed binding energies of 0^+ and 1^+ states were not well reproduced with the use of ND and NF. Therefore, we are much interested in using a new version of Nijmegen soft-core YN interaction, NSC97 [9] and in examining this interaction by four-body calculation of ${}^4_\Lambda$H and ${}^4_\Lambda$He including the $\Lambda N - \Sigma N$ coupling explicitly. This calcualtion is in progress.

Acknowledgement

This work was supported by the Special Postdoctal Researchers Program at RIKEN (E. H.) and by the Grant-in-Aid for Scientific Research in Priority Areas.

References

1. K. Miyagawa, H.Kamada, W. Glöckle and V. Stoks, *Phys. Rev.* C **51** 2905 (1995) .
2. M. Kamimura, *Phys. Rev.* A **38** 621 (1988).
3. H. Kameyama, M. Kamimura and Y. Fukushima, *Phys. Rev.* C **40** 974 (1989).
4. M. Kamimura and H. Kameyama, *Nucl. Phys.* A **508** 17c (1990).
5. E. Hiyama and M. Kamimura, *Nucl. Phys.* A **588** 35c (1995).
6. E. Hiyama, M. Kamimura, T. Motoba, Y. Yamada and Y. Yamamoto, **Phys. Rev.** C **53** 2075 (1996).
7. E. Hiyama, M. Kamimura, T. Motoba, Y. Yamada and Y. Yamamoto, *Prog. Theor. Phys.* **97** 881 (1997).
8. E. Hiyama, M. Kamimura, K. Miyazaki and T. Motoba, *Phys. Rev.* C **59** 2351 (1999).
9. T. A. Rijken, V. G. J. Stoks, Y. Yamamoto, *Phys. Rev.* C **59** 21 (1999).

QUARK PAULI PRINCIPLE IN Λ HYPERNUCLEAR SYSTEMS

Y. SUZUKI

Department of Physics, Niigata University,
Niigata 950-2181, Japan
E-mail: suzuki@nt.sc.niigata-u.ac.jp

H. NEMURA

Graduate School of Science and Technology, Niigata University,
Niigata 950-2181, Japan
E-mail: nemura@nt.sc.niigata-u.ac.jp

Y. FUJIWARA

Department of Physics, Kyoto University,
Kyoto 606-8502, Japan
E-mail: fujiwara@ruby.scphys.kyoto-u.ac.jp

C. NAKAMOTO

Suzuka National College of Technology,
Suzuka, Mie 510-0294, Japan
E-mail: nakamoto@genl.suzuka-ct.ac.jp

By taking into account the quark structure of baryons we study the NN and YN interactions as well as the quark Pauli effect on s-shell Λ hypernuclear systems. The resonating group formulation of the spin-flavor SU_6 quark model gives a reasonable fit to available scattering data when an effective quark-quark potential is generated from the mechanism of one-gluon exchange and scalar and pseudo-scalar nonet-meson exchange as well. By excluding a special five-baryon configuration which is forbidden by the Pauli principle at the quark level, it is shown that the Λ separation energy of ${}^{5}_{\Lambda}\text{He}$ or ${}^{\;\,6}_{\Lambda\Lambda}\text{He}$ is significantly reduced for a baryon size of 0.86 fm, while it is unchanged for a smaller size of 0.6 fm.

1 Introduction

The quark model is successful for the spectroscopy of baryons and mesons. The purpose of this contribution is to review our recent works which exploit the compositeness underlying the quark model of a nucleon (N) and a hyperon (Y). The first one is an attempt at a simultaneous description of the baryon-baryon (NN, YN, YY) interactions in the resonating group method (RGM) formulation of the spin-flavor SU_6 quark model.[1,2,3,4] The second is to examine the extent to which the Pauli principle at the quark level affects the properties of the baryon dynamics, especially the binding energies of Λ hypernuclei.[5,6]

In the RGM formulation of the spin-flavor SU_6 quark model the effective quark-quark ($q-q$) interaction is built by combining a phenomenological quark-confining potential and the color analogue of the Fermi-Breit interaction with a minimum effective meson-exchange potential (EMEP) of scalar and pseudo-scalar meson nonets directly coupled to quarks. The flavor symmetry breaking for the YN system is explicitly introduced through the quark-mass dependence of the Hamiltonian, as well as the flavor dependence of the exchanged meson masses. An advantage of introducing the EMEP at the quark level is that it gives a stringent constraint on the flavor dependence of the various NN and YN interactions. In this way rich information on the NN interaction can be utilized to correlate the YN interactions which are still at an ambiguous stage because of scarce experimental data.

The Pauli principle plays a vital role in the binding mechanism for systems of identical particles. Its importance is well known in atoms and atomic nuclei. However, for example in Λ hypernuclei, where the Λ-particle is embedded in nuclei, the role of the Pauli principle has not yet been explored from the point of view of the binding mechanism. Though Λ can be distinguished from the nucleon N at the baryon level, they are subjected to the Pauli principle at the quark level. We will estimate how much the Λ separation energy of $A = 5, 6$ Λ hypernuclei changes by the quark Pauli effect.

2 Baryon-baryon interaction in an SU_6 quark model

We assume that the octet baryon is a $3q$ system, and the orbital motion of each quark is assumed to be described by a (0s) harmonic-oscillator function, $\exp(-\boldsymbol{\rho}^2/2b^2)$, with a size parameter b. The orbital part of the baryon wave function is expressed as

$$\begin{aligned}\phi^{(\mathrm{o})} &= (\pi b^2)^{-\frac{9}{4}}\exp\left(-\frac{1}{2b^2}(\boldsymbol{\rho}_1^2+\boldsymbol{\rho}_2^2+\boldsymbol{\rho}_3^2)\right)\\ &= \left(\frac{\pi b^2}{3}\right)^{-\frac{3}{4}}\exp\left(-\frac{3}{2b^2}\boldsymbol{r}_1^2\right)\phi^{(\mathrm{int})},\end{aligned}\tag{1}$$

where $\boldsymbol{r}_1 = \frac{1}{3}(\boldsymbol{\rho}_1+\boldsymbol{\rho}_2+\boldsymbol{\rho}_3)$ is the baryon's position vector.

The Hamiltonian for a two-baryon system consists of the non-relativistic kinetic energy, the quadratic confinement potential, the Fermi-Breit interaction with explicit quark-mass dependence, and the scalar- (S-) and pseudoscalar- (PS-) meson exchange potentials acting between quarks:

$$H = \sum_{i=1}^{6}\left(m_i c^2 + \frac{\boldsymbol{p}_i^2}{2m_i}\right) + \sum_{i<j}^{6}\left(U_{ij}^{Cf} + U_{ij}^{FB} + \sum_{\beta} U_{ij}^{S\beta} + \sum_{\beta} U_{ij}^{PS\beta}\right). \tag{2}$$

It is important to treat the Galilean non-invariant terms of U^{FB} carefully and restore the empirical reduced mass of the YN system in the RGM framework.[1,2] The EMEP's, $U^{S\beta}$ and $U^{PS\beta}$, are constructed from the quark-meson vertex Lagrangian by the standard procedure. No vector mesons are included because the short-range vector-meson exchange potentials could be already accounted for by the quark-exchange mechanism.

Three different versions called RGM-F [1,2], FSS [3,4] and RGM-H [4] have been constructed. They have different features in the treatment of the EMEP. The model RGM-F takes into account only the tensor component generated from the π- and K-meson exchanges besides the S-meson nonet, and uses some approximations in evaluating the spin-flavor factors of the quark-exchange RGM kernel. On the other hand, in the FSS and RGM-H the spin-flavor factors are calculated exactly at the quark level, and the spin-spin terms originating from all members of the PS-meson nonet are included. By this improvement, one can set a strong constraint on the flavor dependence of each of NN and YN interactions, keeping the correct threshold energy of the ΣN channel in the ΛN - $\Sigma N(I = 1/2)$ coupled-channel system. We do not ignore the δ-function type contact term contained in the spin-spin term of the PS mesons, but introduce it with a common reduction factor c_δ, which is chosen to be $c_\delta = 0.381$ (FSS) or 0.339 (RGM-H) for all channels. The SU_3 relation of the coupling constants emerges as a natural consequence of the SU_6 quark model. For scalar mesons, the $F/(F + D)$ ratio turns out to take the SU_6 value of purely electric type. This is too restrictive to reproduce the experimental data for the low-energy YN cross sections. We will relax this restriction in two ways; one is to change the mixing angle of the flavor-singlet and octet scalar mesons only for the $\Sigma N(I = 3/2)$ channel, and the other is to employ the same approximation as RGM-F solely for the isoscalar S-mesons, ϵ and S^*. We call these models FSS and RGM-H, respectively. Predictions of these two models are not very different except for the roles of the $LS^{(-)}$ force in the ΛN - $\Sigma N(I = 1/2)$ coupled-channel system. The SU_3 parameters of EMEP, scalar meson masses, and the quark model parameters are determined to fit the NN S-wave and P-wave phase shifts under the constraint that the deuteron binding energy and $^1S_0$ scattering length are properly reproduced. The low-energy cross section data for YN scattering is also employed to fix the parameters in the strangeness sector, especially the up-down to strange quark-mass ratio $\lambda = m_s/m_{ud}$. For the explicit values of model parameters, the original papers [3,4] should be referred to.

The models FSS and RGM-H reproduce the empirical NN phase shifts up to $J = 4$ within an accuracy of a couple of degrees, although a slight overestimate in the $^3D_2$ state still remains. [4] The spin-spin term of the one-pion exchange is particularly important to reproduce the low-energy behavior of the NN $^1S_0$ phase shift as well as the higher partial waves. The η, η' and δ mesons play the essential role in correcting the well-known imbalance [1,2] of the NN ^1S and ^3S central quark model potentials. The deuteron properties and effective range parameters are also reasonably reproduced.

Figure 1 shows the elastic differential cross sections of the np scattering, compared with the experimental data.[7] In Fig. 1 the solid curve indicates predictions by RGM-H, the dashed curve by FSS. No imaginary potential is introduced in this calculation. Although the cross sections overestimate the experimental data at backward angles, characteristic behavior of the energy dependence and the angular distribution is reasonably well reproduced in the wide energy. The differential cross sections and the polarization for the pp elastic scattering are also examined up to 800 MeV, and a behavior similar to that for the np system is obtained.

The low-energy scattering and reaction "total" cross sections for Σ^+p and Σ^-p

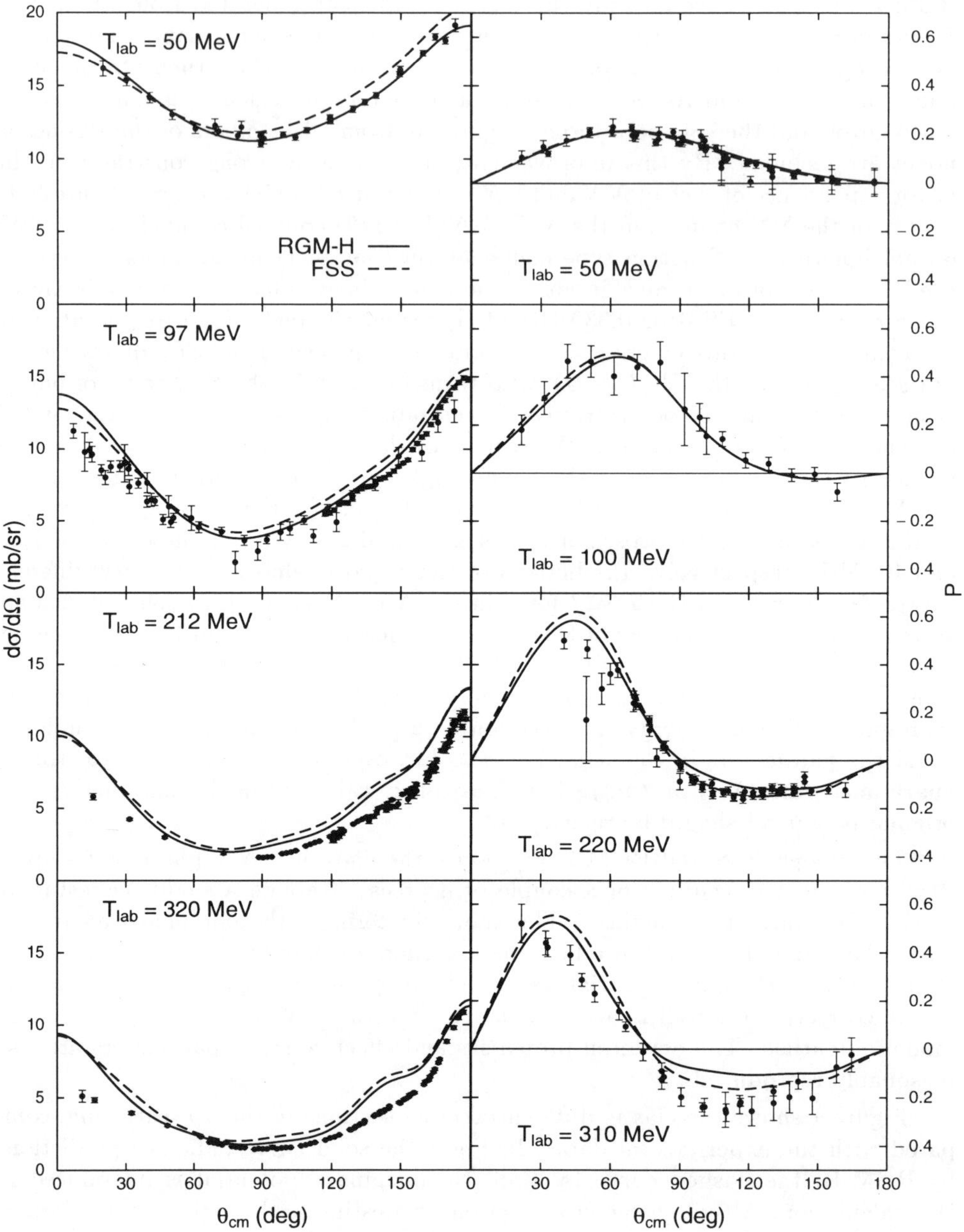

Figure 1: The differential cross sections and the polarization for the np scattering at $T_{lab} \leq 320$ MeV. Solid and dashed curves denote RGM-H and FSS results, respectively. Experimental data are from Ref. 7.

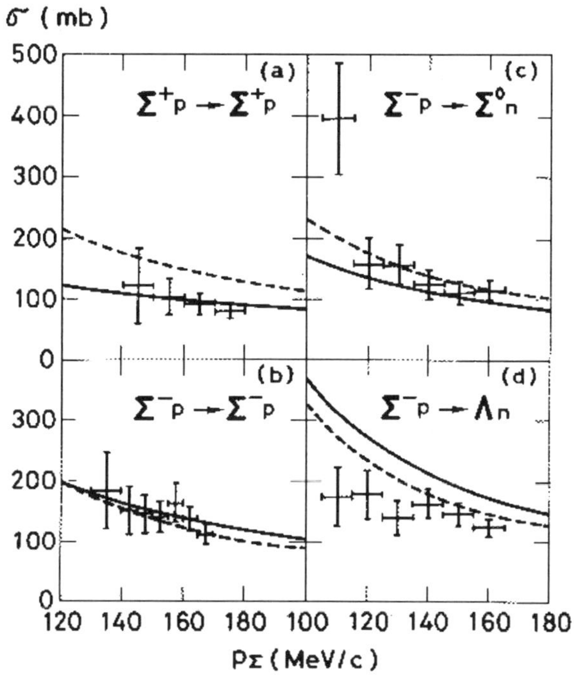

Figure 2: The low-energy Σ^+p and Σ^-p "total" cross sections in FSS (solid curve) and RGM-H (dashed curve), compared with experiment [8,9].

systems predicted in FSS (solid curves) and RGM-H (dashed curves) are compared in Fig. 2 with the experimental data. The improvement over the RGM-F result[2], especially in the Σ^-p elastic cross sections, is an outcome of the weaker central attraction of the $\Sigma N(I = 1/2)$ channel in these models. On the other hand, $\Sigma^-p \to \Lambda n$ reaction cross sections are slightly overestimated especially in FSS.

The calculated total cross sections for the Λp scattering are compared with the experimental data in Fig. 3. Since the threshold energy of the ΣN channel is properly reproduced both in FSS and in RGM-H, a cusp structure of the ΛN $^3S_1$ state appears at the incident-momentum of $p_\Lambda = 638$ MeV/c. This is generated by the strong ΛN $^3D_1$ - ΣN $^3S_1$ coupling due to the one-pion tensor force. In FSS (solid curve) the cusp structure is strongly enhanced by the bump structure predicted in RGM-F, which is now spread out over a wide energy region of $p_\Lambda = 500 \sim 800$ MeV/c. This is the consequence of the P-wave channel-coupling effect with the $\Sigma N(I = 1/2)$ channel due to the $LS^{(-)}$ force. On the other hand, RGM-H (dashed curve) predicts a rather moderate rise of the cross sections at the cusp region, which seems to be consistent with the experimental data.[12] Since the $LS^{(-)}$ force in the present models originates entirely from the Fermi-Breit interaction, experimental confirmation of the enhanced cusp structure is of great interest in establishing the role of this force.

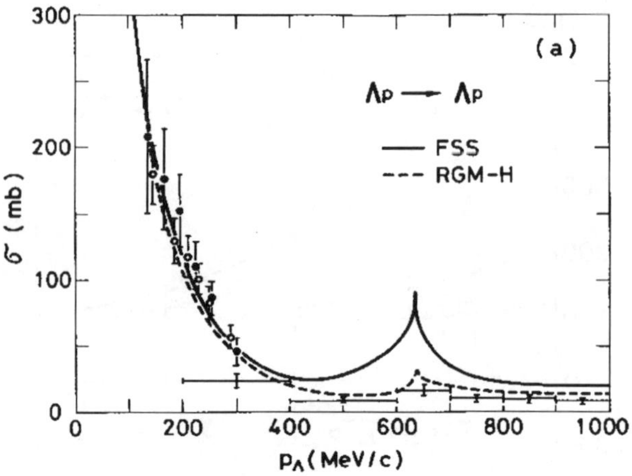

Figure 3: Calculated Λp elastic total cross sections in FSS (solid curve) and RGM-H (dashed curve), compared with experiment [10,11,12].

3 Quark Pauli effects in Λ hypernuclei with $A = 5$ and 6

Though the quark model description of the NN and YN interactions has given us a reasonable fit to the experimental data, we are still at a qualitative stage of the understanding of the interactions. In Fig. 4 we show S-wave phase shifts predicted by the model FSS and the Nijmegen model F potential[13]. Both of them reproduce the Λp total cross sections of Fig. 3, but they exhibit rather different characteristics in the phase shift behavior. To consider the effect of the quark structure of baryons on the binding energy of Λ hypernuclei, it is desirable to use such a ΛN potential that reproduces the Λ separation energies of $A = 3, 4$ Λ hypernuclei where no Pauli effect is expected at the quark level. The solid curve of the figure denotes the phase shift calculated by the potential determined in this way.[5]

To consider the effect of the quark structure of baryons in Λ hypernuclear system, let us write a $15q$, five-baryon wave function for $ppnn\Lambda$, $ppn\Lambda\Lambda$, or $pnn\Lambda\Lambda$

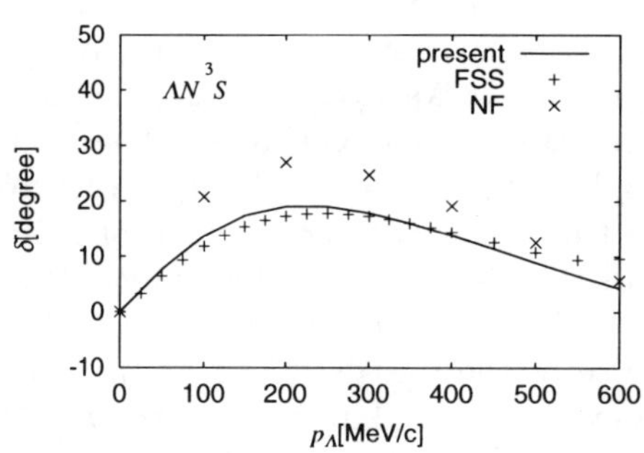

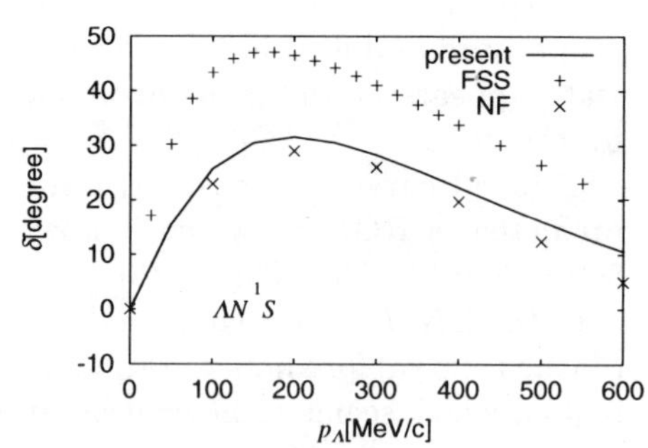

Figure 4: Comparison of ΛN S-wave phase shifts calculated by the effective potential of Ref. 5 (solid line), the model FSS (+) and the Nijmegen model F ($\times$) potential, respectively.

system in the form

$$\Phi = \mathcal{A}_{\rm q}\left\{\left(\prod_{i=1}^{5}\phi^{(\rm int)}(i)\phi^{(\rm sfc)}(i)\right)\Psi(\boldsymbol{r}_1,...,\boldsymbol{r}_5)\right\}, \tag{3}$$

where $\phi^{(\rm sfc)}$ is the spin-flavor-color part of the baryon, $\Psi(\boldsymbol{r}_1,...,\boldsymbol{r}_5)$ the relative motion function between the baryons and $\mathcal{A}_{\rm q}$ the antisymmetrizer of the quarks. There exist functions $\Psi(\boldsymbol{r}_1,...,\boldsymbol{r}_5)$, called Pauli-forbidden, for which Φ vanishes identically. It is easy to conclude that

$$\psi_{\rm F} = \left(\frac{\pi b^2}{3}\right)^{-3}\exp\left(-\frac{3}{2b^2}\sum_{i=1}^{5}(\boldsymbol{r}_i-\boldsymbol{R})^2\right) \tag{4}$$

is manifestly Pauli-forbidden, where $\boldsymbol{R}=\frac{1}{5}\sum_{i=1}^{5}\boldsymbol{r}_i=\frac{1}{15}\sum_{i=1}^{15}\boldsymbol{\rho}_i$ is the center-of-mass coordinate of the five baryons.

The state (4) corresponds to a unique configuration where the five baryons are on top of one another. Note that the existence of this forbidden state is never expected from the Pauli principle at the baryon level where Λ is distinguished from N. The consequence of the Pauli-forbidden state is that the relative motion function Ψ between the baryons be subjected to remain in the subspace which is orthogonal to the forbidden state, that is, $\langle\psi_{\rm F}|\Psi\rangle = 0$. The constraint applies to the hypernuclei with $A \geq 5$.

Since the quark is confined in the baryon, the parameter b determining its spatial extent has to be chosen appropriately.[14] A reasonable guess for b is made from the charge radius of the proton, leading to $b \approx 0.86$ fm. Actually the baryon structure is complicated and the value of $b = 0.86$ fm could be a maximum as an acceptable value. In the extreme limit of the quark shell model[15], b would be determined from the size of the hypernucleus. On the contrary, the choice of $b = 0$ would correspond to another extreme limit that the baryon has no quark structure and thus there is no quark Pauli effect.

In order to evaluate quantitatively the binding energy reduction due to the quark Pauli effect, the variational trial function has to be chosen flexibly enough. We assumed that the trial function with the total orbital angular momentum $L = 0$ is given by

$$\begin{aligned}\Psi_{JMIM_I} &= \sum_k C_k\Big(\Psi_k - \sum_a \Gamma_a\langle\Gamma_a|\Psi_k\rangle\Big),\\ \Psi_k &= \mathcal{A}\{G(\boldsymbol{x},A_k)\chi_{kJM}\eta_{kIM_I}\},\end{aligned} \tag{5}$$

with the correlated Gaussian G

$$G(\boldsymbol{x},A_k) \equiv \exp\Big(-\frac{1}{2}\boldsymbol{x}A_k\boldsymbol{x}\Big), \tag{6}$$

where Γ_a is an orthonormal set of functions constructed from the Pauli forbidden space. $\mathcal{A}$ is the operator antisymmetrizing identical baryons, $\boldsymbol{x} = (\boldsymbol{x}_1,...,\boldsymbol{x}_{A-1})$ stands for a set of relative (e.g., Jacobi) coordinates, and χ_{kSM_S} (η_{kIM_I}) is the spin (isospin) function of the system.

Table 1: The Λ separation energies, in MeV, of $A = 5$ systems as a function of the baryon size b (in fm). The case of $b = 0$ indicates that no quark Pauli effect is imposed in the calculation.

	Theory			Expt.
	$b = 0$	0.7	0.86	
$B_\Lambda({}^5_\Lambda\mathrm{He})$	4.98	4.70	3.63	3.12±0.02
$B_{\Lambda\Lambda}({}^{\ 5}_{\Lambda\Lambda}\mathrm{H})$	5.63	5.52	5.12	
$B_{\Lambda\Lambda}({}^{\ 5}_{\Lambda\Lambda}\mathrm{He})$	5.53	5.45	5.05	

The basis element is increased one by one in the stochastic variational method[16,17] until the solution reaches a practical convergence. Here the basis set Ψ_k is chosen so as to minimize the energy, whereas the set Γ_a is chosen to maximize the energy. The accuracy of solutions is vital in the present calculation because we have to take account of very small components produced by the quark Pauli principle.

Table 1 compares the Λ separation energies calculated for $A = 5$ systems (${}^5_\Lambda\mathrm{He}$, ${}^{\ 5}_{\Lambda\Lambda}\mathrm{H}$, ${}^{\ 5}_{\Lambda\Lambda}\mathrm{He}$). Spin, parity and isospin (J^π, I) are ($\frac{1}{2}^+$, 0), ($\frac{1}{2}^+$, $\frac{1}{2}$), and ($\frac{1}{2}^+$, $\frac{1}{2}$), respectively. Experimental search for ${}^{\ 5}_{\Lambda\Lambda}\mathrm{H}$ is under way.[18] The NN and ΛN potentials were the same as employed in Ref. 5. The NN potential reproduces the binding energies and sizes of ${}^3\mathrm{H}$, ${}^3\mathrm{He}$, and ${}^4\mathrm{He}$ very well[17], and likewise the ΛN potential is set to well reproduce the energies of $A = 3 - 4$ hypernuclei. The $\Lambda\Lambda$ potential was taken from Ref. 19.

As shown in Table 1, the quark Pauli effect leads to a significant reduction of the binding energy for ${}^5_\Lambda\mathrm{He}$ at $b = 0.86$ fm. As discussed in Ref. 5, this effect plays an important clue in resolving the anomaly in the ${}^5_\Lambda\mathrm{He}$ binding energy (see, e.g., Refs. 21 and 22) provided this choice of baryon size is acceptable. The quark Pauli effect becomes much less in ${}^{\ 5}_{\Lambda\Lambda}\mathrm{H}$ and ${}^{\ 5}_{\Lambda\Lambda}\mathrm{He}$, which reflects the fact that they are spatially more extended than ${}^5_\Lambda\mathrm{He}$. In fact, the root mean square (rms) radius of ${}^{\ 5}_{\Lambda\Lambda}\mathrm{H}$ (and also ${}^{\ 5}_{\Lambda\Lambda}\mathrm{He}$) turns out to be 2.0 fm, while the rms radius of ${}^5_\Lambda\mathrm{He}$ is 1.6 fm.

The results of Table 1 can be qualitatively understood as follows. Let Ψ_0 and E_0 denote the wave function and the energy of the ground state which is obtained when no quark Pauli effect is imposed. When the Pauli effect is taken into account, the wave function may be written according to the perturbation theory as

$$\Psi = \frac{1}{\sqrt{1-\alpha^2}}(\Psi_0 - \alpha\psi_F) \quad \text{with} \quad \alpha = \langle\psi_F|\Psi_0\rangle. \tag{7}$$

Assuming that α is small, the energy difference is given by

$$E - E_0 = \langle\Psi|H|\Psi\rangle - E_0 \approx \alpha^2\langle\psi_F|H|\psi_F\rangle. \tag{8}$$

The energy difference is determined by the product of α^2, the probability of finding the forbidden component in Ψ_0, and the energy expectation value of the Pauli forbidden state. Let us use the above equation to estimate how much the energy difference is produced by the quark Pauli principle. In the case of ${}^5_\Lambda\mathrm{He}$, $\alpha^2 \approx 0.004$, $\langle\psi_F|H|\psi_F\rangle \approx 500$ MeV for $b = 0.86$ fm, and thus $E - E_0 \approx 2$ MeV, while in the case of ${}^{\ 5}_{\Lambda\Lambda}\mathrm{H}$ $\alpha^2 \approx 0.001$, $\langle\psi_F|H|\psi_F\rangle \approx 800$ MeV, and $E - E_0 \approx 1$ MeV. This estimation is in good agreement with the result of Table 1.

We have calculated the binding energy of ${}^{6}_{\Lambda\Lambda}$He in a complete six-body treatment. Only the central potential is included. This system has three types of forbidden states (4), containing $ppnn\Lambda$, $ppn\Lambda\Lambda$, or $pnn\Lambda\Lambda$ five baryons. Table 2 lists the energies obtained by excluding the forbidden states. The energy obtained without the quark Pauli effect ($b = 0$) is apparently overbound. The elimination of the forbidden states leads to a surprisingly large reduction in the binding energy. In order to see whether or not this large effect is due to the choice of the $\Lambda\Lambda$ potential, we have repeated the calculation by using another $\Lambda\Lambda$ potential determined by the quark cluster model[23]. The result was very similar. The rms radius of ${}^{6}_{\Lambda\Lambda}$He is calculated to be 1.6 fm, considerably smaller than that (2.0 fm) of ${}^{5}_{\Lambda\Lambda}$H, which leads to a larger quark Pauli effect in ${}^{6}_{\Lambda\Lambda}$He.

Table 2: The Λ separation energy $B_{\Lambda\Lambda}$, in MeV, of ${}^{6}_{\Lambda\Lambda}$He as a function of the baryon size b (in fm).

Theory				Expt.
$b = 0$	0.6	0.7	0.86	
14.3	13.9	12.9	9.4	10.9±0.6

4 Summary

We showed that a simultaneous description of the NN and YN interactions is possible in the standard $(3q)$-$(3q)$ RGM, if the full Fermi-Breit interaction with the explicit flavor symmetry breaking is augmented with the effective meson-exchange potentials including the central, spin-spin, and tensor components generated from the scalar and pseudo-scalar nonet mesons. These effective meson-exchange potentials were calculated microscopically at the quark level. The two versions, FSS and RGM-H thus constructed, gave similar results for the NN and YN scattering observables, except for the Λp cusp region.

We also studied the consequence of the quark structure of the baryons on the binding energies of Λ- and $\Lambda\Lambda$ hypernuclei. We showed that, though the Λ particle can be distinguished from the nucleon, the quark model predicts a special forbidden state involving five baryons. Without taking into account the quark Pauli effect, the ΛN and $\Lambda\Lambda$ interactions that are equivalent to the meson theoretical or the quark cluster model potentials led to a substantial overbinding of the Λ separation energies of ${}^{5}_{\Lambda}$He and ${}^{6}_{\Lambda\Lambda}$He. By excluding the forbidden states we found that the binding energies started to decrease at the baryon size of $b = 0.6$ fm and reduced significantly at the size of free baryons, $b = 0.86$ fm. Since the forbidden states considered here are many-body configurations involving five baryons, the Pauli effect cannot be revealed through a study on the baryon-baryon interaction alone. In other words, a full understanding of the binding mechanism of the hypernuclei may not be attainable only through the knowledge of the two- and three-body interactions of the baryons.

Acknowledgments

This work was supported in part by Grant-in-Aid for Scientific Research (No. 10640255) of the Ministry of Education, Science and Culture (Japan).

References

1. C. Nakamoto, Y. Suzuki and Y. Fujiwara, Prog. Theor. Phys. 94 (1995) 65.
2. Y. Fujiwara, C. Nakamoto and Y. Suzuki, Prog. Theor. Phys. 94 (1995) 215, 353.
3. Y. Fujiwara, C. Nakamoto and Y. Suzuki, Phy. Rev. Lett. 76 (1996) 2242.
4. Y. Fujiwara, C. Nakamoto and Y. Suzuki, Phys. Rev. C54 (1996) 2180.
5. H. Nemura, Y. Suzuki, Y. Fujiwara and C. Nakamoto, Prog. Theor. Phys. **101** (1999) 981.
6. Y. Suzuki and H. Nemura, submitted to Prog. Theor. Phys.
7. Scattering Analysis Interactive Dial-up (SAID), Virginia Polytechnic Institute, Blacksburg, Virginia (R.A. Arndt, private communication).
8. F. Eisele, H. Filthuth, W. Föhlisch, V. Hepp and G. Zech, Phys. Lett. 37B (1971) 204.
9. R. Engelmann, H. Filthuth, V. Hepp and E. Kluge, Phys. Lett. 21 (1966) 587.
10. G. Alexander, U. Karshon, A. Shapira, G. Yekutieli, R. Engelmann, H. Filthuth and W. Lughofer, Phys. Rev. C173 (1968) 1452.
11. B. Sechi-Zorn, B. Kehoe, J. Twitty and R.A. Burnstein, Phys. Rev. 175 (1968) 1735.
12. J.A. Kadyk, G. Alexander, J.H. Chan, P. Gaposchkin and G.H. Trilling, Nucl. Phys. B27 (1971) 13.
13. M. M. Nagels, T. A. Rijken and J. J. de Swart, Phys. Rev. D**20** (1979), 1633.
14. S. Takeuchi, K. Shimizu and K. Yazaki, Nucl. Phys. A**449** (1986), 617.
 S. Takeuchi and K. Shimizu, Phys. Lett. B**179** (1986), 197.
15. E. V. Hungerford and L. C. Biedenharn, Phys. Lett. B**142** (1984), 232.
16. V. I. Kukulin and V. M. Krasnopol'sky, J. Phys. G**3** (1977), 795.
17. K. Varga and Y. Suzuki, Phys. Rev. C**52** (1995), 2885.
 Y. Suzuki and K. Varga, *Stochastic Variational Approach to Quantum-Mechanical Few-Body Problems*, Lecture Notes in Physics, Vol. m54, (Springer-Verlag, Berlin Heidelberg, 1998).
18. T. Fukuda for the E906 collaboration, Nucl. Phys. A**639** (1998), 355c.
19. E. Hiyama, M. Kamimura, T. Motoba, T. Yamada and Y. Yamamoto, Prog. Theor. Phys. **97** (1997), 881.
20. S. Nakaichi-Maeda and Y. Akaishi, Prog. Theor. Phys. **84** (1990), 1025.
21. R. H. Dalitz, R. C. Herndon and Y. C. Tang, Nucl. Phys. B**47** (1972), 109.
22. B. F. Gibson and E. V. Hungerford III, Phys. Rep. **257** (1995), 349.
23. C. Nakamoto, Y. Fujiwara and Y. Suzuki, *Proceedings of the KEK-Tanashi International Symposium on Physics of Hadrons and Nuclei*, Dec. 14-17, 1998, Tokyo, Japan, to be published in Nucl. Phys. A.

TWO-FREQUENCY SHELL MODEL FOR HYPERNUCLEI

YIHARN TZENG,[1] S.Y. TSAY TZENG,[2] T.T.S. KUO,[1,3] AND T.-S.H. LEE[4]

[1]*Institute of Physics, Academia Sinica, Taipei, Taiwan, R.O.C.*

[2] *National Taipei University of Technology, Taipei, Taiwan, R.O.C.*

[3]*Department of Physics, State University of New York, NY 11794, USA*

[4]*Physics Division, Argonne National Laboratory, Argonne, IL 60493, USA*

We discuss a two-frequency shell model for hypernuclei and the derivation of its effective interaction starting from a hyperon-nucleon potential in free space. Hyperons in a hypernucleus are generally less bound and spatially more extended than nucleons. To accomodate this feature, we have allowed in our model that the oscillator frequency $\hbar\omega_Y$ for the hyperon to be smaller than $\hbar\omega_N$ for the nucelon. In a folded-diagram calculation for ${}^{16}_{\Lambda}O$ using the Nijmegen potential, the Λ single particle energy is found to have a saturation minimum at an oscillator frequency $\hbar\omega_\Lambda \approx 10$ MeV which is considerably smaller than $\hbar\omega_N = 14$ MeV. The $0s^\Lambda - 0p^N$ spin-dependence parameters have been calculated using our two-frequency approach. The results given by the Nijmegen(1989) potential are significantly different from those by the *Jülich* $- B$ potential. The ΛNN three-body interactions induced by $\Lambda N - \Sigma N$ transitions are found to be important for the spin parameters.

In microscopic calculations for hypernuclei, one would like to calculate the various nuclear properties, such as the nuclear spectra, starting from a fundamental free hyperon-nucleon (YN) interaction V_{YN}. One usually starts from a nuclear many-body problem of the form $H\Psi = E\Psi$, with $H = T + V$ where T denotes the kinetic energy and V denotes both the nucleon-nucleon (NN) and YN potentials. This many-body problem can be formally reduced to a model space (usually referred to as the P-space) problem given as [1]

$$PH_{eff}P\Psi = EP\Psi; \ H_{eff} = H_0 + V_{eff}. \tag{1}$$

Here H_0 is the unprturbed Hamiltonian and V_{eff} the model-space effective interaction. It is given as a folded-diagram series

$$V_{eff} = \hat{Q} - \hat{Q}' \int \hat{Q} + \hat{Q}' \int \hat{Q} \int \hat{Q} - \hat{Q}' \int \hat{Q} \int \hat{Q} \int \hat{Q} \cdots, \tag{2}$$

where $\int$ denotes a generalized fold, and $\hat{Q}'$ and $\hat{Q}$ represent the irreducible vertex function $\hat{Q} - box$ [1]. In the above approach, clearly the central task is the derivation of the effective interaction V_{eff}, and it is often not an easy task. In this report, we would like to discuss a two-frequency shell model approach for calculating the above effective interaction for hypernuclei, based on realistic YN potentials.

The calculations of effective YN interactions for shell-model studies have been performed by many authors[2,3,4,5,6,7]. A common feature of all of these previous approaches is to assume that the harmonic oscillator basis wavefunctions for nucleons and hyperons are of the same frequency; i.e. $\hbar\omega_Y = \hbar\omega_N$. Hyperons in a nucleus are in general much less bound than nucleons, and the wave functions for the former are spatially more extended than those for the latter. Consequently, it would seem to be natural that one should use different basis functions for nucleons and hyperons, with $\hbar\omega_\Lambda \leq \hbar\omega_N$. In a previous paper [8] we have applied this type of two-frequency approach to halo nuclei.

Let us first discuss the YN G-matrix within the framework of the above two-frequency approach. Like the nucleon-nucleon potentials, the YN potential V_{YN} also has a strong repulsive core. Hence V_{YN} itself is not suitable to be used directly in perturbative microscopic calculations of hypernuclei. Rather, we should first convert it to a hyperon-nucleon G-matrix, and then perform the many-body calculation in terms of the G-matrix interactions. The G-matrix should be calculated in accordance with the model space which one would choose to describe the hypernucleus.

Let us take ${}^{16}_{\Lambda}O$ as an example. For this nucleus, one may choose a model space consisted of nucleons in the $(0s, 0p, 0d1s)$ shell and hyperons (Λ) in the $(0s, 0p)$ shell. The model-space G-matrix is defined by the integral equation [9]

$$G(\omega) = V + VQ\frac{1}{\omega - QTQ}QG(\omega), \tag{3}$$

where ω is an energy variable. Q is a two-body Pauli exclusion operator. Its role is to prevent double counting, namely the intermediate states allowed for G must be outside of the chosen model space. Thus the Pauli operators Q is dependent on the model space chosen, and consequently the corresponding G matrix would also be the model-space dependent. T is the two-nucleon kinetic energy. Note that our G-matrix has orthogonalized plane-wave intermediate states. The above is a coupled-channel G-matrix, with V, G and Q given respectively as

$$\begin{bmatrix} V_{\Lambda N-\Lambda N} & V_{\Lambda N-\Sigma N} \\ V_{\Sigma N-\Lambda N} & V_{\Sigma N-\Sigma N} \end{bmatrix}, \tag{4}$$

$$\begin{bmatrix} G_{\Lambda N-\Lambda N} & G_{\Lambda N-\Sigma N} \\ G_{\Sigma N-\Lambda N} & G_{\Sigma N-\Sigma N} \end{bmatrix}, \tag{5}$$

and

$$\begin{bmatrix} Q_{\Lambda N-\Lambda N} & 0 \\ 0 & Q_{\Sigma N-\Sigma N} \end{bmatrix}. \tag{6}$$

The presence of the Pauli operator Q has added difficulty to the solution of the above G-matrix equation. The Q operator is defined in terms of oscillator wave functions. For ${}^{16}_{\Lambda}O$, we employ a model space consisted of Λ and nucleons only. (Σ is not included in our model space.) Hence for this case, we have $Q_{\Sigma N-\Sigma N}$=1. [1] $Q_{\Lambda N-\Lambda N}$ may be specified in terms of the orbit numbers n_1, n_2, $n_{2\Lambda}$, $n_{3\Lambda}$ [5,6], which represent the boundaries of the chosen model space.

A convenient way to solve the above G-matrix equation is by way of the matrix identity

$$Q\frac{1}{QAQ}Q = \frac{1}{A} - \frac{1}{A}P\frac{1}{P\frac{1}{A}P}P\frac{1}{A}, \tag{7}$$

where P and Q are projection operators, with P + Q =1, and A is a matrix in the (P + Q) space. With this identity, the exact solution of Eq.(3) is

$$G(\omega) = G_F(\omega) + \Delta G(\omega) \tag{8}$$

with

$$\Delta G(\omega) = -G_F(\omega)\frac{1}{e}P\frac{1}{P[\frac{1}{e}+\frac{1}{e}G_F(\omega)\frac{1}{e}]P}P\frac{1}{e}G_F(\omega) \tag{9}$$

where e represents $(\omega - T)$. The free-space G matrix is given by

$$G_F(\omega) = V + V\frac{1}{\omega - T}G_F(\omega). \tag{10}$$

Note that G_F does not contain the Pauli operator Q. Hence its calculation is relatively easy. The matrix elements of $\frac{1}{e}G_F$, $G_F\frac{1}{e}$ and $\frac{1}{e}G_F\frac{1}{e}$ can also be calculated conveniently. From these matrix elements, the Pauli correction term ΔG of Eq.(9) can be readily obtained. The above method has been applied [5,6] to the YN G-matrix for the one-frequency case, namely the oscillator frequency for the hyperon wave function is the same as that for the nucleon.

There is a technical difficulty in extending the above G-matrix method to the two-frequency case, where the oscillator frequencies $\hbar\omega_Y$ and $\hbar\omega_N$ are different. This is because we do not have a handy computer program for calculating the two-frequency Moshinsky expansion coefficients for the transformation between the laboratory $(n_1l_1n_2l_2)$ and relative and center-of-mass $(NLnl)$ oscillator wave functions. To circumvent this difficulty, we have adopted an expansion (brute force) procedure, namely expanding the oscillator wave functions with $\hbar\omega_Y$ in terms of those with $\hbar\omega_N$, or *vice versa*. Typically, a high accuracy is obtained when we include about 8 terms in the expansion. In short, we have been able to calculate the YN two-frequency G-matrix in an accurate way.

Next we calculate the model-space effective interaction of Eq.(2). Here the choice of model space may play an important role. The effective interaction method described above above is general, and is applicable for any model space one chooses to employ. For any P, the model-space effective interaction V_{eff} exists and can be formally derived. There is, however, an important point which may be emphasized. Namely, although formally one can choose to use any model space P, it is nevertheless very helpful to use a "good-quality" P-space, by which we mean that there is a large overlap between our chosen P-space and the true nuclear wave function Ψ. The calculation of V_{eff} is considerably simplified [8] if the model space is of "good quality". For hypernuclei, hyperons are generally less bound and spatially more extended than nucleons. Clearly the quality of the shell-model P-space would be improved if a two-frequency approach is employed where $\hbar\omega_Y$ is suitably smaller than $\hbar\omega_N$.

Let us review the procedure of our calculation for V_{eff}. We first calculate the vertex function $\hat{Q}-box$ using the two-frequency G-matrix. For a given $\hat{Q}-box$, we summed up the folded-diagram series of Eq.(2) using the Lee-Suzuki iteration method [10,11]. We note that although the G-matrix of Eq.(3) is energy (ω) dependent, the effective interaction given by Eq.(2) is energy independent. For the $\hat{Q}-box$ we include only the irrducible diagrams first and second-order in G [12]. Clearly there is the question of convergence which should be addressed. Namely, what is the effect of the higher-order diagrams which we have left out in our $\hat{Q}-box$?

For halo nuclei, this convergence problem is alleviated when a two-frequency shell model is employed [8]. Here the halo nucleons are much less bound than the core nucleons. Hence the oscillator frequency $\hbar\omega_{halo}$ for the halo nucleons is chosen to be smaller than $\hbar\omega_{core}$ for the core nucleons. In this way the coupling between the halo and core nucleons is weakened, and thus the higher order diagrams of the $\hat{Q} - box$ become significantly less important. The situation for hyperons in a hypernucleus is quite similar. Sample calculations have indicated that the core polarization diagrams for hyperons are similarly unimportant, when a two-frequency approach is employed. Although further investigation remains to be carried out, we feel that the two-frequency shell model for hypernuclei may have an important advantage, namely with this approach the high-order core polarization diagrams of the YN $\hat{Q} - box$ are likely to become unimportant and consequently it can be fairly accurately calculated using the low-order approximation mentioned earlier. This behavior is indeed consistent with the P-space quality consideration mentioned earlier.

Now we come to a central issue of our approach: Namely how do we determine the oscillator frequencies $\hbar\omega_\Lambda$ and $\hbar\omega_N$? In this work, the frequency for nucleons is determined by using the well known empirical formula $\hbar\omega_N = 45A^{-1/3} - 25A^{-2/3} =$ 14 MeV for the considered $A = 16$ system. This formula was based on the empirical nuclear radii, for non-strange nuclei. For hypernuclei, it seems that we do not yet have such information and thus a similar empirical formula is not yet available. So we will have to treat the frequency $\hbar\omega_\Lambda$ "essentially" as a free parameter in the calculation.

To try it out, we have performed calculations for the Λ single particle energies for ${}^{17}_{\Lambda}O$, with respect to a ${}^{16}O$ core, using the Nijmegan(1989) potential [13]. They are calculated by using a folded-diagram method, with the $\hat{Q}$-box consisted of the 1-body diagrams first- and second-order in G-matrix. A range of values for $\hbar\omega_\Lambda$ have been used, all with $\hbar\omega_N$ fixed at 14 MeV. Results show a saturation behavior, namely we have a concave-upward parabolic-like curve when plotting the single particle energy as a function of $\hbar\omega_\Lambda$. For the $s^{\Lambda}_{1/2}$ orbit, the saturation point is at $\hbar\omega_\Lambda \approx 10\ MeV$ with a binding energy of about 13 MeV.

Although the above saturation behavior is encouraging, an important point may be pointed out. We are calculating the energy difference between ${}^{17}_{\Lambda}O$ and ${}^{16}O$, and for it the usual variational principle does not apply. In other words, our result does not constitute an upper bound. It is necessary, however, that our calculation must be stable with respect to a small variation of $\hbar\omega_\Lambda$ (i.e. vanishing derivative with respct to variation of Λ). This criterion is met at the above saturation point. As mentioned earlier, an important step in our approach is the determination of $\hbar\omega_\Lambda$, and the above stability requirement is perhaps a best available way for its determination.

We have carried out calculations for ${}^{16}_{\Lambda}O$ using $\hbar\omega_\Lambda$=10 MeV and $\hbar\omega_\Lambda$=14 Mev. Because of space limitation, we shall report our results very briefly here. (A detailed report will be published elsewhere.) Generally speaking, our lowest four J^- states $(1^-, 0^-, 1^-, 2^-)$ are in good agreements with experiments [15,16,17] and the results given by the Nijmegen and *Jülich*-B [14] potentials are fairly similar. But the next four J^+ states given by these two potentials are noticeably different, and detailed experimental information about them seem to be not yet available.

We have also calculated the $s - p$ shell spin dependence parameters, and let us briefly report our results. The spin dependence of the $\Lambda - N$ effective interaction is a topic of much current interest. As proposed by Millener et al. [18,19], we may generally write the $\Lambda - N$ effective interaction as

$$V_{\Lambda N}(r) = V_o(r) + V_\sigma(r)\mathbf{s_N}\cdot\mathbf{s_\Lambda} + \mathbf{V_\Lambda(r)l_{N\Lambda}}\cdot\mathbf{s_\Lambda} + \mathbf{V_N(r)l_{N\Lambda}}\cdot\mathbf{s_N} + \mathbf{V_T(r)S_{12}}. \quad (11)$$

For the $\Lambda s_{1/2}$-$Np_{3/2,1/2}$ interaction, the above effective interaction is completely represented by 5 parameters: V, D, S_Λ, S_N, and T. They are just certain average radial integrals involving, respectively, the radial potentials V_o, V_σ, V_Λ, V_N and V_T.

In our calculation for ${}^{16}_{\Lambda}O$ our model space consists of only nucleons and lambda. Thus the model space effective interaction should contain ΛNN 3-body force with intermediate states having a Σ particle. For the particle-hole interaction, the lowest-order diagram which contributes to the 3-body force is the second-order diagram whose initial, intermediate and final states are respectively (Λ-particle N-hole), (Σ-particle N-hole), and (Λ-particle N-hole). We have calculated this diagram, within the framework of our two-frequency approach.

This ΛNN 3-body-force diagram is observed to have an important effect on the spin parameters. For example, the D parameters obtained from the Nijmegen (*Jülich*) are respectively 1.45 (-1.63) MeV, while the corresponding contribution from the 3-body-force diagram are 0.53 (0.22) MeV. The V parameters obtained from these two potentials are respectively -0.78 (-1.19) MeV, while the contribution from the 3-body-force diagram are weaker, being 0.15 (0.06) MeV respectively. Clearly this diagram has important effect on the parameters V and D for the Nijmegen potential. But its effect for the *Jülich* potential is considerably less, a reflection that it has a weaker $\Lambda - \Sigma$ transition potential than the Nijmegen potential.

In summary, we have investigated a two-frequency shell model for investigating the structure of hypernuclei starting with a meson-exchange hyperon-nucleon potential in free space. In a hypernucleus, hyperons are generally less bound than nucleons, the former being more spatially extended. It seems to be natural that one should allow the oscillator frequency for the hyperon wave function to be different from that for the nucleon wave function. We have applied this approach to a folded-diagram G-matrix calculation of ${}^{16}_{\Lambda}O$. The predicted Λ single particle energy exhibits a saturation minimum at $\hbar\omega_\Lambda \approx 10$ MeV which is much smaller than the value $\hbar\omega_N = 14$ MeV for nucleons. Within the framework of our two-frequency shell model, we have calculated the spin-dependence parameters for the $\Lambda - N$ effective interactions.

A central parameter in our approach is the oscillator frequency $\hbar\omega_\Lambda$ for the Λ orbits. The rms radius for the Λ particle is directly related to $\hbar\omega_\Lambda$. If this rms radius were experimentally known, it would lay a strict constraint on the $\hbar\omega_\Lambda$ value. Further studies in this regard, both experimentally and experimentally, would be very useful. We believe that the two-frequency shell-model approach provides a more realistic framework than the conventional (one-frequency) shell-model approach for understanding the structure of hypernuclear nuclei in terms of the elementary hyperon-nucleon interactions.

Acknowledgements

The authors are very gratefuul to J. Millener for many helpful discussions. This work was supported partly by the NSC(Taiwan) Grant NSC87-2112-M-001-004Y, the US NSF Grant #INT9601361, the US DOE grant DE-FG02-88ER0388, and the US DOE Nuclear Physics Division contract No. W-31-109-ENG-38.

References

1. T.T.S. Kuo and E. Osnes, *Springer Lecture Notes in Physics* **364**,1 (1990).
2. Y. Yamamato, H. Bandō, and J. Žofka, Prog. Phys. **80**, 757 (1988)
3. T. Motoba, H. Bandō, R. Wünsch, J. Žofka, Phys. Rev. C **38** 1322 (1988).
4. Y. Yamamoto and H. Bandō, Prog. Theor. Phys. **73**, 905 (1985); T. Yamada, T. Motoba, K. Ikeda and H. Bando, *ibid.* Suppl. **81**, 104 (1985); H. Bandō, T. Motoba, J. Žofka, Int. J. Mod. Phys. **A5**, 4021 (1990).
5. Jifa Hao, T.T.S. Kuo, A Reuber, K. Holinde, J. Speth, and D.J. Millener, Phys. Rev. Lett. **71**, 1498 (1993).
6. T.T.S. Kuo and Jifa Hao, Prog. Theor. Phys. Suppl. **117**, 351 (1994).
7. Yiharn Tzeng, S.Y. Tsay Tzeng, T.T.S.Kuo, T.-S.H.Lee, Nucl. Phys. A639 (1998) 165c.
8. T.T.S.Kuo, F.Krmpotic, Y.Tzeng, *Phys. Rev. Lett* **78**, 2708 (1997).
9. E.M. Krenciglowa, C.L. Kung, T.T.S. Kuo, and E. Osnes, Ann. Phys. (N.Y.) **101**, 154 (1976).
10. S.Y. Lee and K. Suzuki, Phys. Lett. **B91**, 173 (1980).
11. K. Suzuki and S.Y. Lee, Prog. Theor. Phys. **64**, 2091 (1980).
12. M.F. Jiang, R. Machleidt, D.B. Stout, and T.T.S. Kuo, Phys. Rev. C **46**, 910 (1992).

13. M.M. Nagel, T.A. Rijken and J.J. de Swart, Phys. Rev. D **15**, 2547 (1977); *ibid.* **20**, 1633 (1979); P.M.M. Maessen, T.A. Rijken and J.J. de Swart, Phys. Rev. C **40**, 2226 (1989).

14. B. Holzenkamp, K. Holinde, and J. Speth, Nucl. Phys. **A500**, 485 (1989); A. Reuber, K. Holinde, and J. Speth, Czech. J. Phys. **42**, 1115 (1992); Nucl. Phys. **A570**, 543 (1994).

15. R.E. Chrien, Nucl. Phys. **A478**, 705c (1988).

16. P. H. Pile *et al.*, Phys. Rev. Lett. **66**, 2585 (1991).

17. T. Hasegawa *et al.*, Phys. Rev. **C53**, 1210 (1996).

18. D.J. Millener, A. Gal, C.B. Dover, and R.H. Dalitz, Phys. Rev. C **31**, 499 (1985).

19. D.J. Millener, C.B. Dover, and A.Gal, Phys. Rev.**C38** 2700(1988)

HYPERNUCLEAR PROPERTIES DERIVED FROM G-MATRIX INTERACTIONS

Y. YAMAMOTO

Physics section, Tsuru University, Tsuru, Yamanashi 402-8555, Japan
E-mail: yamamoto@tsuru.ac.jp

G-matrix calculations are performed systematically with the various OBE models, and discuss the specific differences among them. It is demonstrated that the spin-dependent features of the G-matrices can be seen in the contributions to U_Λ from the spin-independent, spin-spin, LS and tensor components of the G-matrices, and the various OBE models are discriminated characteristically. On the basis of the local density approximation, the structure calculations for $^{89}_{\Lambda}$Y are done with use of the G-matrix interactions. The Λ effective mass is extracted by adjusting the calculated energy levels so as to reproduce the recent experimental data precisely.

1 Introduction

The properties of hypernuclear systems link closely to underlying hyperon-nucleon (YN) interactions. Our knowledge of YN interactions is quite incomplete due to the lack of free-space scattering data. It is quite important to extract the information on the interactions from hypernuclear phenomena. The G-matrix gives a good starting point for such approach [1]. On the ΛN interactions, we are now in the stage to establish quantitatively their spin-dependent components on the basis of the coming high-resolution data: The γ-rays detections will provide precise spectroscopic data of light Λ hypernuclei systematically. Then, particular components of the free-space ΛN interactions such as spin-spin and spin-orbit terms, should be revealed when we can eliminate model-dependent uncertainty by performing precise structure calculations of hypernuclei with use of G-matrix interactions.

In the past we performed the G-matrix calculations systematically with the various OBE models by Nijmegen and Jülich groups, and found the specific differences among them [1]. In the case of Nijmegen models there were three models: the hard-core models (NHC-D and NHC-F) [2] D and F and the soft-core model (NSC89) [3]. Recently the revised versions (NSC97) of the soft-core model have been proposed [4]: There are six versions (a, b, c, d, e, f) specified by the different strengths of the magnetic $F/(F+D)$ ratio α_V^m for the vector mesons. Although these six ones give an equally good description of the YN scattering data, their spin-spin terms are very different from each other.

Another topic in this report is related to the recent data of the (π, K) spectrum for $^{89}_{\Lambda}$Y which has been observed at KEK (E369) with nice energy resolution of 1.65 MeV (FWHM). We demonstrate that this data provides the valuable information on the well depth and the effective mass of Λ in nuclear medium.

2 Features of G-matrix interactions derived from OBE models

Let us derive the ΛN G-matrix interactions from various OBE models in nuclear matter. The calculations are done in the same framework as the past work [1]:

We adopt the simple QTQ prescription for the intermediate-state spectrum, which means that no potential term is taken into account in the off-shell propagation.

Table 1. Partial wave contributions to the Λ potential energy $U_\Lambda(k_\Lambda = 0)$ at $k_F = 1.35$ fm^{-1} for the various OBE models. All entries are in MeV.

	$U_\Lambda(S)$	($^1S_0$, $^3S_1$)	$U_\Lambda(P)$	SUM
NSC97				
a	–34.5	(–3.8 , –30.7)	0.6	–33.9
b	–35.5	(–5.5 , –30.0)	1.4	–34.1
c	–37.5	(–7.8 , –29.7)	2.1	–35.3
d	–38.7	(–11.0 , –27.7)	3.5	–35.1
e	–38.8	(–12.8 , –26.0)	4.5	–34.3
f	–37.3	(–14.4 , –22.9)	6.2	–31.1
NSC89	–24.0	(–15.3 , –8.7)	0.2	–23.8
NHC-D	–32.5	(–7.4 , –25.1)	–8.0	–40.5
NHC-F	–30.7	(–10.0 , –20.7)	–0.9	–31.6
JA	–30.8	(–3.6 , –27.2)	1.8	–29.0
JB	–34.9	(–0.5 , –34.4)	3.6	–31.3

Table 2. Decomposition of U_Λ ($k_F = 1.35$ fm^{-1}) to spin-independent (U_0), spin-spin ($U_{\sigma\sigma}$), LS (U_{LS}), and tensor (U_T) parts. All entries are in MeV.

	NSC89	97f	97e	97d	F	D
-S-						
U_0	–6.00	–9.33	–9.70	–9.67	–7.67	–8.13
$U_{\sigma\sigma}$	3.10	1.68	1.04	0.43	0.77	–0.24
-P-						
U_0	0.27	0.92	0.72	0.61	0.13	–1.08
$U_{\sigma\sigma}$	–0.43	–0.50	–0.44	–0.42	–0.39	0.46
U_{LS}	–0.53	–0.47	–0.46	–0.43	–0.49	–0.44
U_{ten}	0.14	0.22	0.17	0.15	0.14	0.09

In Table 1 we show the potential energies $U_\Lambda(k_\Lambda = 0)$ at normal density (k_F=1.35 fm^{-1}) and their S-state ($^1S_0$ and $^3S_1$) and P-state contributions in the cases of various OBE models. Here, the results denoted by JA and JB are in the cases of Jülich models $\tilde{A}$ and $\tilde{B}$ [5] It should be noted that the relative ratios of $U_\Lambda(^1S_0)$ and $U_\Lambda(^3S_1)$ are very different from each other, indicating the different spin-spin components. In the cases of NSC97 models the spin-spin parts varies continuously from a ($\alpha_V^m = 0.4447$) to f ($\alpha_V^m = 0.3647$).

In order to see the spin-dependent features of the G-matrices clearly, we obtain the contributions to U_Λ from the spin-independent, spin-spin, LS and tensor components of the G-matrices, denoted as U_0, $U_{\sigma\sigma}$, U_{LS} and U_{ten}, respectively, by the following transformation:

$$\begin{aligned}
U_0(S) &= (U(^3S_1) + U(^1S_0))/4\ , \\
U_{\sigma\sigma}(S) &= (U(^3S_1)/3 - U(^1S_0))/4\ , \\
U_0(P) &= ((U(^3P_0) + U(^3P_1) + U(^3P_2))/3 + U(^1P_1))/4\ , \\
U_{\sigma\sigma}(P) &= ((U(^3P_0) + U(^3P_1) + U(^3P_2))/9 - U(^1P_1))/4\ , \\
U_{LS}(P) &= (-2U(^3P_0) - U(^3P_1) + U(^3P_2))/12\ , \\
U_{ten}(P) &= (-10U(^3P_0) + 5U(^3P_1) - U(^3P_2))/72\ .
\end{aligned} \tag{1}$$

In Table 2, the obtained values are shown for NSC89, NSC97(d, e, f), NHC-F and NHC-D. It should be careful that U_{LS} includes effectively the effect of the Anti-symmetric LS (ALS) interaction, not only of the LS one. The experimental manifestation of the ΛN spin-spin interaction is the $0^+ - 1^+$ doublet states of $^4_\Lambda$H, where the $J_< = 0^+$ state is below the $J_> = 1^+$ one by about 1 MeV. The analysis of $^4_\Lambda$H with the G-matrix interactions indicates that the adequate strength of spin-spin interaction is between those of NHC-F and NSC89 [1]: Those of NSC97e and NSC97f seem to be adequate. A complementary indication can be obtained from the three-body calculations of $^3_\Lambda$H. Miyagawa performed the calculations using the NSC97 models [6]: NSC97f reproduces the reasonable Λ binding energy. NSC97e gives rise to only a very weakly bound state, and no bound states are obtained for NSC97a–d.

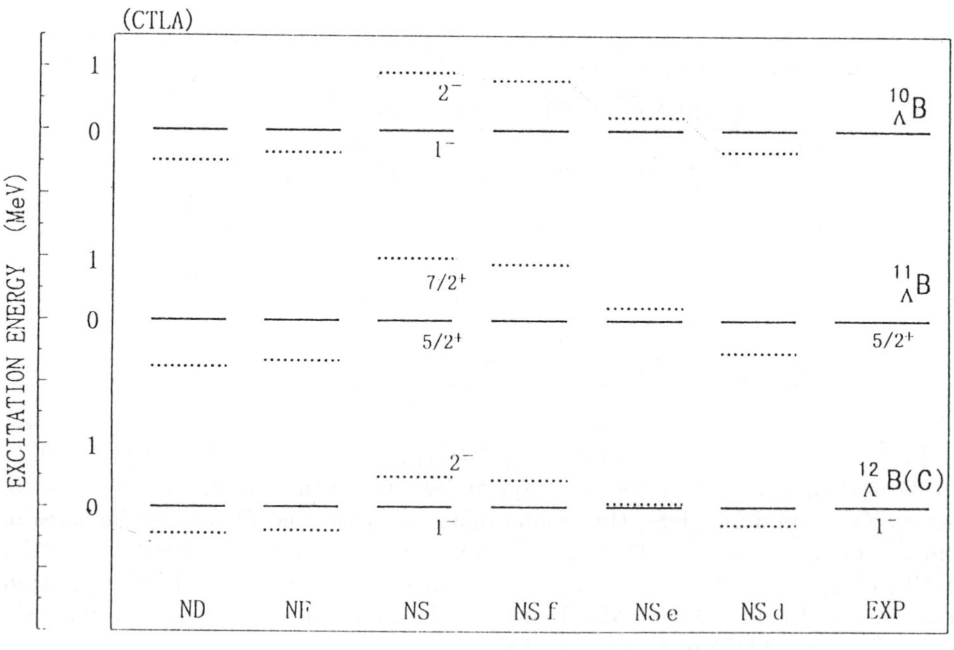

Fig.1

In contrast with the spin-spin components, the spin-orbit ones of Nijmegen models turn out to be of similar strengths. The reason can be understood in terms of the SU(3) irreducible representation (irrep). The ^3P NN interactions belong to the $\{27\}$ irrep, which is also the dominant part of the ^3P ΛN ones. This means that the latters are strongly restricted by the NN scattering data under the assumption of SU(3) symmetry.

For applications to various hypernuclear problems, it is convenient to construct effective local potentials which simulate the spin-independent, spin-spin, spin-orbit (LS and ALS) and tensor parts of the G-matrices. They are dependent on k_F. When these effective interactions are used in the shell-model calculations, the parameters k_F are adjusted so as to reproduce the experimental values of Λ binding energies in ground states. It should be noted that the level structures are not so affected by the variation of k_F, because the k_F dependences of the spin-dependent terms are weak.

The ground-state doublet splittings of some light p-shell hypernuclei are indicative of the spin-spin interactions. In Fig.1 we show the recent result for the ground spin-doublet states of $^{10}_{\Lambda}$B, $^{11}_{\Lambda}$B, $^{12}_{\Lambda}$C and $^{12}_{\Lambda}$B, where the all spin-dependent terms are taken into account [7]. Experimentally the ground-state spins of $^{11}_{\Lambda}$B and $^{12}_{\Lambda}$B are $J_< = 5/2^+$ and 1^-, respectively. The repulsive spin-spin interactions make the $J_<$ states lower than the $J_>$ ones. (The notations NS, NSd, NSe and NSf in the figure mean NSC89, NSC97d, e and f, respectively.) On the other hand, the spin-orbit interaction works more attractively on the $J_>$ states against the spin-spin interaction. The spin-spin parts of NSC97 (f and e) turn out to be repulsive enough to give the right order suggested by the experimental data, though they are less repulsive than that of NSC89. The spin-spin parts of the other versions (a,b,c,d) lead to the reversed orders of the spin-doublets as well as NHC-D and -F. There still remains ambiguities because the strengths of spin-orbit (LS+ALS) interactions are not established experimentally. As stated above, however, they are more or less similar among the various Nijmegen models, and therefor the difference is mainly due to the spin-spin parts. There is another interesting possibility: In the quark-model approach the far weaker spin-orbit interaction is predicted because of the stronger cancellation between the LS and ALS terms. This situation may allow the less repulsive spin-spin interactions than those of the OBE models.

One can see another feature in the above Tables: The P-state contributions (spin-independent parts) are found to be very different among these models. They range from the strongly attractive (NHC-D) to the strongly repulsive (NSC97f). For instance, the values of U_Λ for NHC-F, NSC97f and JB are very similar, but their ratios of S- and P-state contributions are very different from each other. Here we demonstrate that the difference of odd-state components is revealed drastically in high-density matter, which may be related to the Λ mixing in neutron star matter. In Fig.2 we give the U_Λ values in neutron matter and their partial-wave (S, P and D) contributions for NHC-D (solid) and NHC-F (dashed). Here, they are drawn as a function of ρ/ρ_0, ρ_0 being the normal density. It should be noted that the remarkable difference between D and F in the high-density region is responsible to that of the P-state contributions. The elaborate G-matrix calculations for the neutron star matter including hyperons are now in progress by Nishizaki, Takatsuka and Yamamoto [8].

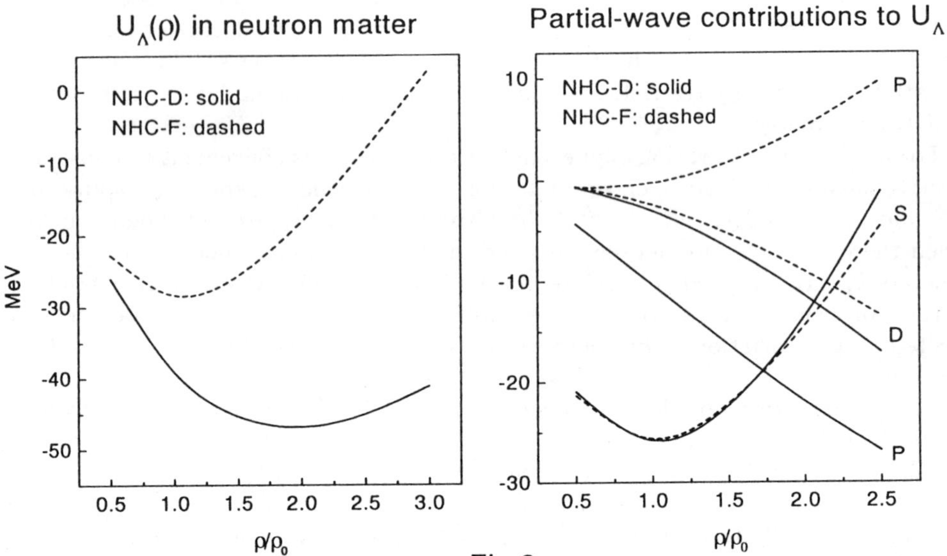

Fig.2

3 U_Λ and M^*_Λ in heavy hypernuclei

Recently, the excitation energy spectrum of ${}^{89}_{\Lambda}$Y has been measured via the (π^+, K^+) reaction at KEK (E369) with the high resolution [9]. The main motivation of this experiment was to observe the possible signature of the spin-orbit (s.o.) splitting. Here, we show that the valuable information on the effective Λ mass also can be obtained from this new data.

Our basic assumption is the Local Density Approximation (LDA): Performing the G-matrix calculations for ΛN pairs in nuclear matter at density ρ, we obtain the Λ single particle potential $U_\Lambda(\rho)$ as a function of ρ. Then, the Λ potential in a finite system is constructed by using the density distribution $\rho(r)$ calculated with the Skyrme-Hartree-Fock method. We adopt here the improved LDA [10] where the finite-range effect of ΛN interactions is taken into account by a parameter t. Moreover, additional parameters α and β for the potential depth and the effective mass, respectively, are introduced for fine fitting to the experimental spectrum.

$$U_\Lambda(r) = \alpha(t\sqrt{\pi})^{-3} \int d\mathbf{r}' U_\Lambda(\rho(r')) \exp(-|\mathbf{r} - \mathbf{r}'|^2/t^2)\rho(r') \ . \tag{2}$$

where α is an adjustable parameter.

The Λ effective mass M^*_Λ is defined by

$$\frac{M^*_\Lambda}{M_\Lambda} = \left(\frac{\tilde{M}_\Lambda}{M_\Lambda}\right) \times \left(\frac{\bar{M}_\Lambda}{M_\Lambda}\right) ,$$
$$\frac{\tilde{M}_\Lambda}{M_\Lambda} = \left[1 + 2M_\Lambda \frac{\partial U_\Lambda}{\partial p^2_\Lambda}\right]^{-1} ,$$

$$\frac{\bar{M}_\Lambda}{M_\Lambda} = 1 - \frac{\partial U_\Lambda}{\partial \epsilon_\Lambda} , \tag{3}$$

where $\tilde{M}_\Lambda$ and $\bar{M}_\Lambda$ are non-locality and energy masses, respectively. Applying LDA to the Λ effective mass, we have

$$M^*_\Lambda(r) = \beta M^*_\Lambda(\rho(r)) + (1-\beta) M_\Lambda \ . \tag{4}$$

Here this adjustment assures $M^*_\Lambda(r) \leq M_\Lambda$ for any value of a parameter β, when $M^*_\Lambda(\rho) \leq M_\Lambda$ is satisfied.

Then, the equation for Λ in a finite system is given by

$$\frac{\hbar^2}{2M^*_\Lambda(r)} \left[-R''_\alpha(r) + \frac{l(l+1)}{r^2} R_\alpha(r) \right] - \left(\frac{\hbar^2}{2M^*_\Lambda(r)} \right)' R'_\alpha(r) + U_\Lambda(r) R_\alpha(r) = \varepsilon_\alpha R_\alpha(r) \ . \tag{5}$$

The adjustable parameters α and β for U_Λ and M^*_Λ, respectively, are determined by the χ^2 fitting to the the experimental spectrum. The range parameter t is fixed adequately in [10], the results being insensitive to its value. In the present case, however, we find that the choice of t has a non-negligible effect on the resultant value of M^*_Λ, namely $M^*_\Lambda(r)$ is rather dependent on $U_\Lambda(r)$. Considering t as a parameter for fine tuning of $U_\Lambda(r)$, we determine it also by the χ^2 fitting.

In order to derive $U_\Lambda(\rho)$ and $M^*_\Lambda(\rho)$ via G-matrix calculations, we adopt here NHC-F and NSC97f (the latter is abbreviated as NSC-f hereafter). The most important condition for the ΛN interactions is to reproduce $U_\Lambda \sim -30$ MeV at normal density, as shown later: We can consider it as a necessity condition which links hypernuclear data to underlying ΛN interactions. As seen in Table 1, $U_\Lambda \sim -30$ MeV is reproduced in the cases of NHC-F and NSC-f. It is a problem, of course, that our results are obtained with the simple QTQ approximation. There is an energy gain of a few MeV [1], if the Λ potential is taken into account continuously in the intermediate spectrum. However, this gain is fairly canceled by considering the nucleon rearrangement effect, which means that U_Λ should be multiplied by $(1-\kappa_N)$. Here κ_N (~ 0.1) is a nucleon correlation probability. Thus, we can say that the QTQ results simulate roughly the more elaborate ones. The values of effective masses are shown in Table 3, which are calculated at normal density.

Table 3. Effective masses calculated at $k_F = 1.35$ fm^{-1}.

	$\tilde{M}_\Lambda/M_\Lambda$	$\bar{M}_\Lambda/M_\Lambda$	M^*_Λ/M_Λ
NHC-F	0.59	1.23	0.73
NSC-f	0.59	1.11	0.66

The E369 data of $^{89}_{\Lambda}$Y shows clearly the s, p, d and f excitation spectrum with high statistics. The observed widths of peaks are broader than the experimental energy resolution, and indicate that the p, d and f peaks are of doublet structures. Here, we assume that the double-peak structures are originated from the ΛN s.o.

interaction. With use of the Scheerbaum approximation [11] the one-body s.o. potential of Λ in a nucleus is related to the two-body LS and ALS interactions as follows:

$$U_\Lambda^{s.o.}(r) = K_\Lambda \frac{1}{r}\frac{d\rho}{dr}\vec{l}\vec{\sigma} \ , \quad K_\Lambda = -\frac{1}{3}\pi(S_{LS}+S_{ALS}) \ ,$$
$$\text{with} \quad S_{LS,ALS} = \frac{3}{\bar{q}}\int_0^\infty r^3 j_1(\bar{q}r) G_{LS,ALS}^{^3O}(r)dr \ , \tag{6}$$

where $G_{LS}^{^3O}(r)$ and $G_{ALS}^{^3O}(r)$ are the LS and ALS components of the ΛN G-matrix interaction in the ^3O state. The calculated values of K_Λ [4] are 19 and 18 (MeV fm^5) for NHC-F and NSC-f, respectively, where $\bar{q} = 0.7$ fm^{-1} was taken[11]. In the present analysis we find that the larger value of K_Λ=27 MeV fm^5 is adequate to reproduce the observed double-peaked structures. However, there still remains a problem for our assumption that these sub-structures are fully originated from the ΛN s.o. interaction: The observed doublets seems to be composed of two peaks of almost equal heights. In the theoretical calculations for spin-doublet states, however, heights of higher-spin states are fairly lower than those of lower-spin ones [12]. Therefore, it is possible that the above large value of K_Λ includes another coupling effect. Our following discussion is on the Λ effective mass, which has to be related to the gross structure of the observed peaks and to be fairly free from the problem of sub-structures.

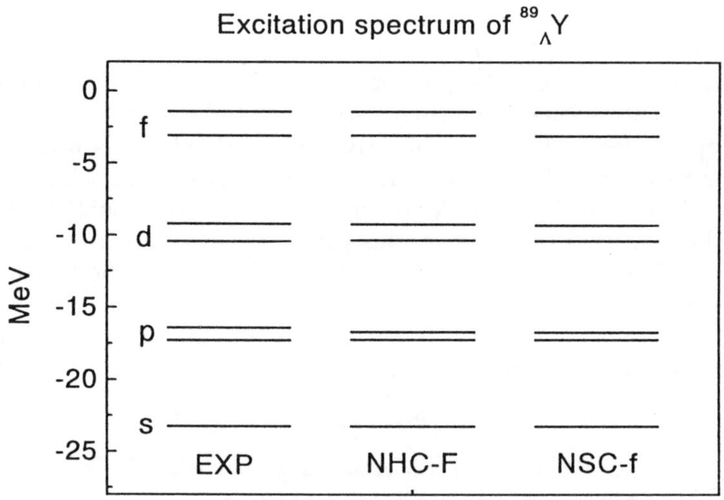

Fig.3

In Fig.2 the calculated spectra in the cases of NHC-F and NSC-f are compared with the experimental one. Here, we can see that the quality of fitting to the data is almost complete. In Table 4 there are given the determined values of parameters and the values of $U_\Lambda(\rho_0)$ and $M_\Lambda^*(\rho_0)$ corrected by α and β, respectively, at normal density. It is found that the corrected values of U_Λ are similar to each other, but the difference of the M_Λ^* values is not negligible.

Table 4. Results of χ^2 fitting. $U_\Lambda(\rho_0)$ and $M^*_\Lambda(\rho_0)$ are corrected by α and β, respectively, at normal density.

	t	α	β	$U_\Lambda(\rho_0)$	$M^*_\Lambda(\rho_0)/M_\Lambda$
NHC-F	0.6	1.008	1.020	−30.5	0.72
NSC-f	0.6	0.981	0.6964	−30.8	0.76

We can explain the reason why the fitting procedures lead to the different values of M^*_Λ for NHC-F and NSC-f as follows: The Λ potentials $U_\Lambda(r)$ derived from NHC-F and NSC-f are of different ranges, because the ρ-dependences of $U_\Lambda(\rho)$ are different from each other. Let us demonstrate them simply by $U_\Lambda(\rho/2) - U_\Lambda(\rho)$, which are 7.1 MeV and 7.7 MeV for NHC-F and NSC-f, respectively. This means that the former is more attractive than the latter at the low-density region, namely the obtained potential $U_\Lambda(r)$ is of longer range. Thus, it turns out that the Λ effective mass derived from the experimental data is generally dependent on the adopted ΛN interaction: One should not consider U_Λ and M^*_Λ separately. In case of NHC-F the corrected value 0.72 of $M^*_\Lambda(\rho_0)$ is very close to the non-corrected one 0.73. On the other hand, the large correction from 0.66 to 0.76 is found in the case of NSC-f. Then, we can say that the U_Λ and M^*_Λ derived from NHC-F are more consistent with the experimental data. It should be noted that the small value of M^*_Λ for NSC-f is related to its strongly repulsive p-state interactions.

Recently the possibility of superfluidity of Λ's mixed in neutron star matter is investigated [13], where the problem of M^*_Λ in dense neutron matter plays an essential role. Our result indicates that a reasonable way in calculations of M^*_Λ in such an exotic situation is to use the ΛN interaction such as NHC-F which reproduces U_Λ and M^*_Λ consistently in normal situations of hypernuclei.

References

1. Y. Yamamoto, T. Motoba, H. Himeno, K. Ikeda, and S. Nagata, Prog. Theor. Phys. Suppl. **No.117** 361 (1994).
2. M.M. Nagels, T.A. Rijken, and J.J. deSwart, Phys. Rev. **D15** 2547 (1977); **D20** 1633 (1979).
3. P.M.M. Maessen, T.A. Rijken, and J.J. deSwart, Phys. Rev. **C40** 2226 (1989).
4. T.A. Rijken, V.G.J. Stoks, and Y.Yamamoto, Phys. Rev. **C59** (1999) 21.
5. A. Reuber, K. Holinde, and J. Speth, Nucl. Phys. **A570** (1994) 543.
6. K. Miyagawa, private communication.
7. T. Motoba, private communication.
8. S. Nishizaki, T. Takatsuka, and Y. Yamamoto, in progress.
9. T. Nagae, in this proceedings.
10. J.P. Jeukenne, A.Lejeune, and C.Mahaux, Phys. Rev. **C16** (1977) 80.
11. R.R. Scheerbaum, Nucl. Phys. **A257** (1976) 77.
12. T. Motoba, H. Bandō, R. Wünsch, and J. Zofka, Phys. Rev. **C38** (1988) 1322.
13. T. Takatsuka, in this proceedings.

ESTIMATE OF INFLUENCE OF ΛNN FORCES AND THE NUCLEAR CORE SIZE ON THE $^7_\Lambda Li$ LEVEL SPECTRUM

V. N. FETISOV

Lebedev Physical Institute, Russian Academy of Sciences, Moscow 117924, Russia
E-mail: fetisov@sci.lebedev.ru

The effective zero-range ΛNN interaction with the Λ-spin dependence pecular to dispersive ΛNN forces is introduced to estimate its influence on the shell model spectra of light hypernuclei. The parameters of three-body and pair interactions are evaluated using the values of B_Λ for s-shell hypernuclei with A=4 and 5. It is shown that the description of the $^7_\Lambda Li_{g.s.}$ doublet energy $\Delta E \simeq 689$ keV fitted with other hypernuclei gives the reduced radius $R \simeq 2.2$ fm for ^6Li bound in $^7_\Lambda Li$ (for the free nucleus ^6Li $R \simeq 2.4$ fm) and the dispersive ΛNN force contribution to this energy ~ 250 keV. A more refined version of the $^7_\Lambda Li$ level spectrum including ΛN and ΛNN forces and the compression of the ^6Li core is proposed. .

1 Introduction

The current shell model calculations of hypernuclear spectra are based on the pair ΛN interaction and in doing so, the coupling of Λ and Σ channels is disregarded[1,2]. However, it is of interest to estimate the contributions of ΛNN forces to the level spacing with the KEK-BNL hypernuclear γ-spectroscopy program in mind[3].

This work was stimulated by results[4] where the large contribution ($\sim 30\%$) of dispersive ΛNN (DF) forces to the $1^+, 0^+$ spin splitting $\Delta E_4(1^+, 0^+) \simeq 1.1$ MeV in $^4_\Lambda H(He)$ has been obtained and by the discussion[5] of two forms of DF and their role in light hypernuclei as well.

Recently the $^7_\Lambda Li_{g.s.}$ doublet splitting $\Delta E_7(3/2^+_1, 1/2+) \simeq 689 \pm 1.1$ keV has been measured in the $(\pi^+, K^+\gamma)$ reaction in KEK[6]. This value exceeds the expected energy ~ 440 keV[a] evaluated in the shell model with the spin-spin parameter of ΛN interaction $\Delta(B) \simeq 0.3$ MeV[2] describing hypernuclear levels in the vicinity of $^{10}_\Lambda B$. From this KEK result it is inferred that the empirical value of $\Delta(Li)$ in $^7_\Lambda Li$ is nearly twice as large as that for more heavy hypernuclei.

To elucidate the reasons of this difference we study in the next sections the dependence of Δ and the $^7_\Lambda Li$ level splitting from the host nucleus (or hypernucleus) sizes and ΛNN forces.

2 Nuclear sizes and the ΛN spin-spin interaction

For the p shell hypernuclei four potential parameters Δ, S_Λ, S_N and T (expressed here in MeV) corresponding to spin-spin, two spin-orbit and tensor parts of ΛN interaction determine the level positions[9]. The harmonic oscillator wave functions with a common h.o. parameter r_0 for the Λ particle and nucleons are accepted to estimate a scale of influence of nuclear sizes on Δ that is important parameter to

[a]The identification of the $^{10}_\Lambda B$ secondary γ-line with $E_\gamma = 442 \pm 2.1$ keV[7] ascribed[8] to $^7_\Lambda Li$ is probably an open problem.

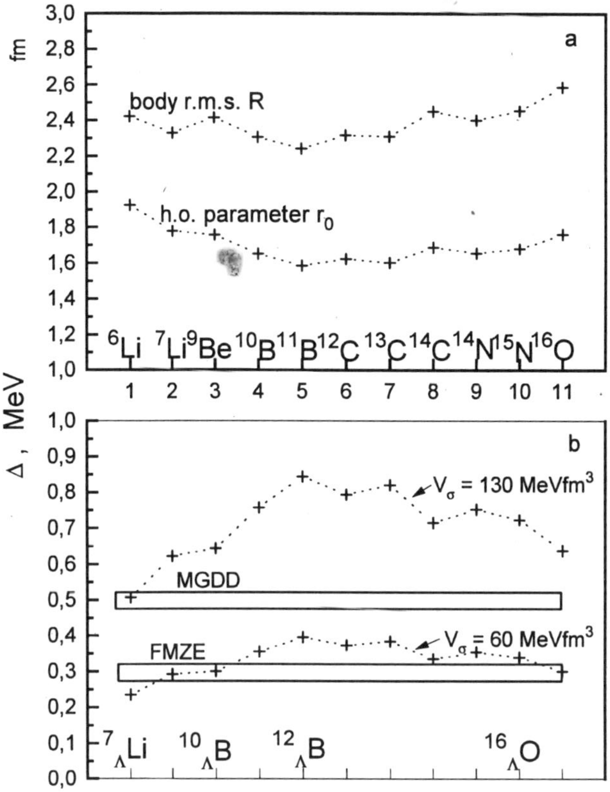

Figure 1: Nuclear body r.m.s. radii, oscillator parameters and Δ for the p shell hypernuclei.

produce a doublet splitting. In this case Δ is of the form

$$\Delta(r_0) = V_\sigma (a/\pi)^{3/2} \frac{(1 + ar_0^2 - \varepsilon)}{(1 + 2ar_0^2)^{5/2}}, \tag{1}$$

where $V_\sigma = V_s - V_t$ is a difference of singlet and triplet volume integrals for the ΛN Gaussian potential $\sim exp(-ar^2)(1 - \varepsilon + \varepsilon P_x)$ with $a = 0.9376$ fm^{-2} [10] and $\varepsilon = 0.25$ [4].

The body r.m.s. radii R and the values of $r_0(^AZ)$ are shown for the p shell nuclei in Fig.1a. They were calculated with the Tassie-Barker correction on the center-of-mass motion [11] using nuclear charge radii and the proton radius ~ 0.8 fm [12]. One of last estimates of $\Delta \simeq 0.5$ have been obtained from a description of $\Delta E_4(1^+, 0^+)$ (the MGDD set of parameters) [1]. Within the Λ-nucleus model [10] and using ΔE_4 as well we find $V_\sigma \simeq 130$ MeVfm^3 and too large $B_\Lambda({}^5_\Lambda He) \simeq 6.9$ MeV. For this case eq. (1) gives the upper curve shown in Fig.1b. But these large values of Δ have been excluded [2] early by the BNL data [7]. If one accepts $\Delta(B) \simeq 0.3$ (the FMZE set of parameters) [2] and $r_0(B) \simeq 1.74$ fm then $V_\sigma \simeq 60$ MeVfm^3 and we have the lower curve in Fig.1b with $\Delta(Li) \simeq 0.23$. The value $\Delta(Li) \simeq 0.23$ is in contradiction with the KEK data. Really the ${}^7_\Lambda Li$ doublet splitting can be written with a good accuracy as

$$\Delta E_7(3/2^+_1, 1/2^+) \simeq \frac{3}{2}\beta^2\Delta(Li), \tag{2}$$

where $\beta = 0.992$ is the amplitude of the dominant ${}^{13}S$-state in the ${}^6Li_{g.s.}$ wave function [13]. For $\Delta(Li) = 0.23$ obtained for the free nucleus ${}^6Li$ ($r_0 = 1.92$ fm) the value $\Delta E_7 = 340$ keV is too small in comparison with $\Delta E_7^{exp.} \simeq 689$ keV. Since Δ is a decreasing function of r_0 to reproduce $\Delta E_7^{exp.}$ the bound nucleus ${}^6Li$ should be compressed upto unrealistic small sizes ($R \simeq 1.8$ fm, $r_0 \simeq 1.45$ fm) in a strong contradiction with a scale of compression ($\Delta R/R \simeq 15\%$) predicted by the cluster model calculations [14]. These difficulties of the ${}^7_\Lambda Li$ shell model with the pair ΛN forces leads us to consideration of the role of ΛNN forces.

3 The influence of ΛNN forces and the compression of the nuclear core on the ${}^7_\Lambda Li$ spectrum

Two types of ΛNN forces arise from a strong $\Lambda N \leftrightarrow \Sigma N$ coupling. First of them $V^{2\pi}_{\Lambda NN}$ is generated due to the sequential double pion exchange accompanied by $\Lambda\Sigma$ conversion [15] and the other one, so called dispersive forces $V^D_{\Lambda NN}$, occurs when one of baryons of the intermediate ΣN pair interacts with an additional nucleon (3π-exchange) [4,5]. The direct employment of explicit expressions for ΛNN forces derived by the pion exchange formalism and given in ref. 5 is not well grounded in the shell model calculations especially due to their indeterminate behaviour at small distances between baryons that is strongly dependent from short range correlations. Because of the short range nature of three-body forces in a scale of hypernuclear sizes it seems to be reasonable to write them in a zero range approximation conserving the Λ-spin dependence pecular to dispersive forces

$$V_{\Lambda NN} = \delta(\vec{r}_\Lambda - \vec{r}_1)\delta(\vec{r}_\Lambda - \vec{r}_2)(t + t^s\vec{\sigma}_\Lambda(\vec{\sigma}_1 + \vec{\sigma}_2)), \tag{3}$$

Here a constant t includes Λ-spin independent parts of the dispersive and 2π-exchange three-body forces which, as calculations show, have a tendency to compensate each other. The spin-dependent term of the potential (3) is not zero only in ${}^{13}S$-state of a nucleon pair. Obviously it gives no contribution to matrix elements for nuclear cores with the total spin S=0. These properties of eq.(3) reflects probably the suppression of $\Lambda\Sigma$ coupling seen explicitly for ${}^5_\Lambda He$ in the two-channel formalism of the s shell hypernucleus description [16].

Four volume integrals $V_{s(t)}$ and t, t^s were determined using B_Λ in the $0^+, 1^+$ states of ${}^4_\Lambda H(He)$ and $B_\Lambda({}^5_\Lambda He)$ within the Λ-nucleus model [10] with the Gaussian nuclear wave functions from equations

$$-\frac{\hbar^2}{2\mu}\frac{d^2\chi_0}{dr^2}+\{Sexp(-\frac{6}{R_3^2}r^2)-V(S)exp(-\frac{3}{2R'^2_3}r^2)\}\chi_0=(-2.22MeV)\chi_0, \quad (4)$$

$$-\frac{\hbar^2}{2\mu}\frac{d^2\chi_1}{dr^2}+\{Pexp(-\frac{6}{R_3^2}r^2)-aV(S)exp(-\frac{3}{2R'^2_3}r^2)\}\chi_1=(-1.1MeV)\chi_1, \quad (5)$$

$$-\frac{\hbar^2}{2\mu_\alpha}\frac{d^2\chi}{dr^2}+\{\frac{81}{16}(P+\frac{1}{3}S)\left(\frac{R_3}{R_\alpha}\right)^6 exp(-\frac{4}{R_\alpha^2}r^2)$$
$$-b\left(\frac{R'_3}{R'_\alpha}\right)^3 V(S)exp(-\frac{3}{2R'^2_\alpha}r^2)\}\chi=(-3.12MeV)\chi, \quad (6)$$

where χ are the radial functions of the Λ-partice;

$$S=\lambda(t-2t^s),\quad P=\lambda(t+\frac{2}{3}t^s),\quad \lambda=\frac{3^{5/2}}{\pi^3R_3^6},\quad \alpha=V_t/V_s;$$

$$V(S)=\frac{3}{2}(1+\alpha)V_s\left(\frac{3}{2\pi R'^2_3}\right)^{3/2},\quad a=\frac{1+5\alpha}{3(1+\alpha)},\quad b=\frac{2(1+3\alpha)}{3(1+\alpha)};$$

$$R_3=\bar{R}(^3H,^3He)=1.6\text{ fm},\quad R_\alpha=R(^4He)=1.46\text{ fm};$$
$$R'_{3(\alpha)}=(R^2_{3(\alpha)}+c)^{1/2},\quad c=1.6026\ ^2.$$

The volume integrals as functions of the ratio $\alpha=V_t/V_s$ are shown in Fig.2. For $\alpha\simeq 0.67$ one finds t^s=0, $V_\sigma\simeq 130$ MeVfm^3 and therefore all the value of ΔE_4 results from spin-spin ΛN forces. If $\alpha=1$ then $V_\sigma=0$, $t^s\simeq 440$ MeVfm^6 and the total doublet splitting is described by DF. For the acceptable value $V_\sigma\simeq 60$ MeVfm^3 derived from $\Delta(B)=0.3$ we have $\alpha=0.83$ and $t^s=250$ MeVfm^6.

Now the additional term originates from DF in the final formula for ΔE_7

$$\Delta E_7=\beta^2\left\{\frac{3}{2}V_\sigma(a/\pi)^{3/2}\frac{(1+ar_0(Li)^2-\varepsilon)}{(1+2ar_0(Li)^2)^{5/2}}+\frac{41}{3^{7/2}\pi^3r_0(Li)^6}\times t^s\right\}. \quad (7)$$

The curves marked by a in Fig.3 shows the separate contributions to $\Delta E_{tot.}$ from ΛN and ΛNN forces as functions of $r_0(Li)$ (or R) at $V_\sigma=60$ MeVfm^6. In this case the DF contribution to ΔE_4 is about $\frac{1}{2}\Delta E_4^{exp.}\simeq 500$ keV. As Fig.3 suggests there is a need to reduce the ^6Li radius up to $R\simeq 2.2$ fm ($\Delta R/R\simeq 8\%$) to reproduce $\Delta E_7^{exp.}$. The parameters used for the curve b give somewhat larger values of R and $\Delta(B)\simeq 0.4$ and the DF contribution to ΔE_4 is $\simeq\frac{1}{3}\Delta E_4^{exp.}$ which is just the value predicted in ref.4. Lastly, the curve c corresponds to the case when the whole of splitting ΔE_4 results from spin-spin ΛN forces and then one needs to increase (!) the ^6Li radius up to $R\simeq 2.5$ fm to be in agreement with the KEK data.

All the $^7_\Lambda Li$ low-lying levels displayed in Fig.4 have been obtained with the Barker's NN interaction [13] using $V_\sigma=60$ MeVfm^3, $r_0=1.74$ fm (for compressed

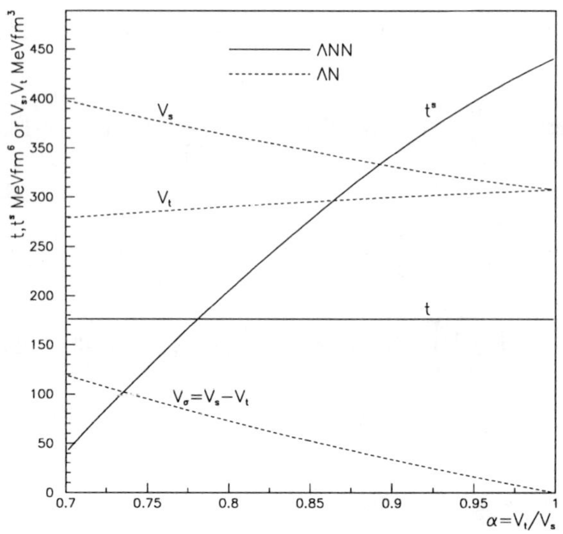

Figure 2: The volume integrals as the functons of α.

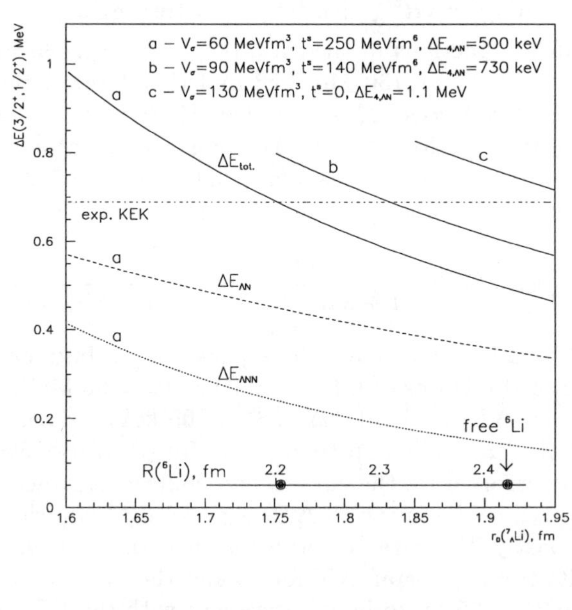

Figure 3: The dependence of the $^7_\Lambda Li$ doublet splitting on r_0 or $R(^6Li)$.

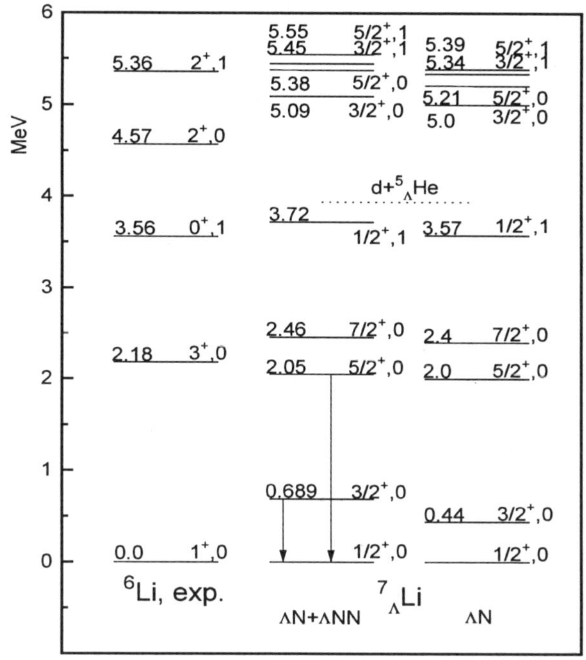

Figure 4: The measured γ-transitions and the new version of the ${}^{7}_{\Lambda}Li$ spectrum.

${}^{6}Li$, $\Delta = 0.3015$), $S_N = -0.4$, $S_\Lambda = -0.02$, $T = 0.02$ [2], $t = 176$ MeVfm^6 and $t^s = 250$ MeVfm^6. The spectrum marked by ΛN is given for $t = t^s = 0$.

The γ -deexcitation of the 2.05 MeV $5/2^+$ level observed in KEK also is well reproduced with the spin-orbit parameter $-S_N = 0.4$ exceeding in four times $-S_N$ found for other p shell hypernuclei. The spectrum depends only on the sum $a + S_N$ where $a = -1.584$ is the Barker's single nucleon spin-orbit constant taking here for the free nucleus. It suggests that increasing $|S_N|$ should be rather associated with increasing $|a|$ ($\Delta a/a \simeq 20\%$) due to a decrease of the diffuseness of a single-nucleon potential under compression of ${}^{6}Li$. This effect has been recognized recently for ${}^{12}_{\Lambda}Be$ and ${}^{16}_{\Lambda}C$ in the Skyrme-Hartree-Fock calculations [17].

4 Summary

In the framework of nuclear and hypernuclear models acceped here it is impossible to reproduce in consistent way the ${}^{7}_{\Lambda}Li_{g.s.}$ doublet splitting observed and the level positions in the s and p shell hypernuclei only with the pair ΛN interaction.
A shell model description of this doublet splitting is possible

i) if one introduces the Λ-spin dependent dispersive ΛNN forces and *ii*) if the nuclear core response manifests itself in the compression of ^6Li. This conclusion is in line with the cluster model calculations in which a strong "glue-like" role of the Λ particle has been revealed [18,14] and the dynamical contraction of ^6Li ($\sim$ 15%) has been predicted. In this work the r.m.s. radius of the bound ^6Li extracted from the doublet splitting is about 2.2 fm that correponds to its overall radial compression $\sim$ 8%.

The DF contribution to the T=0 ground state doublet of $^7_\Lambda Li$ is largest and it is equal to $\sim$ 250 keV versus the 50 keV contribution to the T=1 doublet in $^7_\Lambda He$. A large value of the induced spin-orbit parameter $|S_N|$ may be a signal of increasing the nucleon spin-orbit constant $|a|$ ($\sim$ 20%) due to the compression of ^6Li.

Acknowledgments

Author is very indebted to Profs. H. Tamura and K. Tanida for providing him with the new KEK data and to Profs. D.E. Lanskoy, L. Majling and T. Motoba for useful discussions and to Prof. E. Hiyama for sending the papers before publications.

References

1. D.J. Millener, A. Gal, C.B. Dover and R.H. Dalitz, *Phys. Rev.* C **31**, 499 (1985).
2. V.N. Fetisov, L. Majling, J. Žofka and R.A. Eramzhyan, *Z. Phys.* A **339**, 399 (1991).
3. H. Tamura, *Nucl. Phys.* A **639**, 83c (1998); H. Tamura, in these proceedings.
4. A.R. Bodmer and Q.N. Usmani, *Nucl. Phys.* A **477**, 621 (1988).
5. A. Gal, *Proc. of the LAMPF* (π, K) *Workshop*, Los Alamos, Oct. 11-13, 1990.
6. H. Tamura, private communication, January 1999; K. Tanida, in these proceedings.
7. R.E. Crien, S. Bart, M. May et al., *Phys. Rev.* C **41**, 1062 (1990).
8. L. Majling, J. Žofka, V.N. Fetisov and R.A. Eramzhyan, *Z. Phys.* A **337**, 337 (1990).
9. A. Gal, J.M. Soper and R.H. Dalitz, em Ann. Phys. **63**, 53 (1971).
10. R.H. Dalitz and B.W. Downs, *Phys. Rev.* **111**, 967 (1958).
11. L.J. Tassie and F.C. Barker, *Phys. Rev.* **111**, 940 (1958).
12. R.C. Barrett and D.F. Jackson, *Nuclear Sizes and Structure*, 1977.
13. F.C. Barker, *Nucl. Phys.* **83**, 418 (1966).
14. E. Hiyama, M. Kamimura, K. Miyzaki and T. Motoba, *Gamma transitions in A=7 hypernuclei and a possible derivation of hypernuclear size, 1998,* to be submitted in *Phys. Rev.*
15. R. Bhaduri, B. Loiseau and Y. Nogami, *Ann. Phys.* **44**, 67 (1967).
16. B.F. Gibson, I.R. Afnan, J.A. Carlson and D.R. Lehman, *Prog. Theor. Phys. Suppl.* No. **117**, 339 (1994).
17. T.Yu. Tretyakova and D.E. Lanskoy, to be pulished in *Eur. Phys. J.*.
18. T. Motoba, H. Bandō and K. Ikeda, *Prog. Theor. Phys.*, **70**, 189 (1983).

Hypernuclear spectroscopy for $^{12}B_\Lambda$ through (K^-_{stop}, π^0) reaction using NMS.

M. Youn
(for the AGS E907 Collaboration [a])
Department of Physics, University of Houston, Houston, TX 77204, USA
E-mail: myoun@bnl.gov

We have studied the energy level structure of the hypernucleus $^{12}B_\Lambda$, through the reaction $^{12}C(K^-_{stop}, \pi^0)^{12}B_\Lambda$. We used the Neutral Meson Spectrometer (NMS) to detect the π^0 from the reaction, and an active target chamber(ATC) to detect the reaction vertex. The basic physics ideas of the experiment are discussed. The performance and characteristics of the NMS and other devices are presented, together with a preliminary analysis of π^0 spectroscopy from the reaction.

1 INTRODUCTION

The hypernucleus that is made out of adding a hyperon to a nucleus exhibits many aspects that are not found in conventional nuclear systems; non-mesonic weak decay of hyperons, outstanding charge symmetry breaking(CSB) effect, and induced three-body forces are examples of the advantages of studying hypernuclear systems.

One of the basic ideas of the experiment, AGS-E907, is to observe the effect of CSB by comparing known energy levels [1] of $^{12}C_\Lambda$ and the energy levels of the mirror nucleus $^{12}B_\Lambda$ that has not yet been investigated using counter techniques. To excite the energy levels of $^{12}B_\Lambda$ we performed the experiment using the reaction $^{12}C(K^-_{stop}, \pi^0)^{12}B_\Lambda$.

The reaction (K^-_{stop}, π^0) offers several advantages that may not be found in other kinds of reactions. The (K^-, π^0) reaction can form nuclei that are mirrors to those from the (K^-, π^-) reaction. In (K^-, π^0) reaction, a Λ is produced from a proton rather than from a neutron; the resulting Λ-hypernucleus is left in a proton hole state. Using a stopped K^- beam makes the initial state kinematics simple and well-defined because the reaction takes place at rest. A good resolution can be achieved with (K^-_{stop}, π^0) reaction [2], owing to the well-defined initial state kinematics and owing to the fact that the outgoing photons from the decay of π^0 suffer no energy loss. Compared to the typical energy resolution of 2 - 4 MeV (FWHM) of the reactions (K^-, π^-) or (π^+, K^+), (K^-_{stop}, π^0) reaction can provide resolution of around 1 MeV (FWHM). On the other hand the background can be worse than the usual (π^+, K^+) reaction, for example.

[a] List of collaboration: M. Ahmed, X. Cui, A. Empl, E. Hungerford, A. Lan, B. Mayes, M. Youn (Univ. of Houston), R. Chrien, R. Gill, P. Pile, A. Rusek, R. Sawafta, R. Sutter(Brookhaven National Lab.), R. Boudrie, C. Edwards, C. Morris, J. Peng, H. Thiessen(Los Alamos National Lab.), D. Dehnhard, J. Gerald, J. O'Donnell, H. Juengst, J. Bjoraker, J. Liu, T. Vidnovic(Univ. of Minnesota), D. Androic, D. Bonsar, M. Furic, T. Petkovic, M. Planinic(Univ. of Zagreb), I. Supek(Ruder Boskovic Institute), L. Tang(Hampton Univ.), V. Zeps(Univ. of Kentucky), K. Johnston(Lousiana Tech Univ.), O. Hashimoto(Tohoku Univ.), J. Comfort, C. Gaulard, A. Ramirez(Arizona State Univ.), M. Terigan(Weizmann Institute), W. Briscoe(George Washington Univ.)

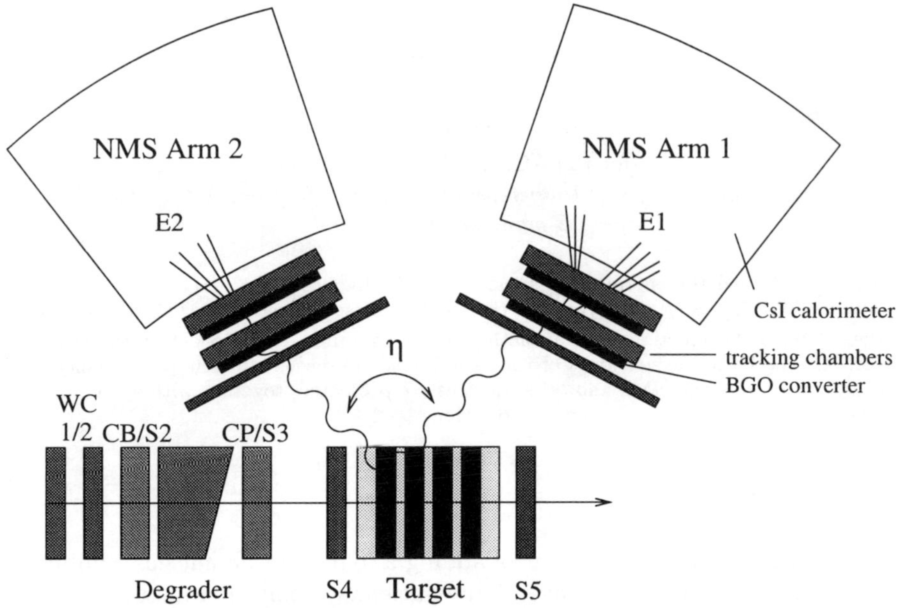

Figure 1: A schematic for detectors setup of AGS-E907, showing the beamline trigger elements, the ATC, and the NMS. WC1, 2 are beamline chambers for K-beam. S2 and S3 are the trigger counters for primary and degraded K beam. CP and CB are the Čerenkov counters.

2 Experiment layout and detectors

The experiment AGS-E907 was set up at AGS-C8 beamline, using the low energy separated beam(LESB-II) for the experiment. The production target, located far upstream of the beamline, produced a secondary beam from the primary proton beam. LESB-II consists of magnets, separators, and slits that deliver K beam to experimental setup.

Figure 1 shows a schematic layout of detectors and target for this experiment. In the figure, WC 1,2 are beamline chambers to detect the tracks of the incident K beam. A set of scintillators S1(not shown in the figure), S2, and S3 are located at each side of degrader to form a trigger for the energy-degraded K-beam incident on target. Using the Čerenkov counters CB and CP, we triggered beam kaons and descriminated against other beam particles.

We used the ATC as our target. It consists of 4 layers of ^{12}C target sandwiched between layers of chambers to detect the tracks of particles before and after the reaction. The analysis of the ATC determines the coordinates of the reaction vertex.

The NMS has two identical arms. Each arm consists of a CsI calorimeter, layers of BGO converters associated with planes of tracking chambers, and a layer of thin plastic scintillators that is used as a charged particle veto.

A π^0 is known to decay into two photons with a branching ratio of 98.8%. The NMS detects two photons in coincidence, one in each arm. The veto counters in front of each arm ensures that no charged particle entered into NMS. Layers of

BGO converters provide chances for the photon to convert into charged particle shower. Layers of tracking chambers behind the BGO converters provide information about the tracks of the shower. The conversion vertex of the arm is obtained by extrapolating the track into BGO converter.

Knowing the conversion vertices from the tracking chambers and the reaction vertex from the ATC, one can calculate the opening angle η between two photons. The BGO converters and the CsI calorimeter provide the energy of the photon at each arm, E_1 and E_2. From energy-momentum conservation, one gets the relations;

$$\begin{aligned} M_{\pi^0} &= \sqrt{2E_1E_2(1-cos\ \eta)}\ , \\ E_{\pi^0} &= m_{\pi^0}\sqrt{\frac{2}{(1-cos\ \eta)(1-X^2)}}\ , \\ where\ X &\equiv \frac{E_1-E_2}{E_1-E_2}\ . \end{aligned} \tag{1}$$

In the equation (1), the quantity X characterizes the energy asymmetry between the two photons.

Important cuts for the analysis involve X and the invariant mass of the photon pair M_{π^0}. Selecting the events that show $M_{\pi^0} \approx 134.97$ MeV(known mass of π^0) ensures that the pair of photons originated from the decay of a π^0. Applying an analysis cut on the absolute value of X improves energy resolution, although it reduces solid angle. Because the cut on X affects both resolution and solid angle, the calorimetry resolution remains important though the contribution to E_{π^0} in Eq. (1) is a second order correction.

3 Performance of the detectors

Among the decay modes of K^+, the channel $K^+ \to \pi^+\pi^0$ provides a source of monoenergetic π^0 of total energy 245.11 MeV, when the K^+ decays at rest. Using the monoenergetic π^0 from the decay of K^+_{stop}, we performed detectors calibration as well as resolution study.

Figure 2 shows energy sum from the calorimeters when the photon in each arm of NMS converted in the BGO. The absolute calibration is obtained by putting the centroid of the peak at 245.1 MeV, while the relative calibration among the crystals is obtained from the cosmic ray data analysis. In this figure, we obtain the FWHM of the peak to be 13.5 MeV; the relative resolution of NMS calorimetry is about 5.5% FWHM.

The reconstructed invariant mass of the π^0 is shown at Figure 3 (a). Figure 3 (b) shows the E_{π^0} from the analysis of the decay of K^+_{stop}, with the cut of $\mid X \mid \leq 0.1$. The peak centroid is at 245.1 MeV, and the width of the peak is 1.3 MeV FWHM which is the resolution of the NMS detector system for the experiment.

The acceptance of NMS depends both on the cut on X, and on the geometric solid angle. We find that the cut $\mid X \mid \leq 0.1$ is a reasonable compromise between the resolution and statistics. Tightening this cut improves resolution, but it reduces the acceptance of NMS. The acceptance of the NMS obtained from a simulation is

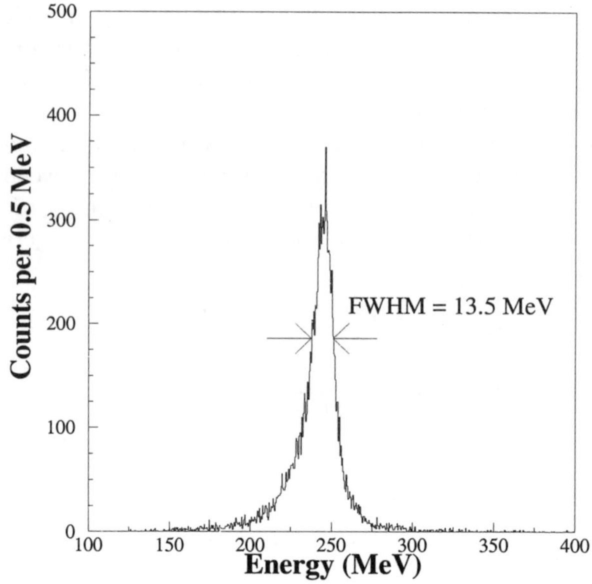

Figure 2: Sum of energy from calorimeters, including NMS CsI and BGOs.

shown in Figure 4. The figure shows two curves that correspond to the settings for K^+_{stop} decay data under the condition $\mid X \mid \leq 0.1$, and $^{12}C(K^-_{stop}, \pi^0)$ under $\mid X \mid \leq 0.3$.

4 Preliminary result and discussion

Figure 5 shows a preliminary analysis of E_{π^0} spectrum from $^{12}C(K^-_{stop}, \pi^0)$ data. For this spectrum, we used a cut on $\mid X \mid \leq 0.1$, and a cut on reconstructed invariant mass. The statistics analyzed for this spectrum was obtained from about 2.0×10^9 K^- triggers after the degrader. Due to the low momentum of K^- beam after it passes through the degrader, we estimate 50% of them arrive at the target. The data shown here amount to 40% of total statistics we obtained.

The peak at $E_{\pi^0} = 232$ MeV is from the decay of $\Sigma^+ \to p\pi^0$ at rest, where the Σ^+ is produced from the elementary reaction $K^-p \to \Sigma^+\pi^-$ or $K^-pp \to \Sigma^+n$. If the Σ^+ decays before it stops, it contributes to the continuum background. For the case of Σ^+ production that does not involve a π-emission, the momentum of the Σ^+ is higher, and provides more chance for an in-flight decay of the Σ^+ rather than for the decay at rest. The π^0 from the in-flight decay can be boosted up to $E_{\pi^0} \approx 350$ MeV to form a broad continuum above the region $E_{\pi^0} \geq 290$ MeV, where we expect to find the signatures from $^{12}B_\Lambda$.

The peak at $E_{\pi^0} = 309$ MeV is believed to be from $^{12}C(K^-_{stop}, \pi^0)^{12}B$ forming the ground state of $^{12}B_\Lambda$, and the peak at $E_{\pi^0} = 299$ MeV is believed to be associated with $^{12}B_\Lambda$ in p-shell, although the accumulated statistics are as yet inadequate.

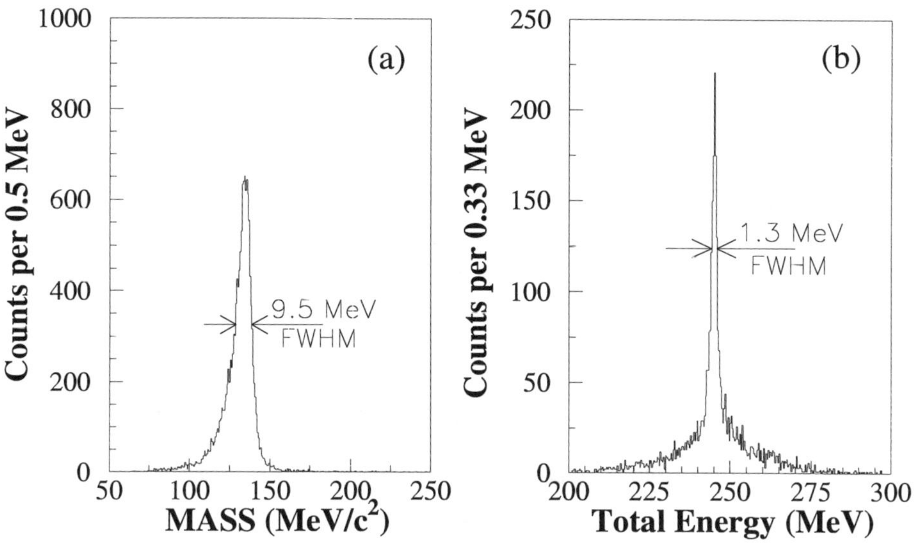

Figure 3: (a) Invariant mass of the pair of photons at each arm of NMS. (b) Total energy of π^0 from the decay of K^+_{stop}, that shows 1.3 MeV FWHM resolution.

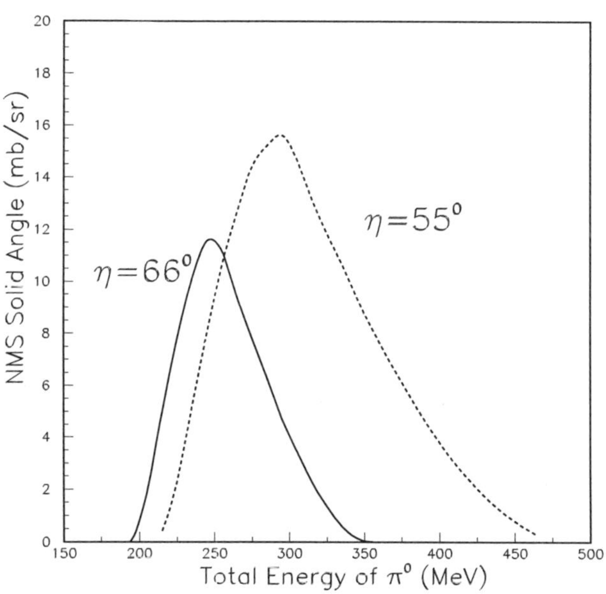

Figure 4: Acceptance of NMS in unit of solid angle (mb/sr), for the settings to detect π^0 from decay of K^+_{stop} (solid line), and $^{12}C(K^-_{stop}, \pi^0)$ (dotted line). The curves shown were obtained with the conditions $| X | \leq 0.1$ for K^+_{stop} setting, and $| X | \leq 0.3$ for $^{12}C(K^-_{stop}, \pi^0)$.

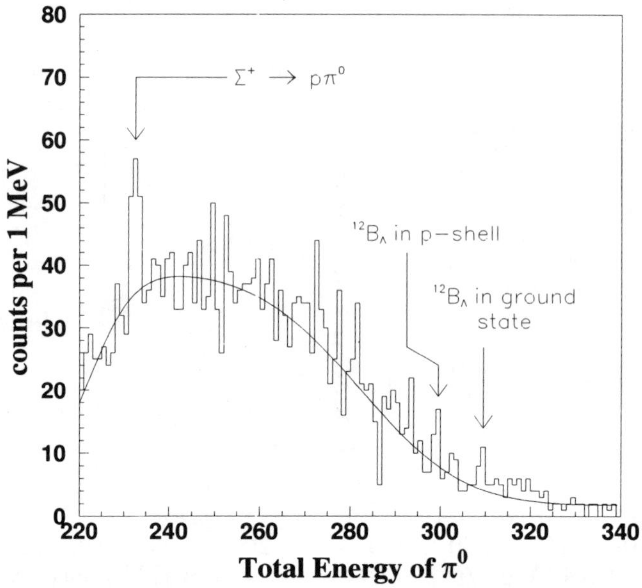

Figure 5: A preliminary total energy spectrum of π^0 from the reaction $^{12}C(K^-_{stop}, \pi^0)$. The curve represents a phenomenological fit for the continuum background.

We are attempting to extend the analysis to include all the recorded data. From a crude estimation of statistics, we expect around 60 counts in $^{12}B_\Lambda$ ground state.

References

1. T. Hasegawa *et al.*, *Phys. Rev. Lett.* **74**, 224 (1995).
2. D. F. Measday, *Czech. J. Phys* **420**, 1049 (1992).

STRUCTURE OF P-SHELL HYPERNUCLEI

D. J. MILLENER
Brookhaven National Laboratory, Upton, NY 11973, USA
E-mail: millener@bnl.gov

New γ-ray data for ${}^{7}_{\Lambda}$Li from KEK E419 and new (π^+, K^+) data on ^{12}C, ^{13}C, and ^{16}O targets from KEK E336 and E369 is used to update Millener, Gal, Dover and Dalitz's 1985 analysis of the spin dependence of the effective ΛN interaction.

1 Introduction

With only very limited YN two-body scattering data available, baryon-baryon potential models based on meson exchange rely heavily on the extension of NN potential models to the strangeness -1 sector by use of symmetry schemes to relate the coupling constants at different baryon-baryon-meson vertices, cutoff parameters in form factors, and so on. When applied to hypernuclei in the form of G-matrix interactions derived from the potentials by solving the ΛN-ΣN coupled channels problem, the YN interactions give a qualitative description of the overall binding energies of Λ hypernuclei characterized by the depth of the Λ-nucleus potential well required to describe the binding energies of Λ single-particle states for a wide range of orbits and masses. However, the spin-dependence of the YN interactions, particularly the central spin-spin interaction, varies greatly[1].

In this situation, it has long been recognized that the energy spectra of Λ hypernuclei can provide rather direct information on the spin dependence of the effective ΛN interaction. Given that the ΛN interaction is weaker than the NN interaction and that the Pauli principle is absent for the Λ in a Λ hypernucleus, the properties of the in-medium ΛN effective interaction are more simply related to the properties of the free YN interaction than in the corresponding case for ordinary nuclei. For the particular case of p-shell hypernuclei with the Λ in a $0s$ orbit, there are only five radial intergrals which characterize the $p_N s_\Lambda$ effective interaction. Of these one characterizes the strength of the spin-independent central interaction while the remaining four characterize the overall strengths of the central spin-spin, symmetric spin-orbit, antisymmetric spin-orbit, and tensor interactions[2].

Experimentally, an important step forward has been made with the development of a 14 element array (Ge ball) of BGO suppressed Ge detectors[3] for hypernuclear γ-ray spectroscopy. The results of a very successful first experiment on ${}^{7}_{\Lambda}$Li, performed at KEK in the summer of 1998, are reported at this meeting by Tanida. Two γ rays with energies of 2.05 MeV and 0.69 MeV, corresponding to the ground-state transitions from the first $5/2^+$ and $3/2^+$ states of ${}^{7}_{\Lambda}$Li, have been observed; the latter transition represents the first measurement of an s_Λ doublet splitting in a p-shell hypernucleus. In addition, the lifetime of the $5/2^+$ level has been measured.

After a brief review of the parametrization of the $p_N s_\Lambda$ effective interaction in Sec. 2, it is shown in Sec. 3 how the new data on ${}^{7}_{\Lambda}$Li from KEK E419 affects the values of the parameters used by Millener, Gal, Dover, and Dalitz[2] (MGDD) in their 1985 fit to the then available data on p-shell hypernuclei. The remaining sections examine the consistency of the parametrization which fits the ${}^{7}_{\Lambda}$Li data with recent (π^+, K^+) data on ^{12}C, ^{13}C, and ^{16}O targets from KEK E336 (presented at this meeting by Hashimoto) and from KEK E369 for a ^{12}C target (presented at this meeting by Nagae).

2 Effective Interaction Parameters

By taking the appropriate linear combinations of the symmetric and antisymmetric spin-orbit interactions, the ΛN effective interaction may be written as

$$V_{N\Lambda}(r) = V_0(r) + V_\sigma(r)\mathbf{s_N}.\mathbf{s_\Lambda} + V_\Lambda(r)\mathbf{l_{N\Lambda}}.\mathbf{s_\Lambda} + V_N(r)\mathbf{l_{N\Lambda}}.\mathbf{s_N} + V_T(r)\mathbf{S}_{12}.$$

In shell-model calculations for $s^4p^{A-5}s_\Lambda$ configurations, the two-body ΛN interaction can be expressed in terms of five radial integrals, one associated with each term in the above expression and conventionally denoted by $\bar{V}$, Δ, S_Λ, S_N, and T. If the same harmonic oscillator (HO) wave functions are used for the N and the Λ, then $\bar{V}$ and Δ are the average of radial integrals in relative $0s$ and $0p$ states while only relative p-state integrals contribute for the non-central interactions[2]. If, on the other hand, it is assumed that $\hbar\omega_N = \hbar\omega_\Lambda$, the relative s-state and p-state intensities are weighted by factors $m_N/(m_N + m_\Lambda)$ and $m_\Lambda/(m_N + m_\Lambda)$, respectively. The choice does not affect the phenomenological parametrization of the $p_N s_\Lambda$ interaction but does affect the relationship to matrix elements which involve p_Λ orbits.

A selection of parametrizations of the $p_N s_\Lambda$ interaction is given in Table 1. The first line contains the MGDD parametrization in which the spin-spin parameter Δ is chosen to be consistent with the average 0^+, 1^+ separation in the $A = 4$ hypernuclei while the remaining spin-dependent parameters are chosen to be roughly consistent with estimates based on the Nijmegen model D potential[4]. The parametrization of Fetisov et al.[5] takes into account the non-observation of a γ ray between the members of the ground-state doublet of ${}^{10}_{\Lambda}$B, and the assignation of a 440-keV γ ray to a transition between the members of the ground-state doublet in ${}^{7}_{\Lambda}$Li (this assignation is now invalidated by the results of KEK E419). The next three lines of Table 1 show parameters deduced from the finite-nucleus G-matrix calculations[6,7]

Table 1. $p_N s_\Lambda$ interaction parameters, radial integrals in MeV.

	V	Δ	S_Λ	S_N	T
MGDD		0.50	−0.04	−0.08	0.04
Fetisov		0.30	−0.02	−0.10	0.02
G(JA)	−0.992	0.020	−0.005	−0.327	0.063
G(JB)	−1.099	−1.316	0.014	−0.324	0.066
G(JB)*	−1.183	−1.590	0.044	−0.413	0.056
G(NSC89)	−0.901	1.106	−0.200	−0.190	0.058
3-body	0.131	0.470	0.013	0.009	0.002
V_{eff}	−0.840	1.414	−0.160	−0.230	0.055

for the energy-independent versions of the Jülich A and B potential models[8] ($\hbar\omega_N = \hbar\omega_\Lambda = 11$ MeV, except for the case marked by an asterix, for which $\hbar\omega_N = 14$ MeV). The next line shows similar G-matrix results[9] for the Nijmegen soft-core potential[10], followed by the contribution of the lowest-order 3-body diagram in which a Λ converts to a Σ one one nucleon and converts back on another nucleon. The final line shows the result of summing the folded-diagram series to obtain an energy-independent effective interaction.

As is well-known[1], the JB and NSC89 interactions represent extremes for the spin-spin interaction, which is basically unconstrained by the existing YN data. As discussed by Rijken at this meeting, the strength of the spin-spin interaction is strongly correlated with value of the magnetic $F/(F+D)$ ratio α_V^m for the vector mesons, a fact which has been exploited in the construction of a new set of six NSC97 potential models[11]. For three of these (NSC97a,b,c) $\Delta < 0$ while for the remaining three (NSC97d,e,f) $\Delta > 0$, with models e and f favored by the empirical value of Δ required to fit the data from light hypernuclei. It is interesting to note that for NSC89 the 3-body diagram makes a repulsive contribution to $\bar{V}$ and a substantial contribution to Δ. The 3-body contribution in this case should represent an upper limit because it is believed[1] that the $\Lambda - \Sigma$ coupling for NSC89 is too strong. There is a need to calculate finite-nucleus effective interactions for NSC97 models using the methods represented by the last three lines of Table 1. The NSC97 models also exibit substantial odd-state repulsion in the central interaction, a fact which is of importance when the Λ is in a $0p$ orbit.

The Jülich interactions have the feature that the symmetric and antisymmetric spin-orbit forces are comparable in magnitude and opposite in sign, leading to a small value for S_Λ and a substantial value for S_N. As will be seen, this feature seems to be consistent with the existing data on p-shell hypernuclear spectra. In the Nijmegen interactions, the same type of coherence exists but the ordinary (symmetric) spin-orbit interaction is dominant. Finally, the tensor matrix element T is about the same for all interactions.

3 $^7_\Lambda$Li

A spectrum of $^7_\Lambda$Li which reflects the new information on γ-ray transitions from KEK E419, namely the energies of 2050 keV and 690 keV for the $5/2^+ \to gs$ and $3/2^+ \to gs$ transitions, is shown in Fig. 1. Fig. 1 is an update of the corresponding figure of MGDD in two respects. First, a p-shell effective interaction (hereafter labelled DJM) from a fit to energy levels of the $A = 6-9$ nuclei is used in place of the CK616 interaction (see later for the rationale for this change). Second, the value of S_N for MGDD in Table 1 is increased in magnitude to -0.47 MeV to fit the experimental excitation energy of 2.05 MeV for the $5/2^+$ level; the need for a similar but smaller change in S_N was noted by MGDD for the case of the CK616 interaction. Fig. 1 also incorporates information from a very recent $^5_\Lambda\mathrm{He} + N + N$ three-body calculation[12], reported on at this meeting by Kamimura, in that the γ-ray branching ratios use the B(M1) and B(E2) values of Hiyama et al.[12] and the "experimental" energies. The calculated (π^+, K^+) cross sections (in μb) integrated from $0° - 15°$ are taken from the appendix of Hiyama et al. From the information

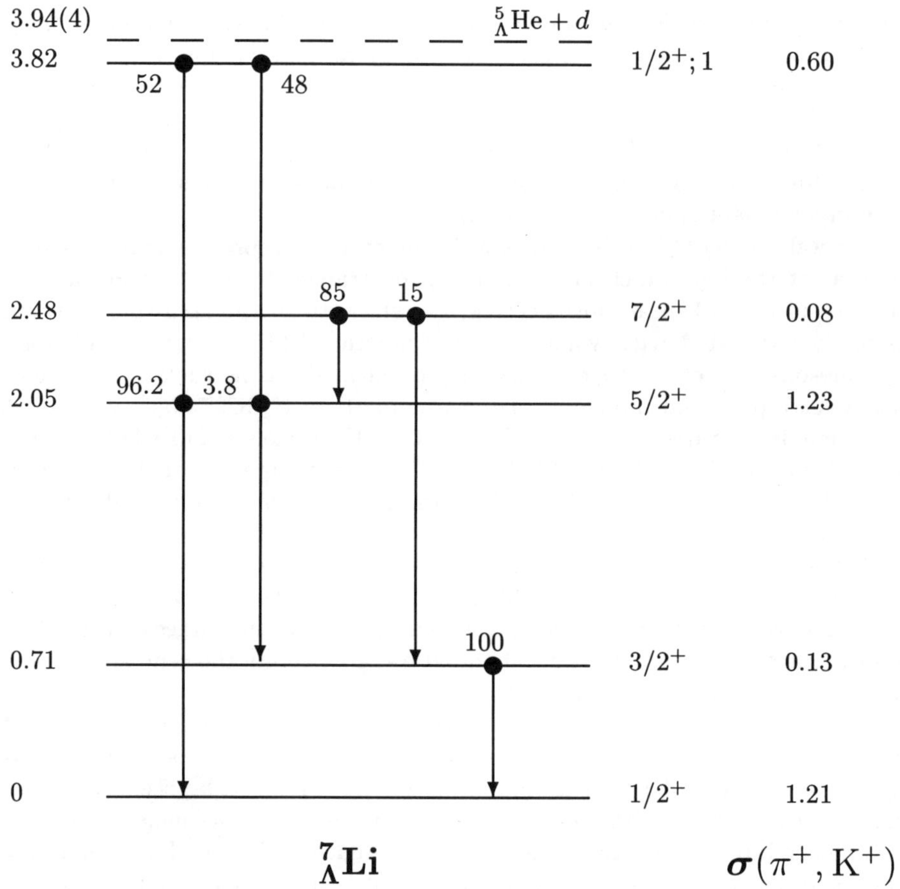

Figure 1. $^7_\Lambda$**Li spectrum and** γ**-ray transitions.**

in Fig. 1, one can calculate the relative yields of γ-rays (ignoring a small $\sim$ 2% weak-decay branch for the $5/2^+$ level)

$$5/2^+ \rightarrow gs = (1.23 + 0.08 \times 0.85) \times 0.962 = 1.249$$

$$3/2^+ \rightarrow gs = 0.13 + 1.298 \times 0.038 + 0.08 \times 0.15 + 0.60 \times 0.52 = 0.503$$

$$1/2^+ \rightarrow 3/2^+ = 0.60 \times 0.52 = 0.312$$

$$1/2^+ \rightarrow gs = 0.60 \times 0.48 = 0.288$$

Taking into account the fact that the detector efficiency for 0.69 MeV γ rays is 2.3 times higher than for 2.05 MeV γ rays, it can be seen that the yields of the two γ rays should be comparable, which is indeed the case. This is strong evidence that the $T = 1$ $1/2^+$ state γ decays, whether or not it lies above the $^5_\Lambda$He + d threshold.

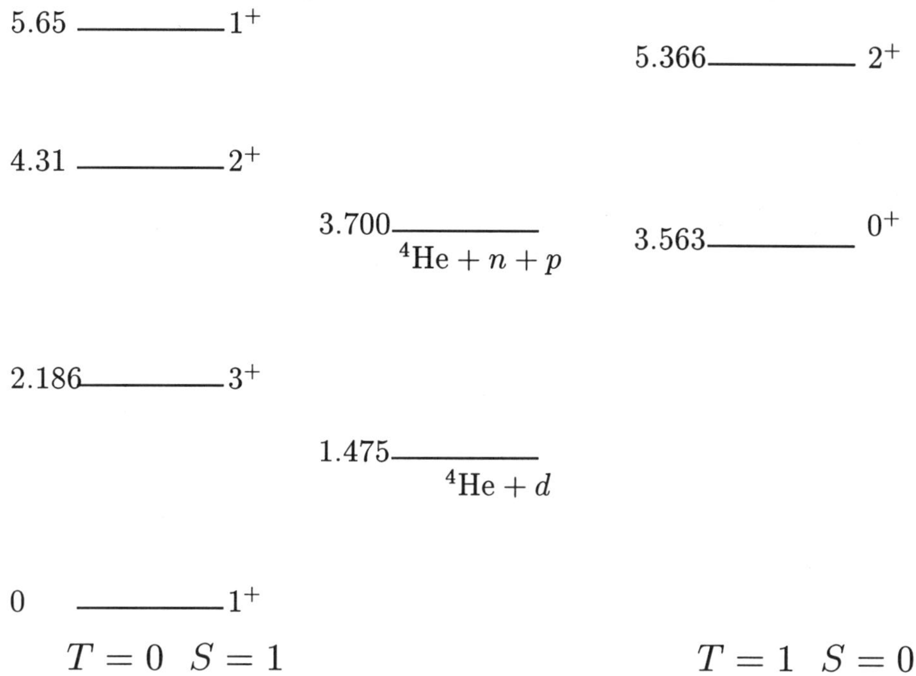

Figure 2. Spectrum of ^{6}Li.

It also means that Doppler-broadened γ-ray lines with energies of ~ 3.3 MeV and ~ 4 MeV should be present in the spectrum but with intensities somewhat below the detection efficiency of E419.

The spectrum of the core nucleus ^{6}Li, to which the s_Λ is coupled to form $^7_\Lambda$Li, is shown in Fig. 2. The Cohen and Kurath interactions[13] do not fit the ^{6}Li spectrum particularly well and exhibit the peculiarity that the single-particle spin-orbit splitting at $A = 5$ is small but grows to ~ 6 MeV to fit $A = 15$. This is achieved by having a one-body spin-orbit interaction masquerade as a two-body interaction, its effective strength therefore increasing linearly with the number particles; all 6 two-body spin-orbit matrix elements (4 ALS) can be substantial in contrast to the dominant role of the odd-state spin-orbit interaction in a G-matrix. The DJM interaction represents a fit to 32 energy levels for $A = 6 - 9$, much in the spirit of Kumar[14] ($A = 5$ spin-orbit splittings of ~ 3.5 and ~ 4 MeV, respectively). The ^{6}Li LS coupling ground-state wave functions for the CK616 and DJM interactions are shown in Table 2.

The small quadrupole moment of ^{6}Li ($Q = -0.083$ fm^2) provides a tight constraint on the wave function, as discussed by Elliott[15] in 1953 (then, $|Q| < 0.1$ fm^2), via

$$Q(^6\text{Li}) = e^0\sqrt{2/5}(-\sqrt{8}\alpha\beta + \sqrt{5/2}\gamma^2 - 7/2\sqrt{10}\beta^2)b^2$$

Table 2. Ground-state wave functions for ^{6}Li.

^{6}Li(gs) =	$\alpha\ ^3S_1$	$+\ \beta\ ^3D_1$	$+\ \gamma\ ^1P_1$
CK616	0.9576	0.2777	0.0761
DJM	0.9873	0.0422	0.1532

where $e^0 = (1 + \delta e_p + \delta e_n)/2$ and b is the harmonic oscillator length parameter. Independent of e^0 and b, CK616 gives an order of magnitude larger Q than DJM (which gives $Q = -0.133$ fm^2 for $e^0 = 0.815$ and $b = 1.77$ fm). The point is that direct mixing of $^3S_1$ and $^3D_1$ via the tensor force gives $\beta/\alpha > 0$ while indirect mixing via the one-body spin-orbit interaction gives the opposite sign. This is in exact analogy to the famous cancellation for ^{14}C β decay; a similar interplay leads to small mixings between the dominantly ^{2}F and ^{4}P $5/2^-$ states in ^{7}Li and ^{7}Be.

The smaller ^{3}D admixture in the DJM wave function has a substantial effect on the energy separation of the ground-state doublet in $^7_\Lambda$Li, as can be seen from Table 3 where the coefficients of the ΛN interaction parameters are listed for both MGDD and DJM. The separation is raised from 610 keV for MGDD to 712 keV for DJM, which is close to the measured separation of $\sim$ 690 keV. As the last line of Table 3 shows, the spin-spin contribution is dominant.

Table 3. Splitting of $^7_\Lambda$Li gs doublet $\Delta E_{\Lambda N}$.

	Δ	S_Λ	S_N	T	$\Delta E_{\Lambda N}$
MGDD	1.348	0.152	−0.059	−1.301	610 keV
DJM	1.444	0.054	0.016	−0.271	717 keV
	722	−2	−7	−11	712 keV

Table 4 shows a similar breakdown for the excitation energy of the $5/2^+$ state. Here S_N is the only contribution which lowers the excitation energy (unless $S_\Lambda > 0$), and a value of -0.47 MeV is required to obtain agreement with the experimental excitation energy of 2.05 MeV for the DJM case if the other parameters remain at their MGDD values. The enhanced nuclear spin-orbit interaction due to S_N increases the separation of states in the $L = 2$, $S = 1$ triplet in the ^{6}Li core (see Fig. 2), thus lowering the 3^+ level.

Table 4. $5/2^+ - 1/2^+$ separation $\Delta E = 2186 + \Delta E_{\Lambda N}$.

	Δ	S_Λ	S_N	T	ΔE
MGDD	0.075	−1.003	0.952	0.226	2235 keV
DJM	0.154	−1.105	0.678	1.095	2326 keV
	77	44	−319	44	2053 keV

An interesting question to ask is whether the $5/2^+$ and $3/2^+$ states based on the

$2^+;1$ core state are likely to γ decay in ${}^{7}_{\Lambda}$Li. To isospin mix the ${}^1D_2$ and ${}^3P_2$ basis states with the ${}^3D_2$ $2^+;0$ state requires the electromagnetic spin-orbit interaction, and an estimate by Bray et al.[16] gives an admixture of 2×10^{-4}. Using an $\alpha + d$ potential well which accounts for the widths of the 3^+ and 2^+ levels in ^{6}Li, gives an estimate of 28 eV for Γ_d in ${}^{7}_{\Lambda}$Li, a value which is much greater than the estimated Γ_γ of ~ 2 eV. The DJM calculation gives a separation of only 10 keV for the $5/2^+$ and $3/2^+$ $T = 1$ levels so that, in this case, both levels would be expected to weak decay in ${}^{7}_{\Lambda}$He.

In this section, it has been shown that the calculated spin-dependent splittings in ${}^{7}_{\Lambda}$Li are quite sensitive to the choice of core wave functions. If the DJM p-shell interaction (non-central interactions fixed, central and one-body spin-orbit interactions fitted to $A = 6 - 9$) is used, the only change needed in the MGDD parametrization of the ΛN effective interaction is to increase S_N from -0.08 to -0.47 MeV (the necessity for a change to ~ -0.25 MeV when using the CK616 interaction was noted by MGDD). This change lowers the excitation energy of the $5/2^+$ level in ${}^{7}_{\Lambda}$Li. The new set of parameters which define the modified MGDD interaction are

$$\Delta = 0.50 \quad S_\Lambda = -0.04 \quad S_N = -0.47 \quad T = 0.04$$

In the next sections, it is asked whether the change in S_N is consistent with recent (π^+, K^+) data from KEK on ^{12}C, ^{13}C, and ^{16}O targets.

4 ${}^{13}_{\Lambda}$C

Although the spectrum of ${}^{13}_{\Lambda}$C observed in (K^-, π^-) and (π^+, K^+) reactions is rather complex[17] because the neutron removal parentage from the ^{13}C ground state goes to widely-separated states of both [44] and [431] spatial symmetry in ^{12}C, the $1/2^+$ ground state and the $3/2^+$ state based on the 4.44-MeV 2^+ level of ^{12}C are strongly populated and well separated from each other and higher states. From the high-statistics (π^+, K^+) data from KEK E336[18] it is possible to obtain an accurate value for the separation energy of theses two states, the preliminary value from E336 being 4.89(7) MeV. This excitation energy is considerably larger than the unperturbed separation energy of 4.44 MeV and the MGDD value of 4.49 MeV, and is primarily sensitive to S_N, as may be seen from

$$E_x(3/2^+) = 4.44 - 0.044\Delta - 1.455 S_\Lambda - 0.803 S_N - 1.142 T$$

with the modified MGDD interaction (CKPOT for the core) which gives an excitation energy of 4.81 MeV for the $3/2^+$ state.

5 ${}^{12}_{\Lambda}$C

In this case, the neutron removal spectrum is rather simple with the parentage to the ^{11}C ground state being strongest and the only other significant parentage being to the 2.00-MeV $1/2^-$ and 4.80-MeV $3/2^-$ states, leading to three 1^- states which can be populated by non-spin-flip transitions in the (π^+, K^+) reaction. The energies of the 1^- excited states are again sensitive to S_N and are given by

$$E_x(1_2^-) = 2.00 + 0.336\Delta + 1.130S_\Lambda - 0.909S_N + 0.703T$$

$$E_x(1_3^-) = 4.80 + 0.383\Delta - 0.397S_\Lambda - 1.336S_N + 0.488T$$

Results for the excitation energies and formation strengths, represented by the square of the L=1, S=0 one-body density-matrices, are given in Table 5 for the weak-coupling limit, the MGDD interaction, and the modified MGDD interaction. As noted in MGDD, $\Delta > 0$ leads to constructive admixtures for the ground-state cross section and destructive admixtures for the excited states.

Table 5. Excitation energies and formation strengths for 1^- states in ${}^{12}_{\Lambda}$C.

S_N	$\rho^2(1_1^-)$	$E(1_2^-)$	$\rho^2(1_2^-)$	$E(1_3^-)$	$\rho^2(1_3^-)$
WC	0.475	2.00	0.125	4.80	0.063
−0.08	0.540	2.19	0.075	5.12	0.048
−0.47	0.527	2.54	0.083	5.66	0.054

The increase in the excitation energies shown in the last line of Table 5 gives much better agreement with the results from three KEK experiments, namely 2.58(17) MeV from E140a[19], 2.71(13) MeV from E336[18], and 2.83(9) MeV from E369 (Nagae, this meeting) for the second 1^- level. This is also true for the energy of the third 1^- level, although in this case there is no clear separation from higher positive-parity levels based on $1\hbar\omega$ postive-parity states of the ^{11}C core.

6 ${}^{16}_{\Lambda}$O

The splitting between the lowest two 1^- states, based on the $p_{1/2}$ and $p_{3/2}$ neutron hole states, in ${}^{16}_{\Lambda}$O is substantially increased with the modified MGDD interaction, as can be seen from the expression below

$$E(1_2^-) - E(1_1^-) = 6.176 - 0.24\Delta - 1.25S_\Lambda - 1.50S_N - 0.73T\ .$$

In this case, the predicted energy separation of 6.75 MeV is considerable larger than the prelimary value of 6.26 MeV from KEK E336, which is puzzling given the evidence from ${}^{12}_{\Lambda}$C and ${}^{13}_{\Lambda}$C. The energy centroid of the peak associated with the 1_2^- level should be lowered somewhat by the population of a 2^+ level built on the the 5.24-MeV $5/2^+$ level of ^{15}O but this effect should modest because the cross section estimated from the spectroscopic factor deduced from the ^{16}O$(e,e'p)$ reaction[20] is only 10% of that for the 1_2^- level.

As is well known, the splitting of the ground-state doublet is very sensitive to the ΛN tensor interaction

$$E(1_1^-) - E(0^-) = -0.38\Delta + 1.37S_\Lambda + 7.87T\ .$$

and, if the splitting is as small as predicted (86 keV), will have to be measured by observing the difference between the energies of the γ rays deexciting the 1_2^-

level. Of course, this measurement will also provide an accurate measurement of the separation of the 1^- levels.

7 Summary

Experiment E419 at KEK has made the first observation of a γ-ray transition between the members of an s_Λ doublet in a p-shell hypernucleus, namely the ground-state doublet in ${}^{7}_{\Lambda}$Li. This doublet separation (690 keV) is primarily sensitive to the central spin-spin component of the effective ΛN interaction, although the transition energy is in fact quite sensitive to the fine details of the ^{6}Li core wave function, and, through a simple shell-model analysis, is in quantitative agreement with the $0^+ - 1^+$ energy separation in the $A = 4$ hypernuclei. It is clear that only (e) and (f) of the NSC97 potential models have a central spin-spin interaction of the correct sign and approximately the correct magnitude (the Jülich interactions are not satisfactory in this regard). In the above cases, the effective ΛNN interaction which arises from the ΛN−ΣN coupling contributes to the energy separations (see Sec. 2) and needs to be evaluated by many-body effective interaction theory for finite nuclei, as discussed by Kuo in these proceedings.

The Λ-spin dependent component of the effective ΛN spin-orbit interaction is known to be small, the best upper limit coming from the separation of the $5/2^+$, $3/2^+$ doublet in ${}^{9}_{\Lambda}$Be (new data from BNL E930, which uses the Ge ball, is currently under analysis). The energy separations of states belonging to different s_Λ doublets are generally sensitive to the nucleon-spin dependent component of the ΛN spin-orbit interaction. This is the case for the $5/2^+ \to 1/2^+$ transition in ${}^{7}_{\Lambda}$Li, measured previously with NaI detectors at BNL but more precisely in KEK E419. Again, the value of S_N required to fit the transition energy is sensitive to details of the ^{6}Li core wave functions. An interesting observation is that the large value $S_N = -0.47$ MeV required to fit the the $5/2^+ \to 1/2^+$ transition energy raises the energies of the 1^- and $3/2^+$ excited states in ${}^{12}_{\Lambda}$C and ${}^{13}_{\Lambda}$C, respectively, bringing them into much better agreement with the energies deduced from recent (π^+, K^+) experiments at KEK. For unknown reasons, the Jülich potentials, but not the Nijmegen potentials, qualitatively reproduce the empirical feature of S_Λ small and S_N large (comparable magnitudes for the symmetric and antisymmetric spin-orbit interactions).

The influence of the tensor force is strongest at the end of the p-shell, the proposed BNL E930 measurement being that of the $1^-_2 \to 0^-_1$ and $1^-_2 \to 1^-_1$ transitions, with intensities in the ratio of $\sim 3:1$, in ${}^{16}_{\Lambda}$O. In addition, the $1^-_2 \to 1^-_1$ transition energy provides another check on the value of S_N.

In summary, a rather consistent picture of the spin dependence of the effective ΛN interaction is emerging, although the measurement of several more key γ-ray transition energies in p-shell hypernuclei is required to overdetermine the spin dependence and provide strict consistency checks. Nuclear many-body theory is probably reliable enough to use the measured spin dependence to put very good constraints on ΛN − ΣN potential models. Of course, the empirically determined $p_N s_\Lambda$ matrix elements relate only to an integrated strength for various components of the ΛN interaction. Sensitivity to odd-state components and quadrupole components of the central force requires the consideration of $p_N p_\Lambda$ and other matrix elements, as noted for p_Λ states in p-shell hypernuclei[17]. Full $1\hbar\omega$ shell-model calculations are needed to properly eliminate spurious center-of-mass states and to acount for all the low-lying states which arise from the coupling and an s_Λ to low-lying non-normal-parity states of the core nucleus. The codes should also include both the effective ΛN and ΛNN interactions.

Acknowledgments

This research was supported by the U. S. Department of Energy under Contract No. DE-AC02-98CH10886 with Brookhaven National Laboratory.

References

1. Y. Yamamoto, T. Motoba, H. Himeno, K. Ikeda, and S. Nagata, *Prog. Theor. Phys. Suppl.* **117**, 361 (1994).
2. D.J. Millener, A. Gal, C.B. Dover, and R.H. Dalitz, *Phys. Rev.* C **31**, 499 (1985).
3. H. Tamura, *Nucl. Phys.* A **639**, 83c (1985).
4. M. M. Nagels, T. A. Rijken and J. J. de Swart, *Phys. Rev.* D **15**, 2547 (1977) (model D); *ibid.* **20**, 1633 (1979) (model F).
5. V.N. Fetisov, L. Majling, J. Žofka, and R.A. Eramzhyan, *Z. Phys.* A **339**, 399 (1991).
6. J. Hao, T.T.S. Kuo, A. Reuber, K. Holinde, J. Speth, and D.J. Millener, *Phys. Rev. Lett.* **71**, 1498 (1993).
7. T.T.S. Kuo and J. Hao, *Prog. Theor. Phys. Suppl.* **117**, 351 (1994).
8. B. Holzenkamp, K. Holinde, and J. Speth, *Nucl. Phys.* A **500**, 485 (1989); A. Reuber, K. Holinde, and J. Speth, *Czech. J. Phys.* **42**, 1115 (1992); *Nucl. Phys.* A **570**, 543 (1994).
9. T.T.S. Kuo, in these proceedings.
10. P. M. M. Maessen, T. A. Rijken and J. J. de Swart, *Phys. Rev.* C **40**, 2226 (1989).
11. Th. A. Rijken, V. G. J. Stoks, and Y. Yamamoto, *Phys. Rev.* C **59**, 21 (1999).
12. E. Hiyama, M. Kamimura, K. Miyazaki, and T. Motoba, *Phys. Rev.* C **59**, 2351 (1999).
13. S. Cohen and D. Kurath, *Nucl. Phys.* **73**, 1 (1965).
14. N. Kumar, *Nucl. Phys.* A **225**, 221 (1974).
15. J.P. Elliott, *Proc. Roy. Soc.* A **218**, 345 (1953).
16. K.H. Bray, J.M. Cameron, H.W. Fearing, D.R. Gill, and H.S. Sherif, *Phys. Rev.* C **8**, 881 (1973).
17. E.H. Auerbach, A.J. Baltz, C.B. Dover, A. Gal, S.H. Kahana, L. Ludeking, and D.J. Millener, *Ann. Phys.* (NY) **148**, 381 (1983).
18. O. Hashimoto, *Nucl. Phys.* A **639**, 93c (1998).
19. T. Hasegawa et al., *Phys. Rev. Lett.* **74**, 224 (1995).
20. M. Leuschner et al. *Phys. Rev.* C **49**, 955 (1994).

THREE-BODY MODEL STUDY OF γ TRANSITION IN $A = 7$ HYPERNUCLEI [a]

E. HIYAMA
Institute for Physical and Chemical Research (RIKEN), Wako, Saitama 351, Japan

M. KAMIMURA and K. MIYAZAKI
Department of Physics, Kyushu University, Fukuoka 812, Japan

T. MOTOBA
Laboratory of Physics, Osaka Electro-Communication University, Neyagawa 572, Japan

We calculate the strengths of γ transitions in ${}^7_\Lambda$Li and ${}^7_\Lambda$He on the basis of the ${}^5_\Lambda$He+N+N three-body model which has been applied to a systematic study of the energy and nucleon halo structure of the $A = 7$ isotriplet hypernuclei. The calculation makes firm that the $B(E2;5/2^+ \to 1/2^+)$ value of the ground-state band in ${}^7_\Lambda$Li is remarkably reduced in comparison with $B(E2;3^+ \to 1^+)$ of the corresponding transition in the core nucleus ^{6}Li. We propose a prescription for how to derive the size of the ground state of the hypernucleus ${}^7_\Lambda$Li using the emprical value of $B(E2;5/2^+ \to 1/2^+)$ together with that of $B(E2;3^+ \to 1^+)$ in ^{6}Li and the size of the ground state of ^{6}Li.

1 Introduction

One of the interesting aspects in hypernuclei physics is to investigate nuclear responses to an added Λ hyperon which is free from the nuclear Pauli principle. In p-shell hypernuclei, the remarkable role of the Λ particle has been pointed out in Refs.[1,2]: the participation of the Λ particle gives rise to more bound states and appreciable contraction of the system (this stablization is called the 'glue like' role of Λ). This feature of light hypernuclei has been studied mostly in systems composed of a stable nucleus and a Λ particle.

On the other hand, a new type of Λ addition to an unstable nucleus having a neutron having or proton halo was investigated by the present authors and their collaborators in the case of the $A = 7$ isotriplet hypernuclei [9]. In Ref. [9], a ${}^5_\Lambda$He $+$ $N + N$ three-body model for the $A = 7$ hypernuclei (${}^7_\Lambda$He,${}^7_\Lambda$Li,${}^7_\Lambda$Be) together with an $\alpha + N + N$ model for the $A = 6$ nuclei (^{6}He,^{6}Li and ^{6}Be) and made a unified study of those systems successfully.

In order to study the structure of the $A = 7$ hypernuclei and obtain information on the ΛN interaction, an experimental project (E419) with high-resolution γ-ray spectroscopy is under way at KEK [3]. The first purpose of thhis paper is to predict $E2$ and $M1$ γ-transition probabilities of $A = 7$ hypernuclei with the present wavefunction before experimental results are reported.

It is of particular interest to experimentally measure the size of any hypernucleus for the first time and compaire it with that of the core nucleus ^{6}Li.

The second purpose is to encouradge the measurement of $B(E2; 5/2^+ \to 1/2^+)$ in ${}^7_\Lambda$Li and to propose a prescription to derive the side of the ground state of ${}^7_\Lambda$Li

[a] presented by M. Kamimura

from that $B(E2)$ value with the aid of the empirical $B(E2; 3^+ \to 1^+)$ and the size of the ground state of ^{6}Li.

2 Model and Method

We employ a $^5_\Lambda$He $+ N + N$ model for the $A = 7$ hypernuclei and an $\alpha + N + N$ model for the $A = 6$ nuclei (Fig. 1 of Ref. [9]).

In order to solve four-body problem accurately, we employ a coupled variational method[5,6] with the use of Jacobian coordinate Gaussian basis functions which span all the rearrangement channels in Fig. 1. This method has been succesfully applied to the three- nucleon bound states[6,7] with realistic two-body and three-body force and to muonic molecules[5], unstable nuclei[8] and hypernuclei[9,10]. The method works excellently in describing both the short-range correlations and the long-range tail behaviour. We take a realistic NN Bonn-A potential. As for the αN interaction, the one proposed in Ref. [4] is adopted; it reproduces precisely the αN scattering phase shifts at low energies. The Pauli principle between the valence nucleons and the α-core nucleons is taken into consideration by introducing a projection operator which rules out the amplitude of the Pauli-forbidden αN state $(0s)$ from the total wave function. As for the ΛN interaction, we employ a one-range Gaussian (ORG) potential proposed in Ref. [1] The $\alpha\Lambda$ interaction is constructed by folding the ΛN interaction into the nucleon density of the α particle. Similarly, the interaction between the $^5_\Lambda$He cluster and a valence nucleon is obtained by folding the αN and ΛN interactions into the α and Λ densities of $^5_\Lambda$He.

The total wave function of the $A = 7$ hypernucleus $^7_\Lambda$Z is described as a sum of the amplitudes of the three rearrangement channels $c = 1 \sim 3$ in Fig. 1.

$$\Psi_{JM}(^7_\Lambda Z) = \sum_{c=1}^{3} \sum_{I,S} \left[\Phi_{1/2}(^5_\Lambda \mathrm{He}) \left[\phi_I^{(c)}(\mathbf{r}_c, \mathbf{R}_c) [\chi_{1/2}(N_1)\chi_{1/2}(N_2)]_S \right]_{J_0} \right]_{JM} \tag{1}$$

Here, $\phi_I^{(c)}(\mathbf{r}_c, \mathbf{R}_c)$ is the spatial coordinate amplitude with angular momentum I, and $\chi_{1/2}(N_1)$ and $\chi_{1/2}(N_2)$ are the two-nucleon spin wave functions coupled to spin S; I and S being coupled to J_0. $\Phi_{1/2}(^5_\Lambda\mathrm{He})$ denotes the wave function of $^5_\Lambda$He with spin 1/2 which couples with J_0 to the total angular momentum $J = J_0 \pm 1/2$. Similar wave functions are adopted for ^{6}Li.

3 Results

In Table 1 (a), we list the calculated $E2$ strenght in $^7_\Lambda$He together with a shell-model estimation[11]. The $E2$ strength is rather small since the motion of the valence neutrons with respect to the c.m. of $^7_\Lambda$He does not contribute to the strength, while only that of the α cluster contributes. The $B(E2)$ values by the shell model, which are much smaller than those by the present model, were derived with the use of an effective charge $(-0.08 \pm 0.01)e$ for the valence neutrons. The $M1$ strength of the $5/2^+ - 3/2^+$ transition is almost the same between the present model and the shell model since the strength merely reflects the radial overlap between the initial- and final-state wavefunctions even though the shape of the wavefunctions

Table 1: Calculated $E2$ transition based on ${}^{5}_{\Lambda}\mathrm{He}+N+N$ model of (a)${}^{7}_{\Lambda}\mathrm{He}$ and (b)${}^{7}_{\Lambda}\mathrm{Li}$ ($T=0,1$) and ${}^{6}\mathrm{Li}(T=0)$.

(a) ${}^{7}_{\Lambda}\mathrm{He}$

Transition	$B(E2)(e^2\mathrm{fm}^4)$	$T(E2)(\mathrm{sec}^{-1})$
$3/2^+ \to 1/2^+$	0.059	0.99×10^9
$5/2^+ \to 1/2^+$	0.068	2.91×10^9

(b) ${}^{7}_{\Lambda}\mathrm{Li}$

	Transition		$B(E2)(e^2\mathrm{fm}^4)$	$T(E2)(\mathrm{sec}^{-1})$
${}^{7}_{\Lambda}\mathrm{Li}(T=0)$	$5/2^+ \to 1/2^+$	present calculation	2.42	1.49×10^{11}
		$\alpha+d+\Lambda$ [1]	2.5	0.95×10^{11}
		shell model [11]	8.6	
	$5/2^+ \to 3/2^+$	present calculation	0.74	3.75×10^9
		$\alpha+d+\Lambda$ [1]	0.4	0.27×10^{11}
	$7/2^+ \to 3/2^+$	present calculation	3.69	2.02×10^{11}
		$\alpha+d+\Lambda$ [1]	3.0	0.73×10^{11}
		shell model [11]	3.1	
${}^{7}_{\Lambda}\mathrm{Li}$ ($T=1$)	$5/2^+ \to 1/2^+$	present calculation	4.58	1.64×10^{11}
	$3/2^+ \to 1/2^+$	present calculation	4.02	5.70×10^{10}
${}^{6}\mathrm{Li}$ ($T=0$)	$3^+ \to 1^+$	present calculation	9.96	1.13×10^{12}
		$\alpha+d$ model	6.6	3.58×10^{11}
		exp.	9.3 $\pm$2.1	5.79×10^{11}

are rather different between the models. Since the decay rate of the free Λ particle is $3.8 \times 10^9 s^{-1}$, it may not be impossible to observed the $E2$ γ rays in ${}^7_\Lambda$He.

In Table 1 (b), calculated $B(E2)$ strength in ${}^7_\Lambda$Li are shown together with those given by the shell model [11] and by the $\alpha + d + \Lambda$ model [1]. The $M1$ strengths are almost the same among the three types of the models because of the same reason as mentioned in ${}^7_\Lambda$He. However, the calculated $B(E2)$ values in ${}^7_\Lambda$Li are quite different between the models.

Firstly, the shell-model values are eminently larger than those by the other models. The reason is as follows: The shell-model calculation [11] took no consideration of the nuclear contraction due to the addition of a Λ particle and assumed the normalization that the sum of $B(E2 : 5/2^+ \to 3/2^+)$ and $B(E2 : 5/2^+ \to 1/2^+)$ is taken to be the same as the observed $B(E2 : 3^+ \to 1^+)$ value in the core nucleus ^{6}Li. On the other hand, the results by the two types of the cluster models exhibit strong reduction of the $B(E2)$ strength due to the shrinkage of the nuclear size in ${}^7_\Lambda$Li. Degree of the shrinkage, namely, the ratio $B(E2 :{}^7_\Lambda \text{Li})/B(E2 :{}^6 \text{Li})$, is listed in Table 2; note that, in the comparison with $B(E2 : 5/2^+ \to 3/2^+)$ and $B(E2 : 5/2^+ \to 1/2^+)$ in ${}^7_\Lambda$Li, the core transition strength $B(E2 : 3^+ \to 1^+)$ is divided into two parts with factors 7/9 and 2/9 multiplied, respectively (see Eq. (3.9) of Ref. [1] for the factors which account for the splitting of $B(E2)$ due to the coupling of spin 1/2 of the Λ particle to the spin 3^+ of ^{6}Li). In the present calculation, the reduction of $B(E2)$ is most evident. Slight deviation in the reduction ratios among the three transitions is simply because the higher-lying state has the larger extension of wavefunction which increases the $B(E2)$ strength. In the $\alpha + d + \Lambda$ cluster model [1], the reduction of $B(E2)$ is weaker than in the present calculation. This is considered to be due to that the spatial extension of the 3^+ resonant state in ^{6}Li is not satisfactorily described by the basis functions with the short-range tails [1] and therefore $B(E2, 3^+ \to 1^+)$ in ^{6}Li was underestimated by about 30 % compared with the observed value.

Table 2: Calculated ratio $B(E2,{}^7_\Lambda \text{Li})/B(E2,{}^6 \text{Li})$ on the basis ${}^5_\Lambda\text{He} + N + N$ model.

$B(E2,{}^7_\Lambda \text{Li})/B(E2,{}^6 \text{Li})$		ratio
$B(E2, 5/2^+ \to 1/2^+)/B(E2, 3^+ \to 1^+)$	present calculation	0.243
	$\alpha + d + \Lambda$	0.379
$B(E2, 5/2^+ \to 3/2^+)/B(E2, 3^+ \to 1^+)$	present calculation	0.074
	$\alpha + d + \Lambda$	0.061
$B(E2, 7/2^+ \to 3/2^+)/B(E2, 3^+ \to 1^+)$	present calculation	0.370
	$\alpha + d + \Lambda$	0.455

We now discuss about the shrinkage of the nuclear size which is reflected in the above mentioned reduction of the $B(E2)$ strengths. Since the $5/2^+ \to 1/2^+$ transition is the most probable to observe experimentally among the three $E2$ transitions in ${}^{7}_{\Lambda}$Li, we consider only this case in the following. In the $\alpha + d + \Lambda$ cluster model [1], it is assumed that the internal motion of the two clusters α and d does not change in the process of the $E2$ transitin and therefore only the $\alpha - d$ relative motion contributes to the $B(E2)$ strength; note that the angle operator $Y_2(\hat{\mathbf{r}}_{\alpha-d})$ is not affected by the contraction of the $\alpha - d$ distance. Thus, the ratios of the $B(E2)$ strengths in Table 2 can directly be related to the r.m.s distance between the core and d, $\bar{r}_{\text{core}-d}$, though the equation

$$\frac{\bar{r}_{\text{core}-d}({}^{7}_{\Lambda}\text{Li})}{\bar{r}_{\text{core}-d}({}^{6}\text{Li})} = \left[\frac{B(\text{E2}, 5/2^+ \to 1/2^+)}{\frac{7}{9}B(\text{E2}, 3^+ \to 1^+)}\right]^{\frac{1}{4}} \quad . \tag{2}$$

According to Ref. [1], the R.H.S of Eq.(7.5) is estimated to be 0.84, which is almost consistent with the value 3.13 fm/3.80 fm =0.82 given by the direct calculation of the L.H.S of Eq.(2) using the wavefunctions themselves.

As mentioned before, the observed $B(E2 : 3^+ \to 1^+)$ strength is underestimated in Ref. [1], but is reproduced by the present work. Therefore, more reliable value of the R.H.S of Eq.(7.5) may be given by the present $\alpha + n + p$ model (^{6}Li) and ${}^{5}_{\Lambda}\text{He} + n + p$ model (${}^{7}_{\Lambda}$Li) in which deuteron cluster is not assumed a *priori* for the valence neutron and proton (deuteron cluster is broken by approximately 40 % in ^{6}Li and ${}^{7}_{\Lambda}$Li [9]). We first examine whether the shrinkage of ${}^{7}_{\Lambda}$Li occurs in the $n - p$ relative distance or in the $n - p$ c.m. distance with respect to the core. As shown in Fig. 1, the $n - p$ relative density averaged over $R_{\text{core}-np}$, $\rho(r_{n-p})$,defined by

$$\rho(r_{n-p}) = \int |\Psi({}^{6}\text{Li or } {}^{7}_{\Lambda}\text{Li}, \text{g.s.})|^2 \, d\mathbf{R}_{\text{core}-np} \, d\hat{\mathbf{r}}_{n-p} \,/\, 4\pi \quad , \tag{3}$$

exhibits almost the same shape between the ground states of ^{6}Li and ${}^{7}_{\Lambda}$Li; in other words, shrinkage of the $n - p$ distance is negligible small. In order to see the degree of shrinkage in the motion of the c.m. of the $n - p$ pair with respect to the core, we introduce $n - p$ c.m. density as a function of $R_{\text{core}-np}$ by

$$\hat{\rho}(R_{\text{core}-np}) = \int |\Psi({}^{6}\text{Li or } {}^{7}_{\Lambda}\text{Li}, \text{g.s.})|^2 \, d\mathbf{r}_{n-p} \, d\hat{\mathbf{R}}_{\text{core}-np} \,/\, 4\pi \quad . \tag{4}$$

In Fig. 1 (b), the density $\hat{\rho}(R_{\text{core}-np})$ is illustrated for the ground states of ^{6}Li (dashed line) and ${}^{7}_{\Lambda}$Li (solid line). In the case of ${}^{7}_{\Lambda}$Li, one sees significant contraction of the relative wavefunction between the c.m. of the $n-p$ pair and the core compared with the case of ^{6}Li. The r.m.s. distance $\bar{R}_{\text{core}-np}$ is obtained as 3.85 fm for the ${}^{6}\text{Li}(0^+)$ and 2.94 fm for ${}^{7}_{\Lambda}\text{Li}(1/2^+)$. We then conclude that the contraction in ${}^{7}_{\Lambda}$Li occurs between the c.m. of the $n - p$ pair and the core with the $n - p$ relative motion unchanged. In this sense, the L.H.S. of Eq.(7.5) is 2.94 fm/3.85 fm =0.764, while the R.H.S. equals to 0.758 within our model. Therefore, we think that, if observation of $B(E2 : 5/2^+ \to 1/2^+)$ is performed in the near future, a reasonable estimation of the size of the hypernucleus ${}^{7}_{\Lambda}$Li will be possible by using Eq.(7.5) together with the experimental values of the R.H.S quantity and $\bar{r}_{\alpha-np}({}^{6}\mathrm{Li})$ in L.H.S. The experimental value of $\bar{r}_{\alpha-np}({}^{6}\mathrm{Li})$ may be estimated by the often-used relation

$$6 \times \bar{r}^2({}^6\mathrm{Li}) = 4 \times \bar{r}^2(\alpha) + 2 \times \bar{r}^2_{n-p} + 2 \times 4/(4+2)\, \bar{r}_{\alpha-np}({}^6\mathrm{Li}), \tag{5}$$

where $\bar{r}({}^6\mathrm{Li}) = 2.82$ fm and $\bar{r}(\alpha) = 1.61$ fm are the r.m.s. radii of mass distribution of ^{6}Li and α particle, respectively. Here, $\bar{r}_{n-p}$ is the r.m.s. distance between the $n-p$ pair in the ground state of ^{6}Li which may be estimated by our model as $\bar{r}_{n-p} =$ 3.42 fm. From these we have $\bar{r}_{\alpha-np}({}^6\mathrm{Li}) = 3.24$ fm. We then propose a relation to estimate experimental value of $\bar{r}_{\mathrm{core}-d}({}^7_\Lambda\mathrm{Li})$ as

$$\bar{r}_{\mathrm{core}-d}({}^7_\Lambda\mathrm{Li}) = 1.98 \times [B(\mathrm{E2}, 5/2^+ \to 1/2^+)]^{\frac{1}{4}}, \tag{6}$$

where the $B(E2)$ is in unit of $e^2\mathrm{fm}^4$.

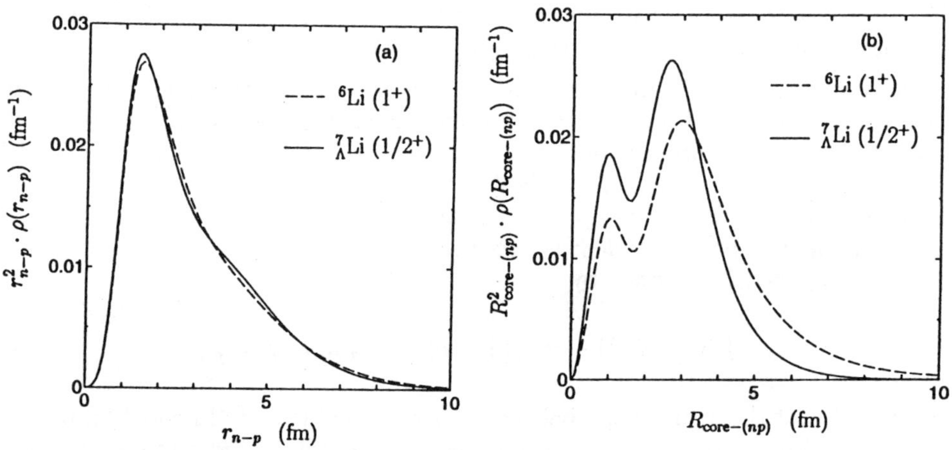

Fig.1. (a) The $(n-p)$ relative distribution $\rho(r_{n-p})$ (b) The np pair c.m. density distribution $\rho(R_{\mathrm{core}np})$.

4 Summary

With the use of ${}^5_\Lambda$He + N + N model, we have calculated $E2$ and $M1$ transitions in ${}^7_\Lambda$He and ${}^7_\Lambda$Li. Observation of the transitions in ${}^7_\Lambda$He and the comparison with the present calculation will contribute to the study of the excitation mechanism of neutron halo of the core nucleus ^{6}He in which gamma transitions cannot be seen due to the prompt neutron decays from the excited states. Strong reduction of $E2$ transitions in ${}^7\Lambda$Li compared with the corresponding transition in the core nucleus ^{6}Li is signature of the shrinkage of the distance between the c.m. of the valence $n-p$ pair and the α cluster. Using this result, we have suggested observation of the $B(E2 : 5/2^+ \to 1/2^+)$ in ${}^7_\Lambda$Li which gives rise to a possibility to estimate experimentally the size of hypernucleus ${}^7_\Lambda$Li.

Acknowledgement

This work was supported by the Special Postdoctal Researchers Program at RIKEN (E. H.) and by the Grant-in-Aid for Scientific Research in Priority Areas.

References

1. T. Motoba, H. Bandō and K. Ikeda, *Prog. Theor. Phys.* **70**, 189 (1983); T. Motoba, H. Bandō, K. Ikeda and T. Yamada, Prog. Theor. Phys. Suppl. No. 81 (1985), Chap. III.
2. H. Bandō, T. Motoba and J.Žofka, *Int. J. Mod. Phys.* A5, 4021 (1990).
3. H. Tamura *et al.*, "Measurement of E2 Transition Rate in ${}^{7}_{\Lambda}$Li Hypernucleus", KEK-PS E419 proposal, 1997.
4. H. Kanada, T. Kaneko, S. Nagata and M. Nemoto, *Prog. Theor. Phys.* **61**, 1327 (1979).
5. M. Kamimura, *Phys. Rev.* A **38** 621 (1988).
6. H. Kameyama, M. Kamimura and Y. Fukushima, *Phys. Rev.* C **40** 974 (1989).
7. M. Kamimura and H. Kameyama, *Nucl. Phys.* A **508** 17c (1990).
8. E. Hiyama and M. Kamimura, *Nucl. Phys.* A **588** 35c (1995).
9. E. Hiyama, M. Kamimura, T. Motoba, Y. Yamada and Y. Yamamoto, *Phys. Rev.* C **53** 2075 (1996).
10. E. Hiyama, M. Kamimura, T. Motoba, Y. Yamada and Y. Yamamoto, *Prog. Theor. Phys.* **97** 881 (1997).
11. R. H. Dalitz and A. Gal, *Ann. of Phys.* **116**, 167 (1978).

HYPERNUCLEAR γ-RAY SPECTROSCOPY EXPERIMENTS WITH GERMANIUM DETECTORS

K. Tanida[a], H. Tamura[b], D. Abe[b], S. Ajimura[c], H. Akikawa[d], K. Araki[b], H. C. Bhang[e], R. E. Chrien[f], P. Eugenio[g], G. B. Franklin[g], Y. Fujii[b], T. Fukuda[h], L. Gan[i], O. Hashimoto[b], H. Hotchi[a], K. Imai[d], Y. Kakiguchi[h], P. Khaustov[g], J. H. Kim[e], Y. D. Kim[j], H. Kohri[c], M. May[f], T. Miyoshi[b], T. Murakami[d], T. Nagae[h], J. Nakano[a], H. Noumi[h], K. Ozawa[b], H. Outa[h], P. H. Pile[f], B. P. Quinn[g], A. Rusek[f], T. Saito[k], J. Sasao[b], Y. Sato[b], S. Satoh[b], R. I. Sawafta[l], R. A. Schumacher[g], M. Sekimoto[h], T. Takahashi[b], T. Tamagawa[a], L. Tang[i], H. H. Xia[n], L. Yuan[i], S. H. Zhou[n], L. H. Zhu[n], X. F. Zhu[n]

(KEK-PS E419/BNL-AGS E930 collaboration)

[a] *Department of Physics, University of Tokyo, Tokyo 113-0033, Japan*
[b] *Department of Physics, Tohoku University, Sendai 980-8578, Japan*
[c] *Department of Physics, Osaka University, Toyonaka 560-0043, Japan*
[d] *Department of Physics, Kyoto University, Kyoto 606-8502, Japan*
[e] *Department of Physics, Seoul National University, Seoul 151-742, Korea*
[f] *Brookhaven National Laboratory, Upton, NY 11973, USA*
[g] *Carnegie Mellon University, Pittsburgh, PA 15213, USA*
[h] *Institute for Particle and Nuclear Physics, KEK, Tsukuba 305-0801, Japan*
[i] *Department of Physics, Hampton University, Hampton, VA 23668, USA*
[j] *Department of Physics, Sejong University, Seoul 143-747, Korea*
[k] *Laboratory of Nuclear Science, Tohoku University, Sendai 980-0826, Japan*
[l] *Physics Department, North Carolina A&T State University, Greensboro, NC 27411, USA*
[n] *Department of Nuclear Physics, CIAE, P.O.Box 275(80), Beijing 102413, China*

We report on the two hypernuclear γ-ray spectroscopy experiments with a germanium detector system(Hyperball) which is recently constructed. In the first experiment at KEK, we observed two γ transitions in ${}^{7}_{\Lambda}$Li, the $E2(5/2^+ \rightarrow 1/2^+)$ transition at 2050 ± 2 keV and the spin-flip $M1(3/2^+ \rightarrow 1/2^+)$ transition at 689 ± 4 keV(preliminary). This is the first observation of well-identified hypernuclear γ-rays with germanium detectors. The measured $B(E2)$ value of the former transition, $4.1 \pm 0.6 \pm 0.5$ e^2fm^4(preliminary), is a clear evidence of the glue-like role of Λ. The strength of the ΛN spin-spin interaction can be derived from the energy of the latter transition. Status of the second experiment at BNL is briefly mentioned.

1 Introduction

Gamma-ray spectroscopy is a powerful means of investivating hypernuclei because of an excellent energy resolution of 2 keV(FWHM) accessible by using germanium (Ge) detectors. In the past, however, only five γ-transitions were identified so far, none of which were observed with Ge detectors.

Recently, we have constructed a detector system which consists of 14 germanium detectors for hypernuclear study(Hyperball). Two experiments, KEK-PS E419 and BNL-AGS E930, have been carried out as the first stage of the detector system[1,2,3]. In this paper, we report on the preliminary results of the former experiment in Section 2 and the status of the latter experiment in Section 3.

2 KEK-PS E419 experiment

KEK-PS E419 is a γ-ray spectroscopy experiment of $^{7}_{\Lambda}$Li produced via the (π^+, K^+) reaction at KEK-PS K6 beamline. Purposes, setup and results of the experiment are described in the following subsections.

2.1 Purposes of the experiment

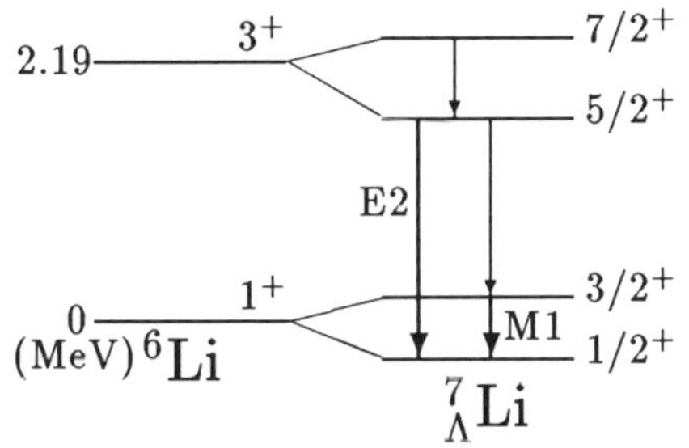

Figure 1: Expected level scheme and γ transitions of $^{7}_{\Lambda}$Li. Corresponding levels of ^{6}Li are also shown.

Figure 1 shows the expected level scheme and γ transitions of $^{7}_{\Lambda}$Li. In the experiment, there are two physics motivations related to the $E2(5/2^+ \to 1/2^+)$ and $M1(3/2^+ \to 1/2^+)$ transitions. One is to verify the "glue-like role" of a Λ particle. A loosely bound nucleus, such as ^{6}Li, is expected to shrink by the presence of a Λ particle in the $0s$ orbit, because of the attractive interaction between a Λ particle and a nucleon. This effect is predicted by Motoba *et al.*[4] and called "glue-like role of Λ". It can be observed by measuring the reduced transition probability , $B(E2)$, of the $E2(5/2^+ \to 1/2^+)$ transition, which is basically the same as the $E2(3^+ \to 1^+)$ transition of the ^{6}Li core nucleus. Since $B(E2)$ is proportional to the fourth power of the nuclear size, the size of the $^{7}_{\Lambda}$Li can be represented as;

$$\frac{R(^{7}_{\Lambda}\mathrm{Li})}{R(^{6}\mathrm{Li})} = \sqrt[4]{\frac{9}{7}\frac{B(E2;^{7}_{\Lambda}\mathrm{Li}\ 5/2^+ \to 1/2^+)}{B(E2;^{6}\mathrm{Li}\ 3^+ \to 1^+)}}, \tag{1}$$

where R denotes the mean distance between α cluster and the center of valence proton and neutron in ^{6}Li and $^{7}_{\Lambda}$Li[5]. Because $B(E2;^6\mathrm{Li}\ 3^+ \to 1^+)$ is known[6] to be $9.3 \pm 2.1\ e^2\mathrm{fm}^4$, we can extract the hypernuclear-size information from $B(E2;^{7}_{\Lambda}\mathrm{Li}\ 5/2^+ \to 1/2^+)$. Experimentally, $B(E2;^{7}_{\Lambda}\mathrm{Li}\ 5/2^+ \to 1/2^+)$ is derived from the lifetime of the $5/2^+$ state which we measured by Doppler attenuation method.

The other purpose is to obtain information on the strength of ΛN spin-spin interaction by measuring the energy spacing of the ground state doublet$(3/2^+, 1/2^+)$ of $^{7}_{\Lambda}$Li. This energy spacing is determined only by the spin-spin interaction because the ground state of the core $^6\mathrm{Li}(1^+)$ has almost pure $^3S_1$ configuration. Thus we can extract the information on the spin-spin interaction from the energy of the γ-ray emitted by the spin-flip $M1(3/2^+ \to 1/2^+)$ transition.

2.2 Experimental setup

The experiment was performed at KEK-PS K6 beamline using the SKS spectrometer and the Hyperball. $^{7}_{\Lambda}$Li hypernuclei were produced by the $^7\mathrm{Li}(\pi^+, K^+)^{7}_{\Lambda}\mathrm{Li}$ reaction. Momenta of beam pions and scattered kaons were measured by the beamline spectrometer and the SKS spectrometer. From these momenta, we can reconstruct missing mass for $^{7}_{\Lambda}$Li, and thus select the production of the bound states of $^{7}_{\Lambda}$Li. The missing mass resolution in this experiment is mainly determined by the thickness of the target(13.5g/cm^2) and is about 5 MeV FWHM. More information about the K6 beamline and the SKS spectrometer is found in Ref.[7]

Gamma-rays were measured by the Hyperball in coincidence with the production of $^7_\Lambda$Li. The Hyperball consists of 14 Ge detectors of 60% relative efficiency with each surrounded by 6 BGO counters for background suppression. Detailed descriptions of the Hyperball are found in Ref.[8]

2.3 Results and discussion

Missing mass spectrum of the ^{7}Li(π^+, K^+)$^7_\Lambda$Li reaction Figure 2 shows the missing mass spectrum of $^7_\Lambda$Li. The peak at around 173 MeV corresponds to bound states of $^7_\Lambda$Li. The "bound state region" gate is set as shown in the figure.

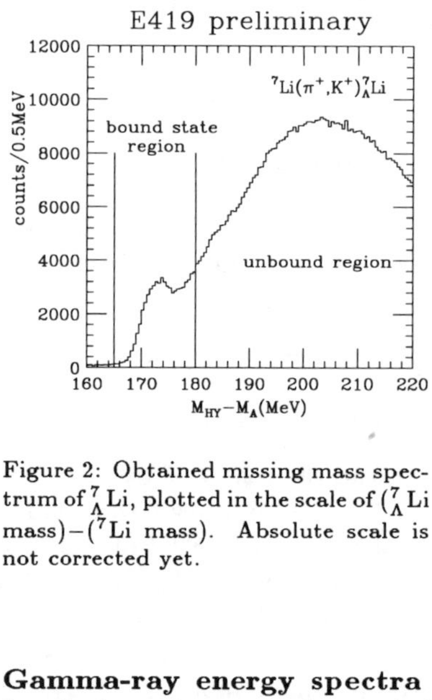

Figure 2: Obtained missing mass spectrum of $^7_\Lambda$Li, plotted in the scale of ($^7_\Lambda$Li mass)−(^{7}Li mass). Absolute scale is not corrected yet.

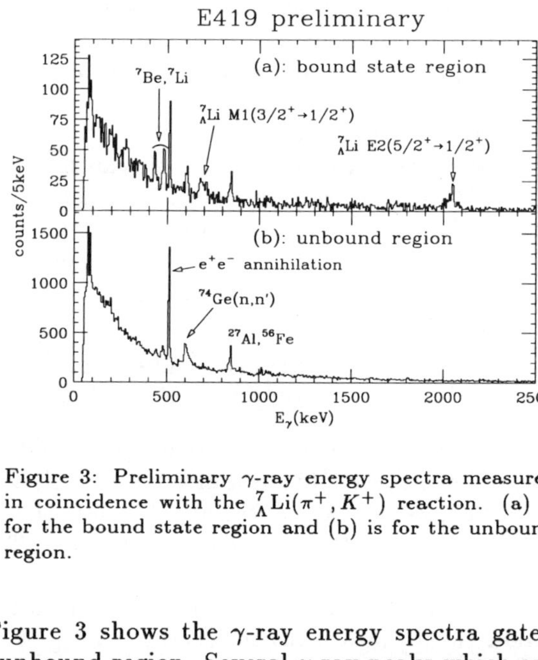

Figure 3: Preliminary γ-ray energy spectra measured in coincidence with the $^7_\Lambda$Li(π^+, K^+) reaction. (a) is for the bound state region and (b) is for the unbound region.

Gamma-ray energy spectra Figure 3 shows the γ-ray energy spectra gated for the bound state region and the unbound region. Several γ-ray peaks which are common to both spectra do not come from $^7_\Lambda$Li hypernucleus but from other sources such as e^+e^- annihilation(511 keV), ^{74}Ge (596 keV), ^{27}Al(844 keV) and ^{56}Fe(847 keV). Two peaks at 429 keV and 478 keV are also seen in both spectra, but are much more prominent in (a) than (b). They are from ^{7}Be*(429 keV) and ^{7}Li*(478 keV) produced by the weak decay of $^7_\Lambda$Li, namely $^7_\Lambda\text{Li} \rightarrow \pi^- + {}^7\text{Be}^*$ and $^7_\Lambda\text{Li} \rightarrow \pi^0 + {}^7\text{Li}^*$.

There are two peaks at 690 keV and 2050 keV which are seen only in (a). This fact indicates that those peaks come from $^7_\Lambda$Li. The peak at 2050 keV is the same as observed in the old experiment with NaI detectors[9], and is attributed to the $E2(5/2^+ \rightarrow 1/2^+)$ transition. The other peak at 690 keV is assigned to the $M1(3/2^+ \rightarrow 1/2^+)$ transition by comparing the obtained yield of the peak with a theoretical calculation in Ref.[5] These transitions are the first well-identified γ transitions from hypernuclei observed with germanium detectors. Detailed descriptions about them are given in the following subsections.

The $E2(5/2^+ \to 1/2^+)$ transition and shrinkage of $^7_\Lambda$Li Figure 4 is the magnification of Figure 3 around the $E2$ peak. The peak has two components, namely, the narrower part and the broader part which is due to the Doppler shift. From the position of the narrower peak, which is free from Doppler shift, the energy of the γ-ray is determined as 2050 ± 2 keV(preliminary). This is consistent with the result of the old experiment with NaI counters(2034 ± 23 keV)[9], but the precision of the present result is much better than that. Indeed, this precision is the world record of hypernuclear experiments.

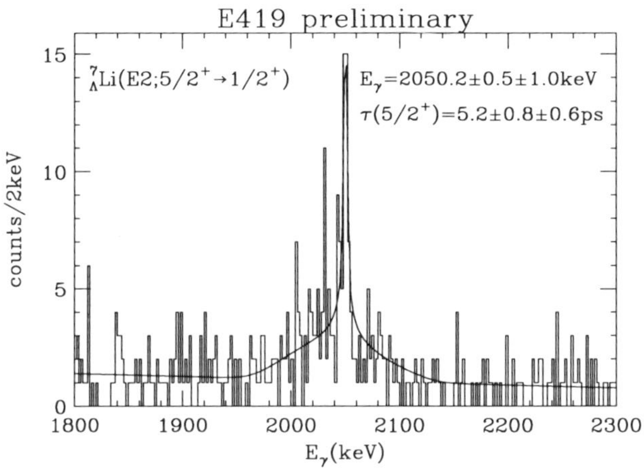

Figure 4: Preliminary γ-ray energy spectrum around the $E2$ peak. The result of the fitting is shown with the solid line.

In order to obtain the lifetime of the $5/2^+$ state, the obtained spectrum was fitted with Monte Carlo simulated spectra which depend on the lifetime of the $5/2^+$ state. A linear background was assumed in the fitting. The result is shown as a solid line in Figure 4. The lifetime of $5/2^+$ obtained from the fitting is $5.2 \pm 0.8 \pm 0.6$ ps(preliminary).

The obtained lifetime is then converted into $B(E2)$ of $4.1 \pm 0.6 \pm 0.5$ $e^2\text{fm}^4$, assuming the branching ratio of the $5/2^+ \to 1/2^+$ transition is 94% [the other branches are the $5/2^+ \to 3/2^+$ transition(4%) and the weak decay(2%)]. The present result is significantly smaller than that of a shell-model calculation by Dalitz and Gal ($B(E2) = 8.6$ $e^2\text{fm}^4$) in which the shrinking effect is not taken into account[10]. Therefore, we conclude that the shrinking effect is observed and that the glue-like role is verified. The present result corresponds to a nuclear-size shrinkage of 13% using Eq. (1).

Cluster model calculations by Motoba *et al.* and recently by Hiyama *et al.* predict $B(E2)$ of 2.5 $e^2\text{fm}^4$ and 2.42 $e^2\text{fm}^4$, respectively[4,5]. These values show better agreement with the present result, but the shrinking effect obtained in our experiment is smaller than their predictions.

The spin-flip $M1(3/2^+ \to 1/2^+)$ transition Figure 5 shows the γ-ray energy spectra around the $M1$ peak. Since we know the direction of the emitted γ-rays and the recoil velocity of ${}^7_\Lambda$Li, Doppler shift can be corrected for event by event. Figure 5 (a) and (b) are the spectra before and after the Doppler correction, respectively. In (b), the $M1$ peak is sharper than in (a), which indicates that the peak is really from ${}^7_\Lambda$Li.

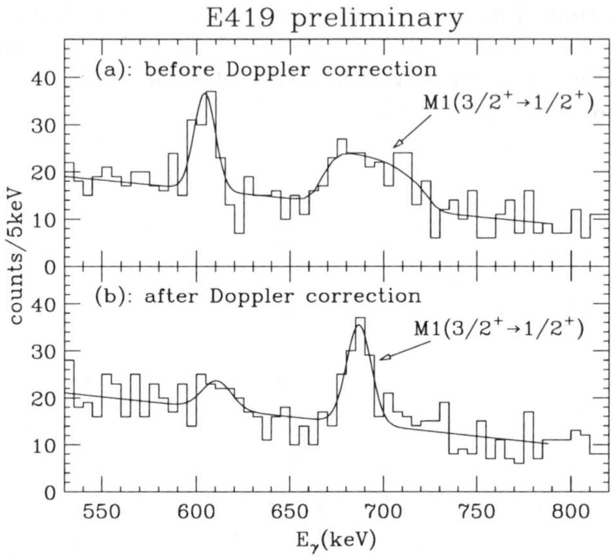

Figure 5: Preliminary γ-ray energy spectra around the $M1$ peak before the Doppler correction(a) and after the Doppler Correction(b). Fitting results are shown with the solid curves.

The M1 peak is fitted to a simulated spectrum in (a) and a gaussian in (b), each with a linear background and a gaussian for the ^{74}Ge peak. The results of the fittings are shown as solid lines in the figure. From the fittings, the energy of the γ-ray is determined to be 689 ± 4 keV(preliminary), where the error is mainly from the ambiguity in the treatment of Doppler shift. The obtained γ-ray energy gives direct information on the strength of the ΛN spin-spin interaction.

3 BNL-AGS E930 experiment

This experiment aims at observing γ-rays from various hypernuclei, namely, ${}^9_\Lambda$Be, ${}^{16}_\Lambda$O, ${}^7_\Lambda$Li and ${}^{12}_\Lambda$C, produced via (K^-, π^-) reactions at AGS-BNL D6 beamline. All of the ΛN spin-dependent interactions, namely, spin-spin,

Table 1: Hypernuclei to be studied in E930 experiment and the related ΛN interactions.

Hypernuclei	Related ΛN interaction
${}^9_\Lambda\mathrm{Be}(3/2^+, 5/2^+)$	spin-orbit
${}^{16}_\Lambda\mathrm{O}(1^-, 0^-)$	spin-spin, spin-orbit, tensor
${}^7_\Lambda\mathrm{Li}(7/2^+, 5/2^+)$	spin-spin, spin-orbit
${}^{12}_\Lambda\mathrm{C}(2^-, 1^-)$	spin-spin, spin-orbit

spin-orbit and tensor interactions are expected to be clarified from this experiment, as shown in Table 1. More descriptions about the experiment are found in Ref.[2]

We have finished data taking for ${}^{9}_{\Lambda}$Be in December 1998, and are analyzing the data now. Preliminary results will be available soon. For other hypernuclei, we are waiting for the next beamtime, which is hopefully available in FY2001.

4 Summary

We performed two hypernuclear γ-ray spectroscopy experiments with our recently constructed Ge detector system(Hyperball). In the first experiment at KEK, we observed well-identified hypernuclear γ transitions with Ge detectors for the first time. The energies of the observed two γ-rays emitted by the $E2(5/2^+ \to 1/2^+)$ and $M1(3/2^+ \to 1/2^+)$ transitions of ${}^{7}_{\Lambda}$Li are 2050$\pm$2 keV and 689$\pm$4 keV(preliminary), respectively. The energy of the latter transition gives direct information on the ΛN spin-spin interaction.

Lifetime of the $5/2^+$ state is measured to be 5.2$\pm$0.8$\pm$0.6 ps, which corresponds to $B(E2; 5/2^+ \to 1/2^+)$ of 4.1$\pm$0.6$\pm$0.5 e^2fm^4(preliminary). This is a clear evidence of the shrinkage of ${}^{7}_{\Lambda}$Li and the glue-like role of Λ is confirmed.

The second experiment at BNL is ongoing. We have finished taking data for ${}^{9}_{\Lambda}$Be and are analyzing them. We are waiting for the next beamtime for the other hypernuclei.

Acknowledgement

We thank the staff of KEK-PS and BNL-AGS for their support to our experiments. We also would like to acknowlegde invaluable discussions with Prof. Motoba, Prof. D. J. Millener and Dr. E. Hiyama. This work is supported by Grand-In-Aid for Scientific Research on Priority Areas(Strangeness Nuclear Physics) from the Ministry of Education, Science, Sports and Culture of Japan.

References

1. H. Tamura *et al.*, KEK-PS Proposal E419, July 1997.
2. H. Tamura *et al.*, BNL-AGS Proposal E930, September 1996.
3. H. Tamura, Nucl. Phys. A639 (1998) 83c.
4. T. Motoba, H. Bando and K. Ikeda, Prog. Theor. Phys. 80 (1983) 189; T. Motoba, H. Bando, K. Ikeda and T. Yamada, Prog. Theor. Phys. Suppl. 81 (1985) 42.
5. E. Hiyama, M. Kamimura, K. Miyazaki and T. Motoba, Phys. Rev. C59, No.5 (1999), in press.
6. F. Ajzenberg-Selove, Nucl. Phys. A490 (1988) 1.
7. T. Fukuda *et al.*, Nucl. Instrum. Methods A361 (1998) 485.
8. K. Tanida, Master's thesis, Univ. of Tokyo, February 1998.
9. M. May *et al.*, Phys. Rev. Lett. 51 (1983) 2085.
10. R. H. Dalitz and A. Gal, Ann. Phys. 116 (1978) 167.

CONTINUUM TAMM-DANCOFF APPROXIMATION CALCULATION OF Λ-HYPERNUCLEUS PRODUCTION CROSS SECTIONS

S. W. HONG, B. T. KIM, D. H. LEE, AND C. Y. RYU

Department of Physics, Sungkyunkwan University, Suwon 440-746, Korea

Calculations of the Λ-hypernucleus production cross section for the (π^+, K^+) reaction are done including $\Lambda - N^{-1}$ correlations in the continuum Tamm-Dancoff approximation. The formalism allows us to calculate the wave function of Λ in nuclei for both bound and continuum states. The quasi-free peak is reproduced by using an energy dependent $\Lambda - N^{-1}$ interaction. The Λ knock-out and the spreading modes are separated and are compared with the coincidence data.

1 Introduction

Since pioneering experiments for the (π^+, K^+) reaction at BNL AGS[1,2] and a series of further vigorous experiments at the KEK 12-GeV Proton Synchrotron[3,4] were carried out with various target nuclei, spectroscopic data for the shell structure of Λ-hypernuclei were accumulated. Theoretically, early calculations on the (π^+, K^+) reactions were made by Motoba *et al.*,[5] who calculated the hypernucleus production cross section in the distorted wave impulse approximation (DWIA)[6]. Recently, extensive calculations were carried out in Refs. 7 and 8 by using the DWIA and the Tamm-Dancoff approximation with Skyrme-type interactions for the Λ particle - neutron hole $(\Lambda - N^{-1})$ states. Here we report on the DWIA calculations done in a way somewhat similar to the ones in Refs. 5 − 8 but also different in some other respects. We use the distorted wave functions for the π^+ and K^+ for the meson distortion and treat the continuum state of the Λ-hyperon properly while including the $\Lambda - N^{-1}$ correlations for the quasi-free region. We also decompose the cross section into the spreading and the knock-out components, which may correspond to non-mesonic and mesonic decays, respectively.

2 Theory

2.1 Inclusive Cross Sections in the DWIA and the Strength Function

The double differential cross section for the inclusive process $A + a \to B + b$ in the center-of-mass system may be written as[9,10,11]

$$\frac{d^2\sigma}{dE_b d\Omega_b} = \frac{\omega_\alpha \omega_\beta}{(2\pi\hbar^2 c^2)^2} \frac{p_b}{p_a} \mid T_{\beta\alpha} \mid^2 \delta(E_B + E_b - E_A - E_a). \tag{1}$$

In the first-order DWIA, the transition amplitude $T_{\beta\alpha}$ is given by[6,10,11]

$$T^{DW}_{\beta\alpha} = \int d\vec{R} \ \chi_b^{(-)^*}(\vec{k}_b, \vec{R}) \langle \Phi_B \Phi_b \mid \sum_{j=1,A} t_{ij} \mid \Phi_A \Phi_a \rangle \chi_a^{(+)}(\vec{k}_a, \vec{R}), \tag{2}$$

where t_{ij} represents the effective interaction[12] between the projectile meson i and the target nucleon j. The transition amplitude may be rewritten as

$$T^{DW}_{\beta\alpha} = \langle \Phi_B \mid \hat{\rho} \mid \Phi_A \rangle, \tag{3}$$

where $\hat{\rho}$ is the hadronic transition operator defined by

$$\hat{\rho} = \sum_{j=1,A} \langle \Phi_{\nu_b} \mid \chi_b^{(-)^*}(\vec{k}_b, \vec{r}) t_{ij}(\omega, \vec{q}) \chi_a^{(+)}(\vec{k}_a, \vec{r}) \mid \Phi_{\nu_a} \rangle. \tag{4}$$

By inserting Eq. (3) into Eq. (1), the double differential cross section is rewritten as

$$\frac{d^2\sigma}{dE_b d\Omega_b} = \frac{\omega_\alpha \omega_\beta}{(2\pi\hbar^2 c^2)^2} \frac{p_b}{p_a} S, \tag{5}$$

where S is the strength function defined by

$$S = -\frac{1}{\pi} Im \langle \rho \mid G \mid \rho \rangle = -\frac{1}{\pi} Im \langle \rho \mid \Psi \rangle \tag{6}$$

with $\mid \rho \rangle = \hat{\rho} \mid \Phi_A \rangle$ and $G = (\omega - H + i\epsilon)^{-1}$. The Green function G propagates the doorway state $\mid \rho \rangle$ to the final state wave function, $\mid \Psi \rangle = G \mid \rho \rangle$. Once $\mid \Psi \rangle$ is calculated, it is straightforward to calculate the cross section.

2.2 The Correlated Source Function Method

The correlated source function method provides a very efficient way of calculating $\mid \Psi \rangle$. First we split H into $H = H_0 + V_{ph}$, where H_0 is the single particle Hamiltonian and V_{ph} is the residual ph interaction. H_0 is further separated into $H_0 = H_C + H_p$, where H_C is the Hamiltonian of the hole nucleus C and H_p is the Hamiltonian of the excited particle p. The Hamiltonian of the excited particle p is given by $H_p = T_p + U_p$, where T_p is the kinetic energy operator and $U_p = V_p + iW_p$ is a complex one-body potential. Then one can obtain the following integral equation:

$$\mid \Psi \rangle = G_0 \mid \rho \rangle + G_0 V_{ph} \mid \Psi \rangle, \tag{7}$$

where G_0 is the unperturbed Green function defined by $G_0 = (\omega - H_0 + i\epsilon)^{-1}$. Equation (7) can be transformed into an equivalent integral equation

$$\mid \Lambda \rangle = \mid \rho \rangle + V_{ph} G_0 \mid \Lambda \rangle, \tag{8}$$

where $\mid \Lambda \rangle$ is defined by $\mid \Lambda \rangle \equiv G_0^{-1} \mid \Psi \rangle$. Note that $\mid \Lambda \rangle$ plays a similar role as $\mid \rho \rangle$. However, it includes the correlations due to V_{ph}, and thus we call $\mid \Lambda \rangle$ the *correlated source function.*[10]

By integrating over all coordinates except for the radial coordinate of particle p, Eq.(8) is reduced to a set of coupled channel (CC) equations:[10]

$$\lambda_{ph}(r) = \rho_{ph}(r) + \sum_{p'h'} \int dr' dr'' V^{j_t}_{ph,p'h'}(r, r') g^{(+)}_{p'h'}(r', r'') \lambda_{p'h'}(r''), \tag{9}$$

where $g^{(+)}_{ph}$ is the radial optical model Green function and the ph matrix elements $V^{j_t}_{ph,p'h'}(r, r')$ are defined by

$$V^{j_t}_{ph,p'h'}(r, r') = rr'([y_{p'} \Phi_{h'}]_{j_t} | V_{ph} | [y_p \Phi_h]_{j_t}). \tag{10}$$

The merit of solving Eq. (8) lies in the fact that $| \lambda\rangle$ is a function localized (bound) in the nuclear region. This is not the case for $| \Psi\rangle$ when the particle p is in the continuum.

2.3 Decomposition of the Inclusive Cross Section

The inclusive cross section of Eq. (5) can be decomposed into partial cross sections: $S = S_{SP} + S_{KO} + S_{ph}$. Here the partial strength functions are defined by[10]

$$S_{SP} = \frac{1}{\pi}\langle\Psi \mid -W_p \mid \Psi\rangle, \tag{11}$$

$$S_{KO} = \langle\Lambda \mid \Omega^{-}\delta(\omega - H_C - T_p)\Omega^{(-)\dagger} \mid \Lambda\rangle, \tag{12}$$

$$S_{ph} = \frac{1}{\pi}\langle\Psi \mid ImV^{\dagger}_{ph} \mid \Psi\rangle, \tag{13}$$

where $\Omega^{(-)^{\dagger}} = 1 + U_p G_0$ is the Moeller wave operator. S_{SP} denotes the spreading (SP) strength function resulting from non-elastic collisions of Λ in the nuclear medium. S_{KO} is the strength function for the Λ knock-out (KO) process. S_{ph} is the component resulting from the imaginary part of the residual interaction. In the present work, we have employed a delta function interaction for V_{ph}, so S_{ph} vanishes. S_{SP} can be calculated in a straightforward manner. This is not the case, however, for S_{KO}. Thus it is calculated from the relation $S_{KO} = -\frac{1}{\pi}Im\langle\rho \mid \Psi\rangle - S_{SP}$.

3 Results and discussion

The distorted wave functions for the incident π^+ and the outgoing K^+ in Eq. (4) are calculated by solving the Schrödinger equation reduced from the Klein-Gordon equation with local optical potentials[11,13,14]. The calculated K^+ and π^+ elastic differential cross sections for ^{12}C target are compared with the observed data[15,16] in Fig. 1.

For the Λ-nucleus mean field potential, the Woods-Saxon potential with a spin-orbit term is used as usual:[11,17]

$$U_\Lambda(r) = V_\Lambda f_V(r) + iW_\Lambda f_W(r) + V^{so}_\Lambda\left(\frac{\hbar}{m_\pi c}\right)^2 \vec{l}\cdot\vec{S}\frac{1}{r}\frac{df_{so}(r)}{dr}, \tag{14}$$

where $f_i(r) = 1/(1+exp(X_i))$ with $X_i = (r - R_i)/a_i$ and $R_i = r_i A^{1/3} (i = V, W, so)$, and $(\hbar/m_\pi c)^2$ is taken as 2 fm^2. The imaginary potential $W_\Lambda f_W(r)$ is partly to account for the spreading effects on Λ in nuclei, but here it is mainly to simulate the experimental energy resolution for the particular experiments considered. The Λ in the ground state decays partly through mesonic decays ($\Lambda \to N + \pi$) and partly through non-mesonic decays ($\Lambda + N \to N + N$) via weak interactions. On the other hand, the Λ in the excited states decays first via strong or electromagnetic interactions before going through weak decays. Thus, the physical origin of $W_\Lambda f_W(r)$ is different, depending on the states of Λ. Here, however, we have simply tried to reproduce the data by choosing W_Λ as the energy resolution of the particular experiment.

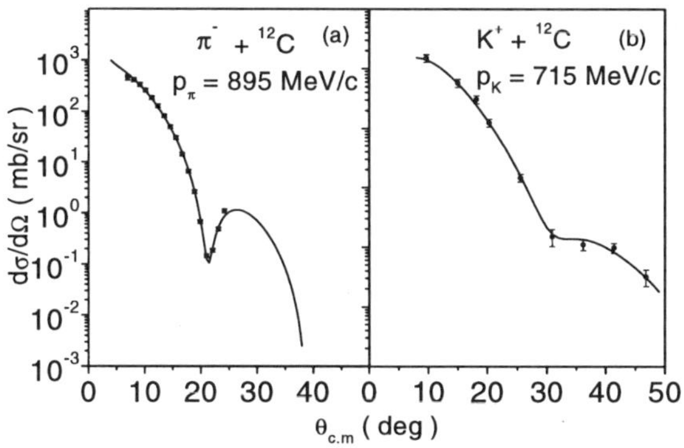

Figure 1: Calculated differential elastic cross sections are compared with the experimental data for (a) $\pi^-+^{12}\text{C}$ and (b) $K^++^{12}\text{C}$. The experimental data are taken from Refs. [15] and [16] for K^+ and π^- scatterings, respectively.

In Fig. 2 we have shown the calculated angular distributions for the $1^-_{g.s.}$ and the 2^+ peaks in comparison with the BNL data[1]. The energy resolution for these data is 3 MeV, so we choose W_Λ to be 3 MeV. The agreement between our calculation and the data is seen to be good.

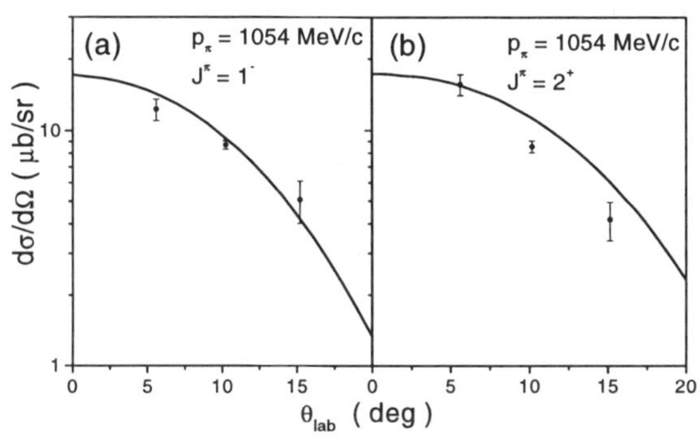

Figure 2: The angular distribution of the differential cross sections for $J^\pi = 1^-_{g.s.}$ and 2^+. The experimental data are from Ref. [1].

In Ref. 18, p and π^- were detected in coincidence with K^+. In that experiment, the energy resolution was about 6 MeV, so we chose $W_\Lambda = 6$ MeV in calculating the cross section. We separated the total strength into spreading and escaping components and compared them with the coincidence data. Figure 3(a) shows the decomposition of the cross section into spreading and knock-out (or escaping) components. The escaping component corresponds to π coincidence events because the knocked-out Λ's will eventually decay into $\pi + N$. Thus, in Fig. 3(b), we compare our calculated escaping cross section with the π^- coincidence data. Since there are no absolute values available for the coincidence cross section, we arbitrarily multiplied the data by a constant factor for comparison. Our calculated spreading cross section may correspond to the proton coincidence data. Thus, in Fig. 3(c), we compare the calculated spreading component with the experimental proton

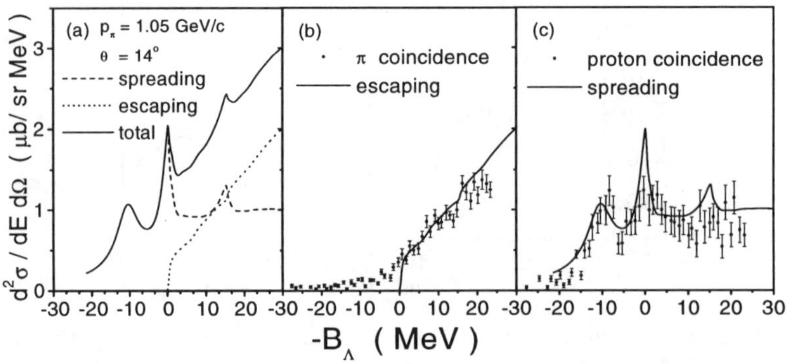

Figure 3: (a) The decomposition of the double differential cross sections into spreading and escaping parts. There are no inclusive cross sections to compare with the calculation. (b) Comparison of the calculated escaping component with the π coincidence cross section. (c) Comparison of the calculated spreading component with the proton coincidence cross section. The experimental data are from Ref. [18].

coincidence cross section multiplied by a constant factor. A fairly good agreement between the calculation and the data is achieved. We obtained a very similar agreement with other coincidence data[19], but a comparison with the data is not repeated here. In the calculations shown in Figs. 2 – 3 we have not included the $\Lambda-N^{-1}$ residual interaction, but we can still more or less reproduce the experimental cross section. In Ref. 3, a characteristic quasi-free peak is found to be located at about $p_K \approx 670$MeV/c. Without including the $\Lambda - N^{-1}$ residual interaction, we cannot fit the quasi-free peak position quite well as shown by the dashed curve in Fig. 4. Thus, we included the energy-dependent residual interaction of the form

$$V_{ph} = V_0\delta(\mathbf{r}_1 - \mathbf{r}_2)(1 + \alpha\sigma_1 \cdot \sigma_2), \tag{15}$$

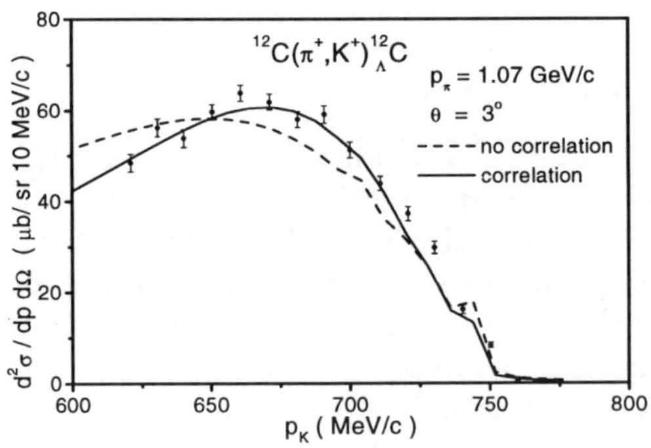

Figure 4: Double differental cross section, $d^2\sigma/dp\, d\Omega$, as a function of the Kaon momentum. The experimental data are taken from Ref. [3].

with $\alpha = -0.05$ and $V_0 = -228\frac{-B_\Lambda}{60}$ MeV fm^3 so that the residual interaction becomes negligible for low excitation energies. By including the energy-dependent $\Lambda - N^{-1}$ residual interation, the quasi-free peak position seems well reproduced. We then repeated the calculations shown in Figs. 2 – 3 with the $\Lambda - N^{-1}$ residual interaction included. The results at low excitation energies change a little but not significantly by including the residual interaction, because it is chosen to be negligible at low excitation energies.

Acknowledgments

This work was supported in part by the Ministry of Education of Korea (BSRI 98-2422).

References

1. C. Milner *et al.*, Phys. Rev. Lett. **54**, 1237 (1985).
2. R. E. Chrien, Nucl. Phys. **A478**, 705c (1988); P. H. Pile *et al.*, Phys. Rev. Lett. **66**, 2585 (1991).
3. M. Akei *et al.*, Nucl. Phys. **A534**, 478 (1991).
4. T. Fukuda *et al.*, Nucl. Inst. Meth. Phys. Res. **A361**, 485 (1995): T. Hasegawa *et al.*, Phys. Rev. Lett. **74**, 224 (1995): S. Ajimura *et al.*, Nucl. Phys. **A585**, 173c (1995): T. Hasegawa *et al.*, Phys. Rev. **C53**, 1210 (1996).
5. T. Motoba, Nucl. Phys. **A479**, 227c (1988); T. Motoba, H. Bando, R. Wünsch, and J. Žofka, Phys. Rev. **C38**, 1322 (1988); K. Itonaga, T. Motoba, O. Richter, and M. Sotona, Phys. Rev. **C49**, 1045 (1994).
6. A. K. Kerman, H. McManus, and R. M. Thaler, Ann. Phys. (N.Y.) **8**, 551 (1959).
7. M. T. López-Arias and F. Fernández, Phys. Lett **B 323**, 273 (1994).
8. M. T. López-Arias, Nucl. Phys. **A582**, 440 (1995).
9. M. L. Goldberger and K. M. Watson, *Collision Theory* (R. E. Krieger Publishing Company, New York, 1975).
10. T. Udagawa, P. Oltmanns, F. Osterfeld, and S. W. Hong, Phys. Rev. **C49**, 3162 (1994).
11. S. W. Hong, B. T. Kim, D. H. Lee, and C. Y. Ryu, J. Korean Phys. Soc. **34**, 9 (1999).
12. M. Sotona and J. Žofka, Prog. Theor. Phys. **81**, 160 (1989).
13. G. R. Satchler, Nucl. Phys. **A540**, 533 (1992).
14. S. W. Hong and B. T. Kim, J. of Physics G: Nucl. Part. Phys. **25**, 1065 (1999).
15. R. Michael *et al.*, Phys. Lett. **B382**, 29 (1996).
16. T. Takahashi *et al.*, Phys. Rev. **C51**, 2542 (1995).
17. H. Bando, T. Motoba, and J. Žofka, Int. J. Mod. Phys. **A5**, 4021 (1990).
18. S. Ajimura *et al.*, Phys. Lett. **282**, 293 (1992).
19. H. Noumi *et al.*, Phys. Rev. **C52**, 2936 (1995).

STRUCTURE OF $^{89}_{\Lambda}$Y MEASURED WITH THE SKS SPECTROMETER

TOMOFUMI NAGAE
High Energy Accelerator Research Organization (KEK),
1-1 Oho, Tsukuba, Ibaraki 305, Japan
E-mail: nagae@post.kek.jp

We have measured an excitation energy spectrum of a Λ-hypernucleus $^{89}_{\Lambda}$Y via the (π^+, K^+) reaction by using the SKS spectrometer at KEK. The best quality data with the energy resolution of 1.63 MeV(FWHM) was obtained in the highest statistics so far. In the spectrum, we can find a splitting of the f-orbital of a Λ particle with the separation of 1.6±0.15 MeV(statistical error only), which may be attributed to the spin-orbit splitting. The (π^+, K^+)$^{51}_{\Lambda}$V spectrum, obtained in the same experiment, also showed a similar structure.

1 SPIN-ORBIT SPLITTING OF Λ-HYPERNUCLEI

The spin-orbit splitting of Λ-hypernuclei is still not well-known. In the literature, the first data which suggested that the spin-orbit splitting of the Λ-hypernuclei would be very small compared with that of normal nuclei was reported by the Sacley-Heidelberg collaboration in 1978[1]. In the ^{16}O(K^-,π^-)$^{16}_{\Lambda}$O reaction, they found the energy difference between ($p^{-1}_{1/2,N}$,$p_{1/2,\Lambda}$) and ($p^{-1}_{1/2,N}$, s_{Λ}) was almost the same as the energy difference between ($p^{-1}_{3/2,N}$,$p_{3/2,\Lambda}$) and ($p^{-1}_{3/2,N}$, s_{Λ}). The efforts to measure the spin-orbit splitting were further continued in the p-shell region: (K^-, π^-)$^{13}_{\Lambda}$C[2] and (K^-,$\pi^-\gamma$)$^{9}_{\Lambda}$Be[3]. The former one set the upper limit of the splitting between ($p^{-1}_{1/2,N}$,$p_{1/2,\Lambda}$) and ($p^{-1}_{1/2,N}$,$p_{3/2,\Lambda}$) to be 0.36±0.3 MeV. The latter one settled down the upper limit to be less than 0.1 MeV through the gamma-ray measurement.

In contrast to these data, there were a few reports which suggested signatures of the spin-orbit splitting. Two data for $^{12}_{\Lambda}$C[4] and $^{16}_{\Lambda}$O[5] were reported in the reanalyses of old emulsion data. The energy resolution was good enough to observe the small splittings of ~0.8 MeV for $^{12}_{\Lambda}$C and 1.56±0.12 MeV for $^{16}_{\Lambda}$O, while the statistics were not so good. Another data[6] was reported in our previous measurement with the SKS spectrometer in the ^{89}Y(π^+, K^+)$^{89}_{\Lambda}$Y reaction. It was suggested[7] that a calculation taking account of a wider spin-orbit splitting than usually expected would explain the spectrum reasonably well. However, the statistics and the energy resolution of 2.2 MeV were not good enough to conclude it.

As compared to the measurement in the p-shell Λ-hypernuclei, there is a great advantage in the measurement of $^{89}_{\Lambda}$Y: the Λ-orbitals with large angular momenta are bound in such a heavy Λ-hypernucleus. Thereby, the spin-orbit splitting would be enhanced, because the magnitude would be proportional to $(\ell+1/2)A^{-2/3}$ as in the case of normal nuclei. The Λ-orbitals up to the f-orbit are bound in $^{89}_{\Lambda}$Y and up to the d-orbit in $^{51}_{\Lambda}$V, respectively. Therefore, even if the spin-orbit splitting for the p-orbit is less than 1 MeV, the splitting could be larger than 1.5 MeV or so for the f-orbit.

On the contrary, the efforts to detect such small spin-orbit splittings in the

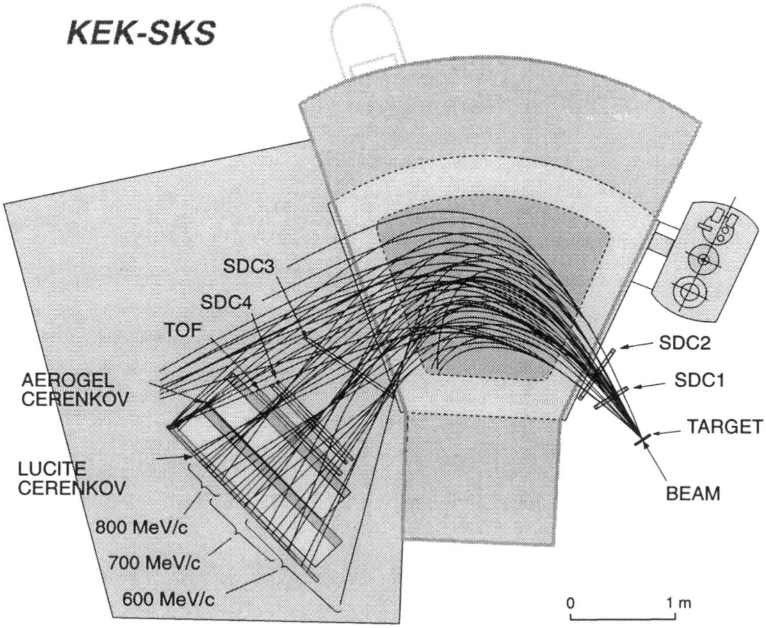

Figure 1. A schematic layout of the SKS spectrometer. Typical tracks are also shown in the figure.

p-shell Λ-hypernuclei through hypernuclear gamma-ray measurements have been carried out at BNL-AGS in the recent experiments E929 and E930.

2 KEK-PS E369

The experiment was carried out at K6 beam line of the 12-GeV KEK proton synchrotron(PS) by using the SKS spectrometer system. In the (π^+, K^+) reaction, the incident momentum of the pion was measured with a beam line spectrometer which comprised of a QQDQQ system, and the momentum of the kaon was measured with the SKS spectrometer. The details of the spectrometer system are described in the reference[8].

The targets used in this experiment were $^{89}Y(2.2\ g/cm^2)$, $^{51}V(2.7\ g/cm^2)$, and $^{12}C(0.9$ and $1.8\ g/cm^2)$. The last target was used for the energy calibration to adjust an absolute energy offset in the binding energy scale and to check the energy resolution.

So far, the best energy resolution achieved in the SKS spectrometer was 1.9 MeV(FWHM) for the $^{12}C(\pi^+, K^+)^{12}_{\Lambda}C$ with the target thickness of 0.9 g/cm^2. In E369, the tracking chamber configuration at the entrance of the SKS magnet was slightly modified; two sets of chambers were installed so as to make them perpendicular to the beam line and their distance to be as large as possible. Also special cares were taken for the tracking-chamber position measurements in the whole spectrometer system in order to reduce systematic errors. In Fig. 2, a new

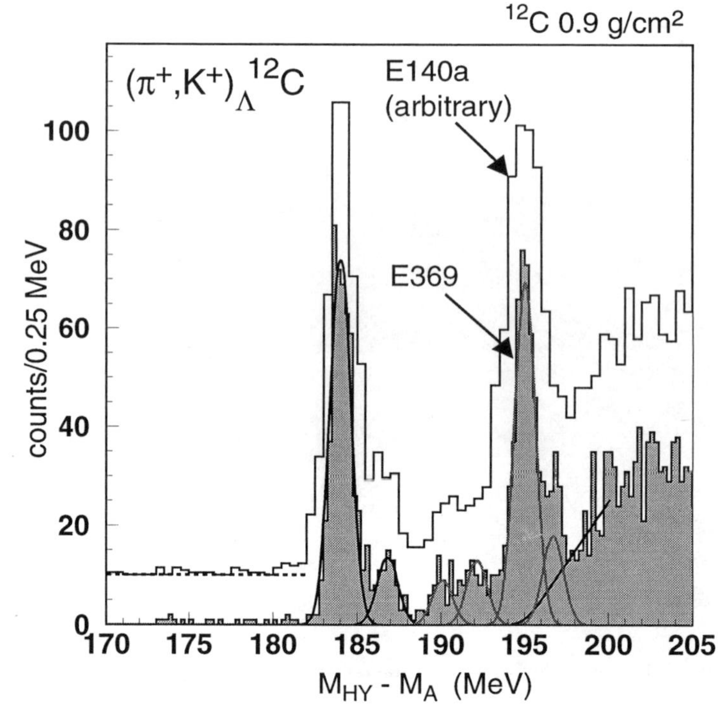

Figure 2. An excitation energy spectrum of ${}^{12}_{\Lambda}$C obtained in E369. The old ${}^{12}_{\Lambda}$C spectrum[9] is superimposed in an arbitrary scale for comparison. The energy resolution is significantly improved from 1.9 MeV to 1.45 MeV(FWHM).

${}^{12}_{\Lambda}$C spectrum obtained in this experiment is shown. For comparison, the old ${}^{12}_{\Lambda}$C spectrum[9] is superimposed in an arbitrary scale. The energy resolution in the new data is 1.45 MeV(FWHM), which is significantly better than before. We could resolve some new peaks for the first time as shown in the figure.

3 RESULTS

The $(\pi^+, K^+)^{89}_{\Lambda}$Y spectrum was obtained in very high statistics: more than 7000 events were accumulated in the bound region. In Fig. 3, a preliminary spectrum is shown as a function of the excitation energy of ${}^{89}_{\Lambda}$Y. The bump structures reflect the major shell structure of the Λ-orbitals coupled to the $g^{-1}_{9/2}$ neutron-hole state. The widths for the p-, d-, and f-orbits are significantly wider than the energy resolution, and the f-orbit looks split into two peaks. In between the major bumps, there exist significant amounts of contributions from the deeper neutron-hole states $f^{-1}_{5/2}$, $f^{-1}_{7/2}$, etc.

In the figure, a fitting was applied based on the calculation by Motoba[7]; the neutron-hole energies were fixed and the hole widths were set to be 3 MeV and 6

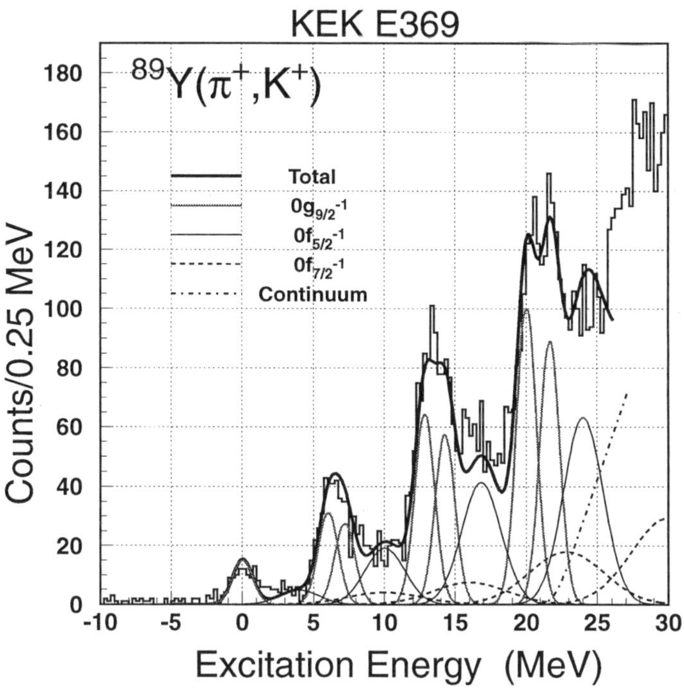

Figure 3. An excitation energy spectrum of ${}^{89}_{\Lambda}$Y obtained in E369. The energy resolution is estimated to be 1.63 MeV(FWHM). The bump structures correspond to the major shell structure of the Λ-orbits, s_Λ, p_Λ, d_Λ, and f_Λ. Two peaks in the f-orbit would be attributed to the spin-orbit splitting.

MeV for $f^{-1}_{5/2}$ and $f^{-1}_{7/2}$, respectively. The only difference was the peak intensity ratios between the two peaks in each Λ-orbit; the ratio was fixed to be 1.1:1 in this fitting, while in the calculation they were 1:1 $\sim$ 1:2. From the fitting, the energy separation between the two peaks in the f-orbit was obtained to be 1.6±0.15 MeV(statistical error only). The separations became smaller for the orbits with smaller angular momenta as shown in the figure. The systematic error on the energy separation seems to be small once we assume there were two peaks with the width fixed as the experimental energy resolution.

The energy resolution of 1.63 MeV(FWHM) was estimated from the ${}^{12}_{\Lambda}$C spectrum obtained from the (π^+, K^+) events coming from a plastic scintillation counter 30-cm upstream of the ^{89}Y target during the data taking. In those events, the kaons from the $(\pi^+, K^+){}^{12}_{\Lambda}$C reaction also passed through the ^{89}Y target. Therefore, we could surely estimate the energy resolution for the $(\pi^+, K^+){}^{89}_{\Lambda}$Y reaction.

A preliminary spectrum of the $(\pi^+, K^+){}^{51}_{\Lambda}$V was also obtained in this experiment as shown in Fig.4. In the spectrum, four peaks with the estimated spectrometer energy resolution of 1.79 MeV(FWHM) are plotted in arbitrary normalizations. It shows a similar structure with the ${}^{89}_{\Lambda}$Y spectrum up to the d-orbit of a lambda particle. It should be noted that the observed bump structures are significantly wider than the spectrometer resolution, which excludes an interpretation of the bump structure as a single peak.

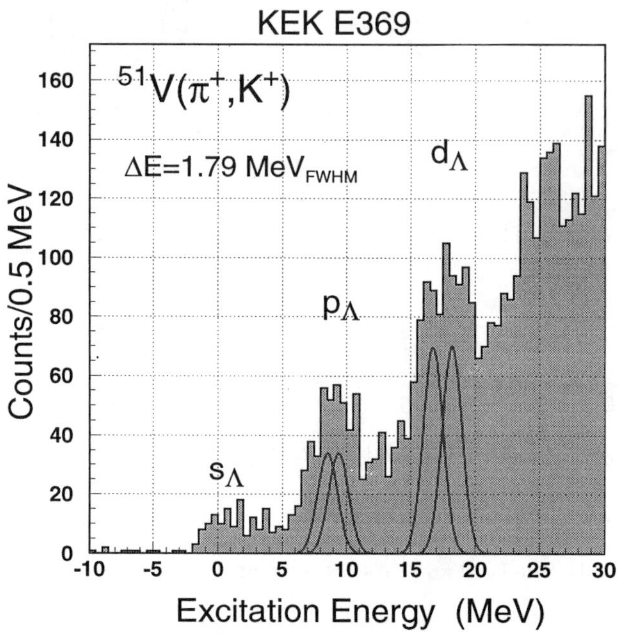

Figure 4. An excitation energy spectrum of ${}^{51}_{\Lambda}$V obtained in E369. The energy resolution is estimated to be 1.79 MeV(FWHM). The bump structures corresponding to the p-orbit and d-orbit are significantly wider than the energy resolution.

4 SUMMARY

The high statistics data of the excitation energy spectrum of ${}^{89}_{\Lambda}$Y was obtained in a good energy resolution of 1.63 MeV(FWHM). The observed bump structures corresponding to the Λ major-shell structure were significantly wider than the experimental resolution. The widest bump of the f-orbit seems split into two peaks, which could be attributed to the spin-orbit splitting. In a preliminary analysis of the ${}^{89}_{\Lambda}$Y spectrum, the splitting for the f-orbit is 1.6±0.15 MeV(statistical error only); it corresponds to V_{LS}=4.3±0.4 MeV according to the Motoba's calculation [7], where V_{LS} is the strength of the spin-orbit potential expressed in the Woods-Saxon form as $U_\Lambda(r) = V_0 f(r) + V_{LS}(\frac{\hbar}{m_\pi c})\frac{df(r)}{rdr}\vec{\ell}\cdot\vec{s}$, $f(r) = \{1+\exp[(r-R)/a]\}^{-1}$. The $(\pi^+, K^+)^{51}_{\Lambda}$V spectrum also showed a similar structure which was consisitent with the interpretation of spin-orbit splittings for these peak structures.

Acknowledgments

The author would like to thank all the collaborators of the KEK-PS E369 experiment;H. Hotchi[a], H. Outa[a], H. Noumi[a], M. Sekimoto[a], T. Fukuda[a], H.C. Bhang[b], Y.D. Kim[b], J.H. Kim[b], H. Park[b], L.I. Tang,[c] R. Sawafta[d], K. Tanida[e], O. Hashimoto[f], H. Tamura[f], T. Saito[f], T. Takahashi[f], Y. Sato[f], T. Endo[f], S. Satoh[f], J. Nishida[f], T. Miyoshi[f], T. Kishimoto[g], A. Sakaguchi[g], S. Ajimura[g], Y. Shimizu[g], T. Mori[g], S. Minami[g], M. Sumihama[g],
a) KEK, b) Seoul National University, c) Hampton University, d) North Carolina A&T, e) University of Tokyo, f) Tohoku University, g) Osaka University.

References

1. W. Brückner *et al.*, *Phys. Lett.* **79B**, 157 (1978).
2. M. May *et al.*, *Phys. Rev. Lett.* **47**, 1106 (1981).
3. M. May *et al.*, *Phys. Rev. Lett.* **51**, 1983 (2095).
4. R.H. Dalitz, D.H. Davis and D.N. Tovee, *Nucl. Phys.* *A*450, 311c (1986).
5. R.H. Dalitz, D.H. Davis, T. Motoba, D.N. Tovee, *Nucl. Phys.* *A*625, 71 (1997).
6. T. Nagae, *Proc. of the 23rd INS Int. Symp.*, Tokyo, Japan, March 1995, Eds.S. Sugimoto and O. Hashimoto, p.175.
7. T. Motoba, *Proc. of the 23rd INS Int. Symp.*, Tokyo, Japan, March 1995, Eds.S. Sugimoto and O. Hashimoto, p.187.
8. T. Fukuda *et al.*, *Nucl. Inst. Meth. A* **361**, 485 (1995).
9. T. Hasegawa *et al.*, *Phys. Rev. Lett.* **74**, 224 (1995).

REACTION SPECTROSCOPY OF Λ HYPERNUCLEI

Osamu Hashimoto

Department of Physics, Tohoku University, Sendai 980-8578
E-mail: hashimot@lambda.phys.tohoku.ac.jp

Reaction spectroscopy as a tool to investigate Λ hypernuclear structure and ΛN interaction is discussed showing recent spectroscopic data obtained through the (π^+, K^+) reaction using a large acceptance high resolution spectrometer (SKS) at KEK 12 GeV PS. High quality spectra of light Λ hypernuclei in the p-shell region are compared with the recent DWIA calculation, in which a good agreement are observed. Analysis of the spectral shapes for ${}^{16}_{\Lambda}O$ and ${}^{13}_{\Lambda}C$ and their comparison with (K^-, π^-) data provided information on the splitting of the spin-orbit partner states, suggesting they are small. Future prospects of hypernuclear reaction spectroscopy are also presented.

1 Introduction

A Λ hypernucleus provides a unique playground for the investigation of hadronic many-body system with strangeness -1 and new aspects of the strong and weak interaction in nuclei. Recent progress in the spectroscopic study of Λ hypernuclei demonstrated the value of the Λ hypernuclear spectroscopy and established a new pathway towards understanding nuclear system with strangeness -1 in a quantitative way. A Λ hypernuclear state is formed in the reaction as a nucleon-hole Λ-particle state. Those hypernuclear states in which a Λ hyperon is bound above the p-shell orbital are often nucleon unbound. The spreading widths of such high-lying states are calculated to be narrower than a few 100 keV even they are excited above the nucleon emission thresholds, which is contrary to the case of an ordinary nucleus.[1,2] In addition to the narrowness of the spreading widths, perturbation by the presence of a Λ hyperon in the nuclear system is small because of the weakness of the interaction between a nucleon and a Λ hyperon. Therefore, information on Λ hypernuclear structure can be related to the ΛN interaction rather straightforwardly. Thus, the Λ hypernuclear reaction spectroscopy is indispensable for the investigation of the ΛN interaction. Such characteristics are realized only in Λ hypernuclei and are hardly seen in other hypernuclei, and thus the spectroscopy of Λ hypernuclei has a unique value in the strangeness nuclear physics.

A Λ hypernucleus has been studied mostly with K^- and π^+ meson beams and so far reactions used for the study of Λ hypernuclei are (K^-, π^-) and (π^+, K^+) reactions. Between them, the reaction spectroscopy by the (π^+, K^+) reaction has been playing an increasingly important role recently, since it efficiently populates bound hypernuclear states due to the large momentum transfer to the recoil hypernuclei. In the present paper, the status of the reaction spectroscopy will be discussed based on the recent progress with the (π^+, K^+) reactionat KEK-PS, and the future prospects with other reactions.

2 Reaction spectroscopy by the (π^+, K^+) reaction with the SKS spectrometer

The reaction spectroscopy by the (π^+, K^+) reaction was first applied to a carbon target at BNL, and later to heavier ones at BNL and KEK.[3,4] Although the hypernuclear cross sections are about 2 orders of magnitude smaller than the case of the (K^-, π^-) reaction, it can be easily compensated by high intensity pion beams at 1 GeV/c. With an intention to fully take an advantage of the (π^+, K^+) reaction for the hypernuclear reaction spectroscopy, a superconducting kaon spectrometer(SKS) was constructed at KEK 12 GeV PS.[5,6] It has the best resolution of 1.5-2MeV FWHM and simultaneously a large solid angle of 100 msr, accepting 60 % of ${}^{12}_{\Lambda}$C gound-state yield in the (π^+, K^+) reaction.[7] Large acceptance of the SKS spectrometer allows us to perform coincidence experiments such as study of hypernuclear weak decay and γ ray spectroscopy. A series of experiments for the investigation of hypernuclei using the SKS spectrometer has been carried out in the past years and they are listed in Table 1

Table 1: Hypernuclear physics experiments with the SKS spectrometer

Experiment	Subject	Reference
E140a (Hashimoto, Tohoku)	Systematic spectroscopy of Λ hypernuclei ${}^{10}_{\Lambda}B$, ${}^{12}_{\Lambda}C$, ${}^{28}_{\Lambda}Si$, ${}^{89}_{\Lambda}Y$, ${}^{139}_{\Lambda}La$, ${}^{208}_{\Lambda}Pb$	7,8
E278 (Kishimoto, Osaka)	Nonmesonic weak decay of polarized ${}^{5}_{\Lambda}$He	9
E307 (Bhang, Seoul)	Lifetimes and weak decay widths of light and medium – heavy Λ hypernuclei	10,These proceedings
E336 (Hashimoto, Tohoku)	Light Λ hypernuclear spectroscopy ${}^{7}_{\Lambda}Li$, ${}^{9}_{\Lambda}Be$, ${}^{12}_{\Lambda}C$, ${}^{13}_{\Lambda}C$, ${}^{16}_{\Lambda}O$	11
E369 (Nagae, KEK)	Spectroscopy of ${}^{89}_{\Lambda}$Y	These proceedings
E419 (Tamura, Tohoku)	γ ray spectroscopy of ${}^{7}_{\Lambda}$Li	These proceedings
E438 (Noumi, KEK)	Σ hypernuclei by the (π^-, K^+) reaction	run in fall,1999
proposed (Outa, KEK)	Decay widths of ${}^{209}_{\Lambda}$Bi	

Among them, the three experiments, E140a, E336 and E369, are particularly for the spectroscopy of Λ hypernuclei and studied Λ hypernuclear structure and ΛN interaction through the measurement of high quality excitation spectra.

The E140a, the first of the series of experiments, aimed to investigate the nature of a Λ hyperon deeply bound to heavy nuclei. The binding energies of a Λ hyperon bound by a wide range of nuclei as heavy as ^{208}Pb are deduced and it was revealed that the single-particle nature of a Λ hyperon persists even in heavy nuclei.[8] Much improved statistics and resolution were obtained specifically for ${}^{89}_{\Lambda}$Y in the most recent E369 experiment, which is reported in this conference. In the E336

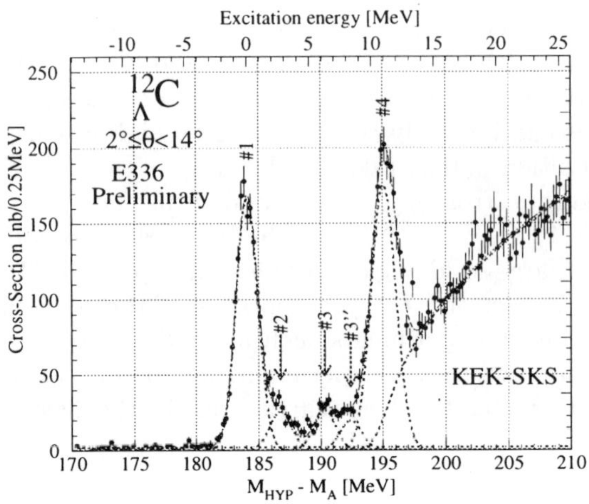

Figure 1: High statistics excitation spectrum of $^{12}_{\Lambda}$C obtained by ^{12}C(π^+,K$^+$)$^{12}_{\Lambda}$C reaction with the SKS spectrometer system. The spectrum has statistics 5 times better than the previous E140a experiment.

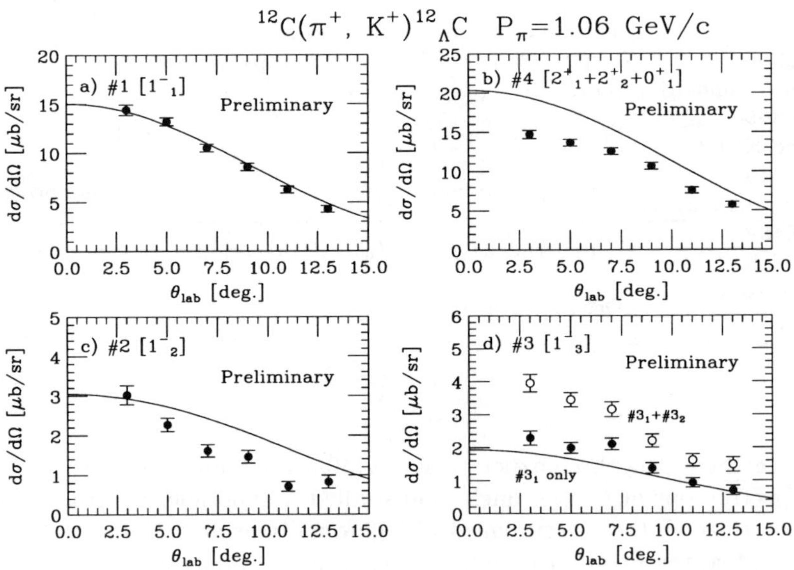

Figure 2: Angular distribution of the ^{12}C(π^+,K$^+$)$^{12}_{\Lambda}$C reaction leading to the low-lying hypernuclear states.

experiment, the p-shell Λ hypernuclei as listed in Table 1 were intensively studied with high statistics reaction spectra. In the following, we will report on the recent analysis of the E336 experiment which was carried out with the upgraded SKS spectrometer. Typical intensity of the pion beam was 3×10^6 per spill with spill repetition of every 3-4 sec. The yield rate of the $^{12}_{\Lambda}$C ground state was about 5 events/g/cm^2/10^9 pions, that is, about 1000 $^{12}_{\Lambda}$C ground states are detected by the spectrometer per day. The overall energy resolution was confirmed to be 2 MeV FWHM by the carbon excitation spectrum as previously reported in E140a, although the resolution varies from target to target. The statistics of the E336 experiment for each target were 5-10 times improved compared to the previous ones.

In the analysis of the E336 spectra, the angular distributions and the absolute cross sections are deduced taking an advantage of high statistics data. Since the angular acceptance of the SKS spectrometer is large, the integrated cross sections over the angular range from 2 to 14 degrees were derived by correcting the ϕ acceptance of SKS at each scattering angle θ event by event. Thus, the present absolute cross sections can be directly compared with theoretical calculations integrated over the same angular range. It is noted that the cross sections given in E140a spectra were obtained by integrating events over the angular range with ϕ acceptance of the SKS spectrometer not corrected. The present cross sections for $^{12}_{\Lambda}$C are, however, consistent with the E140a result if the difference in the derivation is taken into account. In the following two sections, we will discuss the obtained spectra of $^{12}_{\Lambda}$C, $^{16}_{\Lambda}$O, $^{13}_{\Lambda}$C in more detail.

3 Core excitation and the $^{12}_{\Lambda}$C hypernucleus

In the $^{12}_{\Lambda}$C spectrum of the KEK E140a experiment, two satellite peaks were observed near the two prominent peaks (#1 and #4) which are originated from configurations of $\nu p^{-1}_{3/2} \otimes \Lambda s_{1/2}$ and $\nu p^{-1}_{3/2} \otimes \Lambda p_{1/2,3/2}$, respectively. The satellite peaks carry about 1/10 cross sections and were interpreted as originated from an $s_{1/2}$ Λ hyperon coupled to excited states of the ^{11}C core. The excitation energies and the cross section of the states were discussed in conjunction with a spin-dependent part of the ΛN interaction.[7] In the present E336 experiment, we have accumulated statistics for the $^{12}\mathrm{C}(\pi^+,\mathrm{K}^+)^{12}_{\Lambda}\mathrm{C}$ reaction 5 times better than the previous E140a experiment. In the present higher-quality spectrum shown in Fig.1, the controversial #3 peak at $\mathrm{E}_x \approx 6.9$ MeV clearly shows its compositeness. There are arguments that the peak is not necessarily due only to a simple ^{11}C(4.8 MeV)xs_{Λ} configuration but could have some contribution from intershell configuration mixing.[14,15] Thus, the present spectrum was fitted with 5 gaussian peaks plus quasifree contribution and the obtained excitation energies are listed in Table 2 The excitation energy of the #3 peak, 6.9 MeV, is considerably greater than the corresponding core excitation energy of 4.8 MeV, which poses a puzzle to theoretical interpretation.

The angular distributions were obtained by fitting the excitation spectra in every 2 degree bin and are shown in Fig. 2. In the figure, the DWIA calculation by Itonaga[16] is also shown by solid lines. It is seen that the agreement between the data and calculation is very good. For the (#3+#3') peak, we also see quite good

Table 2: Excitation energies and cross sections of the ${}^{12}_{\Lambda}$C states measured by the ^{12}C(π^+,K$^+$)${}^{12}_{\Lambda}$C reaction.

Statistical errors only

Peak#	E140a	E336(Preliminary)	
	Ex(MeV)	Ex(MeV)	Cross Section(2°-14°)(μb)
#1(1_1^-)	0	0 MeV	1.47±0.05
#2(1_2^-)	2.58±0.17	2.71±0.13	0.23±0.03
#3(1_3^-)	} 6.89±0.42	6.05±0.18	0.22±0.03
#3′		8.10±0.38	0.17±0.03
#4(2^+)	10.68±0.12	10.97±0.05	1.76±0.07

agreement when only the peak #3 is compared, supporting the assumption that the peak is composite. The overall agreement will imply that the present DWIA calculation are quite reliable in predicting the absolute cross sections to each state.

4 The ${}^{16}_{\Lambda}$O and ${}^{13}_{\Lambda}$C hypernuclei and *ls* splitting

Determining the magnitude of ΛN spin-orbit splitting is one of the current central issues in Λ hypernuclear spectroscopy. From the CERN ${}^{12}_{\Lambda}$C and ${}^{16}_{\Lambda}$O data,[17] and BNL ${}^{13}_{\Lambda}$C data[18], very small values for the splitting were deduced. However, it was recently suggested that the splitting could be much larger based on reanalysis of the emulsion ${}^{16}_{\Lambda}$O data[19] and analysis of ${}^{89}_{\Lambda}$Y spectrum by the (π^+, K$^+$) reaction.[20]

In the ${}^{16}_{\Lambda}$O hypernuclei, we expect to observe 4 peaks in the bound region by the (π^+, K$^+$) and (K$^-$, π^-) reactions at the forward angles as shown in Table 3. The 2_1^+ and 0_1^+ states are the spin-orbit partners and the two states are split by the spin-orbit interaction. The 2_1^+ state is excited preferentially by the (π^+, K$^+$) reaction, while the 0_1^+ state are strongly populated by the recoilless (K$^-$, π^-) reaction.[21,22] The energy splitting of the two states is then closely related to the strength of the spin-orbit potential as discussed based on a calculation that took into account particle-hole interaction.[19]

From the present high quality (π^+, K$^+$) spectrum shown in Fig. 3, the excitation energy of the 2_1^+ state are determined with a good precision to be 10.6 ± 0.1 MeV. The excitation energy of the *ls* partner 0_1^+ can be determined from the CERN (K$^-$, π^-) in-flight spectrum,[17] although the original paper did not give the excitation energy. The fitting to the CERN ${}^{16}_{\Lambda}$O spectrum was recently carried out by Tovee and Davis.[23] From these excitation energies shown in Table 3, the energy difference between 2_1^+ and 0_1^+ is preliminarily derived to be 0 $\pm 0.3^{stat} \pm 0.2^{sys}$ MeV. We can also estimate the energy difference by comparing the stopped (K$^-$,π^-) spectrum in which both 2_1^+ and 0_1^+ states can be populated[24] with the present (π^+, K$^+$) spectrum. Assuming the yield ratio of the two states in the stopped (K$^-$,π^-) spectrum to be 2 to 1 as calculated by Motoba *et. al.*, the energy difference was evaluated to be 0.72 ± 0.5 MeV. These small splittings of the *ls* partners are consistent with the old CERN and BNL data.

Table 3: Comparison of excitation energies of the ${}^{16}_{\Lambda}$O low-lying states as derived by the (π^+, K^+), in-flight (K^-, π^-) and stopped (K^-, π^-) reactions.

Peak#	Spin-Parity configuration	Excitation energy(MeV)		
		(π^+,K^+) E336 Preliminary	in-flight(K^-,π^-) CERN [23]	stopped(K^-,π^-) KEK [24]
#1	$1_1^-(p_{\frac{1}{2}}^{-1}\otimes\Lambda s_{\frac{1}{2}})$	0	0	0
#2	$1_2^-(p_{\frac{3}{2}}^{-1}\otimes\Lambda s_{\frac{1}{2}})$	6.3±0.1	6.0±0.4	6.37±0.44
#3	$2_1^+(p_{\frac{1}{2}}^{-1}\otimes\Lambda p_{\frac{3}{2}})$	10.6±0.1	—	} 10.82±0.44
	$0_1^-(p_{\frac{1}{2}}^{-1}\otimes\Lambda p_{\frac{1}{2}})$	—	10.6±0.3	
#4	$2_2^+(p_{\frac{3}{2}}^{-1}\otimes\Lambda p)$	16.7±0.1	—	} 16.32±0.44
	$0_2^+(p_{\frac{3}{2}}^{-1}\otimes\Lambda p)$	—	17.2±0.3	

The ${}^{13}_{\Lambda}$C is believed to be one of the most promising candidates for the determination of spin-orbit splitting in the Λ p orbit, that is $p_{1/2}$ and $p_{3/2}$ states. From the recent observation of hypernuclear γ ray from the $p_{1/2}$ state to the ground state, the excitation energy of the $p_{1/2}$ state is now known to be $10.95 \pm 0.1 \pm 0.2$ MeV.[25] In the present (π^+, K^+) ${}^{13}_{\Lambda}$C spectrum, we can extract the excitation energy of the $p_{3/2}$ state which is the spin-orbit partner of the $p_{1/2}$ state since the $p_{3/2}$ state is favorably excited by the (π^+, K^+) reaction. In the spectrum shown in Fig 4, a shoulder is clearly observed to the left hand side of the prominant peak at 10 MeV. The shoulder can be interpreted to consist of two peaks, #3 and #4, as predicted by the DWIA calculation, one the $p_{3/2}$ states and the other the state $s_{1/2}$ coupled to the 1^+ state in ^{12}C. By fitting the acceptance corrected spectrum, the excitation energies of the bound region peaks are derived as listed in Table 4. For the excitation energies of #3 and #4 peaks, we temporarily assigned their spin-parities to be $3/2^-$ and $1/2^+$ based on the theoretical calculation.[21] Systematical errors of the excitation energies were evaluated in the table taking into account the possible range of the cross section ratio between the two states. The energy difference of the two peaks will then be derived to be $\Delta E = 1.4 \pm 0.3^{stat} \pm 1.2^{sys}$ MeV.

The experiment which aims to observe the γ ray transitions from the $p_{3/2}$ and $p_{1/2}$ states to the ground state simultaneously was performed at BNL and its analysis is underway as reported in this conference[26], The present ${}^{13}_{\Lambda}$C spectrum also gives a constraint to the magnitude of the splitting.

In the case of both ${}^{16}_{\Lambda}$O and ${}^{13}_{\Lambda}$C, excitation spectra of the reactions that excite a different member of the ls partner are compared to deduce the energy differences. It is reliably done since only the excitation energies are our concern and the both reactions populate the ground state. The reaction spectroscopy offers an alternative way to determine the splitting and such procedure would be greatly benifitted when the resolution is much improved.

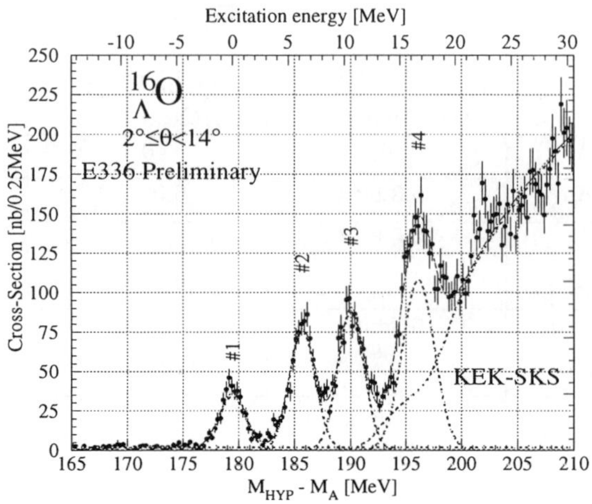

Figure 3: Preliminary excitation spectrum of $^{16}O(water)(\pi^+, K^+)^{16}_{\Lambda}O$ reaction obtained by the SKS spectrometer system.

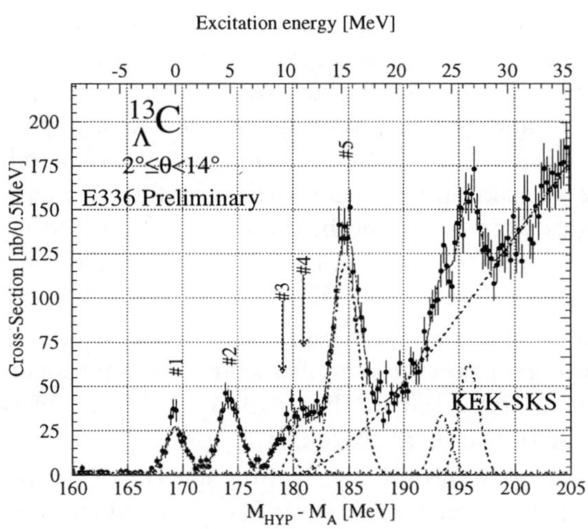

Figure 4: Preliminary excitation spectrum of $^{13}C(\pi^+, K^+)^{13}_{\Lambda}C$ reaction obtained with the SKS spectrometer system.

Table 4: Excitation energies and cross sections of the ${}^{13}_{\Lambda}$C states measured by the ^{13}C(π^+,K$^+$)${}^{13}_{\Lambda}$C reaction

Peak #	configuration $[{}^{12}\mathrm{C}(\mathrm{J}^{\pi}_{\mathrm{c}},\mathrm{T}_{\mathrm{c}}) \otimes \Lambda l_j]\mathrm{J}^{\pi}_{\mathrm{n}}$	E336 Preliminary	Ex(MeV)
#1	$[{}^{12}\mathrm{C}(0^+,0) \otimes \Lambda s_{\frac{1}{2}}]\frac{1}{2}^+_1$		0
#2	$[{}^{12}\mathrm{C}(2^+,0) \otimes \Lambda s_{\frac{1}{2}}]\frac{3}{2}^+_1$		4.9±0.1
#3	$([{}^{12}\mathrm{C}(0^+,0) \otimes \Lambda p_{\frac{3}{2}}]\frac{3}{2}^-)$		**9.6±0.3±1.0***
#4	$([{}^{12}\mathrm{C}(1^+,0) \otimes \Lambda s_{\frac{1}{2}}]\frac{1}{2}^+_2)$		11.6±0.2±1.0*
#5	{ $[{}^{12}\mathrm{C}(2^+,0) \otimes \Lambda s_{\frac{1}{2}}]\frac{1}{2}^+_4$ $[{}^{12}\mathrm{C}(1^+,1) \otimes \Lambda p_{\frac{1}{2}}]\frac{5}{2}^-_2$ $[{}^{12}\mathrm{C}(2^+,1) \otimes \Lambda p_{\frac{1}{2}}]\frac{3}{2}^+_4$ }		15.4±0.1

∗ A systematical error evaluated by considering possible contamination from the #4(#3) peak is quoted.

5 Future prospects and summary

We have learned that the hypernuclear reaction spectroscopy with MeV resolution, which has been intensively carried out at KEK via the (π^+, K$^+$) reaction, become invaluable to in the investigation of the Λ hypernuclear structure. The improvement of the resolution from 3 MeV to 2 MeV(FWHM) greatly contributed to the progress of Λ spectroscopy. It has demonstrated that spectroscopy has a power to the investigation of strangeness nuclear physics. As reporteed in this conference, we also saw the impact of the resolution improvement from 2 to 1.5 MeV in ${}^{89}_{\Lambda}$Y spectrum.

Table 5: Comparison of Λ hyperon production reactions

$\Delta Z = 0$ *neutron to Λ*	$\Delta Z = -1$ *proton to Λ*	comment
(π^+, K^+)	(π^-, K^0)	stretched,high spin
in-flight (K^-, π^-)	in-flight (K^-, π^0)	substitutional at low momentum
stopped (K^-, π^-)	stopped (K^-, π^0)	large yield, via atomic states
$(e, e'K^0)$	$(e, e'K^+)$	spin flip,unnatural parity, virtual (γ,K)
$(p, p'K^0)$	$(p, p'K^+)$	virtual (π,K)
(p, K^0)	(p, K^+)	very large momentum transfer

In the coming years, we expect that the spectroscopic studies of Λ hypernuclei will be further explored with higher resolution and by employing new reactions. Various reactions that can be used for the Λ hypernuclear spectroscopy are listed in

Table 5 In addition to the (π^+, K^+) reaction and (K^-, π^-) reaction which change a neutron to a Λ hyperon, the reactions that convert a proton to a Λ hyperon will become available. It will allow us to produce neutron-rich Λ hypernuclei not investigated so far and the structure study will greatly advanced. Among them, the $(e,e'K^+)$ reaction will play a significant role in the near future as also reported in this conference. It will populate spin-flip states of Λ hypernuclei because of the electron beam. Furthermore, the reaction will give us an opportunity to realize the best resolution in the hypernuclear reaction spectroscopy, sub-MeV or even 2-300 keV(FWHM), since high emittance beams can be used.[27] As shown in the Table 5, neutral meson detection can also expand the possibility of spectroscopic study. The first result of the (K^-, π^0) reaction was also presented in the conference.[28] In addition, the sub-MeV reaction spectroscopy by the (π^+, K^+) is planned at JHF50 GeV with a new high resolution pion beam line.[29] These efforts altogether will form one of cores for the progress of the strangeness nuclear physics . It should be noted, however, that the high resolution and also a high detection-efficiency experiments are of vital importance for the spectroscopy in order to obtain high quality spectra.

As reported in this workshop, the γ ray spectroscopy of Λ hypernuclei will also play a key role with its unprecedented energy resolution.[30] It is a powerful tool for those states below the particle emission threshold, though a Λ hyperon in *p* orbital are hardly bound in many of Λ hypernuclei.

The reaction spectroscopy, which can access to the hypernuclear states above the nucleon emission threshold, and the γ-ray spectroscopy which has almost 3 orders of magnitude better energy resolution will play complimentary roles in the spectroscopic investigation of Λ hypernuclei.

6 Summary

The reaction spectroscopy of Λ hypernuclei made a great progress in the past years, particularly taking the advantage of the (π^+, K^+) reaction. Thanks to the good resolution of 1.5-2 MeV and the high detection efficiency, good quality spectra were measured for wide variety of Λ hypernuclei. The information on the hypernuclear structure and ΛN interaction are obtained as examples of ${}^{12}_{\Lambda}C$, ${}^{16}_{\Lambda}O$ and ${}^{13}_{\Lambda}C$ showed.

There are now substantial effort towards higher energy resolution spectroscopy of Λ hypernuclei. Within a year we expect the first sub-MeV resolution spectroscopy by the $(e,e'K^+)$ reaction at Jefferson Laboratory. We foresee the frontier of the hypernuclear spectroscopy will be expanded through upgrading the existing experimental facilities and employing new reactions. In this connection, γ-ray spectroscopy with a few keV resolution, already has a significant impact to the Λ hypernuclear spectroscopy. In the turning of the century, these next generation experiments with sub-MeV resolution and/or with new hyperon production reactions will further explore the significance of Λ hypernuclear spectroscopy in the strangeness nuclear physics.

The E336 experiment were carried out in collaboration of Tohoku, KEK, Seoul, Tokyo, Osaka, Hampton and Hampton. The author deeply acknowledges the members of the collaoration, particularly to Mr. T. Endo for preparing the figures and tables. Intensive discussion with Prof. T. Motoba, K. Itonaga, D.H. Davis and J. Millener are also greatly appreciated.

References

1. H. Bandō, T. Motoba and Y. Yamamoto, Phys. Rev. **C31** 265 (1985).
2. A. Likar, M. Rosina and B. Povh, Z. Phys. **A324** 35 (1986).
3. C. Milner *et. al.*, Phys. Rev. Lett. **54** 1237 (1985).
   P. H. Pile *et al.*, Phys. Rev. Lett. **66** 2585 (1991).
4. M. Akei *et al.*, Nucl. Phys. **A534** 478 (1991).
5. O. Hashimoto, *et. al.* Il Nuovo Cimento **102** 679 (1989).
6. T. Fukuda, *et. al.*, Nucl. Instr. Meth. **A361** (1995) 485.
7. T. Hasegawa, *et. al.*, Phys. Rev. Lett. **74** (1995) 224.
8. T. Hasegawa, *et. al.*, Phys. Rev. **C53** (1996) 1210.
9. S. Ajimura, *et. al.*, Phys. Rev. Lett. **80** (1998) 3471.
10. H. Bhang, *et. al.*, Phys. Rev. Lett. **81** (1998) 4321.
11. O. Hashimoto, *et. al.*, Nucl. Phys. **A639** (1998) 93c.
12. T. Nagae, this proceedings
13. S. Ajimura, *et. al.*, Nucl. Phys. **A585** (1995) 173c.
14. Proceedings of the 23rd INS symposium, p.23, 1995, University Academy Press, Eds. S. Sugimoto and O. Hashimoto
15. T. Motoba, Nucl. Phys. **A639** (1998) 135c.
16. K. Itonaga, *et. al.*, Prog. Theor. Phys. **84** (1990) 291 and Private communication
17. W. Bruckner *et. al.*, Phys. Lett. **79B** (1978) 157.
18. M. May *et. al.*, Phys. Rev. Lett. **47** (1981) 1106.
19. R.H. Dalitz *et. al*, Nucl. Phys.**A625** (1997).
20. T. Nagae, Proc. of the 23rd INS int. symp. on Nuclear and Particle Physics with Meson Beams in the 1 GeV/c Region, Tokyo, March, 1995. Eds. S. Sugimoto and O. Hashimoto., and these proceedings 1999.
21. T. Motoba, H. Bandō, R. Wünsch and J. Zofka, Phys. Rev. **C38** 1322 (1988).
22. K. Itonaga, T. Motoba and H. Bando, Prog. Theor. Phys. **84** (1990) 291.
23. D.H. Davis, Private communication 1997.
24. H. Tamura, R.S. Hayano, H. Outa and T. Yamazaki, Prog. Theor. Phys. Suppl. **117** (1994) 1.
25. M. May *et. al.*,Phys. Rev. Lett. **78** (1997) 4343.
26. T. Kishimoto, These proceedings
27. E. Hungerford, Prog. Theor. Phys. Suppl. **117** (1994) 135.
28. M. Youn, These proceedings
29. H. Noumi, Nucl. Phys. **A639** (1998) 121c.
30. H. Tamura, These proceedings

THE STORY OF Σ HYPERNUCLEI–A MODERN FABLE

ROBERT E. CHRIEN
Brookhaven National Laboratory, Upton NY 11973, USA
E-mail: chrien@bnl.gov

The reality of Σ hypernuclei has been the subject of intense concern among experimenters and theoreticians for more than 20 years. The possible existence of Σ hypernuclei was first suggested by by a pioneering experiment on a ^{9}Be target at the CERN PS. There were reported to be two narrow ($\Gamma < 8$ MeV) peaks in the continuum region. This finding was quite unexpected since the widths of Σ states were believed to be large due to the strong conversion process. It is obvious that if such relatively long-lived systems were confirmed unambiguously by experiment, their masses and widths provide important constraints on the ΣN effective interaction and its relation to the ΛN and NN interactions. Since the Σ carries isospin, the role of isospin and isospin conservation in hadronic reactions could be explored. This report stimulated a number of subsequent experiments at the BNL-AGS and KEK, along with further experiments with a specially created short kaon beam at the CERN PS. Experimental data were reported for different targets at different momenta and at different conditions. Various tagging techniques were employed to suppress backgrounds, but always at the expense of a reduction in statistical quality. Because of problems with resolution and statistics, contradictions among the different sets of data resulted more often in clouding the issues than clarifying them. Thus, up until a few years ago, there was no statistically clear confirmation of this surprising finding. In the last few years, however, a series of definitive experiments has been performed at the BNL-AGS in an effort to resolve the discrepancies and settle the controversy. Besides repeating the initial experiment on Be, a target of ^{6}Li was run. The earlier stopped kaon data on ^{4}He was repeated in an in-flight experiment, and the suggestion of a Σ bound state, with virtually pure isospin, confirmed. We now have a better understanding of the role of isospin in the nucleon-hyperon interaction and the importance of three body forces in hypernuclei. According to that understanding, it is unlikely that Σ states for $A > 5$ will ever be seen. In this review, a critique of the past BNL experiments is presented and conclusions on the status of the Σ database drawn. As it is with the fables of old, this modern story also has a moral and it can be derived by paraphrasing and combining two old ones: "Seek and ye shall find; but beware, lest what ye find be only in the eye of the beholder."

Since the advent of in-flight studies of hypernuclei initiated at CERN in the 1970's, they have been used to draw conclusions about the baryon-baryon interaction. After a series of fruitful studies on the nuclear structure of p-shell hypernuclei, experimenters naturally turned to the Σ hyperon to discover details of the spin-isospin character of the interaction. The report of Bertini *et al.*[1], reporting narrow Σ states in beryllium, started a flurry of activity; it was deemed of particular interest to compare the spin-orbit splitting of Λ and Σ hypernuclei.

Of course, the very existence of Σ hypernuclei was always in question because of the presence of the strong conversion from Σ to Λ hyperons in the nuclear medium. Special features would have to exist in the radial and spin- isospin dependence of the hyperon-nucleon potential to suppress that strong conversion. Such considerations made the reported observations, with their possibilities in providing interesting information on the nature of the interaction, very exciting.

In an effort to bolster the experimental evidence for Σ hypernuclei, a modified beamline at CERN was constructed so that the lowest practical K^- beam could be achieved. Because of the increased mass of the Σ over the Λ, the beam momentum needed for minimum momentum transfer is reduced. Thus the CERN-Heidelburg collaboration hoped to take advantage of a new short beamline with which they investigated targets of carbon and oxygen.

In the initial report of Bertini *et al.* [1], two peaks, with FWHM's of less than 8 MeV, were reported at an excitation 80 MeV above the Λ hypernuclear threshold region. The spectrum bore a strong resemblance to the Λ region. The presence of such peaks, in the continuum above Σ binding, stimulated the hope that Σ states may indeed be very narrow.

This work encouraged much theoretical and further experimental effort; the outcome of this effort, however, led to contradictory interpretations of such fundamental quantities such as the Σ-nucleon spin-orbit splitting. Follow-up experiments by Bertini *et al.*[2,3], Yamazaki *et al.* [4], and Piekarz *et al.*[5] advanced claims for observing narrow states for the p-shell nuclear targets of ^{6}Li, ^{12}C, and ^{16}O. They were led to make a number of conflicting predictions of the Σ-Nucleon spin-orbit splitting and well depth. One characteristic of these early works was the necessity to use tagging techniques, or the observation of secondary particles, to enhance the signal above the background of kaon decay. The problem is particularly severe for the π^- experiments; on the other hand, the double charge exchange for (K^-,π^+) suppresses decay signals. Thus most reported Σ states were for the (K^-,π^+) reaction. The use of tagging, however, resulted in an order-of-magnitude reduction in statistics.

The somewhat confusing situation was first summarized in the review of Dover, Millener, and Gal [6]. They point out that a rough estimate for the width of a Σ state localized in a well of imaginary depth W_Σ is 2 W_Σ, or about 15 to 20 MeV, based on the earlier estimates of Batty *et al.* [7].

Figure 1, taken from the review of Dover et al.[6] shows the relevant squares of Fermi-averaged amplitudes for the elementary Σ production processes. The strong momentum dependence was used in the CERN and BNL experiments to draw conclusions about the nature of the apparent peaks in the $(K^-,\pi^\pm)$ data. These cross sections are constructed from the amplitudes for Σ production by Gopal [8], and the elementary cross sections were also used for system efficiency calibrations because of the ease with which a (K^-,π^+) peak on a hydrogenous target could be seen.

A subsequent experiment with stopping kaons was reported[9] for a ^{4}He target. This result supported an earlier He bubble chamber experiment [10], and suggested observed structure near threshold as evidence for a bound Σ state. The stopped kaon data is shown in figure 2, taken from Hayano et al.[9]. The large backgrounds unavoidable in the stopped kaon work make the structure hard to see. An effort was subsequently carried on as part of AGS E774 to provide an in-flight experimental verification of the bound state[11]. The statistical quality of the verification was not sufficient to settle the issue.

In this review, I will not describe in more detail the CERN and KEK results of the earlier years. They were contradictory and could not be reconciled; they were typically characterized by a very small number of events, especially after tagging.

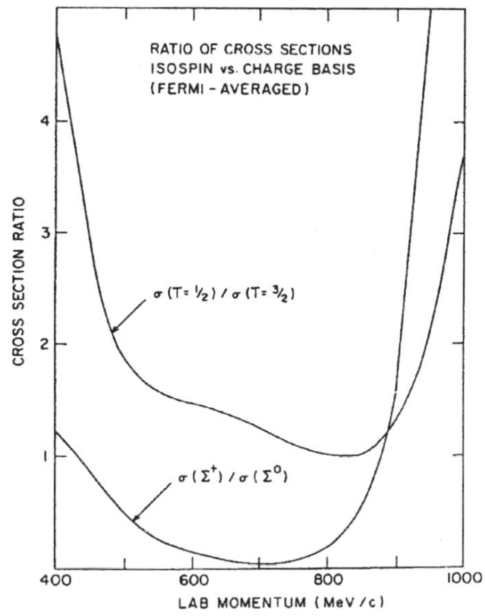

Figure 1. The ratios of Fermi-averaged cross sections in charge and isospin bases, taken from Dover, Gal, and Millener.

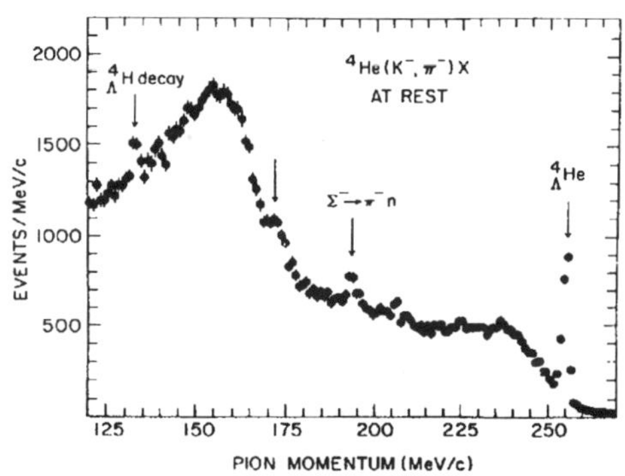

Figure 2. Observation of a suggested bound state in the stopped kaon (K^-, π^-) reaction on ^{4}He at KEK from the report of Hayano *et al.* [9]

No evidence for statistical significance was presented for these experiments. In any case, it is impossible for an outsider to assess possible problems and to offer sensible criticisms of these experiments, done many years ago. Furthermore, the CERN experiments were undertaken at a time when the CERN PS was being phased out for hypernuclear studies. The experimenters were given a very short time to use this short kaon beam line, which was designed to match the low momentum suitable for Σ substitutional state production. This turns out to be a common problem for

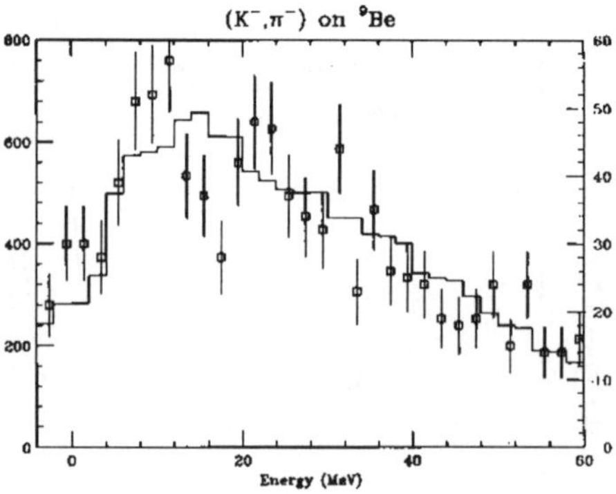

Figure 3. A direct comparison of the CERN and BNL results for ^{9}Be. The statistical errors shown are for the CERN data; the histogram represents the BNL measurement.

hypernuclear work–too little and too late. One longs for a dedicated facility, like the proposed Japanese Hadron Facility, to do this work.

I will therefore confine my remarks to more recent BNL work and the insight it provides. I will also discuss several older published BNL-AGS experiments in an effort to render some consistency to this often confusing subject.

The first figure I show in the set of the new BNL experiments is an overlay of the CERN and BNL data on ^{9}Be [12]. While the CERN experiment was done at 720 MeV/c, the BNL work at 600 MeV/c has a lower momentum transfer, which would be expected to enhance any coherent excitation. While the resolution achieved in the BNL experiment–4 MeV–is slightly broader than the 3.0 MeV achieved in the CERN experiment, the difference is not significant compared to the claimed peak width of 8 MeV. The lack of confirmation of narrow states, and the statistical quality of the BNL experiment, are clear from figure 3.

I now describe how the superior statistics and background suppression were achieved. These most recent experiments were carried out at the C6 beam line of the Low Energy Separated Beam (LESB-2) of the Brookhaven National Laboratory Alternating Gradient Synchrotron (BNL-AGS). Negative kaons of 600 MeV/c were incident on a metallic targets of Be and ^{6}Li, and on a liquid helium target. The π^- and π^+ reaction products were analyzed with the Moby-Dick spectrometer, set at a lab reaction angle of 4 deg. The usual particle identification used with Moby-Dick, consisting of time-of-flight and Čerenkov counters, was supplemented by a muon-range telescope which tagged events, presumably pions, which ranged out in a 20 cm iron block. The effectiveness of these background suppressions are shown in figure 4.

A number of complementary runs were taken to make up for the limited momentum acceptance of the Moby-Dick spectrometer over the range of significant Σ production. These included $(K^-, \pi^\pm)$ at 460 MeV/c, (K^-, π^-) at 438 MeV/c, and (K^-, π^-) at 550 MeV/c. Further, calibration runs were taken with a 2.28 gm/cm^2 polyethylene target to check against previously measure elementary kaon production cross sections on protons, and to study spectrometer acceptance. The

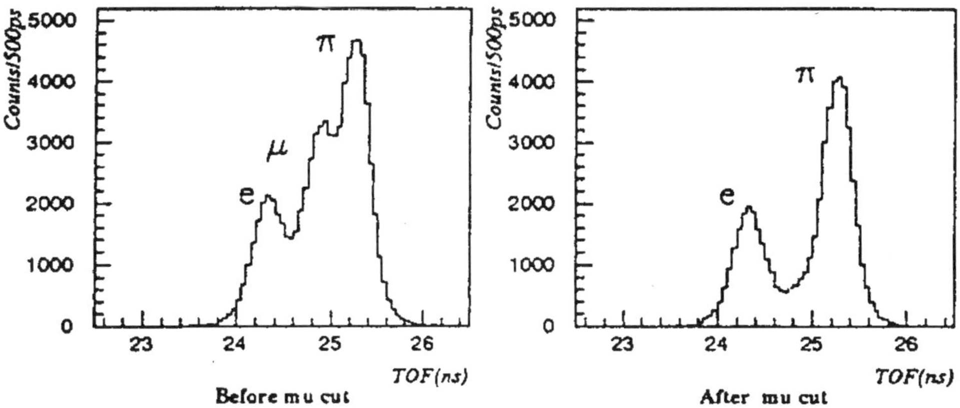

Figure 4. The effectiveness of the muon filter and timing resolution is shown on this plot, taken from the thesis of Miyachi [13].

acceptance was further studied by kinematically isolating decays in the target region. The absolute Moby-Dick solid angle acceptance of 18 msr is believed known to an accuracy of 15 %, based on these and previous studies. The plots to be presented here are missing mass spectra, with the excitation energy zero chosen to correspond to Σ^0 production in the (K^-, π^-) reaction.

Following on the inadequate result of the E774 check on the bound state in He suggested by the stopped kaon experiment at KEK, Nagae *et al.*[14] led the collaboration with an in-flight experiment at BNL with Moby-Dick. It was the last experiment ever done with that venerable spectrometer. With the improved timing and an improved muon filter provided by the Japanese, the in-flight experiment provided a striking confirmation of the KEK stopped kaon experiment.

These last BNL resuts show a consistent picture of the progression of Σ excitations for A=4, 6, and 9. These are displayed in Figure 5. They constitute a data base which is a challenge for theorists to incorporate in their knowledge of the fundamental hyperon-nucleon interaction. The characteristic features of these spectra are remarkable. First, there is an obvious isospin dependence in the comparison of the (K^-, π^-) and (K^-, π^+) spectra. Second, there is a progressive shift in the appearance of an enhancement in the (K^-, π^-) spectrum with mass number, ranging from -10 MeV for He to 14 MeV for Be (note that all excitation energies are referred to 0 on a Σ^0 scale). Finally there is a progressive broadening and shift to higher energies of the (K^-, π^+) spectrum. Any theoretical analysis must, of course, take account of the interference between Λ and Σ amplitudes in a coupled channels calculation for the π^- reaction.

A important contribution to the understanding of ^{4}He has been made by Harada and his collaborators. As they point out[15 16], the results indicate a very large isospin dependence in the Σ -nucleus potential. The ^{4}He π^- results indicate a practically pure T=1/2 bound state in ${}^4_\Sigma$He . Harada suggests a large "Lane" term in the potential; the 1/A dependence in that term indicates that we will not have bound

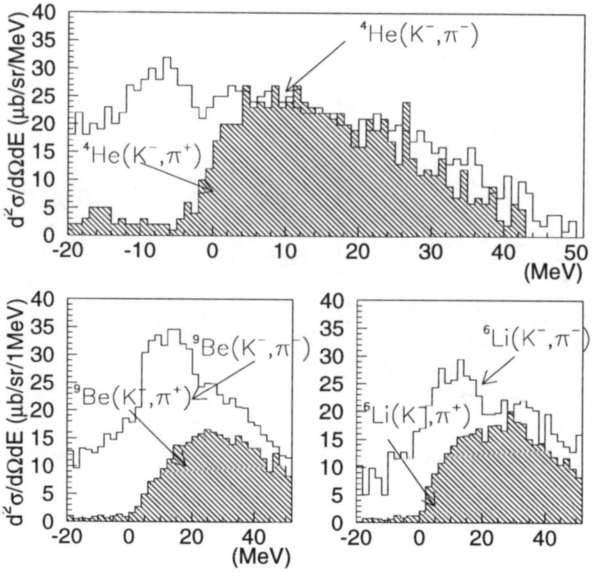

Figure 5. Excitation spectra obtained for targets of He, ^{6}Li, and Be. Note the similarities and trend with mass number in these spectra.

states for A>5. Hence a Σ spectroscopy does not exist for the p-shell and beyond. Harada suggests there is a strong repulsive potential for the $T = 3/2$ part of the potential. This large isospin difference is just what the BNL data indicate in the figures above. Harada further suggests that there is a sizeable three-body force due to the Σ and this shows up in the binding energy systematics of hypernuclei. A complete fit was reported by Harada, covering the region from the Λ threshold to beyond the Σ thresholds.

What is perhaps not so clear is the extension of these ideas to systems heavier than helium and the reconciliation of shell model and cluster descriptions of these heavier hypernuclei.

The shell model calculations have perhaps been, up to now, hampered by lack of a consistent data base of cross sections. This has been made clear to me by my review of earlier BNL experiments. I want here to describe two problem areas and how they have been (maybe) resolved. These problems are well illustrated by a comparison of Piekarz *et al.*[5] and Tang *et al.*[17]. The next figures illustrate the problem.

The first is the very large difference in the cross sections for lithium in the two sets (nearly a factor of 4 for nuclei differing by only one mass unit–that is, ^{6}Li and ^{7}Li). I learn from theoretically minded friends that it is simply not possible to fit the large cross sections reported by Tang with a shell model with reasonable pick-up amplitudes. The answer is that Tang *et al.* normalized to the amplitudes of Gopal,

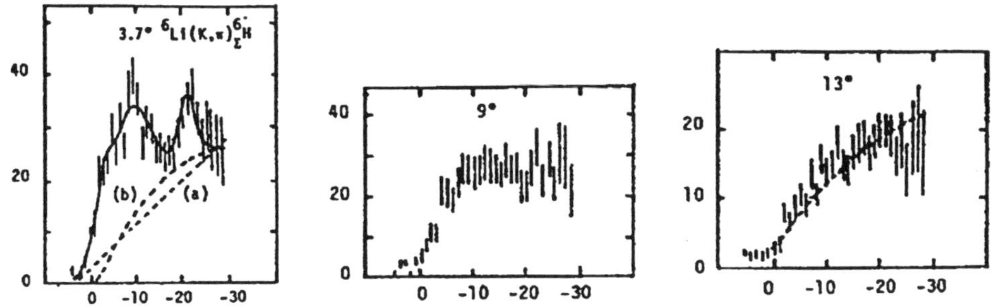

Figure 6. Spectra of Piekarz *et al.* obtained for the reaction (K^-, π^+) on ^{6}Li at 600 MeV/c. The cross section scale is in μbarns/sr.

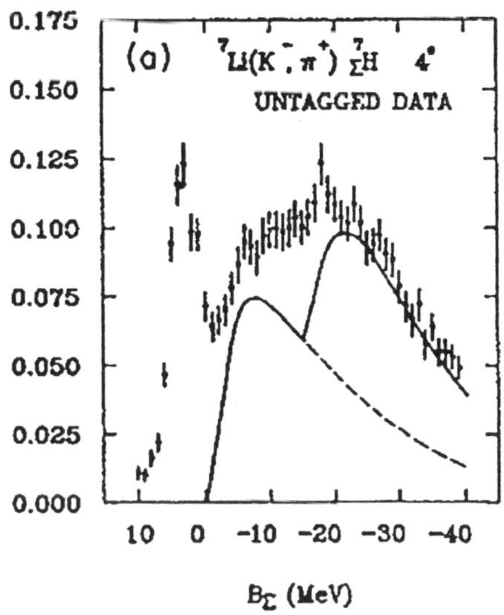

Figure 7. The data of Tang *et al.* using the (K^-, π^+) reaction on a target of ^{7}Li at 715 MeV/c. The cross section scale is in mb/sr. The large bump near zero is a target impurity.

whereas in Piekarz, the cross sections are normalized by reference to Armenteros[18]. While there is no reason to doubt that the Gopal amplitudes show the general trend of the data, near 600 MeV/c they provide cross sections a factor of two larger than the relevant measurements of Armenteros.

Figure 8 demonstrates the very large discrepancy between the measured cross sections and those predicted the the Gopal amplitudes. When that is taken into account, and considering the difference in beam momenta, the sets are consistent.

The second embarrassment (at least to the BNL group) is the report in Piekarz *et al.*[5] of "narrow" states produced by the (K^-, π^+) on ^{6}Li ; the states are seen clearly in fig. 13. The disappearance of those "states" at 9 and 13 degrees was ascribed to the forward angle dominance of the Δ L=0 component and its virtual

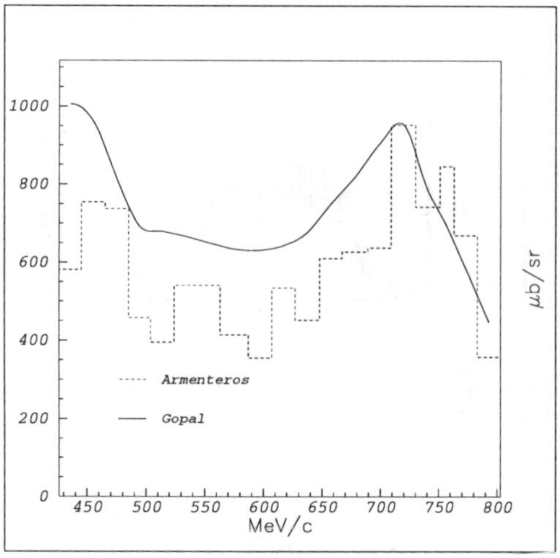

Figure 8. A comparison of the data from Armenteros *et al.* [18] and the global Gopal fit in the region of interest.

disappearance at 9 or 13 degrees. It is difficult to argue that these bumps are not significant statistically.

It is important to point out that the bumps seen in figure 6 are visible also in the spectrum of figure 7. In comparing these, note that the bump near zero in the latter spectrum should be ignored. It is a target impurity.

A clue to explaining these bumps is that the spectrometer setting for both experiments is precisely zero (it is explained in the references that a zero setting corresponds to an average over the spectrometer acceptance, resulting in an effective value of 3.7 degrees). At zero angle, however, it is not possible to make a vertex cut on the tracks of the emitted reaction pions; it therefore not possible to isolate contributions of pions coming from the timing scintillator and Čerenkov veto upstream of the target. We learned in the subsequent Σ experiments described earlier that very large contributions are seen from upstream elements; these are removed by using a non-zero spectrometer angle and a vertex cut to isolate the target contribution.

I suggest therefore that inscattered pions from these elements would have appropriate momenta to account for the broad peaks seen. Such a contribution may as well appear in Tang's spectrum. Data from both of these experiments should be therefore treated with caution because of the above reasons.

The conclusions to be drawn from the present work can be summarized as follows:

1) There are no compelling reasons to believe there are narrow continuum Σ states seen in any hypernuclei above mass number 4.

2) There are sharp and systematic differences between (K^-, π^-) and (K^-, π^+) which demonstrate a large isospin term in the Σ -nuclear potential.

3) There is a reasonable description of a virtually pure I=1/2 isospin state in

helium. That description implies a potential with a 1/A Lane term, suggesting that bound states will not appear for heavier systems

4) There systematic and regular features in the Σ continuum region which call for a systematic treatment, whether in the cluster model or the continuum shell model.

As it is with the fables of old, this modern story also has a moral and it can be derived by paraphrasing and combining two old ones: "Seek and ye shall find; but beware, lest what ye find may be only in the eye of the beholder."

References

1. R. Bertini *et al.*, Phys. Lett. **B90** (1980) 375
2. R. Bertini *et al.* Phys. Lett. **136B** (1984) 29
3. R. Bertini *et al.* Phys. Lett. **158B** (1985) 19
4. T. Yamazaki *et al.* Phys. Rev. Lett. 54 (1985) 102
5. H. Piekarz *et al.* Phys. Lett. **110B** (1982) 428
6. C. B. Dover, D. J. Millener, and A. Gal, Physics Reports **185** 1989
7. C. J. Batty *et al.*, Phys. Lett. **74B** (1978) 27
8. G. P. Gopal *et al.* Nucl. Phys **B119** (1977) 362
9. R. S. Hayano *et al.* Phys. Lett. **231B** (1989) 355
10. R. Roosen *et al.* Nuovo Cimento **49** (1979) 217
11. Y. Shimizu, Master's Thesis, University of Tokyo and private communication (1992)
12. R. Sawafta, Nucl. Phys. **A545** (1995) 103c
13. T. Miyashi, Thesis, University of Tokyo (1997) and private communication
14. T. Nagae *et al.*, Phys. Rev. Lett. **80** (1998) 1605
15. T. Harada, Nucl. Phys. **A507** (1990) 715
16. T. Harada, Phys. Rev. Lett. **81** (1999) 5287
17. L-Q. Tang *et al.* Phys. Rev. **C38** (1988) 846
18. R. Armenteros *et al.*, Nucl. Phys. B21, **B21** (1970) 15-76

THE ISOTOPIC SPIN DEPENDENT (LANE) PART OF THE SINGLE-PARTICLE POTENTIAL OF THE Σ HYPERON IN NUCLEAR MATTER

J. DĄBROWSKI

Theoretical Division, Sołtan Institute for Nuclear Studies, Hoza 69, Pl-00-681 Warsaw, Poland

E-mail: dabrnucl@fuw.edu.pl

The isospin dependence of the real part of the single-particle potential of a Σ hyperon in nuclear matter, $\hat{V}_\Sigma$, is investigated. The isospin dependent part V_1 of $\hat{V}_\Sigma$ is expressed in terms of an effective ΣN interaction in nuclear matter, which depends on proton and neutron densities. With suitable approximations numerical results for V_1 are obtained for four models of the Nijmegen baryon-baryon interaction. A comparison with recent $(K^-, \pi^\pm)$ experiments favors model F as a realistic representation of the ΣN interaction.

1 Introduction

The single-particle (s.p.) potential $V_\Sigma(k_\Sigma)$ of the $\Sigma^\pm$ and Σ^0 hyperons in a spin saturated nuclear matter with Z protons and N neutrons depends on the proton excess parameter $\alpha = (Z - N)/A$. Assuming charge-independence of the baryon-baryon interaction we have in the linear approximation in α:

$$\begin{aligned} V_{\Sigma^\pm}(k_\Sigma) &= V_0(k_\Sigma) \pm \tfrac{1}{2} V_1(k_\Sigma)\alpha \,, \\ V_{\Sigma^0} &= V_0(k_\Sigma) \,. \end{aligned} \tag{1}$$

Similarly as it was noticed first by Lane [1] in the case of the nucleon s.p. potential, the s.p. potentials $V_{\Sigma^\pm}$ and V_{Σ^0} are averaged versions of a more fundamental formula:

$$\hat{V}_\Sigma(k_\Sigma) = V_0(k_\Sigma) + V_1(k_\Sigma)\mathbf{t}_\Sigma \mathbf{T}_A/A, \tag{2}$$

where $\mathbf{t}_\Sigma$ is the Σ isospin ($t_{\Sigma 3} = \pm 1, 0$ for $\Sigma^\pm, \Sigma^0$) and $\mathbf{T}_A$ is the nuclear isospin, $\mathbf{T}_A = \sum \mathbf{t}_N$ [$\mathbf{t}_N$ is the nucleon isospin ($t_{N3} = \frac{1}{2}$ for protons and $-\frac{1}{2}$ for neutrons), $T_{A3} = (Z - N)/2$].

In my talk, I present a calculation of the Lane potential V_1 in nuclear matter. In particular, I consider the contribution to V_1 arising from the intrinsic dependence of the effective ΣN interaction in nuclear matter, $\mathcal{K}$, on two densities, that of protons ρ_p and that of neutrons ρ_n. In the previous simple estimates of V_1, [2], this contribution was disregarded. In the second part of my talk (Sec. 2), I derive the expression for V_1 in terms of $\mathcal{K}$, and introduce the single density approximation which simplifies the expression for V_1. In the last part of my talk (Sec. 3), I present and discuss the result for V_1 obtained with $\mathcal{K}$ calculated in Ref. 3 in the low order Brueckner (LOB) theory for four models of the Nijmegen baryon-baryon interaction. In my talk, I restrict myself to the real part of the Σ s.p. potential.

2 The expression for V_1

To derive the expression for V_1, we start from its definition which follows from Eq. (1):

$$V_1(k_\Sigma) = 2[\partial V_{\Sigma^+}(k_\Sigma)/\partial\alpha]_{\alpha=0} \,. \tag{3}$$

The s.p. potential of Σ^+ in spin saturated nuclear matter, V_{Σ^+}, depends on ρ_p and ρ_n, connected with the respective proton and neutron Fermi momenta $k_F^p \equiv \kappa$ and $k_F^n \equiv \lambda$ by:

$$\left.\begin{array}{l}\kappa = (3\pi^2\rho_p)^{1/3} \\ \lambda = (3\pi^2\rho_n)^{1/3}\end{array}\right\} = k_F(1\pm\alpha)^{1/3} \,, \tag{4}$$

where

$$k_F = [3\pi^2(\rho_p+\rho_n)/2]^{1/3} \,. \tag{5}$$

The dependence of V_{Σ^+} on the two densities, or equivalently on the two Fermi momenta κ and λ, leads to the dependence of V_{Σ^+} on α. To determine this dependence, let us write the expression for V_{Σ^+} in terms of the Σ^+p and Σ^+n effective interactions in spin saturated nuclear matter with a proton excess, $\mathcal{K}_{\Sigma^+p}(\kappa\lambda)$ and $\mathcal{K}_{\Sigma^+n}(\kappa\lambda)$:

$$V_{\Sigma^+}(k_\Sigma) = \sum_{\mathbf{k}_N}^{k_N<\kappa} (\mathbf{k}_N\mathbf{k}_\Sigma|\mathcal{K}_{\Sigma^+p}(\kappa,\lambda)|\mathbf{k}_N\mathbf{k}_\Sigma) + \sum_{\mathbf{k}_N}^{k_N<\lambda} (\mathbf{k}_N\mathbf{k}_\Sigma|\mathcal{K}_{\Sigma^+n}(\kappa\lambda)|\mathbf{k}_N\mathbf{k}_{\Sigma^+}) \,, \tag{6}$$

where spins are suppressed in our notation.

To obtain the expression for V_1, we have to calculate - according to (3) - the derivative of V_{Σ^+} with respect to α, at $\alpha=0$, taking into account the dependence of κ and λ on α, as given by Eq. (4).

The derivative consists of two parts.

The first part comes from the dependence on α of the limits of the sums in Eq. (6). Its contribution to V_1, which we denote by $V_1^{(0)}$, can be easily calculated, and the result is:

$$V_1^{(0)}(k_\Sigma) = \tfrac{1}{4}A\left[\int \frac{d\hat{k}_N}{4\pi}(\mathbf{k}_N\mathbf{k}_\Sigma|\mathcal{K}_{\Sigma^+p}(k_F) - \mathcal{K}_{\Sigma^+n}(k_F)|\mathbf{k}_N\mathbf{k}_\Sigma)\right]_{k_N=k_F} \,, \tag{7}$$

where $\mathcal{K}_{\Sigma^+p(n)}(k_F) = \mathcal{K}_{\Sigma^+p(n)}(\kappa=k_F,\lambda=k_F)$ is defined for $Z=N=A/2$.

The second part comes from the intrinsic dependence of $\mathcal{K}$ on κ and λ. Its contribution to V_1, which we denote by $V_1^{(I)}$, is:

$$V_1^{(I)}(k_\Sigma) = \tfrac{1}{3}\sum_{\mathbf{k}_N}^{k_N<k_F}(\mathbf{k}_N\mathbf{k}_\Sigma|k_F\left[\left(\frac{\partial}{\partial\kappa}-\frac{\partial}{\partial\lambda}\right)\left\{\mathcal{K}_{\Sigma^+p}(\kappa\lambda) + \mathcal{K}_{\Sigma^+n}(\kappa\lambda)\right\}\right]_{\kappa=\lambda=k_F}|\mathbf{k}_N\mathbf{k}_\Sigma) \,. \tag{8}$$

The most difficult part of $V_1 = V_1^{(0)} + V_1^{(I)}$ is $V_1^{(I)}$. Whereas $V_1^{(0)}$ is expressed by $\mathcal{K}_{\Sigma N}(k_F)$, the effective ΣN interaction in nuclear matter with $Z=N$, to determine $V_1^{(I)}$ we should know $\mathcal{K}_{\Sigma N}(\kappa\lambda)$, the effective ΣN interaction in a two-component nuclear matter with $Z \neq N$. A calculation of $\mathcal{K}_{\Sigma N}(\kappa\lambda)$ starting from realistic ΣN interaction would be very tedious and so far has not been performed. On the other hand, results for the simpler effective interaction $\mathcal{K}_{\Sigma N}(k_F)$, obtained with the Brueckner theory, are available in the literature. To be able to use the results obtained for $\mathcal{K}_{\Sigma N}(k_F)$ in our theory of the Lane potential V_1, we introduce the single density (or single Fermi momentum) approximation, applied originally in the problem of spin symmetry energy of liquid ^{3}He [4], and later in the problem of the nuclear isospin [5] and spin [6] symmetry energy. Namely, we introduce the following simlifying assumptions:

$$\mathcal{K}_{\Sigma^+p}(\kappa\lambda) \approx \mathcal{K}_{\Sigma^+p}(\kappa) \,, \tag{9}$$

$$\mathcal{K}_{\Sigma^+n}(\kappa\lambda) \approx \mathcal{K}_{\Sigma^+n}(\lambda) \,. \tag{10}$$

Approximation (9) says that the effect of the proton excess on the Σp effective interaction in nuclear matter is determined primarily by the shift in Fermi momentum of protons. This assumption seems to be physically plausible and it corresponds exactly to the way in which the action of the exclusion principle is altered by the proton excess. The motivation of approximation (10), which applies to the $\Sigma^+ n$ interaction, is analogous.

By applying approximations (9) and (10) in Eq. (8), we get the following approximate expression for $V_1^{(I)}$:

$$\begin{aligned} V_1^{(I)}(k_\Sigma) \approx \tfrac{2}{9} \textstyle\sum_{\mathbf{k}_N}^{k_N<k_F} (\mathbf{k}_N\mathbf{k}_\Sigma | k_F \tfrac{d}{dk_F} \Big\{ \mathcal{K}(\tfrac{3}{2}; k_F) \\ - \mathcal{K}(\tfrac{1}{2}; k_F) \Big\} | \mathbf{k}_N\mathbf{k}_\Sigma) \,, \end{aligned} \tag{11}$$

where $\mathcal{K}(T, k_F)$ denotes the effective ΣN interaction in the state of total ΣN isospin $T = \frac{1}{2}, \frac{3}{2}$ (in nuclear matter with $Z = N$).

3 Results for the YNG interaction and discussion

In calculating $V_1^{(0)}$, Eq. (7), and $V_1^{(I)}$, Eq. (11), I have used the YNG effective ΣN interaction of Yamamoto *et al.* [3]. It is a configuration space representation of the G-matrix calculated in the LOB approximation from the model D [7], model F [8], and the soft-core (SC) model [9] of the Nijmegen baryon-baryon interaction. I use also the YNG interaction obtained from the new soft-core (NSC) model of Rijken, Stoks, and Yamamoto [10].

The results for $V_1^{(0)}$, $V_1^{(I)}$, and $V_1 = V_1^{(0)} + V_1^{(I)}$, calculated at $k_F = 1.35$ fm^{-1} for $k_\Sigma = 0$ are shown in Table 1 which contains also V_0 calculated wih the YNG interaction.

The YNG interaction is constructed so that it simulates the G-matrix in the ground states of the system of a Σ hyperon in nuclear matter, and it is best suited for the case of $k_\Sigma = 0$, presented in Table 1. Calculation for $k_\Sigma > 0$ show that the dependence of $V_1(k_\Sigma)$ on k_Σ is weak (e.g., $V_1(k_\Sigma = 1$ fm^{-1}) differs from $V_1(k_\Sigma = 0)$ by less than 2 %.

Table 1. Different components (in MeV) of $V_\Sigma(k_\Sigma = 0)$ calculated at $k_F = 1.35$ fm^{-1} with the YNG interaction obtained from the indicated models of the ΣN interaction.

Model	V_0	$V_1^{(0)}$	$V_1^{(I)}$	V_1
D	-13.1	51.0	4.1	55.1
F	23.5	67.4	13.1	80.4
SC	-9.6	13.3	17.7	31.0
NSC	-16.6	-26.3	-10.4	-36.7

The contribution $V_1^{(I)}$ increases the magnitude of V_1. Its importance depends on the interaction model. In the case of the SC model $V_1^{(I)}$ is even bigger than $V_1^{(0)}$. In the important case of model F, $V_1^{(I)}$ is also relatively important – it increases V_1 by about 20 %.

The most striking feature which distinguishes model F is that it leads to an repulsive V_0 whereas all the remaining ΣN interaction models lead to an attractive V_0. Furthermore, model F leads to V_1 of a much larger magnitude, than the remaining models.

Let us compare our results with the recent $(K^-, \pi^\pm)$ experiments at BNL on the ^{9}Be target [11,12,13]. The accuracy of these experiments appears to be an order of magnitude batter than in the early CERN experiments. Let us start with the simpler case of the (K^-, π^+) reaction in which only one direct elementary strangeness exchange process $K^-p \to \pi^+\Sigma^-$ occurs. The observed π^+ spectrum is compared in Fig. 1 with the spectrum calculated in [14] in the plane wave impulse approximation with the final state Σ^- s.p. square well potential with the depth $V_{\Sigma 0}$ (positive for an attractive potential). Four values of $V_{\Sigma 0}$ were used: -20, -10, 10, and 20 MeV (curves A, B, C, and D respectively). We see that curve A (obtained with a repulsve potential) shows the best agreement with the experimental data in contradistinction to the C and D curves (obtained with attractive potentials) which fail completely in reproducing the data at higher $-B_\Sigma$. Our clear conclusion is that the interaction of the produced Σ^- hyperon with the nuclear core is repulsive. This obviously favors model F as a realistic representation of the ΣN interaction, and excludes the remaining models.

The π^- spectrum observed in the BNL ^{9}Be(K^-, π^-) experiments indicates that the final state interaction of the Σ hyperon is less repulsive than in the (K^-, π^+) reaction or possibly even attractive [12]. Here, two elementary processes may occur: (A) $K^-p \to \pi^-\Sigma^+$, and (B) $K^-n \to \pi^-\Sigma^0$. The difference between the final state interaction in case (A) and that in the (K^-, π^+) reaction is [see Eq. (1)]: $\Delta V^{(A)} \equiv V_{\Sigma^+} - V_{\Sigma^-} = \alpha V_1 = -\frac{1}{4}V_1$ ($Z = 3, N = 5$ in the nuclear core in the final state). For the same difference in case (B), we have $\Delta V^{(B)} \equiv V_{\Sigma^0} - V_{\Sigma^-} =$

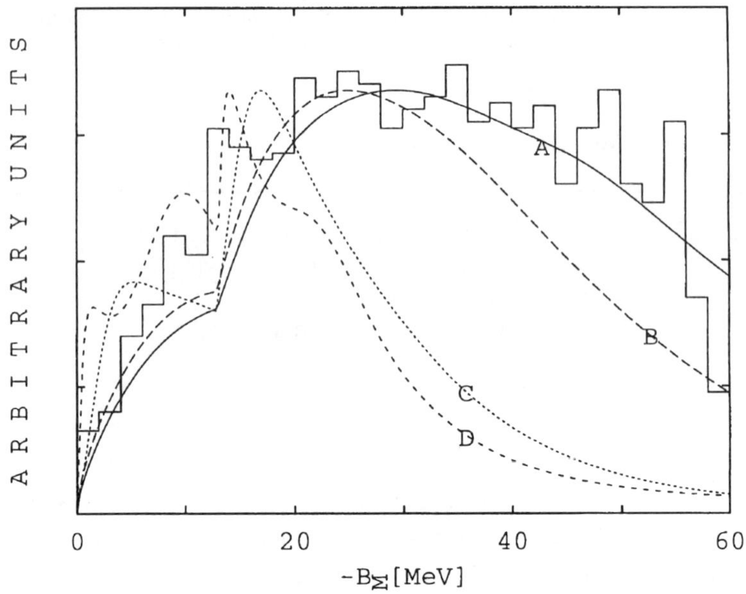

Figure 1. Pion spectrum from (K^-, π^+) reaction on ^{9}Be at $p_K = 600$ MeV/c. B_Σ is the Σ^- binding energy.

$\frac{1}{2}\alpha V_1 = \frac{1}{2}\Delta V^{(A)}$. [a] If we use for V_1 in the nuclear core our nuclear matter results, Table 1, we get $\Delta V^{(A)} \simeq -20$ MeV and $\Delta V^{(B)} \simeq -10$ MeV for model F. These are sizable decreases in the repulsion, required by the comparison of the π^+ and π^- BNL spectra. The effect is smaller for model D and SC, and model NSC leads to an opposite effect: $\Delta V^{(A)} \simeq +9$ MeV, $\Delta V^{(B)} \simeq +4.5$ MeV, i.e., it represents an increase in the repulsion (or a decrease in the attraction). Thus model NSC appears to be incompatible with the BNL data.

We conclude that model F of the Nijmegen baryon-baryon interaction leads to the s.p. potential $\hat{V}_\Sigma$ of a Σ hyperon in nuclear matter which has an repulsive isoscalar part V_0 and a sizable positive Lane potential $V_1 \sim 80$ MeV, and is the only model of the Nijmegen interaction, which is compatible with the recent BNL $(K^-, \pi^\pm)$ data.

Acknowledgments

The author expresses his gratitude to Dr Y. Yamamoto for letting him know Ref. 10 prior to publication, and for supplying him with the parameters of the YNG

[a]Whereas the pion spectra in case (A) and in the (K^-, π^+) reaction depend on the s.p. states of the target protons, the pion sectrum in case (B) depends on the s.p. states of the target neutrons. The effect of this difference requires a more detailed investigation.

interaction for the NSC model. This research was partly supported by Komitet Badań Naukowych under Grant No. PB 2-0956-91-01.

References

1. A. M. Lane, *Nucl. Phys.* A **128**, 256 (1969).
2. C. B. Dover and A. Gal, *Progr. in Particle and Nuclear Phys.* **12**, 171 (1984).
3. Y. Yamamoto, T. Motoba, H. Himeno, K. Ikeda, and S. Nagata, *Progr. Theor. Phys. Suppl.* **117**, 361 (1994).
4. K. A. Brueckner and J. L. Gammel, *Phys. Rev.* **109**, 1040 (1958); **121**, 1863(E) (1961).
5. K. A. Brueckner and J. Dąbrowski, *Phys. Rev.* **134**, B722 (1964).
6. J. Dąbrowski and P. Haensel, *Phys. Rev.* C **7**, 916 (1973); *Canad. J. Phys.* **52**, 1768 (1974); J. Dąbrowski, *Acta Phys. Pol.* B **7**, 657 (1976).
7. N. M. Nagels, T. A. Rijken, and , J. J. de Swart, *Phys. Rev.* D **12**, 744 (1975); **15**, 2547 (1977).
8. N. M. Nagels, T. A. Rijken, and , J. J. de Swart, *Phys. Rev.* D **20**, 1663 (1979).
9. P. M. M. Maessen, T. A. Rijken, and , J. J. de Swart, *Phys. Rev.* C **40**, 226 (1989); *Nucl. Phys.* A **547**, 245c (1992).
10. T. A. Rijken, V. G. J. Stoks, and Y. Yamamoto, *Phys. Rev.* C **59**, 21 (1999).
11. R. Sawafta, *Nucl. Phys.* A **585**, 103c (1995).
12. Y. Shimizu, Doctoral Thesis, University of Tokyo 1996 (unpublished).
13. R. Sawafta, Nucl. Phys. **A639**, 103c (1998).
14. J. Dąbrowski and J. Rożynek, Acta Phys. Pol. **29B** 2147 (1998).

Σ EXCITATION IN NUCLEI AND ITS DECAY

T. HARADA

Department of Social Information, Sapporo Gakuin University, 11-Bunkyodai, Ebetsu, Hokkaido 069-8555, Japan

E-mail: harada@earth.sgu.ac.jp

We examine phenomenologically the ^{4}He(K^-,π^-) spectra in flight and at rest within the distorted-wave impulse approximation by using a coupled ($3N$-Λ)+($3N$-Σ) model. The calculated spectra can explain consistently the available data overall from Λ and Σ regions. The result shows that there exists a pole of the $^4_\Sigma$He unstable bound state on the second Riemann sheet of the complex energy plane, and an asymmetric peak for $^4_\Sigma$He appears because of the Σ branch cut associated with opening the Σ channel. The decay properties of $^4_\Sigma$He are also discussed.

1 Introduction

It has been discussed for about a decade whether there exists a bound state $^4_\Sigma$He or not. Recently, Nagae *et al.*[1] confirmed the existence of the $^4_\Sigma$He bound state in the in-flight ^{4}He(K^-,π^-) reaction at 600 MeV/c (4°) in the E905 experiments at BNL. This state might be interpreted as a strange partner of the α-particle, together with the $^4_\Lambda$He ground state. Therefore, it is of great importance to understand the structure of the $^4_\Sigma$He bound state, because the flavor $SU(3)$ nature seems to survive in Σ-hypernuclear side.[2] Moreover, the ΛN-ΣN coupling is expected to be one of key elements for solving the anomalous binding energies of $^4_\Lambda$He, $^4_\Lambda$H and $^5_\Lambda$He.[3] We believe now that $^4_\Lambda$He and $^4_\Sigma$He are the promising objects to be studied in order to see the importance of the Λ-Σ coupling quantitatively.[4]

In this note, we examine phenomenologically hypernuclear production distributions from K^- capture in flight and at rest on a helium target, with the distorted wave impulse approximation (DWIA) by using a coupled ($3N$-Λ)+($3N$-Σ) model with a spreading potential. Our purpose here is to give an unified understanding for the structure of the $^4_\Sigma$He bound state, in comparison with the available experimental data at BNL,[1,5] ANL[6] and KEK.[7,8]

2 Calculations

Now we consider the K^- capture reactions in a helium target.[9] The inclusive pion spectra for (K^-,$\pi^\mp$) reactions in DWIA are estimated by using the Green's function method.[10] We employ the Fermi-averaged amplitudes $\overline{f}_{(Y\pi)}$ for $K^-N\to\pi Y$ in nuclear medium, and meson distorted-waves for the outgoing $\pi^\mp$ and the initial K^-. The hypernuclear final states are classified as

$$K^- + {}^4\text{He} \to (a)\ \pi^- + {}^3\text{He} + \Lambda,\ (b)\ \pi^- + {}^3\text{H} + \Sigma^+,\ (c)\ \pi^- + {}^3\text{He} + \Sigma^0,$$
$$(d)\ \pi^- + X^{++} + \Lambda, \qquad (1)$$

where X^{++}=d+p or p+p+n for the π^- spectrum. Our calculation has assumed that the $3N$ core-nucleus is rigid, where $3N$ denotes the (^{3}He, ^{3}H)-doublet. Thus

Green's function $\hat{G}$ for the $3N$-Y systems in Eq.(1) is obtained by solving a coupled equation with ^{3}He+Λ, ^{3}H+Σ^+ and ^{3}He+Σ^0 channels, and has natural-parities, 0^+, 1^-, 2^+,$\cdots$, because the spin-flip processes are known to be negligible in $(K^-,\pi^\mp)$ reactions. The $3N$-Y potential matrix with J^π=0^+ on the isospin bases is given by

$$\hat{U} = \begin{pmatrix} U_{\Lambda,1/2} & U_{X,1/2} & 0 \\ U_{X,1/2} & U_{\Sigma,1/2} & 0 \\ 0 & 0 & U_{\Sigma,3/2} \end{pmatrix}, \tag{2}$$

where the Λ-Σ coupling potential $U_{X,1/2}$ is assumed to be real. The diagonal potentials for Y=Σ, Λ and T=1/2, 3/2 are given as

$$U_{Y,T}(R) = V_{Y,T}(R) + iW_{Y,T}(R), \tag{3}$$

where the imaginary part $W_{Y,T}$ is a *spreading* potential, which describes effectively all the $3N$ breakup processes. According to the Green's function method,[10] we can define the strength function as

$$S_\pi = -\frac{1}{\pi}\mathrm{Im}\langle F|\hat{G}|F\rangle, \tag{4}$$

where $|F\rangle$ denotes the production amplitudes excited initially by the $(K^-,\pi^\mp)$ reaction on ^{4}He. The inclusive pion spectra are decomposed into partial (exclusive) spectra corresponding to different physical processes,[10] as classified in Eq. (1). We obtain the following decomposition of the strength function S_π:

$$S_\pi = S^\uparrow_{\Lambda\pi^-} + S^\uparrow_{\Sigma^+\pi^-} + S^\uparrow_{\Sigma^0\pi^-} + S^\downarrow_{\Lambda\pi^-} + S^\downarrow_{\Sigma\pi^-}, \tag{5}$$

where $S^\uparrow_{\Lambda\pi^-}$, $S^\uparrow_{\Sigma^+\pi^-}$ and $S^\uparrow_{\Sigma^0\pi^-}$ describe the strength functions of the quasi-free (escaping) processes with outgoing ^{3}He+Λ, ^{3}H+Σ^+ and ^{3}He+Σ^0 particles, respectively. $S^\downarrow_{\Sigma\pi^-}$ denotes the strength function for spreading through the $\Sigma N \to \Lambda N$ conversion into complicated X^{++}+Λ states with the $3N$ breakup, and $S^\downarrow_{\Lambda\pi^-}$ denotes for spreading through the $\Lambda N \to \Lambda N$ scattering into these X^{++}+Λ states. The decomposition near the Σ^+^{3}H threshold would help us to understand the structure of $^4_\Sigma$He and its decay property. Indeed, the peak below the Σ^+^{3}H threshold is connected with the secondary $3N$ breakup process

$$[^4_\Sigma\mathrm{He}] \to X^{++} + \Lambda, \tag{6}$$

because the Σ interacts subsequently with a second nucleon in $^4_\Sigma$He, and the converted ΛN pair gets large energy from the mass difference $m_\Sigma - m_\Lambda \simeq 80$ MeV.

We have determined parameters of the potential strengths $U_{\Lambda,1/2}$, $U_{X,1/2}$, $U_{\Sigma,1/2}$ and $U_{\Sigma,3/2}$ by making a fit to the experimental data. Here we made $U_{X,1/2}$ satisfy a plausible Σ admixture of about 2% in the $^4_\Lambda$He ground state,[4] and its Λ-binding energy of B_Λ= 2.39 MeV. In addition, since the amplitudes $\overline{f}_{(Y\pi)}$ remain in ambiguities, the relative phase of φ_Λ (φ_Σ) for $\Lambda\pi$ ($\Sigma\pi$ I=1) to $\Sigma\pi$ I=0 channels should be determined from the present fit. As a result, we have obtained the $3N$-Y potential for the best fit, and the suitable relative phases for the spectra.

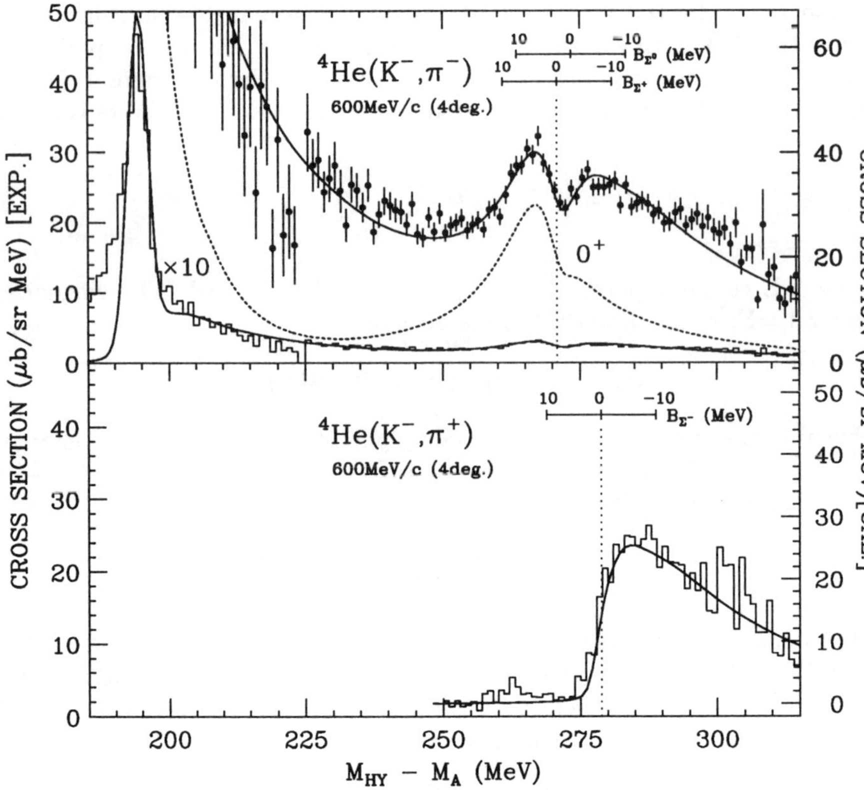

Figure 1: The calculated spectra of the ^{4}He(K^-,$\pi^\mp$) reactions at 600 MeV/c (4°), together with the BNL-E905 data. A peak at 194 MeV corresponds to the $^4_\Lambda$He ground state. The dashed curve denotes the J^π=0^+ component in the π^- spectrum.

3 Results and Discussion

3.1 Inclusive Pion Spectra in the ^{4}He(in-flight K^-,$\pi^\mp$) Reactions

Since the ^{4}He(K^-,π^-) reaction can populate both T=3/2 and T=1/2 states, while the ^{4}He(K^-,π^+) reaction can produce only a T=3/2 state. Figure 1 shows the ^{4}He(K^-,$\pi^\mp$) spectra at 600 MeV/c (4°), in order to be compared with the E905 data at BNL. The resultant spectra are consistent with the $\pi^\mp$ data overall. There exists no bound state in the π^+ spectrum. In the π^- spectrum the peak below the Σ threshold can be reproduced very well. This peak corresponds to a pole of an *unstable* bound state $^4_\Sigma$He with J^π=0^+ in the Σ channel on the *second* Riemann sheet $(-++)^\dagger$ of the complex energy plane. This complex *eigenvalue* represents $E^{(pole)}_{\Sigma^+} = -1.1 - i6.2$ MeV, where the real part of the energy is measured from the $\Sigma^{+\,3}$H threshold. Since the pole resides near the $\Sigma^{+\,3}$H threshold, a position of the

†The sheet is identified by three signs of (Imk_Λ, Imk_{Σ^+}, Imk_{Σ^0}), where k_Y denotes the c.m. momentum of the 3N-Y system in the complex momentum plane.

observed peak is shifted and its shape is asymmetric because of the Σ branch cut associated with opening the Σ channel. (See also Fig. 3). Therefore, we show the peak for ${}^4_\Sigma$He lying at 3.7 MeV below the Σ^{+3}H threshold with the width of about 10 MeV. These values are in agreement with $B_{\Sigma^+}= 4.4 \pm 0.3 \pm 1$ MeV and $\Gamma= 7\pm 0.7 {}^{+1.2}_{-0.0}$ MeV, as reported by Nagae *et al.*[1]

The difference between π^- and π^+ spectra originates from a strong isospin dependence of the Lane term in the $3N$-Y potential. The Lane term plays an essential role in making the $3N$-Y system bound in the Σ channel, and to recover the isospin symmetry to $T\simeq 1/2$ (99%) in ${}^4_\Sigma$He, whereas there is no nuclear bound state in $T=3/2$ for the ^{3}H+Σ^- channel because $U_{\Sigma,3/2}$ is weakly repulsive.[11]

Since the strong $\Sigma N \to \Lambda N$ conversion induces the ${}^4_\Sigma$He decay of Eq.(6), we would expect that the peak for ${}^4_\Sigma$He is clearly seen when events are tagged with an energetic Λ in final states. Hayano *et al.*[5] performed the E774 experiments with multiplicity ≥ 2 condition in the scintillation barrel counter at BNL. The preliminary analysis showed that the spectrum reveals a nearly peak at $B_{\Sigma^+}= 4\pm 1$ MeV with a width of about 10 ± 2 MeV. We find that the shape of the calculated spectrum of $\pi^-+X^{++}+\Lambda$ is in good agreement with the experimental one. The spreading potential $W_{Y,T}$ describing the ${}^4_\Sigma$He decay processes would be acceptable.

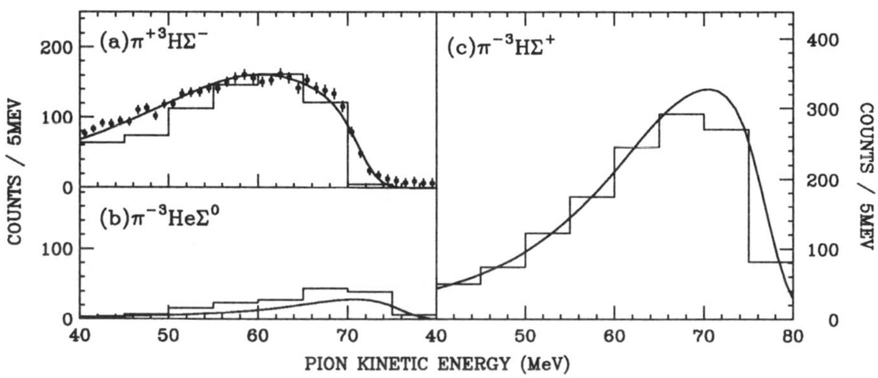

Figure 2: The HeBC T_π distributions of $\pi^{\pm 3}$H$\Sigma^\mp$ and π^{-3}HeΣ^0 final states for the K^--^{4}He absorption at rest. The square dots are the KEK-E167 data for total π^+. The solid curves denote the calculated rates.

3.2 Inclusive and Exclusive Pion Spectra in the ^{4}He(stopped K^-,$\pi^\mp$) Reactions

We calculate the capture rates from the K^{-4}He atom, in order to be compared with the available experimental data.[6,8] By fitting to the these data, we have searched parameters of the relative phases (φ_Λ,φ_Σ) with the $3N$-Y potential we determined. Since atomic orbits where the K^- is absorbed on the K^{-4}He atom are still unknown experimentally, we have assumed the K^- absorptions from the 27% higher ns-orbits and 70% higher np-orbits, as suggested by Koike and Akaishi.[12]

Figure 2 shows the T_π distributions of $\pi^\mp$+^{3}H+$\Sigma^\pm$ and π^-+^{3}He+ Σ^0 final states, together with the data by helium bubble chambers (HeBC) at ANL.[6] The

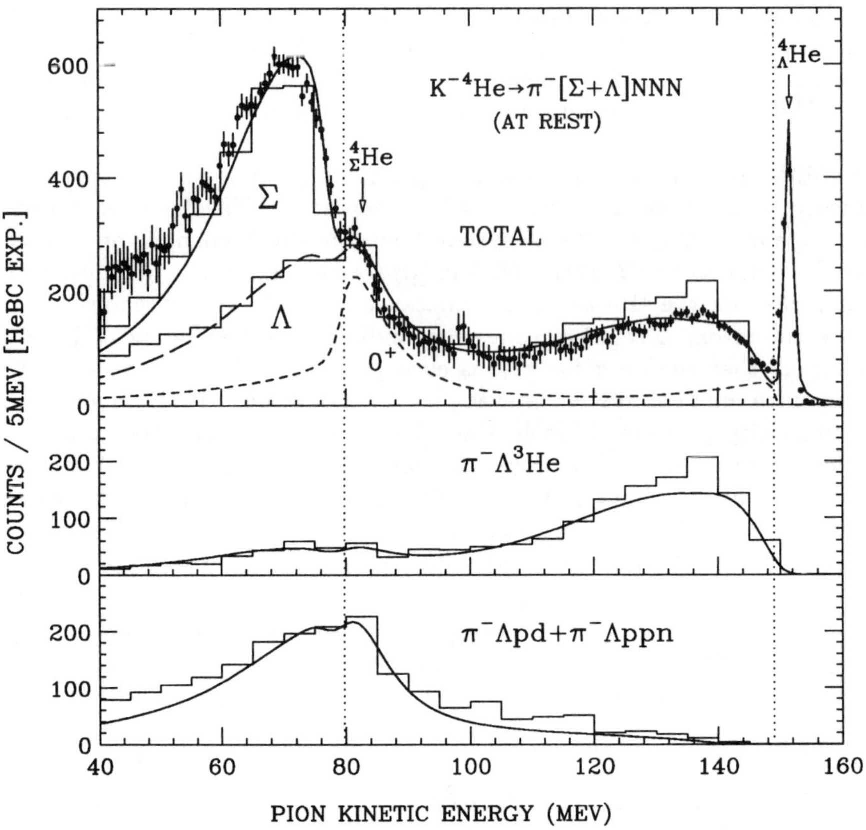

Figure 3: The HeBC T_π distributions of total π^- (top), $\pi^-+{}^3\text{He}+\Lambda$ (middle) and $\pi^-+X^{++}+\Lambda$ (bottom) spectra are compared with the calculated ones (solid lines). The square dots are the KEK-E167 data for total π^-. The long-dashed and dashed curves give full and $J^\pi=0^+$ components, respectively, in the spectrum for total Λ with π^-.

relative phase of φ_Σ can be easily determined from fitting both magnitudes and shapes of the data, so that these spectra are in excellent agreement with the data. In Fig. 3 we shows the T_π distributions for all the π^- events, $\pi^-+{}^3\text{He}+\Lambda$ and $\pi^-+X^{++}+\Lambda$ spectra, in order to compare the calculated curves with the HeBC data. The relative phase of φ_Λ is determined to fit the "dip" region in the π^- spectrum, where the interference between Λ and Σ amplitudes would be enhanced. It is remarkable that our $3N$-Y calculations can reproduce the HeBC data overall from Λ and Σ regions; the ${}^4_\Sigma\text{He}$ decay into $X^{++}+\Lambda$ channels is dominant.

Outa *et al.*[8] showed the inclusive pion data in ${}^4\text{He}$(stopped K^-,$\pi^\mp$) reactions at KEK-E167, reanalyzing the subtraction for the background from hyperonic weak decays. We superimpose the KEK data on the HeBC data in Figs. 2 and 3. As seen in Fig. 3, the peak for ${}^4_\Lambda\text{He}$ is in good agreement with the data. In the Σ region, the spectrum can reproduce a bump behavior near the $\Sigma^+{}^3\text{H}$ threshold very well. The spreading potential plays significant roles in describing the shape and width of the peak for ${}^{4}_{\Sigma}$He even if the pole resides on the sheet $(-++)$. In fact, there appears the asymmetric peak below the Σ^{+3}H threshold in the $J^{\pi}=0^{+}$ component in the spectrum (1-a)+(1-d) for total Λ with π^{-}, as seen in Fig. 3.

On the other hand, Dalitz and Deloff [6] showed that a DWIA calculation can fit qualitatively the $\pi^{\mp}$ distributions of the HeBC data. However, their analysis suggested that the enhancement just below the Σ^{+3}H threshold corresponds to a threshold-cusp state. This result is inconsistent with ours. The reason might be because a spreading potential is not taken into account in their model.

4 Summary

We have examined phenomenologically the ^{4}He(K^{-},π^{-}) spectra in flight and at rest, by using the coupled-channel framework between the ^{3}He+Λ, ^{3}H+Σ^{+} and ^{3}He+Σ^{0} channels with a spreading potential. The analysis including continuum spectra gives us valuable information about the 3N-Y potential. The calculated spectra can explain consistently all the available data at BNL, ANL and KEK. The result shows that a pole of the ${}^{4}_{\Sigma}$He unstable bound state with $J^{\pi}=0^{+}$, $T\simeq 1/2$ (99%) is located on the second Riemann sheet $(-++)$, where the ${}^{4}_{\Sigma}$He decay with complicated 3N breakup occurs.

Acknowledgments

The author thanks Professor Y. Akaishi, Professor K. Yazaki, Professor T. Nagae, Mr. H. Outa and Dr. Y. Shimizu for many valuable discussion and comments. This work was supported by the Grants-in-Aid for Scientific Research of the Japanese Ministry of Education, No. 08239104.

References

1. T. Nagae, *et al.*, *Phys. Rev. Lett.* **80**, 1605 (1998).
2. T. Harada, *Nucl. Phys.* A **547**, 165c (1992).
3. B.F. Gibson, *et al.*, *Prog. Theor. Phys. Suppl.* **117**, 339 (1994), and references therein.
4. Y. Akaishi, these proceedings.
5. R.S. Hayano, *Nucl. Phys.* A **547**, 151c (1992).
6. R.H. Dalitz and A. Deloff, *Nucl. Phys.* A **547**, 181c (1992);A **585**, 303c (1995), and references therein.
7. R.S. Hayano, *et al.*, *Phys. Lett.* B **231**, 355 (1989).
8. H. Outa, *et al.*, *Prog. Theor. Phys. Suppl.* **117**, 171 (1994).
9. T. Harada, *Phys. Rev. Lett.* **81**, 5287 (1998).
10. O. Morimatsu and K. Yazaki, *Nucl. Phys.* A **483**, 1493 (1988).
11. T. Harada. *et al.*, *Nucl. Phys.* A **507**, 715 (1990).
12. T. Koike and Y. Akaishi, *Nucl. Phys.* A **639**, 521c (1998).

THE $\Lambda\Lambda - \Xi N$ COUPLING AND THE BINDING ENERGY OF $^{6}_{\Lambda\Lambda}$He

I. R. AFNAN

Department of Physics, The Flinders University of South Australia
G.P.O. Box 2100, Adelaide 5001, Australia

To understand the difference between the value of the nn and $\Lambda\Lambda$ two-body matrix elements, we examine the role of the coupling between the $\Lambda\Lambda$ and ΞN channels, first within the framework of $SU(3)$, and second in the numerical results for the binding energy of $^{6}_{\Lambda\Lambda}$He. We find that it is essential to include the coupled channel nature of the BB interaction as we proceed from strangeness zero to the strangeness -1 and -2 channels if we are to understand the relative magnitude of the two-body matrix elements in the different strangeness channels.

1 Introduction

The interest in the strangeness $S = -2$ baryon-baryon (BB) interaction was initially motivated by Jaffe's[1] quark model prediction of the existence of a flavor singlet $[uuddss]$ state, known as the H particle. More recently, the interest in the role of strangeness in neutron stars, and the possible production of nuclei with strangeness $S \leq -2$ at RHIC, has given further impetus to extend the BB interaction to the $S \leq -2$ channel.

At present the only experimental data on the $\Lambda\Lambda$ interaction are the three observed $\Lambda\Lambda$ hypernuclei $^{6}_{\Lambda\Lambda}$He,[2] $^{10}_{\Lambda\Lambda}$Be,[3,4] and $^{13}_{\Lambda\Lambda}$B.[5,6] These invariably give effective S-wave matrix elements of

$$\begin{aligned} \Delta B_{\Lambda\Lambda}(^{6}_{\Lambda\Lambda}\mathrm{He}) &= 10.9 \pm 0.5 - 6.14 \pm 0.05 \approx 4.7 \text{ MeV} \\ \Delta B_{\Lambda\Lambda}(^{10}_{\Lambda\Lambda}\mathrm{Be}) &= 17.7 \pm 0.4 - 13.42 \pm 0.08 \approx 4.3 \text{ MeV} \\ \Delta B_{\Lambda\Lambda}(^{13}_{\Lambda\Lambda}\mathrm{B}) &= 27.5 \pm 0.7 - 11.69 \pm 0.12 \approx 4.8 \text{ MeV} \ , \end{aligned} \tag{1}$$

which suggests that the matrix element of the $\Lambda\Lambda$ interaction in light nuclei is $-\langle V_{\Lambda\Lambda}\rangle \approx 4-5$ MeV. This value is to be compared with the corresponding nn matrix element $-\langle V_{nn}\rangle \approx 6-7$ MeV, and the ΛN matrix element $-\langle V_{\Lambda N}\rangle \approx 2-3$ MeV. As a result we have that $-\langle V_{nn}\rangle > -\langle V_{\Lambda\Lambda}\rangle > -\langle V_{\Lambda N}\rangle$. The question we would like to address in the present report is: Can we understand this relative magnitude of the $S = 0, -1, -2$ matrix elements in light nuclei in terms of flavor $SU(3)$ at the meson-baryon level as given in a one-boson-exchange (OBE) potential? Or do we need to include explicit quark-gluon degrees of freedom? We will demonstrate that the relative magnitude of the matrix elements in the $S = 0, -1, -2$ can be understood only if we include the coupling between the ΛN and ΣN in the $S = -1$ channels, and the $\Lambda\Lambda$ and ΞN in the $S = -2$ channel.

2 The Coupled Channel Problem in Flavor SU(3)

Let us assume that flavor $SU(3)$ is a perfect symmetry of strong interaction, then the baryon octet ($B = N, \Lambda, \Sigma, \Xi$) forms a degenerate multiplet, and the BB system will be in one of the six irreducible representations of $\{8\} \otimes \{8\}$, i.e.

$$\{8\} \otimes \{8\} = [\ \{1\} \oplus \{8_s\} \oplus \{27\}\ ]\ \oplus\ [\ \{8_a\} \oplus \{10\} \oplus \{\overline{10}\}\ ]\ , \tag{2}$$

where $\{8_a\}$, $\{10\}$ and $\{\overline{10}\}$ are the flavor anti-symmetric channels, while the $\{1\}$, $\{8_s\}$ and $\{27\}$ are the flavor symmetric channels. The flavor wave function in the $^1S_0$ partial wave for the nn, ΛN and $\Lambda\Lambda$ channels can now be written with the help of the $SU(3)$ Clebsch-Gordan coefficients,[7] as

$$\begin{aligned} |nn\rangle &= |\{27\}\rangle \\ |\Lambda N\rangle &= \sqrt{\frac{36}{40}}\,|\{27\}\rangle - \sqrt{\frac{4}{40}}\,|\{8_s\}\rangle \\ |\Lambda\Lambda\rangle &= \sqrt{\frac{27}{40}}\,|\{27\}\rangle - \sqrt{\frac{8}{40}}\,|\{8_s\}\rangle - \sqrt{\frac{5}{40}}\,|\{1\}\rangle\ . \end{aligned} \tag{3}$$

Here the $\{27\}$ representation is the only component of the nn flavor wave function. Since the nn interaction in the $\{27\}$ representation is strongly attractive to the extent that it almost supports a bound state, and both the ΛN and $\Lambda\Lambda$ wave flavor wave functions have their largest component in the $\{27\}$ representation, we expect the ΛN and $\Lambda\Lambda$ interactions in the $^1S_0$ to also be dominated by the $\{27\}$ representation. As a result the interactions in the $S = 0, -1$ and -2 are expected to satisfy the relation

$$-\langle V_{nn}\rangle > -\langle V_{\Lambda N}\rangle > -\langle V_{\Lambda\Lambda}\rangle\ . \tag{4}$$

This is in contradiction with the experimental observation from the three $\Lambda\Lambda$ hypernuclei which requires that $-\langle V_{\Lambda\Lambda}\rangle > -\langle V_{\Lambda N}\rangle$. The inclusion of the $\{8_s\}$ and $\{1\}$ component of the flavor wave function into our analysis may not improve the situation since there are indications from OBE potentials that the interaction in these two channels are repulsive[8] which would reduce the $\Lambda\Lambda$ matrix element more than the ΛN matrix element.

Since flavor $SU(3)$ is broken in nature to the extent that the masses of the N, Λ, Σ and Ξ are not equal, the BB problem in the $S = -1, -2$ reduces to a coupled channel problem with two or more thresholds in each strangeness channel. Thus for $S = -1$ we have the ΛN and ΣN thresholds which are separated by some 80 MeV; for the $S = -2$ we have the $\Lambda\Lambda$, ΞN and $\Sigma\Sigma$ thresholds with the separation between the $\Lambda\Lambda$ and ΞN channels being ≈ 25 MeV. This suggests that the coupling in the $S = -2$ may be more important than is the case in the $S = -1$ channel. To examine the effect of the baryon mass splitting on ΛN and $\Lambda\Lambda$ matrix elements, we introduce effective potentials that include the coupled channel effects. These potentials, to lowest order in the coupling, are given by

$$\begin{aligned} \langle V_{\Lambda N}^{\rm eff}\rangle &\approx \langle V_{\Lambda N}\rangle - \frac{|\langle \Lambda N|V|\Sigma N\rangle|^2}{\Delta E_{YN}}\ ; \qquad \Delta E_{YN} \approx 80\,\mathrm{MeV} \\ \langle V_{\Lambda\Lambda}^{\rm eff}\rangle &\approx \langle V_{\Lambda\Lambda}\rangle - \frac{|\langle \Lambda\Lambda|V|N\Xi\rangle|^2}{\Delta E_{\Lambda\Lambda}}\ ; \qquad \Delta E_{\Lambda\Lambda} \approx 25\,\mathrm{MeV}\ . \end{aligned} \tag{5}$$

The fact that the thresholds in the $S = -2$ channel are closer than in the $S = -1$ channel suggests that we get more attraction in the $S = -2$ effective interaction due to the coupled channel nature of the problem, than in the $S = -1$ channel. If in addition we assume that the matrix elements that determine the strength of the coupling can be extracted from $SU(3)$, i.e. flavor $SU(3)$ is approximately a good symmetry, then

$$\begin{aligned}\langle\Lambda\Lambda|V|\Xi N\rangle &= -\frac{18}{40}V_{\{27\}} + \frac{8}{40}V_{\{8_s\}} + \frac{10}{40}V_{\{1\}} \\ \langle\Lambda N|V|\Sigma N\rangle &= -\frac{12}{40}V_{\{27\}} + \frac{3}{40}V_{\{8_s\}} \; .\end{aligned} \tag{6}$$

If the interaction in the $\{27\}$ representation is dominant, then the coupling between the $\Lambda\Lambda$ and the ΞN channels is stronger than is the case between the ΛN and the ΣN channels. This again supports the argument that the coupling resulting from the mass splitting gives more attraction in the $\Lambda\Lambda$ channel than in the ΛN channel. Thus even though the ${}^1\mathrm{S}_0$ interaction in the $\Lambda\Lambda$ channel may be weak (small scattering length), the additional attraction resulting from the coupling due to the baryon mass splitting could result in a scattering length that is comparable to the nn scattering length as first suggested by Dover.[9]

Let us now turn to the $\Lambda\Lambda$ interaction having established that the coupling between the $\Lambda\Lambda$ and ΞN channels is more important than is the case in the $S = -1$ channel. Since the experimental data put a constraint on the scattering amplitude that included the additional attraction due to the coupled channel nature of the problem, the $\Lambda\Lambda$ potential in free space, in the absence of coupling, is considerably stronger than is the case when the coupling is included. However in the medium, the coupling is suppressed as a result of the fact that $\Lambda + \Lambda \rightarrow \Xi + N$ is Pauli blocked. From this we may conclude that while in free space

$$\langle V^{\mathrm{eff}}_{\Lambda\Lambda}\rangle_{\text{with coupling}} = \langle V^{\mathrm{eff}}_{\Lambda\Lambda}\rangle_{\text{no coupling}} \, ,$$

in the medium

$$\langle V^{\mathrm{eff}}_{\Lambda\Lambda}\rangle_{\text{with coupling}} < \langle V^{\mathrm{eff}}_{\Lambda\Lambda}\rangle_{\text{no coupling}} \, .$$

In other words, if the $\Lambda\Lambda$ amplitude in free space is attractive enough to almost support a bound state, in the medium the Pauli principle suppresses this amplitude so that the effective matrix element of the $\Lambda\Lambda$ interaction is smaller than is the case for the nn interaction.

The above qualitative analysis is based on the assumption that flavor $SU(3)$ gives a good estimate of the relative magnitude of matrix elements and the breaking of $SU(3)$ is only in the masses of the baryons. To see if these arguments are valid, in the next section we will report on the construction of a potential in the $S = -2$ channel based on the $SU(3)$ rotation of the Nijmegen model D potential,[10] and then proceed in Sec. 4 to use the resulting potential in the calculation of the binding energy of ${}^{6}_{\Lambda\Lambda}\mathrm{He}$. Finally, in Sec. 5 we will present some concluding remarks.

3 The S = −2 BB Potentials

To perform an $SU(3)$ rotation on an OBE potential defined in $S = 0, -1$ channels, we need to write the Lagrangian in terms of the baryon octet with the mesons as either a singlet or a member of an octet. If the interaction is taken to be of the Yukawa type, the Lagrangian takes the form

$$\mathcal{L}_{int} = -\left\{g_{\{1\}}[B^\dagger B]_{\{1\}}M_{\{1\}} + g_{\{8_s\}}[B^\dagger B]_{\{8_s\}}M_{\{8\}} + g_{\{8_a\}}[B^\dagger B]_{\{8_a\}}M_{\{8\}}\right\} , \quad (7)$$

where B and M are the field operators for the baryon and mesons. In writing the above Lagrangian, which is a scalar, we have coupled the initial and final baryons to a flavor singlet or an octet. Since there are two irreducible octet representations, we need a different coupling constant for each of the representations. This Lagrangian has one coupling constant for each singlet meson $g_{\{1\}}$, and two coupling constants $g_{\{8_s\}}$ and $g_{\{8_a\}}$ for each meson octet. These coupling constants can then be determined by fitting the experimental data.

The Nijmegen potential[10] model D takes for the exchange mesons the psedoscalar octet $\{\pi, \eta, \eta', K\}$, the vector octet $\{\rho, \phi, \omega, K^*\}$ and a scalar meson $\{\epsilon\}$. The masses of the mesons and baryons are taken from experiment, while the coupling constants are adjusted to fit the data in the $S = 0, -1$. This, in principle, determines the long range part of the potential which should be described in terms of meson-baryon degrees of freedom. These same coupling constants can be used to construct the OBE potential for $S \leq -2$. For the short range part of the interaction, model D assumes a hard core, which is an additional parameter that could be different in the $S = 0$ and $S = -1$ channels. The breaking of flavor $SU(3)$ is due to the use of physical masses for the baryons and mesons, and the choice of different hard core radii in the $S = 0, -1$.

We expect the short range part of the BB interaction to be expressed in terms of quark-gluon degrees of freedom. In a valence quark model description of this short range interaction the number of strange quarks increases as we go from $S = 0$ to $S = -1$ and then $S \leq -2$. As the number of strange quarks increases, we expect the role of the Pauli principle at the quark level to change. As a result the short range part of the BB interaction should be S dependent, and cannot be determined from the $S = 0, 1$ channels at the meson-baryon level in terms of a flavor $SU(3)$ rotation of the OBE potential. Thus to determine the BB interaction in the $S = -2$ channel, we needs to introduce at least one additional parameter that determines the short range interaction.

This procedure is followed by Carr *et al.*[11] They only considered S-wave interaction and ignored the tensor interaction. Their potential for the exchange of the ith meson is of the form

$$V_i(r) = V_c^{(i)}(r) + \vec{\sigma}_1 \cdot \vec{\sigma}_2 \; V_\sigma^{(i)}(r) \; , \quad (8)$$

where the radial potential $V_\alpha^{(i)}, \alpha = c, \sigma$ for a mason of mass m_i is taken to be

$$V_\alpha^{(i)}(r) = V_0^{(i)} \left[\frac{e^{-m_i r}}{m_i r} - C\left(\frac{M}{m_i}\right)\frac{e^{-Mr}}{Mr}\right] \; ; \quad \alpha = c, \sigma \; . \quad (9)$$

To guarantee a one parameter short range repulsion, a mass $M = 2500$ MeV is chosen in all partial waves. Then the parameter C determined the strength of the short range interaction. This new parameter C is constrained to ensure that the potential for $r \geq 0.7$ fm is unchanged and the short range interaction is always repulsive. In Fig. 1 we present the $\Lambda\Lambda$ potential $V_{\Lambda\Lambda}$ and the ΞN potential $V_{\Xi N}$.

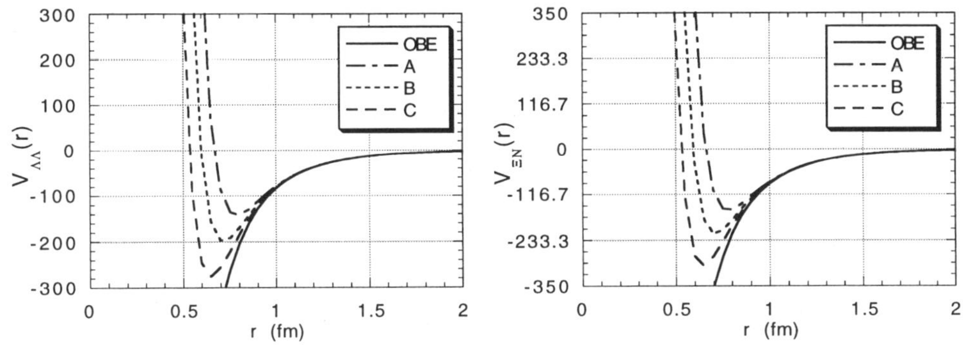

Figure 1. The $S = -2$, $^1S_0$ one-boson-exchange potential $V_{\Lambda\Lambda}$ and $V_{\Xi N}$.

In addition to the OBE potential (solid line), the figure includes three potentials in which the parameter C is adjusted to support a bound state (C), generate an an anti-bound state (B), or have no bound state at all (A).

To examine the role of the coupling between the $\Lambda\Lambda$ and ΞN channels in ${}_{\Lambda\Lambda}^{\ 6}\mathrm{He}$, we need to include this coupling exactly. This can be achieved most simply within a model in which ${}_{\Lambda\Lambda}^{\ 6}\mathrm{He}$ is taken to be a $\Lambda\Lambda\alpha$ three-body system described by the Faddeev equations. These equations can be reduced to coupled one dimensional integral equations if the interactions are assumed to be separable. For this Carr *et al*[11] generated a set of separable potentials that give the same effective range parameters as meson exchange potentials A, B and C. These are referred to as SA, SB and SC. The effective range parameters for both the local and the separable potentials are given in Table 1. Here we observe that the potentials B and SB have a scattering length that is comparable to the *nn* scattering length.

Table 1. The effective range parameters and binding energy of the two-body $\Lambda\Lambda - \Xi N$ system for the local potentials A, B and C and the equivalent separable potentials SA, SB and SC.

Pot.	$a_{\Lambda\Lambda}$	$r_{\Lambda\Lambda}$	$a_{\Xi N}$	$r_{\Xi N}$	B.E. (MeV)
A	-1.91	3.36	-2.12-0.75i	3.45-0.45i	UB
SA	-1.90	3.33	-2.08-0.81i	3.44-0.22i	UB
B	-21.1	1.86	-2.05-6.53i	2.12-0.21i	UB
SB	-21.0	2.54	-2.07-6.52i	2.62-0.15i	UB
C	7.82	1.41	3.08-5.26i	1.74-0.144i	0.71
SC	7.84	1.48	3.05-5.28i	1.45+0.074i	0.71

4 Binding Energy ${}_{\Lambda\Lambda}^{6}$He

In reducing the six body problem of ${}_{\Lambda\Lambda}^{6}$He to a three-body problem we gain the advantage of being able to solve the problem exactly using the Faddeev equations. In particular, the treatment of the coupled channel $\Lambda\Lambda - \Xi N$ is included in full. The main disadvantage of this three-body model is the limitation in the handling of the Pauli principle. This is most pronounced when the two Λs convert to ΞN, as the resulting nucleon should be in an antisymmetric state with respect to the nucleons in the α particle. Since the latter is treated as an elementary particle with no internal degrees of freedom, it is not possible to guarantee that the Pauli principle is satisfied. This effect is most pronounced when the nucleon and α particle are in relative S-wave. There are two ways of overcoming this problem in an approximate way. The first is to assume that the S-wave interaction supports a bound state in the absence of the Pauli exclusion principle. We then can remove this bound state from the spectrum of the Hamiltonian without any modification to the phase shifts. The resultant amplitude has the Pauli forbidden state removed. The second approach is to take the S-wave $N - \alpha$ interaction to be repulsive. The two methods give almost identical results for the binding energy of ^{6}Li.[12] Carr *et al*[11] used the latter approximation as it results in a rank one potential and therefore a simpler set of Faddeev equations for the three-body model of ${}_{\Lambda\Lambda}^{6}$He

The second complication in turning ${}_{\Lambda\Lambda}^{6}$He into a three-body problem is the need to know the $\Lambda - \alpha$ and $\Xi - \alpha$ interactions. These interactions should be consistent with the two-body BB interaction. Here again, Carr *et al*[11] resorted to simple models for these interactions. For the case of the $\Lambda - \alpha$ interaction, they made use of the G-matrix parametrization[13] of the Nijmegen model D potential to fold this interaction with ^{4}He density, while for the $\Xi - \alpha$ interaction they employed a Wood-Saxon optical potential.

Finally, since the conversion of the $\Lambda\Lambda$ to ΞN releases ≈ 25 MeV of energy, that energy could go into the excitation of the α particle. Although it is possible to include such excitation by assuming the α particle can be in one of two states with mass separation comparable to the energy of the first excited state of ^{4}He, Carr *et al*[11] chose not to include this excitation as the effective coupling to the $\Xi N\alpha$ channel is suppressed because the S-wave αN interaction has been taken to be repulsive, i.e. because of Pauli blocking.

To examine the relative importance of the coupling between the $\Lambda\Lambda$ and ΞN in free space, and in the nuclear medium where the Pauli blocking plays an important role, we consider three models for the binding energy of ${}_{\Lambda\Lambda}^{6}$He. These are:

1. In model 1 we include the full coupling between the $\Lambda\Lambda$ and the ΞN. This reduces the three-body model for ${}_{\Lambda\Lambda}^{6}$He to the three-body coupled channel problem $\Lambda\Lambda\alpha - \Xi N\alpha$, which can be solved exactly to give the binding energy for the three-body system under consideration. This is an exact solution to the problem within the limitations stated above. Here we note that this system is not a pure $\Lambda\Lambda\alpha$ system, but has a certain probability of being in a $\Xi N\alpha$ state.

2. In model 2 we include the coupling between the $\Lambda\Lambda$ and ΞN channels in calculating the two-body T-matrix that is the input to the Faddeev equations,

but carry out the three-body calculation within the $\Lambda\Lambda\alpha$ space only. This is equivalent to setting $T_{\Lambda\Xi} = 0$, which gives us a measure of the importance of the coupling between the $\Lambda\Lambda$ and ΞN in nuclear matter, and the effect of the Pauli blocking. In this case the probability of finding the system in $\Xi N\alpha$ state is zero.

3. Finally, in model 3 we assume that the coupling between the $\Lambda\Lambda$ and ΞN can be neglected at the two-body level in free space, i.e. $V_{\Lambda\Xi} = 0$. In this case we construct a two-body $\Lambda\Lambda$ interaction that gives the same $\Lambda\Lambda$ effective range parameters as the coupled channel problem. Thus for model B, the $\Lambda\Lambda$ potential, in the absence of coupling, has an anti-bound state and a scattering length comparable to the nn scattering length.

In Table 2 we present the results of Carr *et al*[11] for the above three models using the potentials SA, SB and SC for the $\Lambda\Lambda - \Xi N$ interaction. We observe that:

- The experimental result of 10.9 ± 0.6 lies between the "exact" results (model 1) for potentials SA and SB, and closer to SB. This suggests that within the present model the short range parameter C should be such that the $\Lambda\Lambda$ scattering length is large and comparable to the nn scattering length if we are to reproduce the binding energy of ${}_{\Lambda\Lambda}^{6}\text{He}$. Therefore we can use the binding energy of one of the $\Lambda\Lambda$ hypernuclei to fix the short range part of the interaction in the $S = -2$ channel and then proceed to use the resulting potential for other $S = -2$ hypernuclei.

- If we now compare the results of model 1 and model 2 in Table 2, we observe that the contribution of $T_{\Lambda\Xi}$ is small (≈ 0.5 MeV) if our potential would give a binding energy comparable to the experimental result. This relatively small reduction in binding energy when the $\Lambda\Lambda - \Xi N$ coupling is included is a measure of the importance of Pauli blocking, even in a nucleus as light as ${}_{\Lambda\Lambda}^{6}\text{He}$. This

Table 2. The binding energy of ${}_{\Lambda\Lambda}^{6}\text{He}$ for the three models under consideration and for the three interactions SA, SB and SC.

	$a_{\Lambda\Lambda}$ (fm)	BE(model 1) ($\Lambda\Lambda\alpha - \Xi N\alpha$) (MeV)	BE(model 2) ($\Lambda\Lambda\alpha$; $T_{\Lambda\Xi} = 0$) (MeV)	BE(model 3) ($\Lambda\Lambda\alpha$; $V_{\Lambda\Xi} = 0$) (MeV)
SA	-1.91	9.738	9.5078	10.007
SB	-21.1	12.268	11.606	14.134
SC	7.82	15.912	14.533	17.842
Exp.		10.9 ± 0.6		

small change in binding energy due to coupling in the nuclear medium does in part justify the exclusion of the excitation of the α particle in the present models.

- On the other hand, if we compare the results of model 1 and model 3 in Table 2, we may conclude that turning off the coupling between the $\Lambda\Lambda$ and ΞN channels at the two-body level, in free space, is in fact a very poor approximation. The difference of ≈ 2 MeV between model 1 and model 3 is a substantial fraction of the value of the $\Lambda\Lambda$ matrix element (a change of $\approx 30\%$). This difference is, to lowest order, the contribution of the coupling between the $\Lambda\Lambda$ and ΞN to the attraction in the effective $\Lambda\Lambda$ matrix element. To see this, we write the binding energy of ${}_{\Lambda\Lambda}^{6}\text{He}$ as the matrix element of the Hamiltonian, i.e.

$$-B.E. = \langle H\rangle = K.E. + 2\langle V_{\Lambda\alpha}\rangle + \langle V_{\Lambda\Lambda}^{\text{eff}}\rangle\ , \tag{10}$$

where $\langle V_{\Lambda\Lambda}^{\text{eff}}\rangle$ is the effective $\Lambda\Lambda$ matrix elements in the medium. If there is no coupling between the $\Lambda\Lambda$ and ΞN channels in free space then

$$\langle V_{\Lambda\Lambda}^{\text{eff}}\rangle = \langle V_{\Lambda\Lambda}\rangle_{\text{no coupling}} \quad \text{for} \quad V_{\Lambda\Xi} = 0\ . \tag{11}$$

On the other hand, the inclusion of the coupling between the $\Lambda\Lambda$ and the ΞN in free space reduces the strength of the $\Lambda\Lambda$ potential, $V_{\Lambda\Lambda}$, so that the $\Lambda\Lambda$ potential and the additional attraction due to the coupling results in an effective matrix element that is fixed by experiment. If we assume that the Pauli blocking removes the contribution of the coupling between the $\Lambda\Lambda$ and ΞN in the nuclear medium, i.e. the difference between the result of model 1 and model 2 in Table 2 is zero, then

$$\langle V_{\Lambda\Lambda}^{\text{eff}}\rangle = \langle V_{\Lambda\Lambda}\rangle_{\text{with coupling}} \quad \text{for} \quad V_{\Lambda\Xi} \neq 0\ . \tag{12}$$

But from Eq. (10) we know that the difference between the binding energy with and without coupling between the $\Lambda\Lambda$ and the ΞN channels in free space is the contribution of the coupling to the attraction, i.e.

$$\begin{aligned} BE(\text{model 1}) - BE(\text{model 3}) &\approx -\langle V_{\Lambda\Lambda}\rangle_{\text{with coupling}} + \langle V_{\Lambda\Lambda}\rangle_{\text{no coupling}} \\ &\approx \frac{|\langle\Lambda\Lambda|V|N\Xi\rangle|^2}{\Delta E_{\Lambda\Lambda}} \approx -2\ \text{MeV}\ . \end{aligned} \tag{13}$$

This result is approximately the difference between the nn and $\Lambda\Lambda$ matrix elements in light nuclei.

This analysis basically establishes the fact that it is essential to include the coupling between the $\Lambda\Lambda$ and ΞN channels in free space even though the coupling is suppressed in the medium due to Pauli blocking.

5 Conclusion

From the above analysis based on the validity of flavor $SU(3)$ for matrix elements of the potential, we can understand how the coupling resulting from mass differences between the members of the baryon octet could justify the fact that $-\langle V_{\Lambda\Lambda}\rangle > -\langle V_{\Lambda N}\rangle$. In fact flavor $SU(3)$ suggests that the coupling in the $S = -2$ is substantially more important than is the case in the $S = -1$ channel. This is a result of the fact that: (i) The $\Lambda\Lambda$ threshold and the ΞN threshold are closer than the ΛN and ΣN thresholds. (ii) The coupling matrix element for the the transition from $\Lambda\Lambda$ to ΞN is larger than the coupling matrix elements for the transition from ΛN to ΣN.

With the help of flavor $SU(3)$ rotation we showed how we can construct the long range part of the potential based on OBE for $S \leq -2$ given that the meson baryon coupling constants are fixed by the data in the $S = 0, -1$ channels. However, this flavor $SU(3)$ rotation does not determine the short range part of the interaction. In fact valence quark models for the BB system would give an S dependence to the short range data, and in an OBE model this requires additional data, e.g. the binding energy of $\Lambda\Lambda$ hypernuclei.

Using this $SU(3)$ rotation Carr *et al*[11] constructed an extension of the Nijmegen model D potential to the $S = -2$ channel with a soft cut-off that does not change the potential outside of 0.7 fm. Finally, Carr *et al*[11] were able to show that a $\Lambda\Lambda$ scattering length comparable to the nn scattering length gives approximately the correct binding energy for ${}_{\Lambda\Lambda}^{6}\mathrm{He}$. This suggests that the experimental binding energy of ${}_{\Lambda\Lambda}^{6}\mathrm{He}$ is consistent with the strength of the BB potential as predicted by flavor $SU(3)$ rotation of the OBE potentials. More important it confirms the prediction of Dover[9] that the $\Lambda\Lambda$ scattering length is comparable to the nn scattering length in the ${}^1\mathrm{S}_0$.

With the inclusion of coupling between the $\Lambda\Lambda$ and ΞN channel there is no need to include repulsive three-body forces to calculate the correct binding energy for ${}_{\Lambda\Lambda}^{6}\mathrm{He}$. This is a result of the fact that the Pauli blocking of the process $\Lambda\Lambda \to \Xi N$ in the nuclear medium reduces the effective attraction in the $\Lambda\Lambda$ channel as compared to the attraction in free space. As a result the $\Lambda\Lambda$ matrix element in light nuclei is less than the nn matrix element by about 2 MeV, which is consistent with the experimental results from $\Lambda\Lambda$ hypernuclei. This is despite the fact that the scattering length for the $\Lambda\Lambda$ is comparable to the nn scattering length.

Finally, we can understand the result that

$$-\langle V_{nn}\rangle > -\langle V_{\Lambda\Lambda}\rangle > -\langle V_{\Lambda N}\rangle \ , \tag{14}$$

on the basis of the coupled channel nature of the BB interaction in the $S = -1, -2$ channels resulting from flavor $SU(3)$ breaking in the mass of the baryon octet.

Acknowledgments

The author would like to thank the Australian Research Council for their financial support. He is indebted to his collaborators S.B. Carr and B.F. Gibson for many enlightening discussions and helpful suggestions.

References

1. A. Jaffe, *Phys. Rev. Lett.* **38**, 195 (1977).

2. D.J. Prowse, *Phys. Rev. Lett.* **17**, 782 (1966).
3. M. Danysz *et al*, *Nucl. Phys.* **49**, 121 (1963).
4. R.H. Dalitz *et al*, *Proc. R. Soc. London* **A 426**, 1 (1989).
5. S. Aoki *et al*, *Phys. Rev. Lett.* **65**, 1729 (1990); *Prog. Theo. Phys.* **85**, 1287 (1991).
6. C.B. Dover, D.J. Millener, A. Gal and D.H. Davis, *Phys. Rev.*, C **44**, 1905 (1991).
7. J.J. de Swart, *Rev. Mod. Phys.* **35**, 916 (1963).
8. Th.A. Rijken, P.M.M. Maessen and J.J. de Swart, *Nucl. Phys.* **A 547**, 245c (1992).
9. C.B. Dover, Properties & Interactions of Hypernuclei, eds. B.F. Gibson, P.D. Barnes and K. Nakai (World Scientific, Singapore, 1994), p 1.
10. M.M. Nagel, Th.A. Rijken and J.J. de Swart, *Phys. Rev.* D **15**, 2547 (1977).
11. S.B. Carr, I.R. Afnan and B.F. Gibson, *Nucl. Phys.* **A 625**, 143 (1997).
12. D.R. Lehman, M. Rai and A. Ghovanlou *Phys. Rev.* C **17**, 744 (1978).
13. Y. Yamamoto and H. Bando, *Prog. Theo. Phys.* **73**, 905 (1985)

HYPERON MIXING IN LIGHT $S = -2$ NUCLEI

T. YAMADA

Laboratory of Physics, Kanto Gakuin University, Yokohama 236-8651, Japan
E-mail: yamada@kanto-gakuin.ac.jp

We discuss hyperon mixing problems in light $S = -2$ nuclei with the [core+Λ+Λ] and [core+Ξ+N] model for the one-boson-exchange-model interactions, and the core+$(3q)$+$(3q)$ model for the quark-cluster-model interactions. In case of the Nijmegen model-D interaction, the Ξ component in ${}^{10}_{\Lambda\Lambda}$Be and ${}^{13}_{\Lambda\Lambda}$B is less than 1%, while in case of the Kyoto-Niigata RGM-F interaction based on the quark cluster model, the hyperon mixing effect is very strong to produce a bound H-nuclear state in ${}^{10}_{\rm YY}$Be.

1 Introduction

The spectroscopic study of strangeness $S = -2$ nuclei is an entrance to the structure study of multi-strangeness hadronic systems and provides the unified understanding of the underlying baryon-baryon interactions. Double-Λ hypernuclei attract special interests because of the following reasons: 1) The system is unique to get important information on the $\Lambda\Lambda$ interaction, and 2) hyperon-mixing states among the baryon octet might appear in some energy region due to the $\Lambda\Lambda$-ΞN-$\Sigma\Sigma$ coupling including the H-dibaryon state. The YN and YY interactions have been proposed with the one-boson exchange model (OBEM)[1,2] and the quark cluster model (QCM)[3,4]. The large qualitative difference between the two models appears in the $S = -2$ sector. For example, 1) the QCM (OBEM) predicts (does not predict) a bound or resonant H-dibaryon state, and 2) the ΞN-$\Lambda\Lambda$ coupling effect is strong in the QCM, while it is weak in the OBEM.

The purpose of this paper is to investigate the hyperon-mixing problem in $S = -2$ nuclei, together with the check of the proposed OBE and QCM interactions. For this purpose, we use the [core+Λ+Λ] and [core+Ξ+N] model for the OBEM interactions, and the core+$(3q)$+$(3q)$ model for the QCM interactions. In the former model, the Nijmegen model-D intereaction[1] is applied to study the energy spectra of ${}^{13}_{\Lambda\Lambda}$B with the core+Λ+Λ model, and then, the $\Lambda\Lambda$-ΞN mixing in ${}^{13}_{\Lambda\Lambda}$B and ${}^{10}_{\Lambda\Lambda}$Be is investigated with the [core+Λ+Λ] and [core+Ξ+N] model. On the other hand, in the core+$(3q)$+$(3q)$ model, we employ the RGM-F interaction proposed by the Kyoto-Niigata group[4] to study the hyperon mixing in ${}^{10}_{\rm YY}$Be.

2 Hyperon mixing in $S = -2$ nuclei with the OBEM interaction

2.1 Energy spectra of ${}^{13}_{\Lambda\Lambda}$B with the core+Λ+Λ model

The total wave function of ${}^{13}_{\Lambda\Lambda}$B is given as

$$\Phi_{JT} = \sum_{c,\nu_r,\nu_R} a_c(\nu_r,\nu_R) \left[\left[\Phi_{J_cT_c}({}^{11}\mathrm{B}) \chi_S(\Lambda\Lambda) \right]_K \left[\phi_{\ell_r}(\nu_r) \phi_{\ell_R}(\nu_R) \right]_L \right]_{JT}, \qquad (1)$$

where $\Phi_{J_cT_c}({}^{11}\mathrm{B})$ denotes a intrinsic part of the shell-model wave function (s^4p^7) for ^{11}B. The $\Lambda - \Lambda$ (core $- (\Lambda + \Lambda)$) relative wave function is expanded in terms of the Gaussian basis $\phi_\ell(\nu)$.

The total Hamiltonian is given as

$$H = h(\text{core}) + T_r + T_R + \sum_{i=1}^{2} \sum_{j \in \text{core}} v_{\Lambda \text{N}}(i,j) + v_{\Lambda\Lambda}, \tag{2}$$

where $h(\text{core})$ denotes the Hamiltonian of the core part with the Cohen-Kurath interaction, and T_r (T_R) is the $\Lambda - \Lambda$ (core $-$ $(\Lambda + \Lambda)$) kinetic energy operator, and $v_{\Lambda\text{N}}$ is the YNG ΛN interaction[5] which reproduces $B_\Lambda^{exp}\left({}^{12}_{\Lambda}\text{B}\,(\text{g.s.})\right)$. As for the $\Lambda\Lambda$ interaction, we use the simulated Nijmegen model-D potential, which is a soft-core potential equivalent to the T-matrix for the original model-D one[1]. The hard-core radius r_c is chosen to reproduce $B_{\Lambda\Lambda}^{exp}\left({}^{13}_{\Lambda\Lambda}\text{B}\right)$ with the core+Λ+Λ model.

We obtain many states of ${}^{13}_{\Lambda\Lambda}$B, which are classified into two groups. First group is that both the two Λ particles are occupied in s shell, and second is that one Λ particle is in s shell and another is in p shell. (The figure of the energy spectra will be given elsewhere.) The energy difference between them is about 12 MeV. The ground state, $(J^\pi, T) = (3/2^-, 1/2)$, and the yrast state belonging to the second group, $(J^\pi, T) = (1/2^+, 1/2)$, are respectively given as

$$|\, 3/2^-,\ 1/2\rangle = \sqrt{0.998}\ |\, {}^{11}\text{B(g.s.)} \otimes \left[{}^1S_0(\Lambda\Lambda) \otimes \ell_R = 0\right]\rangle + \cdots, \tag{3}$$

$$|\, 1/2^+,\ 1/2\rangle = \sqrt{0.995}\ |\, {}^{11}\text{B(g.s.)} \otimes \left[{}^1S_0(\Lambda\Lambda) \otimes \ell_R = 1\right]\rangle + \cdots, \tag{4}$$

where ${}^{2S+1}\ell_j$ and ℓ_R denote, respectively, the angular momentum state of the two Λ particles and orbital angular momentum referring to the core $-$ $(\Lambda + \Lambda)$ relative coordinate. In the excited state, the coupling with other configurations such as with $|\, {}^{11}\text{B(g.s.)} \otimes \left[{}^3P(\Lambda\Lambda) \otimes \ell_R = 0\right]\rangle$ is weak, when we use the simulated Nijmegen model-D potential.

2.2 $\Lambda\Lambda$-ΞN coupling effect in ${}^{13}_{\Lambda\Lambda}B$ and ${}^{10}_{\Lambda\Lambda}Be$

In this section, we study the $\Lambda\Lambda$-ΞN coupling effect in the ground and excited ${}^{13}_{\Lambda\Lambda}$B states given in Eqs. (3) and (4), together with ${}^{10}_{\Lambda\Lambda}$Be, with the $[\text{core} + \Lambda + \Lambda]$, $[\text{core} + \text{p} + \Xi^-]$ and $[\text{core} + \text{n} + \Xi^0]$ model. The $\Sigma\Sigma$ channel is neglected here because of the large mass difference between the $\Lambda\Lambda$ and $\Sigma\Sigma$ channels (160 MeV).

In case of ${}^{13}_{\Lambda\Lambda}$B, the ^{12}C+Ξ^- $\left[= ({}^{11}\text{B} + \text{p}) + \Xi^-\right]$ threshold appears above the ${}^{11}\text{B} + \Lambda + \Lambda$ threshold by about 10 MeV, while the ^{12}B+Ξ^0 $\left[= ({}^{11}\text{B} + \text{n}) + \Xi^0\right]$ threshold appears at about 25 MeV higher than the ^{11}B+Λ+Λ threshold. Therefore, the coupling effect of ${}^{11}\text{B} + \Lambda + \Lambda$ with ${}^{11}\text{B} + \text{n} + \Xi^0$ is considerably weaker than that with ^{11}B+p+Ξ^-. Thus, the $\Lambda\Lambda$-ΞN coupling effect in ${}^{13}_{\Lambda\Lambda}$B is studied with the ^{11}B+Λ+Λ and ^{11}B+p+Ξ^- model in the present paper. On the other hand, in case of ${}^{10}_{\Lambda\Lambda}$Be, both the ^{9}B+Ξ^- and ^{9}Be+Ξ^0 thresholds are approximately degenerate in energy, and appear above the ${}^{11}\text{B} + \Lambda + \Lambda$ threshold by about 25 MeV. Therefore, the $\Lambda\Lambda$-ΞN coupling effect in ${}^{10}_{\Lambda\Lambda}$Be is investigated with use of the isospin bases of ${}^{8}\text{Be} + \Lambda + \Lambda$ and ${}^{8}\text{Be} + \Xi + \text{N}$.

The total wave function is given as

$$\Phi_J = \Phi_{\Lambda\Lambda} + \Phi_{\Xi\text{N}}, \tag{5}$$

Table 1: Ξ components (P_Ξ) and binding energy gains (ΔB) due to the $\Lambda\Lambda$-ΞN coupling in the ground ($J^\pi = 3/2^-$) and excited ($J^\pi = 1/2^+$) states of ${}^{13}_{\Lambda\Lambda}$B and ${}^{10}_{\Lambda\Lambda}$Be.

	${}^{13}_{\Lambda\Lambda}$B		${}^{10}_{\Lambda\Lambda}$Be	
	$3/2^-$	$1/2^+$	$3/2^-$	$1/2^+$
P_Ξ (%)	0.3	0.7	0.3	0.5
ΔB (MeV)	0.3	0.3	0.5	0.4

$$\Phi_{\Lambda\Lambda} = \sum_{c,\nu_r,\nu_R} a_c(\nu_r,\nu_R)\left[[\phi_{\ell_r}(\nu_r)\phi_{\ell_R}(\nu_R)]_L\,\chi_S\right]_J, \tag{6}$$

$$\Phi_{\Xi N} = \sum_{\gamma=1}^{3}\sum_{c_\gamma,\nu_\gamma,\mu_\gamma} a_{c_\gamma}(\nu_\gamma,\mu_\gamma)\left[\left[\phi_{\ell_\gamma}(\nu_\gamma)\phi_{\lambda_\gamma}(\mu_\gamma)\right]_L\,\chi_S\right]_J. \tag{7}$$

Since the nuclear-core part in Eqs. (3) and (4) is almost the ground state of ^{11}B, we treat the nuclear core as a point and spin-less particle for simplicity. This situation is also realized in case of ${}^{10}_{\Lambda\Lambda}$Be. The relative wave function referring to each Jacobi coordinate is expanded in terms of the Gaussian basis $\phi_\ell(\nu)$.

The total Hamiltonian is given as

$$H_{\alpha\beta} = T_\alpha\delta_{\alpha\beta} + V_{\alpha\beta}, \tag{8}$$

$$V_{11} = V_{C\Lambda_1} + V_{C\Lambda_2} + v_{\Lambda\Lambda}, \quad V_{22} = V_{CN} + V_{C\Xi} + v_{\Xi N} + V^{Pauli}, \tag{9}$$

$$V_{12} = V_{21} = \zeta \times v_{\Lambda\Lambda-\Xi N}, \tag{10}$$

where V_{CY} and V_{CN} are the core-baryon folding potentials including the Coulomb potentials, which reproduce the hyperon (proton) binding energy of the ground state for the core+Y (core+N) system. V^{Pauli} is the pseudo-potential to exclude the Pauli-forbidden state between the core and nucleon. As for the valence baryon-baryon interactions, $v_{\Lambda\Lambda}$, $v_{\Xi N}$ and $v_{\Lambda\Lambda-\Xi N}$, we use the simulated Nijmegen model-D potential as mentioned before, and ζ denotes an isospin-spin factor.

The calculated results are shown in Table 1. The Ξ component is as small as $P_\Xi = 0.3 \sim 0.7\%$ and the binding energy gain due to the $\Lambda\Lambda$-ΞN coupling is also as small as $\Delta B = 0.3 \sim 0.5$ MeV. However, the Ξ component in the excited state with the main configuration $\left[\text{core(g.s.)} \otimes [{}^1S_0(\Lambda\Lambda) \otimes \ell_R = 1]\right]_{J=1/2^+}$ is about twice larger than that in the ground state with the main configuration $\left[\text{core(g.s.)} \otimes [{}^1S_0(\Lambda\Lambda) \otimes \ell_R = 0]\right]_{J=3/2^-}$, while the energy gain is almost the same in both the two states. Here, it is instructive to discuss the reason in case of ${}^{13}_{\Lambda\Lambda}$B. Let's first consider the case that the $\Lambda\Lambda$-ΞN coupling is switched off. The Ξ-hypernuclear state coupled mainly with the $\left[{}^{11}\text{B(g.s.)} \otimes [{}^1S_0(\Lambda\Lambda) \otimes \ell_R = 1]\right]_{J=3/2^-}$ configuration is an excited state, $\left[{}^{12}\text{C(g.s.)} \otimes p_\Xi\right]_{J=3/2^-}$, due to the parity conservation. This state appears around the ^{12}C+Ξ^- threshold. Then, the energy difference between the $\Lambda\Lambda$ and Ξ channels is about 40 MeV. On the other hand, in the case of the $\left[{}^{11}\text{B(g.s.)} \otimes [{}^1S_0(\Lambda\Lambda) \otimes \ell_R = 1]\right]_{J=1/2^+}$ configuration with $E_x \simeq$12 MeV (see §2), the Ξ-hypernuclear state coupled mainly with the configuration is the ground state, $\left[{}^{12}\text{C(g.s.)} \otimes s_\Xi\right]_{J=1/2^+}$, whose binding energy is about 8 MeV from the ^{12}C+Ξ^- threshold in the present calculation. Then, the energy difference between the $\Lambda\Lambda$ and Ξ channels is about 20 MeV, which is about half in comparison with the $J^\pi = 3/2^-$ state case.

Next, we switch on the $\Lambda\Lambda$-ΞN coupling. When we assign the matrix element for the coupling potential to $\alpha = \langle s_\Lambda s_\Lambda \mid v_{\Lambda\Lambda-\Xi N} \mid p_\Xi p_N \rangle$, the matrix element of $\langle s_\Lambda p_\Lambda \mid v_{\Lambda\Lambda-\Xi N} \mid s_\Xi p_N \rangle$ is approximately given as $\alpha/\sqrt{2}$ for the dominant Λ-Λ $^1S_0$ state. According to the simple model for the two-channel coupled problem with the energy difference (ΔE) and the coupling matrix element (v with $\mid v/\Delta E \mid \ll 1$), the component of the upper channel in the lower eigenstate obtained after the diagonalization of the Hamiltonian matrix is given as $P_\Xi = (v/\Delta E)^2$, and the energy gain is $\Delta B = v^2/\Delta E$. This is realized in the present problem because α is a few MeV. Then, the rate of the Ξ component in the ground and excited states is given as $P_\Xi(\text{ex.})/P_\Xi(\text{g.s.}) \simeq (\alpha/\sqrt{2}/20)^2/(\alpha/40)^2 \simeq 2$, while the binding energy gain due to the coupling is almost the same in both the states, $\Delta B(\text{ex.})/\Delta B(\text{g.s.}) \simeq \left((\alpha/\sqrt{2})^2/20\right)/(\alpha^2/40) \simeq 1$. These results are consistent with those in Table 1. It is noted that the values of $P_\Xi(\text{ex.})/P_\Xi(\text{g.s.})$ and $\Delta B(\text{ex.})/\Delta B(\text{g.s.})$ do not depend strongly on the strength of the $\Lambda\Lambda$-ΞN coupling potential. We found that the situation in case of $^{10}_{\Lambda\Lambda}$Be is similar to that in case of $^{13}_{\Lambda\Lambda}$B.

3 Hyperon mixing in $S = -2$ nuclei with the QCM interaction

The Kyoto-Niigata RGM-F interaction based on the quark cluster model (QCM) predicts a bound H-dibaryon state with the binding energy of about 19 MeV referring to the $\Lambda\Lambda$ threshold. In this section, we use the RGM-F interaction to study the structure of the $^{10}_{\rm YY}$Be ground state with the core+$(3q)$+$(3q)$ model.

3.1 Formulation of the core+(3q)+(3q) model

In the core+$(3q)$+$(3q)$ model, 1) the valence two baryons are described with the $(3q)$+$(3q)$ cluster model, where the antisymmetrization among the 6 quarks are taken into account exactly and the $(3q)$-$(3q)$ Hamiltonian same as the RGM-F one is used here, 2) the core part is treated as a point model, in which the antisymmetrization between the core and valence quarks is neglected, although we take into account the Pauli forbidden state between the core and valence nucleon, which is removed with the orthogonal condition model (OCM), and 3) the core and valence baryon $(3q)$ interaction is replaced to the phenomenological core–Y or core–N potentials, which procedure corresponds to taking into account only the direct core–$(3q)$ potential.

The total wave function is given as the sum of the three channel wave functions, core+Λ+Λ, core+Ξ+N and core+Σ+Σ, with the resonating group method (RGM),

$$\Phi = \Phi_{\Lambda\Lambda} + \Phi_{\Xi N} + \Phi_{\Sigma\Sigma}, \tag{11}$$

$$= \sum_\alpha \int d\mathbf{r} d\mathbf{R} \left[\mathcal{A}\{\phi_\alpha \delta(\mathbf{r}_\alpha - \mathbf{r})\}\delta(\mathbf{R}_\alpha - \mathbf{R})\right] \chi_\alpha(\mathbf{r}, \mathbf{R}), \tag{12}$$

where $\mathcal{A}$ denotes the antisymmetrization operator among the 6 quarks, and the channel wave function ϕ_α defined through $\phi^{(orb)}(123)\phi^{(orb)}(456)\xi^{SF}_\alpha \xi^C$, and $\chi_\alpha(\mathbf{r}, \mathbf{R})$

Table 2: Calculated binding energies and channel components of the $(3q)+(3q)$ quark cluster model with the original Kyoto-Niigata RGM-F Hamiltonian and modified RGM-F Hamiltonian.

channel	Original RGM-F				Modified RGM-F			
	B (MeV)	component			B (MeV)	component		
		$\langle\Lambda\Lambda\rangle$	$\langle\Xi N\rangle$	$\langle\Sigma\Sigma\rangle$		$\langle\Lambda\Lambda\rangle$	$\langle\Xi N\rangle$	$\langle\Sigma\Sigma\rangle$
$\Lambda\Lambda$	unbound	100 %	0 %	0 %	unbound	100 %	0 %	0 %
$\Lambda\Lambda, \Xi N, \Sigma\Sigma$	18.9	22.6	48.5	28.9	12.2	30.8	43.6	25.6

is the relative wave function with respect to the $(3q)$–$(3q)$ relative coordinate $\mathbf{r}$ and the core–$((3q)+(3q))$ relative coordinate $\mathbf{R}$.

The total Hamiltonian is expressed as

$$H = H_{3q-3q} + \delta_{\alpha\beta}\left[T_{R_\alpha} + V_{CY_1} + V_{CY_2}\right] + \delta_{\alpha 2}\delta_{\beta 2}V_{Pauli} + (\Delta M)_{\alpha\beta}, \tag{13}$$

where H_{3q-3q}, T_{R_α}, V_{CY} and V_{Pauli} denote, respectively, the $(3q)+(3q)$ Hamiltonian, kinetic-energy operator with respect to $\mathbf{R}$, phenomenological potential between the core and valence baryon which reproduces the binding energy of the ground state of the core+baryon system, and pseudo-potential to remove the Pauli forbidden state between the core and valence nucleon. Since the original RGM-F Hamiltonian do not reproduce the experimental threshold energies among the $\Lambda\Lambda$, ΞN and $\Sigma\Sigma$ channels, we introduce the mass difference matrix $(\Delta M)_{\alpha\beta}$ (diagonal matrix) to reproduce them.

The equation of motion is derived from the Schrödinger equation and given as the three-body RGM equation,

$$\sum_\beta \int d\mathbf{r}'d\mathbf{R}'\left[H_{\alpha\beta}(\mathbf{r},\mathbf{R};\mathbf{r}',\mathbf{R}') - EN_{\alpha\beta}(\mathbf{r},\mathbf{R};\mathbf{r}',\mathbf{R}')\right]\chi_\beta(\mathbf{r}',\mathbf{R}') = 0, \tag{14}$$

where $H_{\alpha\beta}$ and $N_{\alpha\beta}$ denote, respectively, the Hamiltonian and norm kernels with respect to the RGM basis function $\mathcal{A}\{\phi_c\delta(\mathbf{r}_c - \mathbf{r})\}\delta(\mathbf{R}_c - \mathbf{R})$ with $c = \alpha$ and β. The RGM equation is solved with the variational method by expanding the relative wave function in terms of the Gaussian basis. For the later discussion, it is convenient to introduce the renormalized wave function $\tilde{\chi}$ and probability density ρ with respect to the $(3q)$-$(3q)$ relative coordinate defined as,

$$\tilde{\chi_\alpha}(\mathbf{r},\mathbf{R}) = \sum_\beta \int d\mathbf{r}'d\mathbf{R}' N^{1/2}_{\alpha\beta}(\mathbf{r},\mathbf{R};\mathbf{r}',\mathbf{R}')\chi_\beta(\mathbf{r}',\mathbf{R}'), \tag{15}$$

$$\rho_\alpha(r) = \int d\hat{\mathbf{r}}d\mathbf{R} \mid \tilde{\chi_\alpha}(\mathbf{r},\mathbf{R}) \mid^2, \quad \sum_\alpha \langle \tilde{\chi_\alpha} \mid \tilde{\chi_\alpha}\rangle = 1. \tag{16}$$

Then, the component of the α channel is given as $\langle\tilde{\chi_\alpha} \mid \tilde{\chi_\alpha}\rangle$, and the H-dibaryon component in the $S = -2$ nuclear state Φ given in Eq. (11) is defined as $\langle\Phi \mid H\rangle\langle H \mid \Phi\rangle$, where $|H\rangle$ is the H-dibaryon wave function.

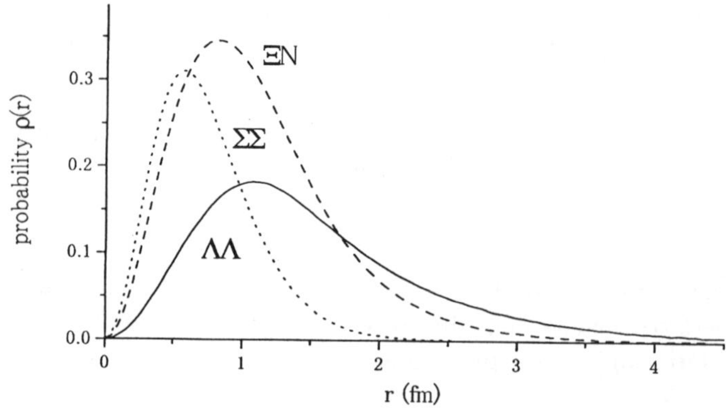

Figure 1: Probability densities $\rho(r)$ for the ΛΛ, ΞN and ΣΣ channels in the H-dibaryon state with the $(3q)+(3q)$ RGM calculation.

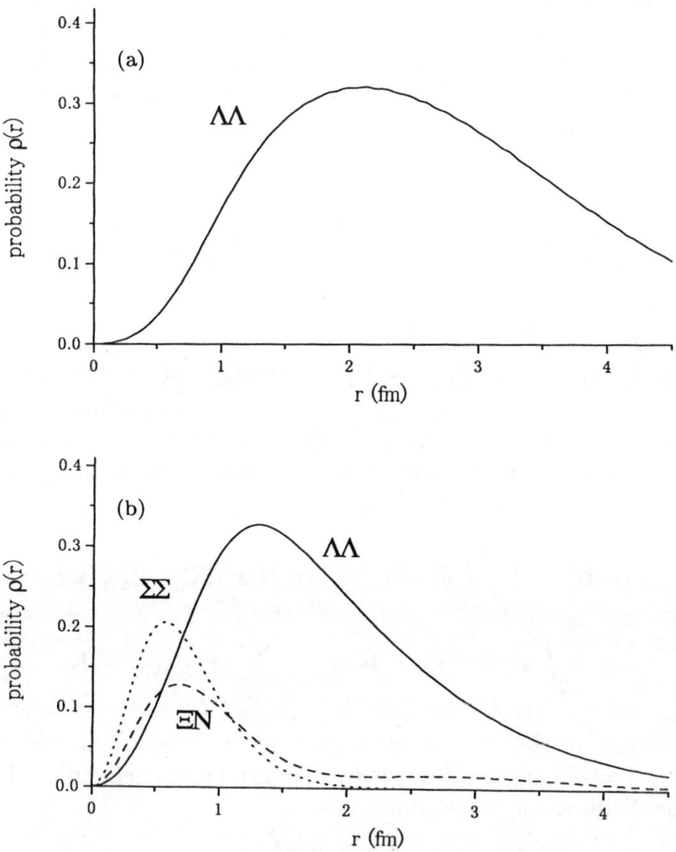

Figure 2: Probability densities $\rho(r)$ with respect to the $(3q)$–$(3q)$ relative part in the ${}^{10}_{\rm YY}$Be ground state with the core+$(3q)$+$(3q)$ model. (a) Results in case of only the core+$\Lambda+\Lambda$ channel, and (b) those in case of the core+Λ+Λ, core+Ξ+N and core+Σ+Σ channels.

Table 3: Calculated binding energies and channel components of the $^{10}_{\rm YY}$Be ground state with the core+(3q)+(3q) model.

channel	$B_{\Lambda\Lambda}$ (MeV)	component			
		$\langle\Lambda\Lambda\rangle$	$\langle\Xi N\rangle$	$\langle\Sigma\Sigma\rangle$	$\langle H\rangle$
$\Lambda\Lambda$	16.0	100 %	0 %	0 %	25 %
$\Lambda\Lambda, \Xi N$	20.7	87	13	0	57
$\Lambda\Lambda, \Xi N, \Sigma\Sigma$	25.3	67	16	17	71

3.2 Results and discussion

First we show the results of the (3q)+(3q) RGM calculation in Table 2. When we use the original RGM-F Hamiltonian H_{3q-3q} and take into account only the $\Lambda\Lambda$ channel, no bound state appears. With the three channel couplings switched on, however, a bound H-dibaryon state appears with about 19 MeV binding energy referring to the $\Lambda\Lambda$ threshold. The channel components shown in Table 2 indicate that the bound state has the characteristic of the flavor SU(3) singlet state. Next we use the modified RGM-F Hamiltonian with the mass difference matrix which reproduces the experimental threshold energies among the $\Lambda\Lambda$, ΞN and $\Sigma\Sigma$ channels, $H_{3q-3q} + (\Delta M)_{\alpha\beta}$. The results are shown in Table 2, and we find a bound state with about 12 MeV binding energy. Although the channel components are a little different with ones in case of the original RGM-F Hamiltonian, they have the characteristic similar to the original H state. Therefore, we call the state as the H state in the present paper. Figure 1 shows the probability density $\rho(r)$ with respect to the (3q)-(3q) relative coordinate, respectively, defined in Eq. (16). We see that all of the $\Lambda\Lambda$-, ΞN- and $\Sigma\Sigma$-channel wave functions concentrate around 1 fm, and the ΞN channel has the largest component.

The core+(3q)+(3q) model is applied to the structure study of the ground state of $^{10}_{\rm YY}$Be, where we use the modified RGM-F Hamiltonian for the (3q)+(3q) part. The calculated results are shown in Table 3. In case of only the core+Λ+Λ channel, the $^{10}_{\rm YY}$Be ground state is bound with $B_{\Lambda\Lambda}$=16 MeV (vs. $B^{exp}_{\Lambda\Lambda}$=17.7 MeV), although the (3q)+(3q) system is not bound in case of the Λ+Λ channel (see Table 2). The reason why the $^{10}_{\rm YY}$Be ground state is bound with only the ^{8}Be+Λ+Λ channel is mainly due to the strong attractive ^{8}Be–Λ potential (B^{exp}_{Λ}=6.7 MeV for the $^9_\Lambda$Be ground state). The probability density $\rho(r)$ for the $\Lambda\Lambda$ channel is shown in Fig. 2(a) and has a peak around 2 fm. Its behavior is much similar to that obtained by the core+Λ+Λ model with the Nijmegen model-D potential. Next, we switch on the three channel couplings. Reflecting the strong coupling effects among the three channels, the binding energy of the $^{10}_{\rm YY}$Be ground state becomes about 25 MeV, which is overbound by about 7 MeV in comparison with the experimental data. The strong coupling effect can be also seen in the channel components shown in Table 3, and it is in contrast to the case of the Nijmegen model-D potential, in which the $\Lambda\Lambda$ and ΞN components in $^{10}_{\rm YY}$Be are, respectively, 99.7 % and 0.3 % (see §2). Since the H component in the ground state of $^{10}_{\rm YY}$Be amounts to about 70 % (see Table 3), we may say that the state is an H-dibaryon state coupled with the nuclear core. However, in comparison of the state with the free H-dibaryon state, we see a little difference between them. The ΞN component is largest in the free H state, while the $\Lambda\Lambda$ component is largest in the hypernuclear state.

The difference can be seen in the probability density $\rho(r)$ for the $\Lambda\Lambda$, ΞN and $\Sigma\Sigma$ in Fig. 2(b). Refecting the fact that the channel couplings occur in the short-range region, the peaks of the ΞN and $\Sigma\Sigma$ channels appear around 0.6 fm. In the $\Lambda\Lambda$ channel, the peak moves to the inside region due to the channel coupling effect in comparison with the case of only the core+Λ+Λ channel (see Fig. 2(a)), and it appears in the intermediate region with $r \sim 1.2$ fm. When we compare them with the free H case in Fig. 1, we find some difference between them. The reasons why the difference appears are given as the follows: 1) When we consider only the $\Lambda\Lambda$ channel, the $(3q)$+$(3q)$ system is not bound, while the core+$(3q)$+$(3q)$ system is bound with $B_{\Lambda\Lambda}$=16 MeV due to the attractive core–Λ potential. This fact pushes down in energy only the $\Lambda\Lambda$ channel and makes large the energy difference between the $\Lambda\Lambda$ and ΞN channels in the core+$(3q)$+$(3q)$ system. 2) Since only the core+Ξ+N channel in the core+$(3q)$+$(3q)$ system has one Pauli-forbidden state between the core and valence nucleon, the nucleon must be occupied in the p state, and the Ξ particle must be also in the p state due to the parity conservation, although the situation is not realized in the $(3q)$+$(3q)$ system. This fact makes weak the $\Lambda\Lambda$-ΞN and ΞN-$\Sigma\Sigma$ couplings in the core+$(3q)$+$(3q)$ system in comparison with the $(3q)$+$(3q)$ system. Consequently, the $\Lambda\Lambda$ channel component in the core+$(3q)$+$(3q)$ system becomes largest in the hypernuclear ground state.

4 Summary

The hyperon mixing problems in light $S = -2$ nuclei have been studied with the [core+Λ+Λ] and [core+Ξ+N] model for the Nijmegen model-D interaction (OBEM), and the core+$(3q)$+$(3q)$ model for the Kyoto-Niigata RGM-F interaction (QCM). In case of the Nijmegen model-D interaction, the Ξ component in the ground and excited states of ${}^{13}_{\Lambda\Lambda}$B and ${}^{10}_{\Lambda\Lambda}$Be are as small as $P_\Xi = 0.3 \sim 0.7$ %, and the binding energy gain due to the $\Lambda\Lambda$-ΞN coupling is $\Delta B = 0.3 \sim 0.5$ MeV. We found the two relations, $P_\Xi(ex.)/P_\Xi(g.s.) \sim 2$ and $\Delta B(ex.)/\Delta B(g.s.) \sim 1$, which do not depend strongly on the strength of the $\Lambda\Lambda$-ΞN conversion potential. On the other hand, with the RGM-F interaction employed, the hyperon mixing effect is strong in the ${}^{10}_{\rm YY}$Be ground state, (67, 16 and 17 % for the $\Lambda\Lambda$, ΞN and $\Sigma\Sigma$ components, respectively), and the ${}^{10}_{\rm YY}$Be state has a H-nuclear-like structure, although it is overbound by about 7 MeV in comparison with the experimental data. Our results indicate that light $S = -2$ nuclei have much different structure in case of the OBEM and QCM interactions.

References

1. M.M. Nagels, T.A. Rijken and J.J. de Swart, *Phys. Rev.* D **12**, 744 (1975); **15**, 2547 (1977); T.A. Rijken, talk in this symposium.
2. K. Tominaga et al., *Nucl. Phys.* A **642**, 483 (1998).
3. M. Oka and K. Yazaki, in *Quarks and Nuclei*, ed. W. Weise (World Scientific, Singapore, 1984), p.241.
4. C. Nakamoto, Y. Suzuki and Y. Fujiwara, *Prog. Theor. Phys.* **97**, 761 (1997); Y. Suzuki, talk in this symposium.
5. Y. Yamamoto et al., *Prog. Theor. Phys.* Suppl.**117**, 361 (1994).

PRODUCTION OF DOUBLE-Λ HYPERNUCLEI : BNL-AGS E906

T. Fukuda, T. Nagae, H. Outa, M. Sekimoto
KEK, Tsukuba, Ibaraki 305-0801, Japan

H. Hotchi, T. Miyachi, J. Nakano, T. Tamagawa, K. Tanida
Department of Physics, University of Tokyo, Tokyo 113-0033, Japan

R. E. Chrien, M. May, E. Meyer, P. Pile, A. Rusek, R. Sutter
Brookhaven National Laboratory, Upton, NY 11973, USA

A. Berdoz, D. Carman, P. Eugenio, G. B. Franklin, P. Khaustov, P. Koran, C. Meyer, K. Paschke, B. P. Quinn, R. A. Schumacher
Department of Physics, Carnegie Mellon University, Pittsburgh, PA 15213, USA

L. Gan, L. Tang, L. Yuan
Department of Physics, Hampton University, Hampton, VA 23668, USA

A. Kurepin, V. Rasin, M. Prokhavatilov, K. Shileev
Institute for Nuclear Research, Russian Academy of Sciences, 117312 Moscow, Russia

J. K. Ahn, H. Akikawa, K. Imai, A. Ichikawa, K. Yamamoto, M. Yosoi
Department of Physics, Kyoto University, Sakyo-ku, Kyoto 606-8224, Japan

S. Ajimura, T. Kishimoto, H. Kori, S. Minami, Y. Shimizu
Department of Physics, Osaka University, Toyonaka, Osaka 560-0043, Japan

Z. Meziani
Department of Physics, Temple University, Philadelphia, PA 19122-6082, USA

H. Fischer, J. Franz, H. Schmitt
Fakultät für Physik, University of Freiburg, D-79104 Freiburg, Germany

C. A. Davis, M. Landry
Department of Physics, University of Manitoba, Winnipeg, Manitoba R3T 2N2, Canada

B. Bassalleck
Department of Physics, University of New Mexico, Albuquerque, NM 87131, USA

We have carried out an experiment at BNL-AGS (E906) to search for double-Λ hypernuclei by observing sequential pionic decays . We will describe the principle of the experiment and report the present status.

1 Introduction

The study of double-Λ hypernuclei will give us information concerning the Λ-Λ force, which is quite important to understand the baryon-baryon interactions in a unified way. Moreover, the existence of double-Λ hypernuclei may place limits on the mass of the predicted stable dibaryon (H), i.e. the observation of a double-Λ hypernucleus will exclude an H mass smaller than $2M_\Lambda$ - (binding energy of the two Λ's in the double- Λ hypernucleus).

However, the experimental studies on double-Λ hypernuclei have been limited so far; in the 1960's only two double-Λ hypernucleus events were reported, one ${}^{10}_{\Lambda\Lambda}$Be[1] and one ${}^{6}_{\Lambda\Lambda}$He[2], although the event in reference 2 is not convincing[3]. These events remained the only ones until recently when a new observation was made at the KEK-PS E176 experiment[4]. However the interpretation of the event is not unique; i.e. either ${}^{10}_{\Lambda\Lambda}$Be or ${}^{13}_{\Lambda\Lambda}$B, and accordingly the extracted Λ-Λ interaction energy is either -4.9 ± 0.7MeV (repulsive Λ-Λ interaction) or $+4.9\pm0.7$MeV (attractive Λ-Λ interaction).

We proposed a new way to study the double-Λ hypernuclei with much more statistics (BNL-AGS E906)[5]. In the following sections, we will discuss the experimental principle and describe the present status of the experiment.

2 PRINCIPLE OF THE EXPERIMENT

The production of double-Λ hypernuclei has been made so far by the stopped-Ξ^- method, where Ξ^- is emitted via a quasi-free process in the (K^-,K^+) reaction and stopped in the target material. Then a Ξ^- -atom is formed and Ξ^- is eventually absorbed by a nucleus, and a proton in the nucleus and the Ξ^- will be converted into two Λ's by the $\Xi^- p \to \Lambda\Lambda$ strong interaction. Then, the double-Λ sticking state is formed with probability of more than ~10 %[6], and it decays into some fragments including a double-Λ hypernucleus or twin single-Λ hypernuclei.

On the other hand, in the (K^-,K^+) reaction, the Ξ^- emission probability is about 70-80 % of the total quasi free events and the rest can be attributed to more complicated processes, such as secondary interaction of Ξ^- inside a nucleus[7,8]. After these complicated processes, there may be a chance for $S = -2$ (eventually two Λ's) to be trapped in a nucleus and to form a double-Λ hypernucleus. A theoretical attempt[9] was made in order to evaluate the production rate of the double-Λ hypernuclei through the quasi-free production of Ξ^- followed by the secondary reaction of Ξ^- inside a nucleus. The estimated rate of the formation of double-Λ compound nucleus is about 0.3 % per quasi free reaction for a ^{12}C target, assuming the Ξ^--nucleus potential of 15 MeV depth. The double-Λ compound nucleus will then decay into hyperfragments such as ${}^{6}_{\Lambda\Lambda}$He and ${}^{5}_{\Lambda\Lambda}$H etc or twin single-Λ hypernuclei.

The double-Λ hypernuclei, which are produced via stopped-Ξ^- process and/or through the quasi-free production of Ξ^- followed by the secondary reaction, can be identified by detecting characteristic mesonic-decaying pions. Let's consider the ${}^{5}_{\Lambda\Lambda}$H case as an example; ${}^{5}_{\Lambda\Lambda}$H can be identified by observing two characteristic pions from the successive pion decays (${}^{5}_{\Lambda\Lambda}\mathrm{H} \to {}^{5}_{\Lambda}\mathrm{He} + \pi^-$ and ${}^{5}_{\Lambda}\mathrm{He} \to {}^{4}\mathrm{He} + p + \pi^-$), provided the widths of the pion spectrum are small enough. A recent calculation [10] shows that the widths of the pion spectrum for ${}^{5}_{\Lambda}$He is about 1.0 MeV, which is consistent with the experimental data of ${}^{5}_{\Lambda}$He decay[11]. From the momentum spectrum of pions in coincidence with the pions with momentum of ~100 MeV/c (${}^{5}_{\Lambda}\mathrm{He} \to {}^{4}\mathrm{He}+p+\pi^-$), we can determine the mass of ${}^{5}_{\Lambda\Lambda}$H. Note that the branching ratios of pion decay rates $\Gamma_{\pi^-}/\Gamma_{tot}$ will be large enough to be observed ; ~ 0.21 and ~ 0.39 for ${}^{5}_{\Lambda\Lambda}$H and ${}^{5}_{\Lambda}$He[12], respectively. It should be also noticed that the

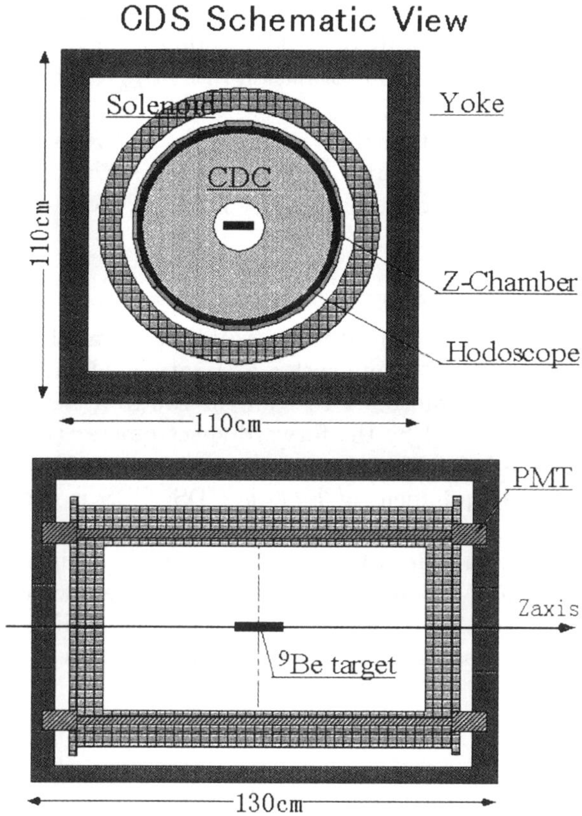

Figure 1: A schematic drawing of the Cylindrical Detector System (CDS)

momenta of decaying pions of $^{5}_{\Lambda\Lambda}$H are expected to be around 130 MeV/c and well separated from other decaying pions. In addition to this decay mode, $^{5}_{\Lambda\Lambda}$H will also decay to $^{4}_{\Lambda}\mathrm{H} + \mathrm{p} + \pi^-$ with a large branching ratio (~ 0.33). The width of this π^- is expected to be small enough ($\sim$ 2 MeV) and $^{4}_{\Lambda}$H will decay with a high energy pion (132.9 MeV/c) and with a large branching ratio (~ 0.36), so that this decay mode can be also used to identify $^{5}_{\Lambda\Lambda}$H . Note that the branching ratio $\Gamma_{\pi^-}/\Gamma_{tot}$ of $^{5}_{\Lambda\Lambda}$H depends only slightly on the binding energy of $^{5}_{\Lambda\Lambda}$H [12].

Decaying pions will be detected by using a Cylindrical Detector System (CDS), which surrounds the target region (Fig.1). The momentum resolution and the solid angle of the CDS will be about 3 MeV/c FWHM (1.7MeV in energy) for 100 MeV/c pions and 65 % of 4π, respectively. We chose ^{9}Be as a target because the available species of the final hyperfragment are limited and thus easy to be identified. In fact, a calculation for the hyperfragment formation rate based on a statistical model shows a concentration of $^{5}_{\Lambda\Lambda}$H (~0.4 per double-Λ compound nucleus) [9,12] for the process through the quasi-free production. Concerning the formation rate of each hyperfragment through stopped Ξ^- process, theories only partly succeed to reproduce the experimental data [13]. For the ^{9}Be case, we expect an absorption of

Ξ^- to the "alpha" in ^{9}Be. This subsystem (Ξ^- plus "alpha") has a large overlap amplitude with the ground state of $^{5}_{\Lambda\Lambda}$H[14], so that it may eventually decay to $^{5}_{\Lambda\Lambda}$H . Here we simply assume a similar formation rate of $^{5}_{\Lambda\Lambda}$H for the stopped Ξ^- process as the quasi-free case.

If we take into account all theoretical inputs and realistic experimental conditions, we expect about 500 counts of decaying pions from $^{5}_{\Lambda\Lambda}$H in coincidence with pions from $^{5}_{\Lambda}$He and $^{4}_{\Lambda}$H for the proposed 2 x 10^{12} K^- on the Be target.

3 EXPERIMENT

The experiment was carried out at the 2 GeV/c K^- beam line (D line) at the AGS. We obtained $(1-2)\text{x}10^6$ K^- /spill with a K/π ratio of about 0.72. The accumulated total number of K^- is about 0.9 x 10^{12} for the run from Sep. to Nov. in 1999. The outgoing K^+ was analyzed by the forward spectrometer (48D48), which was used in AGS experiment 813. The target was a ^{9}Be plate 6" long x 2" wide x 1/2" high. The principal equipment of E906 is CDS. CDS is comprised of three main components, a cylindrical drift chamber (CDC) with a z-chamber, a hodoscope and a solenoidal magnet. In order to reduce the effect of multiple scattering as much as possible, we select low Z materials inside the chamber; we use 80 μm Al wire for the field wire, 20 μm W wire for the sense wire and helium-ethane mixture gas (50 : 50) for chamber gas. The size of the CDC is 298 mm in radius and 920 mm in length.

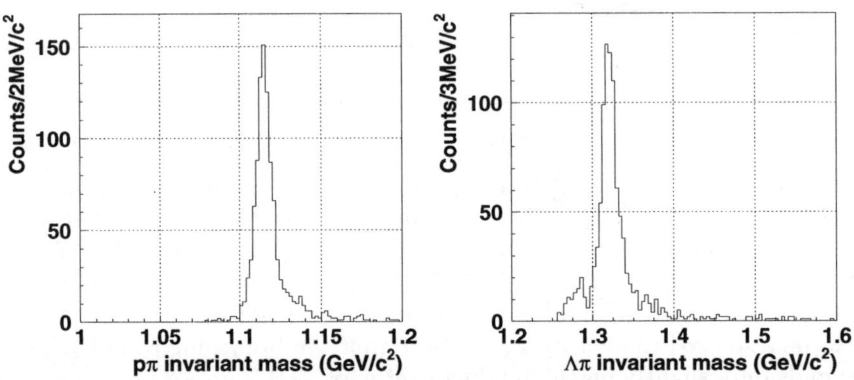

Figure 2: $p\pi^-$ (lefthand side) and $\Lambda\pi^-$ (righthand side) invariant mass spectra for one-positive and two-negative charged track events

The CDC consists of twelve layers; six axial layers and six stereo layers. The total number of the cell is 576. We place a z-chamber outside the CDC in order to get more precise information along the beam direction. The z-chamber has two identical planes, each of which comprises one layer of MWPC with two sheets of cathode readout. The pitch of the anode wires is 2.64 mm. The cathode strip is placed normal to the anode wire which is parallel to the beam direction. The width of the strip and the spacing between adjacent strips are 5 mm and 0.5 mm, respectively. The chamber gas of the z-chamber is an argon-ethane mixture gas (50 : 50). The hodoscope is used for the TOF measurement and the trigger. As the hodoscope is inside the magnetic field, we use a fine-mesh photomultiplier (Hamamatsu H6614). The size of the solenoid magnet is 1180 mm wide, 1180 mm high and 1280 mm long. The maximum field of the magnet is 5 kGauss. The homogeneity of the field is within 0.5 % over the entire CDC volume.

By analyzing the data of the beam line spectrometer and K^+ spectrometer we have clearly seen the quasi-free Ξ^- production peak in the missing mass spectrum. Concerning the performance of CDS, we have observed many clear examples of the Ξ^- decay event; $\Xi^- \rightarrow \Lambda + \pi^-$ and $\Lambda \rightarrow p + \pi^-$. By using these decay events, we have constructed $p + \pi^- \rightarrow \Lambda$ and $\Lambda\pi^-$ invariant mass spectrum, which shows a Ξ^- peak with a rms mass resolution of 8.3 MeV (Fig. 2).

4 PRELIMINARY RESULTS

At first, we analyze "two negatively-charged track events", which are considered to be mainly due to two-π^- coincidence events. Fig. 3 shows a preliminary result for the events; the momentum of pions with higher momentum is plotted when the lower momentum pion is within 90 MeV/c and 110 MeV/c, where most of the lower-momentum pions relevent to the present study should be included, e.g. ${}^5_\Lambda$He (~99 MeV/c). The other cuts for the histograms are rather loose at present; chi squares of each track are less than 5, the distance of closest approach of two tracks is less than 3 cm, and the vertex of the two tracks should be within the target boundary.

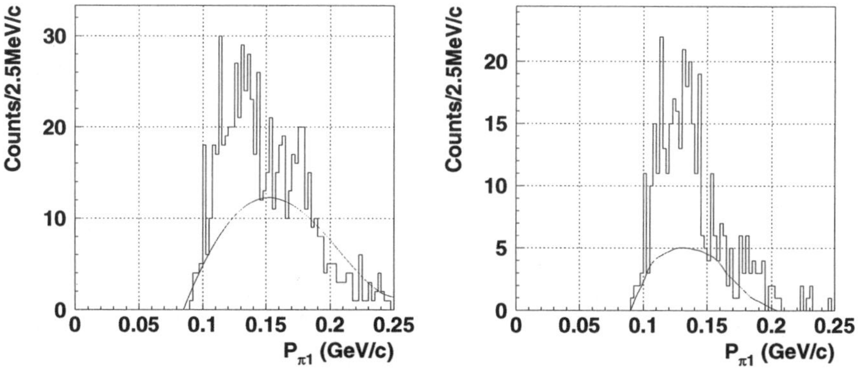

Figure 3: Spectra of the higher momentum pions gated by the lower momentum pions with 90 - 110 MeV/c. The righthand figure is the result by requiring polar angle of two pions to be greater than 60 degrees. See the text for other cuts.

The energy loss in the target has been corrected for by using the information of the vertex of the two tracks. The momentum scale has been calibrated by observing the π^+ from the decay of stopped/in-flight Σ^+ produced in the (K^-, π^-) reaction on CH_2. The main background is thought to come from the decay of the quasi-free cascade, where the proton is missing. In order to reduce this backgrond, pions whose polar angle is greater than 60 degrees are selected because the pions from cascade decay tend to be emitted forward. The righthand figure shows the result with this requirement. We can clearly see that the events less than around 150 MeV/c survive but those with higher momenta are suppressed. In the histogram the shape of the cascade decay by a Monte Carlo simulation is overlaid in order to estimate the background more quantitatively. We tentatively conclude that the events above the solid lines are due to the two-π^- decays from S = -2 nuclear systems. However they may contain both double-Λ hypernuclear decays and twin single-Λ hypernuclear ones. We are now trying to discriminate them by improving the momentum resolution and the vertex resolution etc. At present the momentum resolution of CDS is estimated to be around 5 MeV/c FWHM for 100 MeV/c pions, however ambiguities in the energy loss correction due to the vertex resolution may still exist.

The other interesting subject is an enhanced $\Lambda\Lambda$ production near threshold in the $\Lambda\Lambda$ invariant mass spectrum, which has been discussed by Ahn et al. [15] for the KEK-E224 experiment. It is claimed that the enhancement may indicate either the possible existence of a $\Lambda\Lambda$ resonance, which is referred to as an H-particle, a strong-attractive $\Lambda\Lambda$ final state interaction, or both. Unfortunately their data can't make a definite identification due to the limited statistical significance.

According to a Monte Carlo simulation, our data should contain a larger number of $\Lambda\Lambda$ pairs than E224 by more than one order of magnitude. We analyze "two negatively- and two positively- charged track events", which should include $\Lambda\Lambda$ pairs. At first we find the best candidate for a Λ out of possible combinations and construct the invariant masses for two pairs thus selected. Here we assume negative and positive particles as pions and protons, respectively. A preliminary result of the constructed invariant masses for two pairs of p and π are shown in Fig. 4, which indicates that most of the events are due to two Λ's. Fig. 5 shows the constructed invariant mass spectrum for two Λ's, where the invariant mass regions of 1.09 - 1.14 GeV are selected for two pairs of p and π. The solid curve shows, for reference, a spectral shape of the prediction based on a two-step process model; the $K^-p \to K^+\Xi^-, \Xi^-p \to \Lambda\Lambda$ process in the same nucleus. Other models like an intranuclear cascade model have been also considered in reference 15, but any simple models with no interaction between two Λ's will not predict a larger cross section near threshold. Although the number of analyzed events are smaller than expected by an order of magnitude due to the low track-finding efficiency at present, we clearly see the similar enhancement near the threshold as reference 15. For more quantitative discussion, we are now trying to improve the track-finding efficiency and also to correct for CDS acceptance, which should enhance the cross section for lower invariant mass.

Finally it should be noted that E906 is the first measurement to detect decay particles from a large number of S=-2 nuclear systems (around one million events), so that we can expect to get better information on S=-2 nuclear systems by this new technique, i.e. cascade hypernuclei and their decays, H $\to p\pi^-\Lambda$ decays, Ξ - p

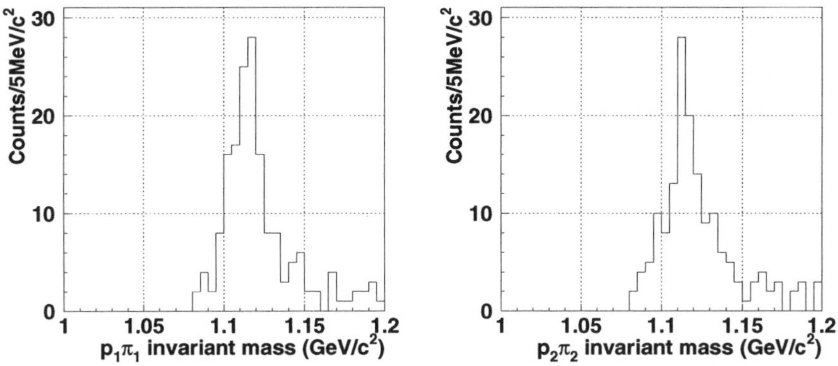

Figure 4: Constructed p π^- invariant mass spectra for p π^- p π^- events.

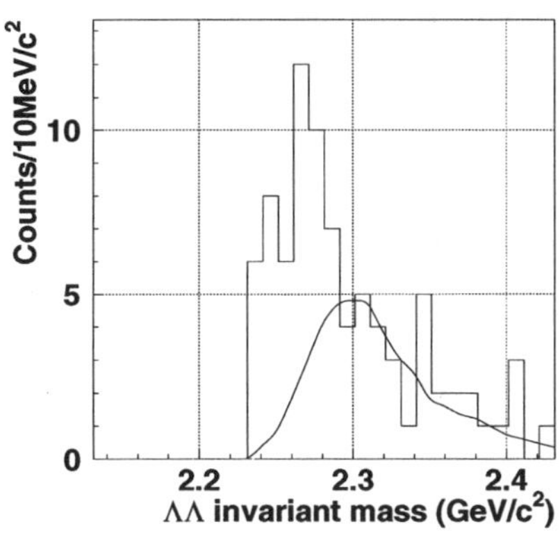

Figure 5: Constructed $\Lambda\Lambda$ invariant mass spectrum.

interactions and so on.

1. M. Danysz et al., Nucl. Phys. 49 (1963) 121.
2. P. J. Prowse, Phys. Rev. Lett. 17 (1966) 782.
3. R. H. Dalitz et al., Proc. R. Soc. Lond. A426 (1989) 1.
4. S. Aoki et al., Prog. Theor. Phys. 85 (1991) 1287.
5. T. Fukuda and R. E. Chrien, AGS proposal E906 (1994).
6. K. Nakazawa, Proc. of the 23rd INS Int. Symp. on Nuclear and Particle Physics with Meson Beams in the 1 GeV/c Region, Tokyo, 1995, p.261.
7. K. Imai, Nucl. Phys. A553 (1993) 667c.
8. T. Iijima et al., Nucl. Phys. A546 (1992) 588.
9. Y. Yamamoto, M. Wakai, T. Fukuda, M. Sano, Prog. Theor. Phys. 88 (1992) 1163.

10. T. Motoba, H. Bando, T. Fukuda, J. Zofka, Nucl. Phys. A534 (1991) 597.
11. W. Gajewski et al., Nucl. Phys. B14 (1969) 11, and P. H. Davis, private communication; Bulletin de I.U.L.B. 26,27 (1966).
12. Y. Yamamoto, M. Wakai, T. Motoba and T. Fukuda, Nucl. Phys. Nucl. Phys. A625 (1997) 107 and Y. Yamamoto, Nucl. Phys. A639 (1998) 393c.
13. T. Yamada, Nucl. Phys. A639 (1998) 385c.
14. Y. Akaishi, private communication.
15. J. K. Ahn et al., Phys. Lett. B444 (1998) 267.

$\Lambda\Lambda$ INTERFEROMETRY IN (K^-, K^+) AND AA REACTIONS

Akira Ohnishi[a], Yuichi Hirata[a], Yasushi Nara[b], Shoji Shinmura[c] and Yoshinori Akaishi[d]
a Division of Physics, Graduate School of Science, Hokkaido University, Sapporo 060-0810, Japan
b Advanced Science Research Center, Japan Atomic Energy Research Institute, Tokai, Ibaraki 319-11, Japan
c Faculty of Engineering, Gifu University, Yanagido 1-1, Gifu 501-11, Japan
d Institute of Particle and Nuclear Study, KEK (Tanashi Branch), Midorimachi 3-2-1, Tanashi 188, Japan

We analyze the invariant mass spectrum of $\Lambda\Lambda$ in ^{12}C$(K^-, K^+\Lambda\Lambda)$ reaction at $P(K^-) = 1.65$ GeV/c by using a combined framework of IntraNuclear Cascade (INC) model and the correlation function technique. The observed enhancement at low-invariant masses can be well reproduced with attractive $\Lambda\Lambda$ interactions with the scattering length either in the range $a = -6 \sim -4$ fm (no bound state) or $a = 7 \sim 12$ fm (with bound state) within the present theoretical treatment. We also discuss $\Lambda\Lambda$ correlation functions in central relativistic heavy-ion collisions as a possible way to eliminate this discrete ambiguity.

1 Introduction

Extracting hyperon-hyperon (YY) interactions is one of the most challenging current problems in nuclear physics. It provides us of an opportunity to verify various ideas on baryon-baryon interactions such as the flavor SU(3) symmetry. YY interactions may also modify the properties of neutron stars, in which abundant hyperons are expected to exist. However, it is very difficult to determine them experimentally. For example, YY scattering experiments, which can directly access YY interactions, require a collider machine with a very compact geometry because of their short lifetimes. On the other hand, only three double Λ nuclei, from which informations of the low-energy ^{1}S$_0$ $\Lambda\Lambda$ interaction are available, are found in these 35 years.[1,2,3] Therefore, other ways to extract YY interactions with various combination of hyperons at various energies have been desired so far.

One of these ways is opened up by a recent measurement of the invariant mass spectrum of $\Lambda\Lambda$ pair in ^{12}C$(K^-, K^+\Lambda\Lambda)$ reaction by Ahn et al.(KEK-E224 collaboration).[4] The measured invariant mass spectrum shows a clear enhancement at around the threshold of $\Lambda\Lambda$ compared to those calculated with the IntraNuclear Cascade (INC) model[5] and to those estimated from the three-body phase volume of ^{11}B$_{g.s.}\Lambda\Lambda$ with the aid of experimentally measured K^+ spectrum.

The invariant mass spectra, or two-particle momentum correlations, have been widely used to evaluate the source size when the final state interactions are negligible or well-known.[6,7,8,9,10,11,12] For (practically) non-interacting bosons such as $\gamma\gamma$, the correlation function tends towards two at small relative momentum, and the enhanced range in the momentum is inversely related to the source size. This enhancement purely comes from the spatially symmetric relative wave function of two bosons. For fermions, the enhancement at low relative momentum goes to one half, if we ignore final state interactions and assume a statistical spin distribution (singlet:triplet = 1:3). Without the final state interaction, the enhancement cannot exceed two, even if we assume that the spin singlet identical fermion pairs are dominant. Therefore, the measured enhancement in the invariant mass spectrum, which seems to be much larger than two, strongly suggests the necessity of the *attractive* final state interaction of $\Lambda\Lambda$.

The two particle momentum correlation between baryons are significantly affected by final state interactions and reaction mechanisms. For example, the proton-proton correlation data shows a depression at small momenta (Coulomb pocket) and an enhancement at around the ^{2}He resonance. In addition, the initial state correlation of spin is known to be important in light-ion induced reactions.[13] Since two-particle momentum correlations are evaluated from the source function and the relative wave function, it is necessary to invoke a well established strong interaction and a dynamical model for the understanding of baryon-baryon correlation. In other words, if we have a *reliable* dynamical model, it would be possible to extract the information on the interaction between the measured particle pair. Actually, precise few-body calculations made it possible to extract neutron-neutron effective range from the particle momentum correlation data.[14]

In this work, first we apply the correlation function technique to the $\Lambda\Lambda$ invariant mass spectrum data, and try to determine $\Lambda\Lambda$ interaction at low energies, by using a source function generated by the IntraNuclear Cascade (INC) model,[5] which well reproduce inclusive (K^-, K^+) spectra. In ^{12}C$(K^-, K^+\Lambda\Lambda)$ reactions, the leading particle creates two Λ at a small distance. This small source size emphasizes the effects of final state interactions at low energies, and a large enhancement is realized. However, we find that the invariant mass spectrum in this reaction is not sensitive to the long range behavior in the continuum wave function, such as the orthogonality to a bound state wave function. Hence the observed enhancement at low energies can be well reproduced with attractive $\Lambda\Lambda$ interactions having the scattering length either in the range $a = -6 \sim -4$ fm (no bound state) or $a = 7 \sim 12$ fm (with bound state).

In order to distinguish whether two lambdas bound or not, it is helpful to utilize reactions with large source size. We then show the calculated results of $\Lambda\Lambda$ correlation functions in relativistic heavy-ion collisions at BNL-AGS, CERN-SPS, and BNL-RHIC energies. The results show that the correlation functions in heavy-ion collisions are sensitive to the existence of bound state in the relative momentum range $q < 40$ MeV/c.

2 Invariant Mass Spectrum in ^{12}C(K^-, K^+) Reaction

The inclusive spectra of (K^-, K^+) reaction on various targets [15] can be explained well by the INC model,[5] as a sum of following contributions; (a) quasi-free $\Xi^{(*)}$ production, (b) heavy-meson $(\phi/a_0/f_0)$ production followed by its decay,[17] and (c) various two-step processes.[16] The predicted reaction mechanism in this model was experimentally verified recently.[16] In this work, we use this INC model to generate a hyperon source function. As for the baryon momentum spectra, however, mean field effects are also important. Thus we include them as external potentials with the depths of $-40, -30, -10$, and -16 MeV for N, Λ, Σ, and Ξ [18] baryons, respectively. In addition, we have adopted the hyperon-nucleon scattering and conversion cross

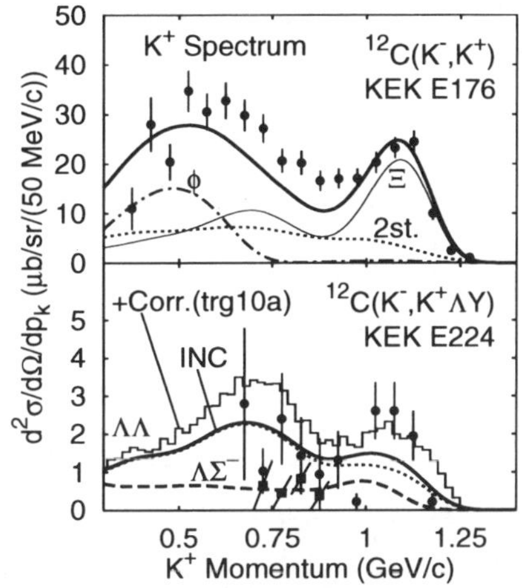

Figure 1: Calculated momentum spectra of K^+ particle in $^{12}\mathrm{C}(K^-,K^+)$ (upper panel) and $^{12}\mathrm{C}(K^-,K^+\Lambda Y)$ (lower panel) reactions are compared with experimental data.[4,15,16]

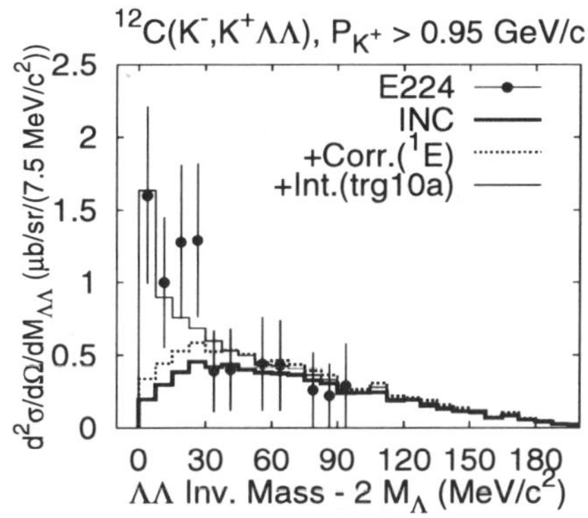

Figure 2: Calculated Λ-Λ invariant mass spectrum by using INC model without (thick) and with (thin) correlation effects are compared with data (solid circles).[4]

sections calculated with the Nijmegen model D potential.[19] For $S = -2$ collisions, we fixed the hardcore radius to be $r_c = 0.5$ fm.

In Fig. 1, we compare the calculated K^+ momentum spectra in inclusive (K^-, K^+) and coincidence $(K^-, K^+\Lambda Y)$ reactions with experimental data.[4,15,16] The calculated results (thick solid curve) well explains the data except for the underestimate of two lambda production at around $P_{K^+} \simeq 1.1$ GeV/c, and the underestimate of inclusive K^+ spectrum at intermediate momenta, $0.5 \leq P(K^+) \leq 0.9$ GeV/c. Since the $\Lambda\Lambda$ invariant mass spectrum is measured in the high momentum range of K^+, $P(K^+) \geq 0.95$ GeV/c, we do not care the latter underestimate in the following discussion of this paper.

The underestimate of coincidence cross section $(K^-, K^+\Lambda\Lambda)$ is concentrated in the low invariant mass region of $\Lambda\Lambda$, as shown in Fig. 2. This is as expected, since the correlation caused by the final state interaction modify the invariant mass spectrum at low mass region most effectively.

The enhancement of the two-particle momentum spectrum has been extensively studied in the context of correlation function. For example, in the correlation function formula,[12] the probability to find particle pair at momenta $\vec{p}_1$ and $\vec{p}_2$ reads,

$$P(\vec{p}_1, \vec{p}_2) = \int d^4x_1\, d^4x_2\, S_{12}(\vec{p}_1, x_1, \vec{p}_2, x_2) \left|\psi^{(-)}(\vec{r}_{12}; \vec{k})\right|^2 , \quad (1)$$

$$\vec{r}_{12} = \vec{r}_1 - \vec{r}_2 + \vec{P}(t_2 - t_1)/2m , \quad (2)$$

$$\vec{P} = \vec{p}_1 + \vec{p}_2 , \vec{k} = \frac{1}{2\hbar}(\vec{p}_1 - \vec{p}_2) , \quad (3)$$

where $\vec{r}_{12}$ is the relative distance of two particles at particle creation, S_{12} denotes the source function, and the wave function $\psi^{(-)}$ is chosen to have the outgoing

Table 1: Two-range Gaussian (trg) $\Lambda\Lambda$ potential parameters which give a local χ^2 minimum for a given range parameter μ_l. The shorter range $\mu_s = 0.45$ fm is fixed for simplicity. For each parameter set, the scattering length a, the effective range r_{eff}, the reduced residual error $\tilde{\chi}^2 \equiv \chi^2/\text{DOF}$, and the binding energy are also shown.

	μ_l (fm)	μ_s (fm)	v_l (MeV)	v_s (MeV)	a (fm)	r_{eff} (fm)	$\tilde{\chi}^2$	B.E. (MeV)
trg06a	0.60	0.45	-900.0	1440.0	-4.4	1.6	0.34	U.B.
trg08a	0.80	0.45	-230.0	470.0	-5.0	1.8	0.36	U.B.
trg10a	1.00	0.45	-105.0	200.0	-6.2	2.0	0.39	U.B.
trg06b	0.60	0.45	-950.0	1310.0	7.5	1.2	0.38	0.72
trg08b	0.80	0.45	-270.0	410.0	8.5	1.3	0.40	0.56
trg10b	1.00	0.45	-135.0	210.0	11.5	1.6	0.43	0.29

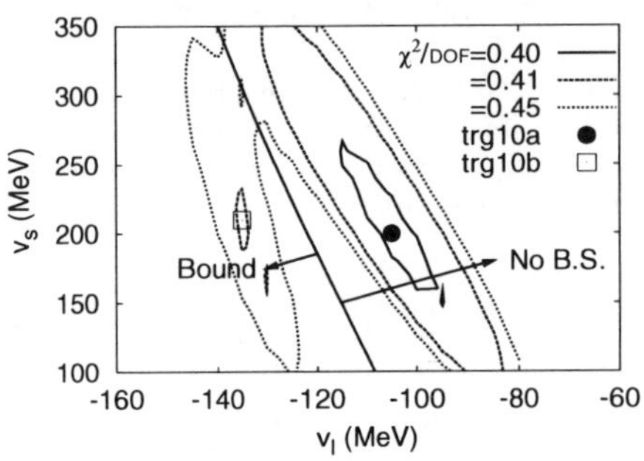

Figure 3: χ^2 surface in the potential strength plane, (v_l, v_s), for $\mu_l = 1.0$ fm. There are two local minima, trg10a and trg10b, shown in a solid point and a open square, respectively.

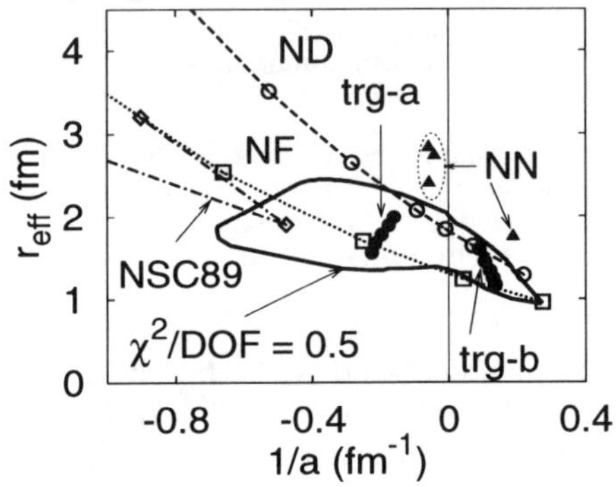

Figure 4: Extracted Λ-Λ scattering parameters. Fitted two-range gaussian potentials (solid circles) are compared with Nijmegen models (open marks). Inside the thick solid line, potential with $\tilde{\chi}^2 \leq 0.5$ exists in the range $0.6 \leq \mu_l \leq 1.0$ fm.

relative momentum. In this work, we limit ourselves to working in a single channel description of $\Lambda\Lambda$, and assume that two Λ particles are in their spin singlet state. Then, the relative wave function of $\Lambda\Lambda$ at small relative momentum is simplified as follows.

$$\psi^{(-)}(\vec{r};\vec{k}) \simeq \sqrt{2}\left[\cos(\vec{k}\cdot\vec{r}) - j_0(kr) + e^{-i\delta_0}u_0(r;k)\right] . \tag{4}$$

Based on these formula, we have calculated the $\Lambda\Lambda$ invariant mass spectrum including the correlation effects. In INC, we have specified the particle "creation" point to be the last point where the Λ particle is created by the leading particle or it collides with other nucleons. Here we have used phenomenological $\Lambda\Lambda$ potentials parametrized in a two-range Gaussian form, $V_{\Lambda\Lambda}(r) = v_l \exp(-r^2/\mu_l^2) + v_s \exp(-r^2/\mu_s^2)$. As shown by the thin histogram in Fig. 2, the enhancement at low invariant masses can be reproduced with an appropriate potential, although the second "peak" around 25 MeV is difficult to explain. At higher invariant masses the correlation effects are washed out by the oscillation of $|\psi|^2$ in the source spreading upto around $r_{12} \simeq 5$ fm.

Some reader may doubt that the enhancement at low energies are too large to explain by the correlation, since the correlation functions (enhancements) are usually less than two, and *pp* correlations are almost zero at low momenta. However, it is the Coulomb interaction that suppresses *pp* correlation at low energies, and the measured *nn* correlation functions in heavy-ion collisions exhibit a strong maximum, $C(q \to 0) > 5$, at zero energies.[20]

We have searched for the χ^2 local minima in the (v_l, v_s) plane for a given longer range parameter, $\mu_l = 0.6 \sim 1.0$ fm (Table 1). There are generally two χ^2 local minima as shown in Fig. 3. In Fig. 4, we compare the extracted $\Lambda\Lambda$ potentials with the Nijmegen models.[19,21,22] From this comparison, model ND [19] (NF [21]) with the hardcore radius $r_c \simeq 0.50(0.46)$ fm, and model NSC89 [22] with the cutoff mass $M_{cut} \simeq 920$ MeV are expected to give reasonable enhancement in the $\Lambda\Lambda$ invariant mass spectrum.

3 Do Two Lambdas Bound ?

As shown in Table 1, all the best fit parameters (trg-a) give negative scattering length ($\delta_0 \sim -ak$), implying that there is no bound state. However, there is another local minimum (trg-b) in the region where there is a bound state, $a > 0$. This double-well structure appears because the enhancement is described by the wave function squared. With the long wave approximation, the wave function becomes essentially a constant, then the correlation effect can be factorized as follows,[14]

$$P(\vec{p}_1,\vec{p}_2) \simeq 2F(k)\, P_c(\vec{p}_1,\vec{p}_2)\ , \quad P_c = \int d^4x_1\, d^4x_2\, S_{12}\ , \quad F(k) = (1-a/b)^2 - ck^2, \tag{5}$$

where b denotes the intrinsic range. There are two solutions of a giving the same enhancement at low energies, $a \sim b\left(1 \pm \sqrt{F(0)}\right)$. When the spectrum is enhanced at low energies, $F(0) > 1$, one of them is negative (no bound state) and the other becomes positive (with bound state).

This situation is shown in the upper panel of Fig. 5, where we show the (momentum integrated) source distribution, and the wave function in the $^1S_0$ channel calculated with the potential parameters of trg10a and trg10b at asymptotic relative momenta of k =0.4, 0.2, 0.1 and 0.01 fm^{-1}. All these momenta are in the range of the lowest energy bin of experimental data. Although there are some differences in the wave functions at extremely small momenta, we cannot distinguish them when integrated out over the momenta in the above range.

In order to overcome this problem and to distinguish these two kind of solutions, it may be helpful to use relativistic heavy-ion collisions, where the source size is much larger than that in (K^-, K^+) reaction and the long wave approximation would not work. For example, the scattering wave function have at least one node in the range $r \leq a$ when a bound state exists, which suppresses the contribution at around $r \sim a$ in two-particle correlations. Since this behavior of wave functions is robust and remains true at low momenta, we expect significant sensitivity to the existence of $\Lambda\Lambda$ bound state if the source covers the range, $3 \leq r_{12} \leq 15$ fm.

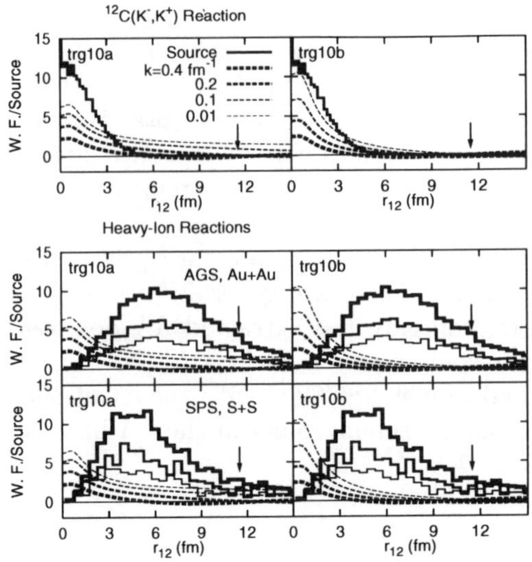

Figure 5: $\Lambda\Lambda$ relative distance distribution in (K^-, K^+) reaction and central relativistic heavy-ion collisions at AGS, SPS, and RHIC energies. The wave functions in $^1S_0$ channel at several momenta are also shown for trg10a (left) and trg10b (right).

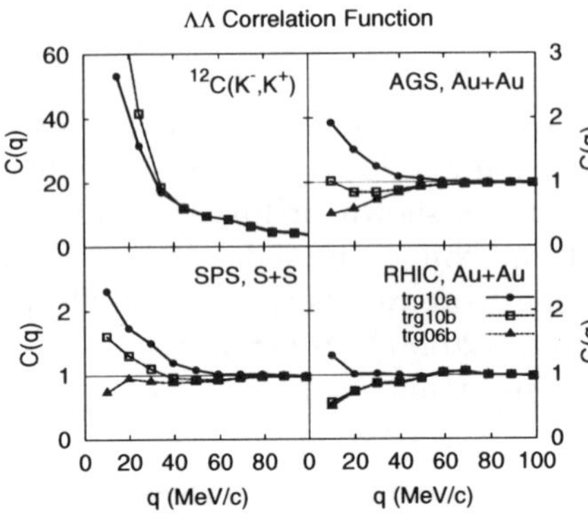

Figure 6: $\Lambda\Lambda$ correlation function in (K^-, K^+) reaction and central relativistic heavy-ion collisions at AGS, SPS, and RHIC energies. The results with trg10a (solid circles) and trg10b (open squares) are compared.

In order to generate relativistic heavy-ion collision events, we have applied a recently developed hadron-string cascade model named JAM[23] to reactions at AGS, SPS and RHIC energies. One of the good features of JAM is its wide energy range of applicability. For example, particle productions are modeled by the resonance production at low energies ($\sqrt{s_{NN}} < 3.5$ GeV), soft string excitation upto around SPS energies, and multiple minijet production at higher energies. At AGS and SPS energies, this model has been shown to work well in describing m_T and rapidity distribution of p, π and K for various combination of projectile and targets.[23]

We have generated central collision events of Au+Au collisions at AGS and RHIC energies and S+S collisions at SPS energy. Generated source distributions at several relative momenta are shown in the lower panel of Fig. 5. These distributions cover the range where the difference of the wave function is clear.

In Fig. 6, we show the calculated $\Lambda\Lambda$ correlation function, $C(q)$, in relativistic heavy-ion collisions with the above source function generated by using JAM. Here statistical spin distribution is assumed. As expected in the above discussion, two-particle correlations are suppressed at small momenta with trg10b and trg06b (with bound state) compared with trg10a (no bound state). When the source size is large enough (Au+Au collisions), the difference becomes larger. Especially, there is a qualitative difference in Au+Au collisions that the correlation function becomes smaller than unity when there is a bound state, while it has a maximum at zero relative momentum for interactions without bound state.

4 Summary

In this paper, first we have analyzed the $\Lambda\Lambda$ invariant mass spectrum in ^{12}C$(K^-, K^+\Lambda\Lambda)$ reaction by using the classical source function generated by using the IntraNuclear Cascade (INC) model combined with the correlation function formula, which takes account of $\Lambda\Lambda$ interaction. Within a single channel description of $\Lambda\Lambda$, and under the assumption that two lambdas are in their spin-singlet state, we limit the allowed range of $\Lambda\Lambda$ scattering parameters. The observed enhancement at low-invariant masses can be well reproduced with attractive $\Lambda\Lambda$ interactions with the scattering length either in the range $a = -6 \sim -4$ fm (no bound state) or $a = 7 \sim 12$ fm (with bound state) within the present theoretical treatment. The origin of this discrete ambiguity is discussed in the context of a long wave approximation.

Next we have discussed a possibility to use relativistic heavy-ion collisions as one of the ways to distinguish the $\Lambda\Lambda$ potentials, which reproduce the invariant mass spectrum of $\Lambda\Lambda$ equally well in (K^-, K^+) but have different sign of the scattering length. The key point is that the continuum wave function has to have at least one node in the range $r \leq a$ if there is a bound state. We find that there appears a qualitative difference in the correlation function $C(q)$, for (really-)heavy-ion collisions, such as Au+Au, which have a large source size.

Although we can also see a similar difference in the correlation function in ^{12}C(K^-, K^+) reaction, we cannot estimate the "uncorrelated" probability in this reaction. Two Λ particles are created by the same leading particle and dynamically correlated in (K^-, K^+) reactions. Therefore, we cannot define the correlation coming from the $\Lambda\Lambda$ interaction separately from this dynamical correlation.

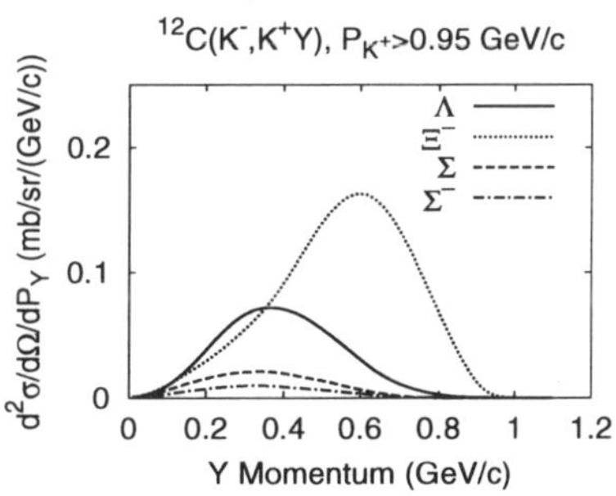

Figure 7: Hyperon momentum spectrum in ^{12}C(K^-, K^+Y) reaction with the condition $P(K^+) > 0.95$ GeV/c.

As discussed at the meeting, there are still large ambiguities to be fixed in the theoretical models which generate source functions. For example, we have to estimate the cross sections of experimentally unknown processes, such as $\eta N \rightarrow KY$, which we estimated by applying the generalized Breit-Wigner formula at KEK energy, $P(K^-)$ =1.65 GeV/c. At higher incident momentum such as in BNL-E906 experiment,[24] where the expected number of $\Lambda\Lambda$ pairs exceeds the current data by around two order of magnitude, strings start to appear and it becomes more elaborate to estimate cross sections. However, we already have careful estimates of K^-N cross sections at this momentum for several channels.[23] In addition, we can in principle reduce the ambiguity by analyzing other cross sections such as (K^-, π) or (K^-, Λ). Especially, it is necessary to analyze the inclusive Λ spectrum (Fig. 7) to verify various inputs of cascade models, before getting any decisive conclusions.

Acknowledgments

The authors would like to thank Dr. J. K. Ahn and Prof. J. Randrup, and all the participants of this workshop for fruitful discussions. This work was supported in part by the Grant-in-Aid for Scientific Research (Nos. 07640365, 08239104 and 09640329) from the Ministry of Education, Science and Culture, Japan.

1. M. Danyz *et al.*, Nucl. Phys. **49** (1963), 121.
2. D. J. Prowse, Phys. Rev. Lett. **17** (1966), 782.
3. S. Aoki *et al.* (KEK-E176 collaboration), Prog. Theor. Phys. **85** (1991), 1287.
4. J.K. Ahn et al. (KEK-E224 collaboration), Phys. Lett. **B444** (1998), 267.
5. Y. Nara, A. Ohnishi, T. Harada, and A. Engel, Nucl. Phys. **A614** (1997) 433.
6. R. Hanbury Brown and R.Q. Twiss, Philos. Mag. **45** (1954), 663; Nature **177** (1956), 27; Nature **178** (1956), 1046.
7. G. Goldhaber et al., Phys. Rev. Lett. **3** (1959), 181; Phys. Rev. **120** (1960), 300.
8. E. V. Shuryak, Phys. Lett. **B44** (1973), 387.
9. S. Pratt, Phys. Rev. Lett. **53** (1984), 1219.
10. S. E. Koonin, Phys. Lett. **B70** (1977), 43.
11. W. G. Gong et al., Phys. Rev. C **43** (1991), 781.
12. W. Bauer, C.K. Gelbke, and S. Pratt, Annu. Rev. Nucl. Part. Sci. **42** (1992), 77.
13. F. Zhu et al., Phys. Rev. **C44** (1991), R582.
14. I. Slaus, Y. Akaishi and H. Tanaka, Phys. Rep. **173**, 257-300 (1989).
15. T. Iijima *et al.* (KEK-E176 collaboration), Nucl. Phys. **A546** (1992) 588.
16. J.K. Ahn et al. (KEK-E224 collaboration), Nucl. Phys. **A625** (1997) 231.
17. C. Gobbi, C. B. Dover and A. Gal, Phys. Rev. **C50** (1994) 1594.
18. T. Fukuda et al. (KEK-E224 collaboration), Phys. Rev. **C58** (1998), 1306.
19. M. M. Nagels, T. A. Rijken, and J. J. de Swart, Phys. Rev. **D15** (1977), 2547.
20. B. Jakobsson et al., Phys. Rev. **C44** (1991), R1238.
21. M.M. Nagels, T.A. Rijken and J.J. de Swart, Phys. Rev. **D20** (1979), 1633.
22. P.M. Maessen, T.A. Rijken and J.J. de Swart, Phys. Rev. **C40** (1989), 2226.
23. Y. Nara, Nucl. Phys. A638 (1998) 555c; Eprint nucl-th/9802016;
    Y. Nara, N. Otuka, A. Ohnishi and T. Maruyama, Prog. Theor. Phys. Suppl. **129** (1997), 33;
    Y. Nara, N. Otuka, A. Ohnishi, K. Niita, S. Chiba, submitted, Eprint nucl-th/9904059.
24. T. Fukuda, in this proceedings.

SEARCH FOR THE H-DIBARYON AND S=-2 NUCLEI VIA THE (K^-, K^+) REACTION (BNL-AGS E885)

K. YAMAMOTO FOR THE BNL-AGS E885 COLLABORATION
Department of Physics, Kyoto University, Kyoto 606-8502, Japan

We have performed an experiment to search for the H-dibaryon and S=-2 nuclei through the (K^-, K^+) reaction at the BNL-AGS (E885). Data analysis is currently in progress. An overview of the experiment and some preliminary results on the direct production of the S=-2 system are presented.

1 Introduction

A study of the double-strangeness systems is one of the most important and challenging subjects in hadron physics in order to improve our understanding on the baryon-baryon interaction and the non-perturbative QCD. Especially, the possible existence of the H-dibaryon has been a long-standing problem, which was originally proposed as a stable 6-quark state against the strong interaction using the MIT bag model [1]. We have approached the above physics interests with the (K^-, K^+) reaction on a carbon target. Since two units of strangeness are exchanged at the reaction vertex, the (K^-, K^+) reaction is an appropriate way to study the S=-2 systems. There are two different processes to produce them. One is the direct production process. The S=-2 nuclei and/or the H-dibaryon are created at the (K^-, K^+) reaction vertex as shown below.

$$K^- +^{12} \mathrm{C} \rightarrow \begin{cases} K^+ + {}^{12}_{\Xi}\mathrm{Be} \\ K^+ + {}^{12}_{\Lambda\Lambda}\mathrm{Be} \\ K^+ + H + X \end{cases} \tag{1}$$

Since the K^+'s from the above reactions could have a higher momentum than those from the quasi-free reaction, the S=-2 objects can be identified as a peak structure in the low missing mass region below free $\Xi^- +^{11} \mathrm{B}$ mass threshold. Another process to produce the S=-2 system is via the atomic capture of a stopped Ξ^-. In this process, Ξ^- hyperons are created first by

$$K^- + \text{'}p\text{'} \rightarrow K^+ + \Xi^-, \tag{2}$$

where 'p' stands for a proton in a carbon nucleus. The Ξ^-'s are decelerated and brought to rest in the target. Then the formation of the S=-2 system is expected from Ξ^- atoms as follows:

$$(\Xi^-, {}^{12}\mathrm{C})_{atom} \rightarrow \begin{cases} H + X \\ {}^{12}_{\Lambda\Lambda}\mathrm{B} + n. \end{cases} \tag{3}$$

In E885, above both processes are objects of the data analysis. However, we stress the direct production process in this paper. Although there exist several previous experiments which aimed at the similar process [2,3,4], we have the better sensitivity owing to our large statistics.

2 Experiment

The experiment was carried out using a 1.8GeV/c K^- beam at the BNL-AGS D6 beam line [5]. The experimental setup is shown in Figure 1. The typical K^-/π^- ratio of the incident beam was 1/1 with the intensity of $1.0 \times 10^6\, K^-$/spill. The discrimination between K^-'s and π^-'s was achieved by an aerogel Čerenkov counter (IC, index 1.03) and time-of-flight.

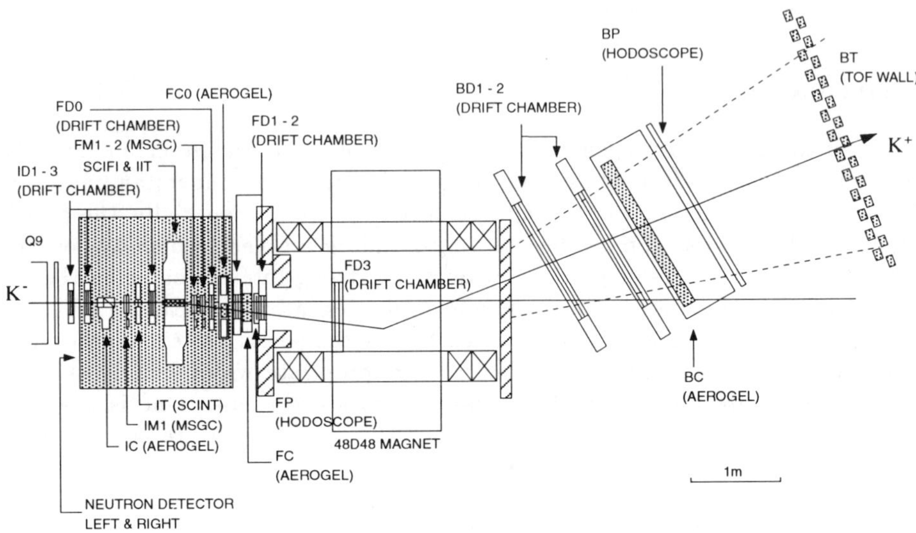

Figure 1: Schematic view of the experimental setup

In order to produce (K^-, K^+) reactions, a diamond target, which is made of artificial diamond powder, was placed at the final focusing point of the beam line. The dimension is 8cm(x) $\times$ 1cm(y) $\times$ 5cm(z) and the density is 3.3g/cm^3. Due to high density, it was also used as a stopping and absorption target of Ξ^-'s so as to increase the probability of the atomic capture process.

The outgoing K^+'s were measured by a K^+ spectrometer at the downstream of the target. The geometric acceptance is about 50msr for the particle with momentum of 1.2GeV/c and the scattering angles from 1 degree to 14 degrees are covered. The momentum of particles was analyzed by a dipole magnet (48D48, 1.4Tm) with the tracking information from drift chambers (FD1-3,BD1-2). One of the main backgrounds in the spectrometer was π^+ from a reaction such as $p(K^-, \pi^+)\Sigma^-$. The separation of K^+'s from π^+'s was accomplished by aerogel Čerenkov counters (FC,BC, index 1.04). Another serious background is the proton resulting from the elastic and inelastic K^-p scattering. In order to suppress these protons, another aerogel Čerenkov counter (FC0) was installed and the second level trigger based on a momentum-velocity correlation was also introduced. The radiator of FC0 has a refraction index of 1.112 and can suppress protons with $P_p < 1.84$GeV/c. The second level trigger is mainly composed of a fast TDC (LeCroy FERET) and a memory look-up unit (MLU). It makes a logical decision using a combination of IT-BT time-of-flight and a FD3 $\times$ BT hit pattern which reflects momentum. The combined information of time-of-flight and momentum is roughly equivalent to the mass reconstruction. Due to this trigger, the typical event rate was reduced to 100/spill, which was small enough for the data acquisition.

Over and under the diamond target, two identical scintillating fiber (SCIFI) blocks were installed in order to observe the charged-particle tracks of decay products coming out of the target. Neutron counters were placed on both the left and right sides of the target. The detailed description on these detectors can be found elsewhere.[6]

3 Data analysis

As mentioned in Section 1, a signal of the direct production of the S=-2 system can be searched for by looking at the missing mass spectrum below the mass threshold of a free Ξ^- and ^{11}B. Figure 2 shows the excitation energy spectrum of the $(\Xi^-, ^{11}\mathrm{B})$ system, which is equivalent to the missing mass spectrum whose origin is moved to the threshold of $\Xi^- + ^{11}\mathrm{B}$. By applying the tight kinematic cuts, most of the backgrounds were rejected and the spectrum was cleaned up below the threshold, especially in the deeply bound region. However, some events still remained near the threshold. A resolution function of the missing mass is essential in this analysis and it turned out to be a single Gaussian with 6MeV r.m.s. It was evaluated from the data analysis of a CH_2 (polyethylene) target. We found the missing mass peak on the Ξ^- production against a free proton, $p(K^-, K^+)\Xi^-$, can be well-fitted with a single Gaussian with 3.8MeV r.m.s., which corresponds to 6MeV r.m.s. for the

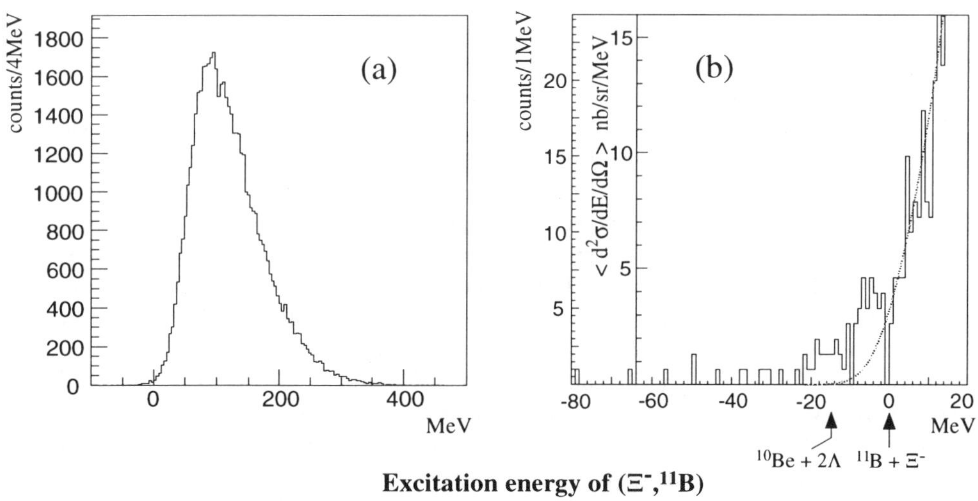

Figure 2: Excitation energy spectrum of the $(\Xi^-, ^{11}\mathrm{B})$ system (a): in the whole range of the quasi-free (K^-, K^+) reactions. (b): around the threshold of $^{11}\mathrm{B} + \Xi^-$.

two-body kinematics on a ^{12}C target. The dotted line in Figure 2(b) represents a polynomial fit to the region of the spectrum above the threshold folded with 6MeV-resolution. There seems some excess in the bound region compared with an expected tail. This may suggest a Ξ^- hypernuclear state which has been discussed previously [7]. But a careful comparison between the theoretical calculation and the data is needed. Since there are few events left below the threshold of ^{10}Be + 2Λ, we set an upper limit on the forward-angle cross section of $^{12}\mathrm{C}(K^-, K^+)^{12}_{\Lambda\Lambda}\mathrm{Be}^*$ as a function of the binding energy of 2Λ's, which is shown in Figure 3. A signal peak with 6MeV r.m.s. is assumed. From a theoretical point of view concerning the direct production of the double-Λ hypernuclei, a calculation for $^{16}\mathrm{O}(K^-, K^+)^{16}_{\Lambda\Lambda}\mathrm{C}^*$ has been made, which was based on the two-step process, $K^-p \to \pi^0\Lambda$ followed by $\pi^0 p \to K^+\Lambda$, in the same nucleus [8]. The forward-angle cross section of a few nanobarn/sr was predicted for a 1.1GeV/c incident K^- beam there.

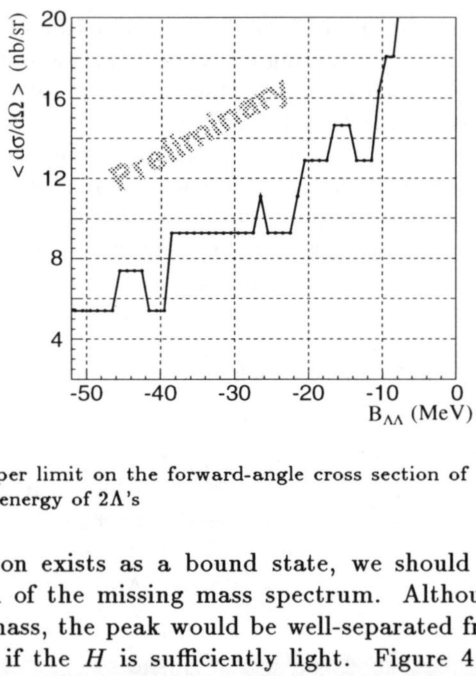

Figure 3: 90% C.L. upper limit on the forward-angle cross section of $^{12}\mathrm{C}(K^-, K^+)^{12}_{\Lambda\Lambda}\mathrm{Be}^*$ as a function of the binding energy of 2Λ's

If the H-dibaryon exists as a bound state, we should see a peak structure in the bound region of the missing mass spectrum. Although the peak position depends on the H mass, the peak would be well-separated from that of the quasi-free Ξ^- production if the H is sufficiently light. Figure 4 shows the excitation energy spectrum with several expected peaks of the corresponding H mass. No distinct peak is seen. Some events left in the deeply bound region may be pions and protons misidentified as kaons. Since no clear signal was found, an upper limit on the forward-angle cross section of the direct H production was set with 90% C.L. as shown in Figure 5. It is in the range of a few nb/sr to 10 nb/sr below the H mass of 2150MeV/c^2. The dashed line represents a theoretical calculation based on the scenario of Aerts and Dover which was proposed for a ^{3}He target originally [9]. It was adopted to a ^{12}C target here. A recent result of Experiment 224 [3] at the KEK-PS is also superimposed with the dash-dotted line. The (K^-, K^+) reactions were measured using a scintillating-fiber live target in that experiment. We could obtain an upper limit which is two orders of magnitude lower than the theoretical

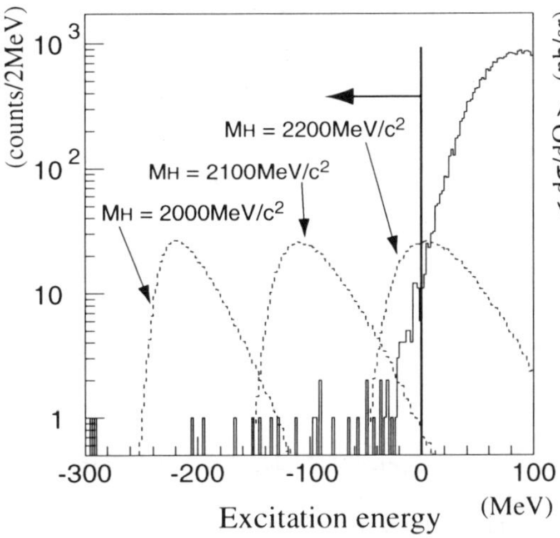

Figure 4: Excitation energy spectrum below 100MeV The dashed lines represent expected peaks of the H of which mass is 2000MeV/c^2, 2100MeV/c^2, and 2200MeV/c^2 respectively.

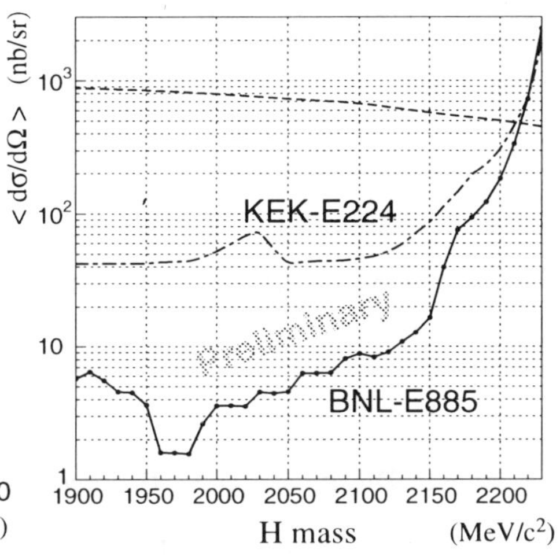

Figure 5: 90% C.L. upper limit on the forward-angle cross section of the direct H production on ^{12}C. The dashed line shows a theoretical calculation based on the scenario of Aerts and Dover (see the text). The dash-dotted line is a result of KEK-E224.

calculation and almost one order of magnitude lower than the result of KEK-E224 in the deeply bound region.

References

1. R.L. Jaffe, Phys. Rev. Lett. **38** (1977) 195.
2. S. Aoki *et al.*, Phys. Rev. Lett. **65** (1990) 1729.
3. J.K. Ahn *et al.*, Phys. Lett. **B378** (1996) 53.
4. R.W. Stotzer *et al.*, Phys. Rev. Lett. **78** (1997) 3646.
5. P.H. Pile *et al.*, Nucl. Instr. Meth. **A321** (1992) 48.
6. M. May, Nucl. Phys. **A639** (1998) 363c., K. Yamamoto *et al.*, Nucl. Phys. **A639** (1998) 371c., P. Khaustov, Nucl. Phys. **A639** (1998) 375c.
7. T. Fukuda *et al.*, Phys. Rev. **C58** (1998) 1306.
8. A.J. Baltz, C.B. Dover and D.J. Millener, Phys. Lett. **B123** (1983) 9.
9. A.T.M. Arets and C.B. Dover, Phys. Rev. **D28** (1983) 450.

STUDY OF DOUBLE-STRANGENESS NUCLEI WITH HYBRID-EMULSION METHOD (KEK-PS E373)

A. ICHIKAWA, J. K. AHN[a], H. AKIKAWA, K. IMAI, H. KANDA,
Y. KONDO, H. TAKAHASHI, J. TOJO, H. TORII, K. YAMAMOTO, Y. YOSOI
Department of Physics, Kyoto University, Kyoto 606-8502, Japan

Y. IWATA, M. KANEKO, K. NAKAZAWA, H. M. PARK, Y. S. RA, S. TORIKAI
Physics Department, Gifu University, Gifu 501-1193, Japan

Y. AKAISHI, T. FUKUDA, M. IEIRI, H. NOUMI, M. SEKIMOTO
KEK,High Energy Accelerator Research Organization, Tsukuba 305-0801, Japan

S. AOKI : *Faculty of Human Development, Kobe University, Kobe 657-8501, Japan*

S. Y. BAHK : *Wonkwang University, Iri 570-749, Korea*

K. M. BAIK, M. S. CHUNG, C. O. KIM, K. S. SIM, J. T. YANG
Korea University, Seoul 136-701, Korea

K. HOSHINO, T. KAWAI
Department of Physics, Nagoya University, Nagoya 464-8601, Japan

J. Y. KIM, S. J. KIM : *Chonnam National University, Kwangju 500-757, South Korea*

T. MOTOBA
Laboratory of Physics, Osaka El.-Com. University, Osaka 572-0833, Japan

C. NAGOSHI : *Higashi Nippon International University, Iwaki 970-8023, Japan*

S. OGAWA, K. OYAMA, H. SHIBUYA
Department of Physics, Toho University, Funabashi 274-0072, Japan

H. OKABE : *Osaka Prefectural Education Center, Osaka 558-0011, Japan*

I. G. PARK, J. S. SONG, C. J. YOON
Gyeongsang National University, Jinju 660-701, Korea

J. T. RHEE : *Konkuk University, Seoul 133-701, Korea*

T. TAKAHASHI : *Faculty of Science, Tohoku University, Sendai 980-8578, Japan*

F. TAKEUTCHI : *Faculty of Science, Kyoto Sangyo University, Kyoto 603-8047, Japan*

H. TANAKA, N. USHIDA : *Aichi University of Education, Kariya 448-8542, Japan*

Y. YAMAMOTO : *Physics Section of Tsuru University, Tsuru 402-0054, Japan*
N. YASUDA : *National Institute of Radiological Science, Chiba 263-0024, Japan*

T. YOSHIDA : *Department of Physics, Osaka City University, Osaka 558-0022, Japan*

[a]Present address: RCNP, Research Center for Nuclear Physics, Ibaraki, 567-0047, Japan

An experiment to search for double-strangeness nuclei in nuclear emulsion is now being carried on as KEK-PS E373. Preliminary results of the data analysis for the K^+ spectrometer and the scintillating fiber detectors are presented. Status of the automatic track finding method for the emulsion is reported.

1 Overview

Repulsive hard core between nucleons has been well explained with the color magnetic interaction between quarks. R.Jaffe predicted that the repulsive hard core originating from the color magnetic interaction did not exist between $\Lambda - \Lambda$, and that the 6-quark state($uuddss$) which was called as H-dibaryon could be stable against the strong interaction[1]. KEK-PS E373 is an experiment to search for nuclei which include two units of strangeness(S=-2). In the previous experiment of KEK-PS E176, the existence of double-Λ hypernuclei which are stable against the strong decay was confirmed[2]. The observation of the double hypernucleus provided the lower limit for the mass of the stable H-dibaryon. However, since there remained two kinds of interpretation of nuclide, it is still unknown whether $\Lambda - \Lambda$ interaction is attractive or repulsive. In E373, we will search the double-strangeness nuclei such as double-Λ hypernuclei, Ξ^- hypernuclei and H-dibaryon at the stopping point of Ξ^- with ten times higher statistics than that of E176.

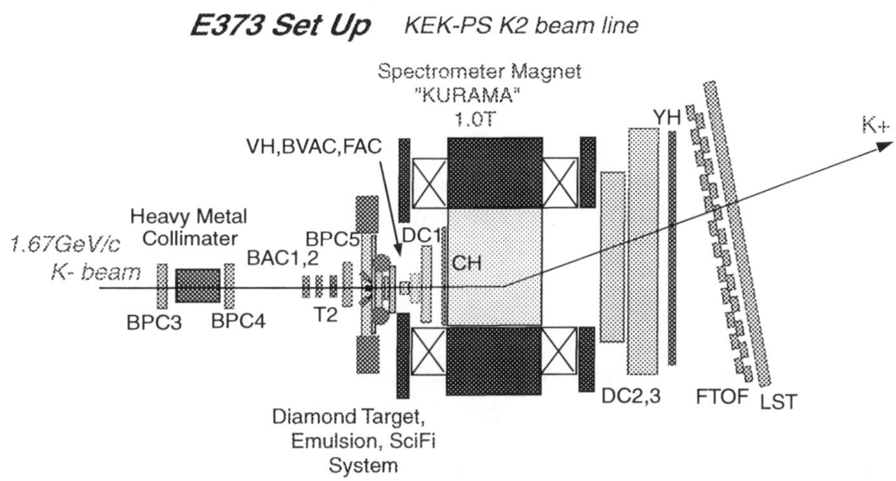

Figure 1: Schematic view of the E373 experimental setup.

Fig.1 shows a scematic view of the experimental setup. Fig.2 shows a schematic view of the target region.

The 1.67GeV/c K^- beam induces a $'p'(K^-, K^+)\Xi^-$ quasi-free reaction in the diamond target. The produced Ξ^- is slowed, stopped in a nuclear emulsion and

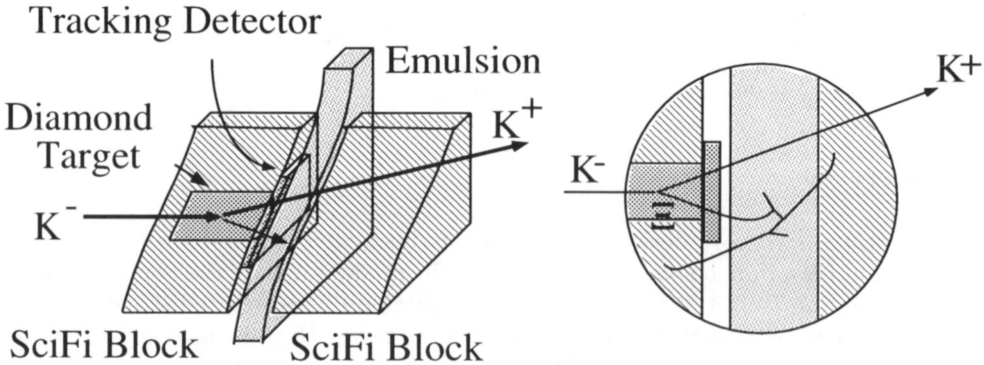

Figure 2: Schematic view of the target region.

expected to produce a double-strangeness nucleus with some probability. The emulsion plates were placed perpendicular to the beam direction. We have developed a high resolution tracking detector using scintillating micro fibers(Fiber-Bundle) to measure the position and the angle of the Ξ^{-3}. In a test experiment using proton beam, the angles of the tracks were determined with an r.m.s precision of 25mrad and the positions were determined with an r.m.s precision of 74μm with this very thin detector of about 4mm thickness. Fast protons and π^-s from the weak decay of double-hypernuclei are detected with the range counter of scintillating mono-fiber blocks(SciFi-Blocks) located on both sides of the emulsion stack. The SciFi-Blocks are also used to determine the vertex point of (K^-, K^+) reaction and to identify the decay of Ξ^-s. Photons from these scintillating fiber detectors are amplified by image intensifier tubes and recorded by CCDs without losing the position information.

In 1998, 45 stacks of emulsion plates made from about 35 liters of nuclear emulsion gel were irradiated with the beam at KEK-PS K2 beam line.

2 Data Analysis

The secondary K^+'s are distinguished from π^+s and protons with the spectrometer. About 62% of the '98's data obtained with the spectrometer were analyzed. For these data, about 31,000 (K^-, K^+) events were obtained. From this result, about 50,000 (K^-, K^+) events are expected for '98's data.

The eye scanning of the images obtained with the scintillating fiber detectors is being done for the events identified as (K^-, K^+) reaction. The events which show the topology of **no thick track in the Fiber-Bundle image, wrong vertex position, or the decay of Ξ^-** are rejected. Fig.3 shows an example of the images obtained with the upstream SciFi-Block. The production and the decay of a Ξ^- can be seen. It was found that about 11% of the (K^-, K^+) events survive this rejection. The corresponding tracks of these events will be scanned in the emulsion plates.

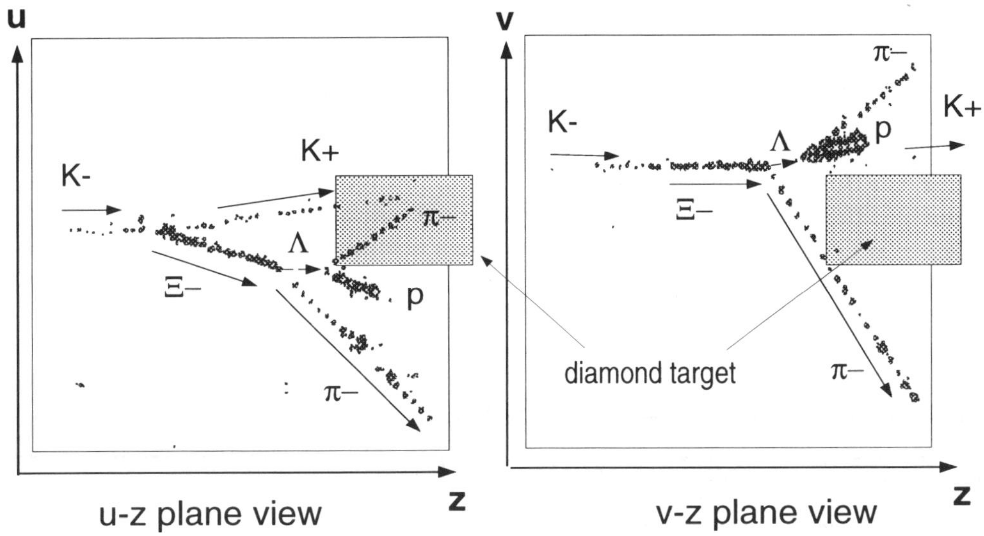

Figure 3: The reconstructed image of the production and the decay of a Ξ^- obtained with the upstream SciFi-Block.

In the emulsion plates, the tracks were recorded with the density of $10^6/cm^2$. Out of thousands of tracks within the area predicted with the Fiber-Bundle, the tracks having predicted angle have to be sought out. It is very tough work to scan these tracks with human eyes. Automatic track finding method is now being developed.

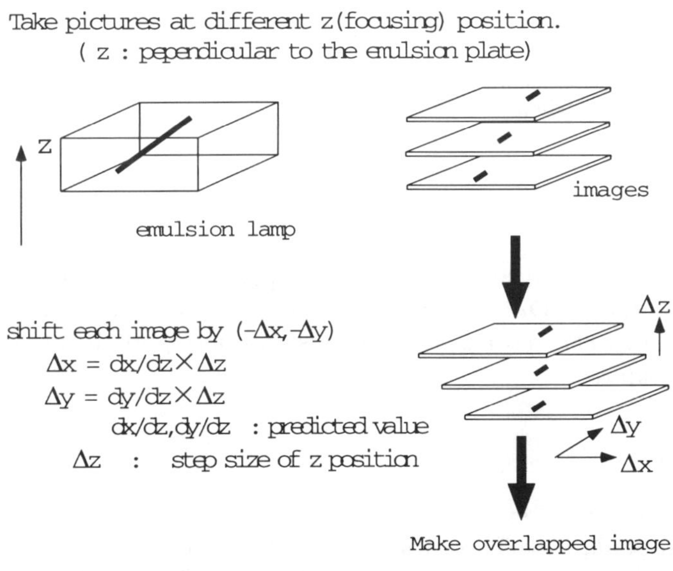

Figure 4: Concept of the emulsion automatic track finding method.

Fig.4 shows the concept of this method. The first step of the method is to take pictures via a microscope and a CCD camera at various depth of the focal plane of the microscope. The obtained tomographic(sliced) images are digitized and stored into the computer memory. Each image is shifted before all images are summed up. The distance of the shift is calculated according to the angle predicted with the Fiber-Bundle so that only the tracks of that angle make brighter spots on the summed image. By searching spots whose brightness is higher than some threshold, the tracks can be sought out. Fig.5(a) shows a summed up image. Notice that this is not obtained by summing up shifted images, but just overlap of tomographic images. Many tracks can be seen in the figure. In Fig.5(b), only the sought-out track is plotted. The efficiency of this method is now being examined.

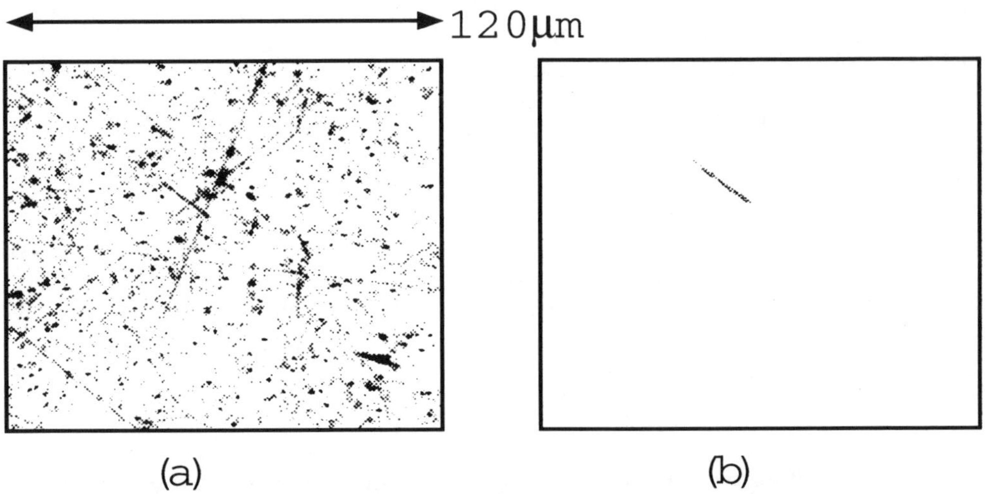

Figure 5: (a)Example of summed up image of the emulsion. (b) Image of the sought out track with automatic track finding method.

3 Summary

We expect that several hundreds of the stopped Ξ^- events will be found in the 45 emulsion stacks. One over ten of these events are expected to form double-hypernuclei. In 1999, another beam time is scheduled and about 35 stacks of emulsion will be exposed.

References

1. R.L.Jaffe, Phys.Rev.Lett. 38 (1977) 195.
2. S.Aoki *et al.*, Prog.Theor.Phys. 85 (1991) 1287.
3. A.Ichikawa *et al.*, Nucl. Instr. and Meth. A417 (1998) 220.

AN EXPERIMENT TO STUDY S=-2 NUCLEAR SYSTEM PS-E373 (KEK) – FOR FULLY AUTOMATED SCANNING OF Ξ^-TRACKS –

H. TANAKA, N. USHIDA
Aichi University of Education, Kariya 448-8542, Japan

J. Y. KIM, S. J. KIM
Chonnam National University, Kwangju 500-757, South Korea

Y. IWATA, M. KANEKO, K. NAKAZAWA, H. M. PARK, Y. S. RA, S. TORIKAI
Physics Department, Gifu University, Gifu 501-1193, Japan
E-mail: nakazawa@phys.ed.gifu-u.ac.jp

I. G. PARK, J. S. SONG, C. J. YOON
Gyeongsang National University, Jinju 660-701, South Korea

C. NAGOSHI
Higashi Nippon International University, Iwaki 970-8023, Japan

Y. AKAISHI, T. FUKUDA, M. IEIRI, H. NOUMI, M. SEKIMOTO
IPNS, KEK, Tsukuba 305-0801, Japan

S. AOKI
Faculty of Human Development, Kobe University, Kobe 657-8501, Japan

J. T. RHEE
Konkuk University, Seoul 133-701, South Korea

K. M. BAIK, M. S. CHUNG, C. O. KIM, K. S. SIM, J. T. YANG
Korea University, Seoul 136-701, South Korea

J. K. AHN, H. AKIKAWA, A. ICHIKAWA, K.-I. IMAI, H. KANDA, Y. KONDO, H. TAKAHASHI, J. TOJO, H. TORII, K. YAMAMOTO, M. YOSOI
Department of Physics, Kyoto University, Kyoto 606-8502, Japan

F. TAKEUTCHI
Faculty of Science, Kyoto Sangyo University, Kyoto 603-8047, Japan

K. HOSHINO, T. KAWAI
Department of Physics, Nagoya University, Nagoya 464-8601, Japan

N. YASUDA
National Institute of Radiological Science, Chiba 263-0024, Japan

T. YOSHIDA
Department of Physics, Osaka City University, Osaka 558-0022, Japan

T. MOTOBA
Laboratory of Physics, Osaka El.-Com. University, Osaka 572-0833, Japan

H. OKABE
Osaka Prefectural Education Center, Osaka 558-0011, Japan

K. OHYAMA, S. OGAWA, H. SHIBUYA
Department of Physics, Toho University, Funabashi 274-0072, Japan

T. TAKAHASHI
Faculty of Science, Tohoku University, Sendai 980-8578, Japan

Y. YAMAMOTO
Physics Section of Tsuru University, Tsuru 402-0054, Japan

D. H. DAVIS, D. N. TOVEE
Department of Physics & Astronomy, University College London, UK

S. Y. BAHK
Wonkwang University, Iri 570-749, South Korea

An experiment to study $S = -2$ nuclear sytems by emulsion-scintillating fiber hybrid method is presented. To search for ten times or more $\Lambda\Lambda$ hypernuclei events than that of the previous experiments, we developed and improved several detectors which will be reported. We will discuss the background tracks for the fully automated scanning system for nuclear emulsion beased on the preliminary analysis.

1 Introduction

The study of the doubly-strangeness nuclear system is very important as the doorway to search for the multi-strangeness nuclear matter. By this study, it is also expected that we can get some informations to know about hyperon-mixing in the nucleus, $\Lambda\Lambda$-ΞN-$\Sigma\Sigma$, including the H-dibaryon. It will be very interesting in a point of view from what kinds of nuclear state existing. [1] On the astrophysical problem, the study of $\Lambda\Lambda$ interaction provides the useful information about properties of the nutron star. [2] The experimental information about $S = -2$ nuclear sytems is, however, very limited. Our knowledge for the ground states of $S = -2$ nuclear sytems is only about ${}^{10}_{\Lambda\Lambda}Be$ and/or ${}^{13}_{\Lambda\Lambda}B$. [3,4]

We have started an experiment (PS-E373 at KEK) [5] to present the information based on ten times higher statistics for various double hypernuclei ($\Lambda\Lambda$-nuclei) than that of the previous E176 experiment [4]. The principle of the $\Lambda\Lambda$-nuclei detection method is the use of nuclear emulsion to produce $\Lambda\Lambda$-nuclei through the Ξ^- capture reaction at rest. In this report, we will introduce the E373 experiment, some properties of emulsion itself and discuss the background tracks by the preliminary analysis.

2 Detection method of $S = -2$ nuclear systems in nuclear emulsion

The nuclear emulsion is the unique detector of $\Lambda\Lambda$-nuclei for more than 30 years. To study for very rare events, emulsion is often used with many counters, chambers, etc. Its hybrid method is the natural style of emulsion use, nowadays.

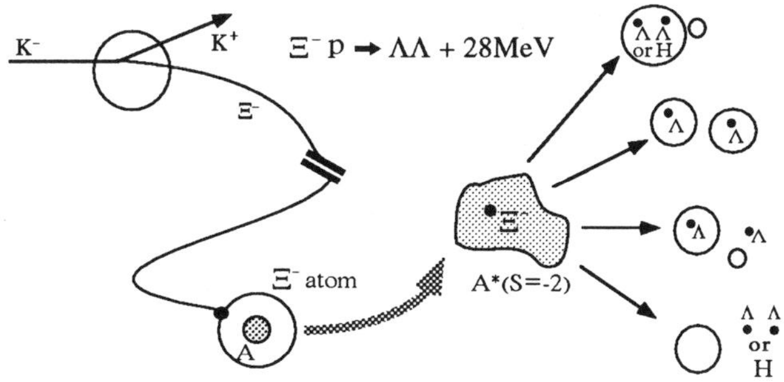

Figure 1: A method for producing $S = -2$ nuclei in the E373

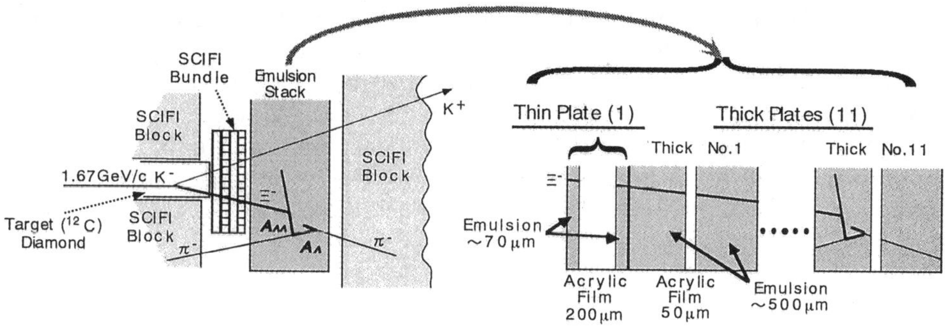

Figure 2: Schematic drawings around the emulsion in the E373

In the E176 experiment, we tagged K^+ particles from the quasi-free 'p'$(K^-,K^+)\Xi^-$ reaction in emulsion. Reconstructed K^+ particles by SSD were scanned in the bottom emulsion plate and followed to upstream emulsion until the reaction vertex. If the Ξ^- candidates track(s) were emitted, we followed them down to their stopping points. At the stopping points, the $\Lambda\Lambda$-nucleus will be produced through the Ξ^- capture at rest in the atomic orbit and the decay of the compound nucleus with two units of strangeness, as shown in Fig. 1. We scanned precisely around the capture point to detect the sequential topology of the $\Lambda\Lambda$-nucleus decay by the weak interaction. By that method, we confirmed the existence of a stable $\Lambda\Lambda$-nuclus against strong interaction [4], and presented the first identification of Ξ^--nucleus bound system by two events of twin single-Λ hypernuclei emission [6].

In the E373 experiment, we changed the above detection scheme to get higher statistics than that of the E176. We, at first, changed the Ξ^- production target from emulsion to diamond as shown in Fig. 2.

One emulsion stack consists of 12 plates, one thin plate and eleven thick plates. On the upstream thin plate, the fully automated scanning [7] will be directly applied for picking the Ξ^- candidates tracks up, not K^+, predicted by the fiber (SCIFI) bundle. The candidate tracks found in the thin plate, we will follow them in the downstream thick emulsion plate by plate. Since the acrylic film which support emulsion shape in the development becomes dead space for the detection of the Ξ^- stopping reaction, we developed the use of 50 μm thick film. We were able to decrease the dead space rate from 6.0 % (E176) to 4.8 %. The analyzing method arround the Ξ^- stopping point is the same as the E176. We have exposed 45 stacks in FY98 beam time.

3 Nuclear emulsion in the E373 experiment

3.1 Uniform AgBrX crystal size of the emulsion

We use the ET-7C emulsion developed by the collaborative R&D between Gifu Univ. and Fuji Photo Film Co.Ltd. The AgBrX cryatal size of the ET-7C, 0.24 $\pm$ 0.02 μm, is more uniform than taht of the ET-7B, 0.24 $\pm$ 0.08 μm, which we used in E176. The reason why we need uniform and small size crystal is that we have to get higher position resolution because the $S = -2$ system production and its sequesntial decay may occur in the tiny volume, *e.g.* $5\times5\times5\mu m^3$, as shown in the previous event [4]. In the beam exposure on FY99, we will use some amount of emulsion of which AgBrX crystal size is more fine, nearly 0.20 μm, than the ET-7C.

3.2 Uniform sensitivity of developed emulsion

In the E176, although we applied kinematical analysis for the $\Lambda\Lambda$-nucleus event, we could not finally identify the nuclide of the $\Lambda\Lambda$-nucleus. [4] After that, we have been developing the method to know the charge of nuclear fragments by the measurement of their ionization loss in emulsion. The preliminary results show the possibility of the charge separation in ${}^1H^+$, ${}^4He^{2+}$ and ${}^7Li^{3+}$ by 4 standard deviations if their ranges are greater than 60 μm. [8] For this charge recognition method, one of the key points is uniform sensitivity by the development along the depth of emulsion plate. We make 500 μm thick emulsion plates in one side. When we develop such as thick emulsion plates, we have to optimize the conditions of the development to obtain uniformity, in usual. From the optimizations of the chemical components, placed time in sveral kinds of chemical water and their temparatures, we can get 30.5 $\pm$ 2 grains/100μm for the full depth of emulsion plates. [a]

4 Preliminary analysis for the fully automated scanning

For the fully automated scanning system, we have to know the angle distribution of the Ξ^- candidate tracks which will be found in the top emulsion plate. We also must know the emulsion thickness distribution after development to measure track angles, precisely. We will discuss the preliminary analysis about them.

[a] Using the tracks of minimum ionized particles, we usually check the uniformity of the developed emulsion by the number of developed grains in a unit length (100μm).

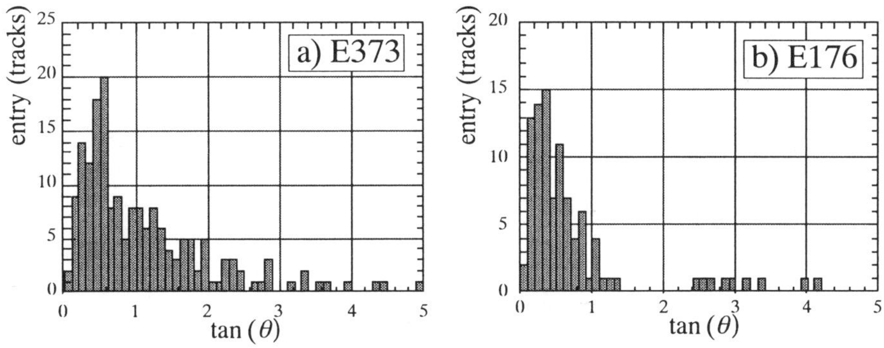

Figure 3: Angle distribution of the Ξ^- candidate tracks at a) SCIFI bundle (E373) and b) (K^-,K^+) reaction points (E176).

4.1 Angles (θ) of the Ξ^- candidate tracks

The distribution of Ξ^- candidate tracks predicted by the reconstructed images in SCIFI bundle is shown in Fig. 3a). The distribution is more broad than that obtained in the E176 which is presented in Fig. 3b). The Ξ^- angle data in the E176 was given by measurement around (K^-,K^+) reaction point of the Ξ^- stopping events. In the E373, the Ξ^- particles were filying 3cm, in maximum, until the SCIFI bundle detector. The angle of the Ξ^- tracks becomes large by the multiple scattering in the diamond target. There are also decay products in the distribution (Fig.3a) such as π^-s and protons from $\Xi^- \rightarrow \Lambda + \pi^-$, $\Lambda \rightarrow p + \pi^-$. We require that the fully automated scanning system must cover effectively for the region of $\tan(\theta) \geq 2$.

4.2 Emulsion thickness after development

To find the Ξ^- candidate tracks by the fully automated scanning system, the thickness of emulsion is requested to be almost uniform as expected, because the system can't pick tracks up if it is very thin. We measured 4 points of tracks ($\tan(\theta) \sim 0.3$) which are shown by a drawing in the center of Fig. 4. Usually, emulsion shrink after development by almost AgBrX escaping. We estimated original thickness by extrapolating two points on both sides acrylic film. The relation of measured and estimated thickness is plotted in Fig. 4a) and 4b) which are for upstream and downstream , respectively. Although the uniformity of upstream is not so bad, the downstream is worse. According to this results, since it may be impossible to get angles by one side emulsion, we need to measure angles by two points on both sides acrylic film. By the additional consideration of the position accuracy between SCIFI bundle and emulsion in FY98 exposure, the background tracks picked by the system will be 10 tracks or more. The number is not so small. The sequential decays of $S = -2$ system are, however, easily understood by their topologies.

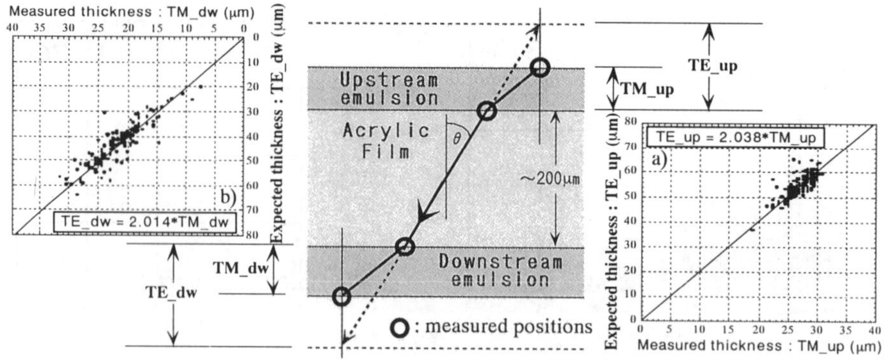

Figure 4: The method for emulsion thickness estimation. Measured thickness and estimated one for the upstream emulsion a) and the downstream b).

5 Summary

From FY98, we have been recording (K^-,K^+) events in nuclear emulsion to present the information about Λ-Λ interaction with ten times or more high statistics than that of the previous experiment. The emulsion handling method was developed and emulsion of itself was improved on its crystal size. The preliminary analyses for Ξ^- track angles, emuslion thickness and position accuracy between SCIFI bundle and emulsion say not so small number of background tracks for Ξ^- candidate. However, we can easily distinguish the production of the $S = -2$ system or not by the decay topology which is sequentially for $\Lambda\Lambda$-nuclei.

References

1. T. Yamada, in this proceedings.
2. T. Takatsuka, in this proceedings.
3. M. Danysz *et al.*, *Phys. Rev. Lett.* **11**, 29 (1963); *ibid. Nucl. Phys.* **49**, 121 (1963).
4. S. Aoki *et al.*, *Prog. Theor. Phys.* **85**, 1287 (1991).
5. K. Nakazawa, *Nucl. Phys.* **A585**, 75c (1995).
6. S. Aoki *et al.*, *Prog. Theor. Phys.* **89**, 493 (1993); *ibid. Phys. Lett.* **B355**, 45 (1995).
7. A. Ichikawa *et al.*, in this proceedings.
8. J. K. Ahn *et al.*, *Annual Report, Tandem V.D.G Lab. Kyoto Univ. 1996-1997*, 141(1998).

KN INTERACTION IN MEDIUM, CHIRAL SYMMETRY AND KAON CONDENSATION

T. TATSUMI and M. YASUHIRA

Department of Physics, Kyoto University Kyoto 606-8502, Japan
E-mail:tatsumi@ruby.scphys.kyoto-u.ac.jp; yasuhira@ruby.scphys.kyoto-u.ac.jp

Relations between KN dynamics and chiral symmetry are discussed, which are closely related with the issue of kaon condensation. We then show how chiral symmetry dictates kaon condensation to us. By the use of the framework we have recently presented quantum and thermal fluctuations around the condensate are treated beyond the mean-field approximation. Some results about the phase diagram and the equation of state are demonstrated.

1 Introduction

Recently there is a large development in the theoretical and experimental study about kaons in medium. Kaon production in heavy-ion collisions, kaonic atoms and the problem of kaon condensation in neutron stars and protoneutron stars are interesting subjects in this context [1,2,3]. Here we would like to discuss the behaviour of kaons in medium and kaon condensation on the basis of chiral symmetry. We shall see that chiral symmetry is very important and useful concept to study kaons in medium and kaon condensation.

The s-wave KN scattering amplitude exhibits quite different nature for the different isospin channels, $I = 0$ and $I = 1$. The $\Lambda(1405)$ resonance is very important in the $I = 0$ channel for low energy scatterings, while it is rather weak in the $I = 1$ channel. We can see that the argument based on chiral symmetry gives a reasonable description of the overall feature of the KN scattering in the vacuum [4].

We are here interested in the K^-N interaction in nuclear matter. When K^- is put in, it propagates while multiply interacting with surrounding nucleons. So the KN scattering amplitude should be modified by them due to the Pauli effects. The nature of Λ resonance should be also modified there. Fortunately, the K^-n interaction is weakly attractive, so that we believe that its modification in nuclear matter should be small. In this case the problem of kaon condensation can be studied without suffering seriously by nuclear correlation effects.

Usually kaon condensation has been treated within the mean-field approximation [3]. However, quantum and thermal fluctuations around the condensate are important to study, e.g., protoneutron stars and their collapse to black holes [5,6], where temperature should become high. We discuss how these fluctuations can be treated with nonlinear chiral Lagrangians by the use of a framework we have recently presented [7]. Thereby we derive the phase diagram in the temperature-density plane and the equation of state (EOS) for the condensed phase.

2 KN interaction in medium

2.1 KN scattering amplitude within the chiral models

Consider the KN forward scattering amplitude D_{KN}:

$$D_{KN}(k) \propto \int d^4x e^{ikx} \langle N|T(\partial_\nu A_5^\nu(x)\partial_\lambda A_5^\lambda(0))|N\rangle$$
$$= \int d^4x e^{ikx} [k_\nu k_\lambda \langle N|T(A_5^\nu(x)A_5^\lambda(0))|N\rangle$$
$$+ ik_\nu \langle N|\delta(x_0)[A_5^0(0), A_5^\nu(x)]|N\rangle - \langle N|\delta(x_0)[A_5^0(x), \partial_\lambda A_5^\lambda(0)]|N\rangle], \quad (1)$$

where A_5^μ is the axial-vector current of the kaon channel. The second and third terms are the Tomozawa-Weinberg term and KN sigma terms (Σ_{KN}), respectively, which give the leading-order (LO) contributions at low energies. The first ("pole") term can be written as

$$(1st.) = \int d^4x e^{ikx} \sum_{Y=\Lambda,\Sigma^0,\Lambda^*,\Sigma^{0*}} \langle N|k_\nu k_\lambda T\{A_5^\nu(x)Y\rangle\langle Y A_5^\lambda(0)\}|N\rangle + k_\nu k_\lambda \alpha^{\nu\lambda} + ... \quad (2)$$

by way of the pole-dominance hypothesis for the divergence of the axial-vector current (PDDAC). The regular terms (the last terms in Eq. (2)) may be regarded to be originated from the higher-order diagrams beyond the tree level.

To summarize, the scattering amplitudes are given, e.g. for K^-,

$$D_{K^-p}(k,k') = \frac{\Sigma_{Kp}}{f^2} + \frac{\nu}{f^2} + d_{1p}\nu^2 + d_{2p} k\cdot k'$$
$$+ \frac{g_\Lambda^2}{2f^2}\left\{\frac{-(M_\Lambda - M_N)(M_N^2 - s)}{M_\Lambda^2 - s - iM_\Lambda\Gamma_\Lambda} - \nu\frac{2M_N(M_\Lambda - M_N) + M_N^2 - s}{M_\Lambda^2 - s - iM_\Lambda\Gamma_\Lambda}\right\}, \quad (3)$$

and

$$D_{K^-n}(k,k') = \frac{\Sigma_{Kn}}{f^2} + \frac{\nu}{2f^2} + d_{1n}\nu^2 + d_{2n} k\cdot k', \quad (4)$$

with $\nu = (p+p')(k+k')/4M_N$. Then the forward amplitude, which is reduced to the scattering length at the threshold, is given by way of the relation

$$f_{KN} = \frac{1}{4\pi(1+\omega_K/M_N)} D_{KN}(k=k'). \quad (5)$$

We have seen that the empirical values for D_{K^-N} [8] can be well reproduced by the following values of the parameters[4],

$$g_\Lambda = 0.7,$$
$$d_p \equiv d_{1p} + d_{2p} = (0.37 - \Sigma_{Kp}/m_K)/f^2 m_K,$$
$$d_n \equiv d_{1n} + d_{2n} = (0.23 - \Sigma_{Kn}/m_K)/f^2 m_K. \quad (6)$$

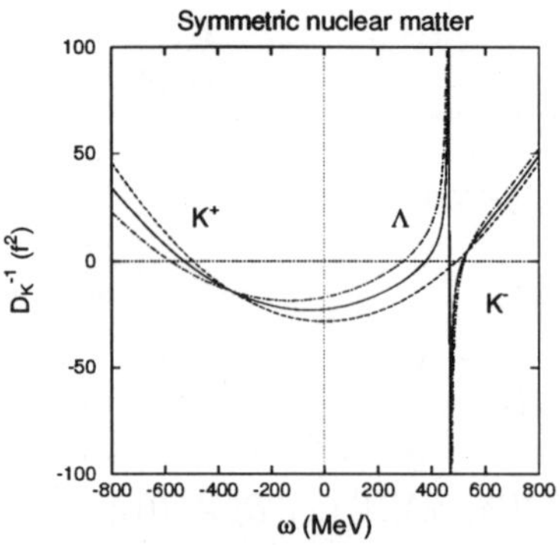

Figure 1: Inverse kaon propagator at $k = 0$. Dashed, solid and dash-dotted lines denote the cases of the vacuum, $\rho = \rho_0$ and $\rho = 2\rho_0$, respectively.

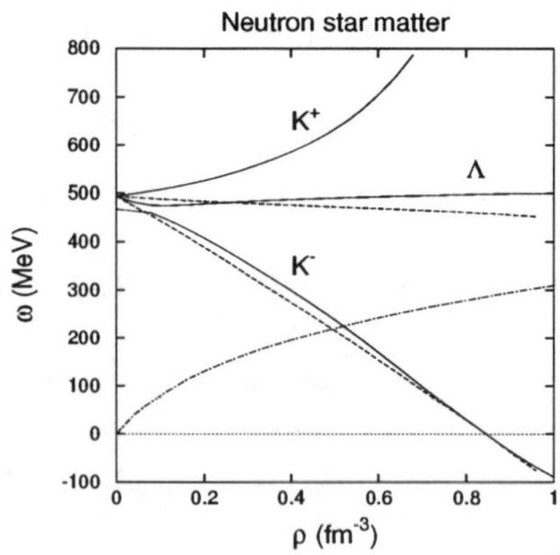

Figure 2: Lowest excitation energies for kaonic modes in the neutron-star matter. Dashed lines are obtained only with LO terms. The difference of the nucleon chemical potentials, $\mu_n - \mu_p$, is also shown for comparison (dash-dotted line). The crossing point of K^- and the dash-dotted line indicates the critical density for kaon condensation.

2.2 K^-N scattering amplitude in medium

KN scattering in the $I = 0$ channel exhibits a peculiar feature due to the presence of the $\Lambda(1405)$ resonance in the intermediate state. In nuclear matter there should exist the collective state, $\Lambda-$ hole state, due to the coupling with K^- [a] ; as a result, the excitation energy of such state is shifted from the original mass difference between nucleon and Λ. Recently this shift has been elaborately studied within the different approaches, still it seems to be controvertial; the upward shift of about 50 MeV has been predicted in ref. 10, while almost unchange has been indicated in ref. 11. In both studies K^- receives an *attractive* effect of about 100MeV, which seems to be consistent with the data analysis of kaonic atoms [2].

The excitation energies of the kaonic modes are determined by way of the kaon propagator D_K:

$$D_K^{-1} = 0, \quad D_K^{-1} = \omega^2 - |\mathbf{k}|^2 - m_K^2 - \Pi(\omega, \mathbf{k}), \quad \Pi = D_{K^-p}\rho_p + D_{K^-n}\rho_n, \tag{7}$$

where we have used the low-density approximation for the self-energy Π. By using the values given in Eq.(6) and $\Sigma_{KN} \equiv 1/2(\Sigma_{Kn} + \Sigma_{Kp}) = 340$ MeV we depict the inverse Green function in Fig. 1. In our case the pole position for the Λ-hole state is shifted *downward* by several tens MeV at the nuclear density $\rho_0 \simeq 0.16 fm^{-3}$, while that for K^- is almost unchanged. [b] Finally K^+ receives a repulsive effect.

[a] This is the state analogous to the well-known Δ-hole state [12].

[b] Our result is apparently inconsistent with the data analysis of kaonic atoms. However, there still remains a possibility that higher-order effects beyond the low density approximation may reconcile it with the empirical data [2].

Our treatment of Λ resonance is completely different from refs. 10,11, where it is regarded as a $KN, \pi\Sigma$ resonant state, *not* an elementaly particle. If we can observe the fate of $\Lambda(1405)$ in nuclear medium, we find whether it is a compact object or not.

On the contrary, KN interaction shows small attraction in the $I = 1$ channel, which is almost constant even below the KN threshold. Hence this small attraction should not be so modified even in medium. When we consider the kaon (K^-) condensation in neutron-star matter, only this channel is relevant. We depict the K^- excitation energy at momentum 0 in Fig.2 , where is also shown the corresponding one with the use of the LO terms for comparison. The critical density for kaon condensation is almost the same, while the behaviour of K^+ energy is remarkable. Hence we can conclude that LO terms theirselves can decribe the phase transition sufficiently.

3 Chiral symmetry and kaon condensation

Tree level study without any fluctuation is possible in a model-independent way [9]. Then kaon condensed state is given by acting the opeartor U_K,

$$|K\rangle = U_K(\langle\theta\rangle)|0\rangle, \quad U_K = \exp(i\langle\theta\rangle Q_4^5), \tag{8}$$

which gives the condensate $\langle K^{\pm}\rangle = \frac{f}{\sqrt{2}}\sin\langle\theta\rangle$ with the chiral angle $\langle\theta\rangle$ as the order parameter. For an effective Hamiltonian $\mathcal{H}_{eff}$ with the kaon chemical potential μ_K,

$$\mathcal{H}_{eff} = \mathcal{H}_{chiral} + \mu_K \rho_Q; \quad \mathcal{H}_{chiral} = \mathcal{H}_{sym} + \mathcal{H}_{SB}, \tag{9}$$

we can evaluate the energy difference

$$\begin{aligned} \Delta\epsilon(\langle\theta\rangle) &\equiv \langle K|\mathcal{H}_{eff}|K\rangle - \langle e|\mathcal{H}_{eff}|e\rangle \\ &= (\cos\langle\theta\rangle - 1)[\mu_K(1+x)\rho/2 - f^2 m_K^2 + \Sigma_{Kp}\rho_p + \Sigma_{Kn}\rho_n] \\ &\quad - 1/2 \cdot \mu_K^2 f^2 \sin^2\langle\theta\rangle, \end{aligned} \tag{10}$$

by way of current algebra and PCAC (LO). The first term in the bracket corresponds to the TW term and the third and forth terms to the KN sigma term.

3.1 Quantum and thermal fluctuations around the condensate

To treat quantum and thermal fluctuations beyond the tree level we must invoke effective Lagrangians, which can reproduce the soft-meson theorem in the tree level. We use here the nonlinear sigma model with meson fields ϕ_a and baryon field B, $\mathcal{L}_{chiral}(U, B) = \mathcal{L}_0(U, B) + \mathcal{L}_{SB}(U, B)$, where $U = \exp[2iT_a\phi_a/f] \in SU(3)$, and $\mathcal{L}_0$ and $\mathcal{L}_{SB}$ are the symmetric and symmetry-breaking (SB) parts, respectively. We shall see how the effect of fluctuations, especially thermal fluctuations, influence the thermodynamic property of the condensed phase.

Consider the partition function Z_{chiral} at temperature T:

$$Z_{chiral} = N \int [dU][dB][d\bar{B}] \exp[S_{chiral}^{eff}], \tag{11}$$

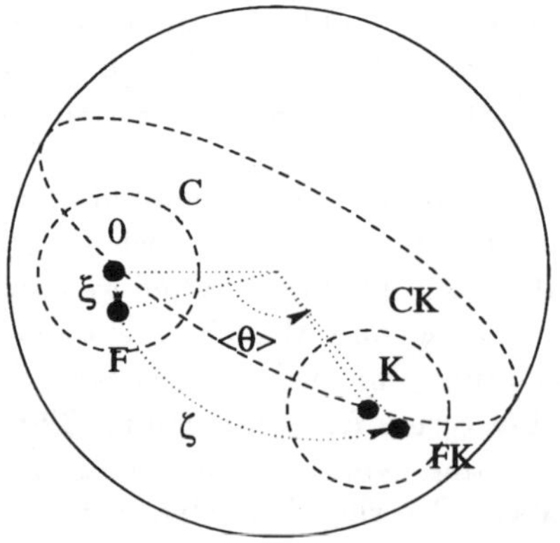

Figure 3: Chiral rotations on the $SU(3)_L \times SU(3)_R/SU(3)_V \simeq SU(3)$ manifold. Circles C, CK indicate the fluctuation areas around the vacuum (O) and kaon condensed state (K), respectively.

where the effective action S^{eff}_{chiral} consists of two terms,

$$S^{eff}_{chiral} = \int_0^\beta d\tau \int d^3x \left[\mathcal{L}_{chiral}(U,B) + \delta\mathcal{L}(U,B)\right] \tag{12}$$

with the imaginary time $\tau = it$ and $\beta = 1/T$; one is the original action and the other is the "induced" SB term besides the SB term inherent in $\mathcal{L}_{chiral}$ as a result of the introduction of chemical potentials μ_K and μ_n to implement the electromagnetic charge and baryon number conservations in the Lagrangian[7],

$$\begin{aligned}\delta\mathcal{L} = &-\frac{f^2\mu_K}{4}\mathrm{tr}\{[T_{em},U]\frac{\partial U^\dagger}{\partial\tau} + \frac{\partial U}{\partial\tau}[T_{em},U^\dagger]\} - \frac{\mu_K}{2}\mathrm{tr}\{B^\dagger[(\xi^\dagger[T_{em},\xi]+\xi[T_{em},\xi^\dagger]),B]\} \\ &- \frac{f^2\mu_K^2}{4}\mathrm{tr}\{[T_{em},U][T_{em},U^\dagger]\} + \mu_n\,\mathrm{tr}\{B^\dagger B\} - \mu_K\,\mathrm{tr}\{B^\dagger[T_{em},B]\}\end{aligned} \tag{13}$$

with $U = \xi^2$.

Generally it is complicated to perform the path integral with the nonlinear Lagrangians by seperating the fileds ϕ_a into the classical fields and fluctuation fields in the standard way, because ϕ_a reside on the curved manifold. Instead we introduce the local coordinates around the condensate to eliminate the curvature in the neighberhood of the condensate, which is equivalent to the following parametrization for U (see Fig. 3);

$$U = \zeta U_f \zeta (\xi = \zeta U_f^{1/2} u^\dagger = u U_f^{1/2}\zeta), \quad \zeta = \exp(\sqrt{2}i\langle M\rangle/f), \tag{14}$$

where $\langle M\rangle$ is the condensate, $\langle M\rangle = V_+\langle K^+\rangle + V_-\langle K^-\rangle$, with $K^\pm = (\phi_4 \pm i\phi_5)/\sqrt{2}$ and $\theta^2 \equiv 2K^+K^-/f^2$, while $U_f = \exp[2iT_a\phi_a/f]$ means the fluctuation field. It may be interesting to see that this procedure can be also regarded as the separation of the zero-mode[14].

Accordingly defining a new baryon field B' by way of

$$B' = u^\dagger B u, \tag{15}$$

we can see eventually that

$$\mathcal{L}_{chiral}(U,B)=\mathcal{L}_0(U_f,B')+\mathcal{L}_{SB}(\zeta U_f\zeta,uB'u^\dagger),\qquad \delta\mathcal{L}(U,B)=\delta\mathcal{L}(\zeta U_f\zeta,uB'u^\dagger) \tag{16}$$

Thus only the SB terms prescribe the KN dynamics in the condensed phase.

The integration measure is given as

$$[dU]=M(\phi)\prod_{a=1}^{8}[d\phi_a], \tag{17}$$

where $M(\phi)$ is the Lee-Yang term [13]; $M(\phi)=\exp\left[\frac{1}{2}\delta^{(4)}(0)\int_0^\beta d\tau\int d^3x\ln g(\phi)\right]$ with the Cartan-Killing metric on the $SU(N)$ manifold by way of the invariant line element, $ds^2=g_{ab}(\phi)d\phi_a d\phi_b=1/2\mathrm{tr}[dUdU^\dagger]$. It is to be noted that the Lee-Yang term is important to make the integration measure chiral invariant, and should be carefully treated[13]. In our case the measure (17) is invariant under the transformation by ζ,

$$[dU]=[d(\zeta U_f\zeta)]=[dU_f], \tag{18}$$

so that the geometry around the condensed point is the same as in the vicinity of the vacuum. Since we are concerned with the fluctuations, the measure can be approximated by the one with flat curvature [14], $[dU_f]\simeq\prod_{a=1}^{8}[d\phi_a]$. Therefore, in this case, we need not be worried about the Lee-Yang term because of $g=1+O(\phi_a^2/f^2)$. Consequently, we find

$$Z_{chiral}\simeq\int\prod_{a=1}^{8}[d\phi_i][dB'][d\bar{B}']\exp\left[\int_0^\beta d\tau\int d^3x\mathcal{L}^{eff}_{chiral}(\zeta,U_f,B')\right]. \tag{19}$$

3.2 *Dispersion relation of kaonic modes*

The integral over ϕ_a is rather easy and we can get the partition function Z_{chiral} up to the one-loop level. In this course, the poles of the thermal Green function for kaons give the excitation spectrum of kaonic modes with energies ($E_\pm$);

$$E_\pm^2=(c_1+c_3+2c_2^2)\pm\sqrt{(c_3+2c_2^2)^2+4c_1c_2^2}, \tag{20}$$

where the coefficients c_i are rather complicated functions of the chiral angle $\langle\theta\rangle$ and kaon chemical potential μ_K [7]. In particular, the mode with E_- is the Goldstone mode and exhibits the Bogoliubov spectrum,

$$E_-^2\sim\frac{c_3}{2C^2}p^2+\frac{p^4}{4C^2}+...,\tag{21}$$

with C being the effective kaon "mass" and c_3 being charge density $\times KK$ scattering length[7]. The origin of the Goldstone mode is as follows. The kaon-condensed state is a broken phase with respect to the rotation around the third axis in the V-spin space (V_3), while the effective Lagrangian is still inavariant. Hence the V- spin symmetry is spontaneously broken in the condensed phase.

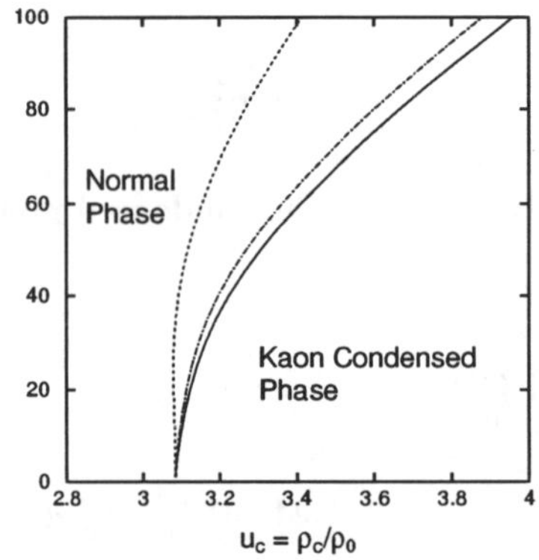

Figure 4: Phase diagram in the temperature and density plane. The dashed line is given in the mean-field level, while the dash-dotted and solid lines include the thermal kaon loops. The solid line is given in the HBL.

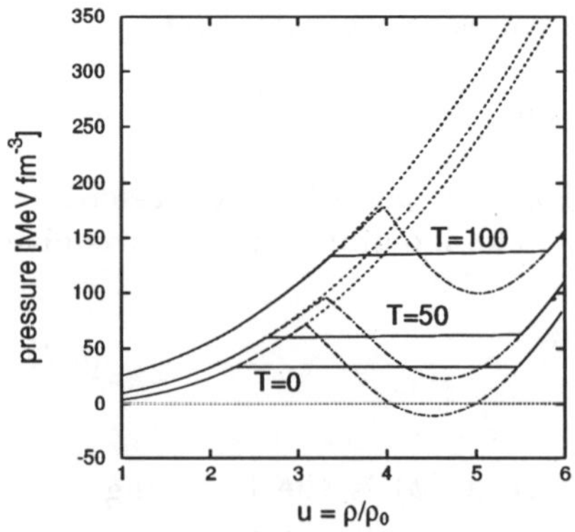

Figure 5: EOS for the isothermal matter in the HBL for a given temperature T in MeV (dash-dotted lines). Dotted lines correspond to those for normal neutron-star matter. Horizontal lines shows the equal-pressure region formed by the Maxwell contsruction.

We use the Hartree approximation to treat the kaon and nucleon loops in a consistent way. As a result all the diagrams are composed of the infinite number of the kaon and nucleon one-loop diagrams. Generally the one-loop diagrams include divergences. We can properly renormalize them and extract the finite contributions [7]. Then the interplay between the kaon and nucleon loops has been analyzed. We have first confirmed that the kaon zero-point fluctuations have little effect on EOS in the heavy-baryon limit (HBL), where the self-energies of a nucleon are neglected. Subsequently it has been shown that the nucleon zero-point fluctuations give sizable effects, while the effects of kaon ones are small. On the other hand, the effects by the thermal kaon and nucleon loops work in a reverse way. This is because one kaonic mode is the Goldstone mode giving a large contribution for the Bose-Einstein distribution function, while the effect of the nucleon loops should be predominantly provided by the density effects as long as temperature is not so high.

3.3 Phase diagram and equation of state

Some results are given for the phase diagram (Fig. 4) and EOS (Fig. 5). First, we can see the role of kaon thermal loops in these figures; thermal effects work against the condensation before the phase transition and harden EOS considerably in the condensed phase. It is to be noted that the result within the mean-field level shows a qualitatively different feature for the critical curve. The resultant EOS shows that the phase transition is of first order, so that we invoke here the Maxwell construction, for simplicity, to obtain the thermodynamically stable EOS. Then we find that the equal-pressure region is shrunk as temperature is increased due to the thermal effects mainly by kaon fluctuations. Nevertheless, there is no critical temperature for the first-order phase transition.

4 Summary and concluding remarks

We first noted the relation between KN interaction and chiral symmetry: we have seen how chiral symmetry describes the KN interaction. We put in $\Lambda(1405)$ by hand to supplement the chiral perturbation theory. Then the excitation energy of the Λ-hole state is shifted *downward* as density increases, about several tens MeV at $\rho = \rho_0$, while that of K^- is almost unchanged. On the other hand, the Λ-hole state and the regular part is not so important for K^-N interaction in neutron matter, so that the in-medium K^-n interaction can be well-described only by the lowest-order chiral pertubation theory.. Kaon condensation looks robust in this context, while effects of hyperon contamination is still open problem[5].

Kaon condensation is discussed from the viewpoint of chiral symmetry: a systematic formulation to treat fluctuations around the kaon condensate is presented by introducing the local coordinate. Using this framework we have obtained the dispersion relations for kaonic modes; one is the Goldstone mode due to the spontaneously breaking of V_3 symmetry, while the other is a very massive mode. Consequently the phase diagram and EOS for the isothermal case are obtained, where the role of thermal kaon-loops should be emphasized. These results may be relevant for the properties of protoneutron stars[15] and their delayed collapse during the initial deleptonization and cooling era[6]. We discuss this subject in detail elsewhere[16].

References

1. G.Q. Li, C.-H. Lee and G.E.Brown, *Nucl. Phys.* **A 625**, 372 (1997).
   G.S. Song, B.-A. Li and C.M. Ko, nucl-th/9809022.
2. C.J. Batty, E. Friedman and A. Gal, *Phys. Reports* **287**, 385 (1997).
3. C.-H. Lee, *Phys. Reports* **275**, 197 (1996).
4. T. Tatsumi, H. Fujii and T. Muto, Proc. of the IV Int. Sympo. on *Weak and Electromagnetic Interactions in Nuclei*, World Scientific, 1995, p.575.
5. M. Prakash, I. Bombaci, M. Prakash, P.J.Ellis, J.M. Lattimer, R. Knorren, *Phys. Reports* **280**, 1 (1997).
6. G.E. Brown and H.A. Bethe, *Astrophys. J.* **423**, 659 (1994).
7. T. Tatsumi and M. Yasuhira, *Phys. Lett.* B **441**, 9 (1998); *Nucl. Phys.* **A**, (1999) to be published (nucl-th/9811067).
8. A.D. Martin, *Nucl. Phys.* B **179**, 33 (1981).
9. As a review, T. Tatsumi, *Prog. Theor. Phys. Suppl.* **120**, 111 (1995).
10. V. Koch, *Phys. Lett.* B **337**, 7 (1994).
    T. Waas and W. Weise, *Nucl. Phys.* **A 625**, 287 (1997)
11. M. Lutz, *Phys. Lett.* B **426**, 12 (1998)
12. T. Suzuki and M. Kohno, *Prog. Theor. Phys.* **68**, 690 (1982).
13. J. Zinn-Justin, *Quantum field theory and critical phenomena*, Oxford Univ. Press, 1996.
14. J. Gasser and H. Leutwyler, *Phys. Lett.* B **188**, 477 (1987).
    F.C. Hansen, *Nucl. Phys.* B **345**, 685 (1990).
15. T. Tatsumi and M. Yasuhira, nucl-th/9904038.
16. M. Yasuhira and T. Tatsumi, in preparation.

STRANGENESS IN THE NUCLEON in the SU(3) CHIRAL QUARK-SOLITON MODEL

HYUN-CHUL KIM
Department of Physics
Pusan National University
609-735 Pusan, Republic of Korea

The recent investigation on strangess content of the nucleon is reviewed within the framework of the SU(3) chiral quark-soliton model. The scalar form factor with the $\sigma_{\pi N}$ term is in good agreement with the analysis of Gasser, Leutwyler, and Sainio. The strange electric form factor is investigated with the kaon effect taken into account. The strange magnetic moment is also studied in a "model-independent" way. The strange part of the tensor charge is presented.

Strangeness in the nucleon has been a hot issue well over decades. The first example is the πN sigma term. Because of the factor-of-two discrepancy between empirical value for the πN sigma term ($\Sigma = 64 \pm 8$ MeV) from the analysis of πN data and naive estimates of the σ term ($\sigma \simeq 25$ MeV) from the baryon mass spectrum [1], which contradicts to the prediction of the low-energy theorem of current algebra, there has been a lot of discussion in the literature [2]. It was once known as "$\Sigma_{\pi N}$ *puzzle*". In the axial channel, deep inelastic muon scattering conducted by European Muon Collaboration (EMC) almost ten years ago indicates that there is a sizeable contribution of the strange quark spin Δs to the proton spin, which was once called "*spin crisis*". This remarkable result has been confirmed by following experiments of the Spin Muon Collaboration (SMC) [3,4], E142 and E143 collaborations [5,6]. The measurement of elastic $\nu p/\bar{\nu}p$ cross section at Brookhaven (BNL experiment 734), from which the strange axial form factor was extracted, came to the more or less same conclusion. In the vector channel the strange magnetic moment presents another key of understanding the strangeness content of the nucleon [7]. Great deal of theoretical effort has been put into the investigation of the strange magnetic moment and strange radius [8–21], which gave the range of possible values of the strange magnetic moment: $(-0.75 \to 2.2)\,\mu_{\rm N}$. This indicates that the strange properties of the nucleon in the vector channel are still poorly understood.

Very recently, the SAMPLE collaboration has announced the first measurement of the strange magnetic form factor G_M^s at $Q^2 = 0.1\ \mathrm{GeV}^2$ [22]:

$$G_{\rm M}^{(\rm s)} = (\,+0.23 \pm 0.37 \pm 0.15 \pm 0.19)\,\mu_{\rm N}. \tag{1}$$

Although the central value for $G_{\rm M}^{(\rm s)}$ is positive, it is still too early to draw any firm conclusion, because of large experimental errors. The HAPPEX collaboration has presented the most recent data for the strange vector form factor [23]:

$$G_E^s + 0.39 G_M^s = 0.023 \pm 0.034(\mathrm{stat}) \pm 0.022(\mathrm{syst}) \pm 0.026(\delta G_E^n), \tag{2}$$

where the last uncertainty arises from the estimated uncertainty in the neutron electric form factor. It contains also rather large errors. One has to wait for other experiments which are presently either being conducted or planned at various electron accelerators. It is therefore of importance to make theoretical predictions for this quantity.

The strangeness in the nucleon is represented by the following nucleon matrix elements of the strange quark operator:

$$\langle N(p')\bar{s}\Gamma s|N(p)\rangle = \langle N(p')\bar{s}\Gamma\lambda_0 s|N(p)\rangle + \frac{1}{\sqrt{3}}\langle N(p')|\bar{s}\Gamma\lambda_8 s|N(p)\rangle, \quad (3)$$

where Γ denotes the 4×4 spin operators: $\Gamma = \mathbf{1}, \gamma_5, \gamma_\mu, \gamma_\mu\gamma_5, \sigma_{\mu\nu}$. Thus, the present aim comes down to the calculation of the above-given matrix elements within the framework of the χQSM. In this talk, results of recent investigations on the strangeness in the nucleon within the framework of the χQSM will be presented. For details about the formalism we refer a review by Christov *et al.* [24].

We start with the scalar channel. The scalar form factor is defined as

$$\sigma(t) = \bar{m}\left\langle N(p')|\bar{u}u + \bar{d}d|N(p)\right\rangle . \quad (4)$$

The analysis of the $\sigma_{\pi N}$ term the momentum dependence of the scalar form factor was carried out by Gasser, Leutwyler and Sainio [2]. The results of Ref. [2] are summarized as $\sigma = 45\pm 8$ MeV and $\Sigma \simeq 60$ MeV, and $y = 2\langle N|\bar{s}s|N\rangle/\langle N|\bar{u}u+\bar{d}d|N\rangle \simeq 0.2$ which means a share of $\langle N|\bar{s}s|N\rangle$ in the $\sigma_{\pi N}$ term.

Figure 1 draws the scalar form factor of the nucleon [25].

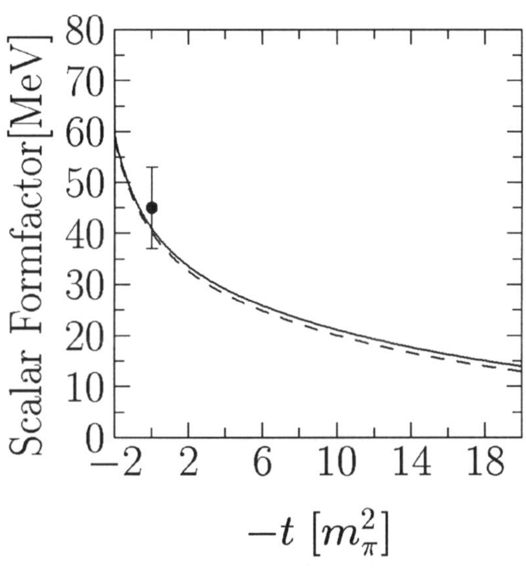

Figure 1: The scalar form factor of the nucleon $\sigma(t)$. The solid curve displays the case of $m_s = 148.49$ MeV, while the dashed curve shows that without m_s.. The error bar denotes the empirical value from Gasser *et al.* [2]

The error bar presented in Fig.1 stands for the empirical analysis due to Gasser, Leutwyler and Sainio. As shown in Fig. 1 the theoretical prediction is in good agreement with Ref. [2]. The m_s corrections seem to be negligible in the scalar form factor. However, they play an important role of suppressing the strangeness contribution: The value of y with the m_s corrections is 0.27 while $y = 0.48$ without the m_s corrections. It implies that though the m_s corrections have a tiny effect on the magnitude of the scalar form factor, it leads to a large suppression of the y.The difference $\Delta\sigma = \sigma(2m_\pi^2) - \sigma(0)$ we have obtained is 18.18 MeV. This value is very close to what Ref.[2] extracted, *i.e.* $\Delta\sigma = \sigma(2m_\pi^2) - \sigma(0) = 15.2 \pm 0.4$ MeV. The prediction of the χQSM for the scalar radius $\langle r^2\rangle_N^S$ is 1.5 fm^2 which is also comparable to the empirical value obtained by Ref. [2] $\langle r^2\rangle_N^S \simeq 1.6\text{fm}^2$.

The strangeness content of the nucleon in the vector channel is one of hot issues these days. While a great deal of theoretical works pile up, there is no clear theoretical consensus and no evident experimental judgment yet. To date there exist only two experimental data [22,23].

Strangeness in the nucleon can be interpreted in terms of the $K\Lambda$ or $K\Sigma$ components coupled to an external probe. So, one can expect that kaons may play a dominant role in describing the strange vector form factors. Thus, we consider the effect of the kaon cloud into account. To do so, we have incorporated the kaonic tails selfconsistently in the profile at the expense of the pion tails [26a].

The effect of the kaon cloud is prominent in the case of the strange electric form factor. As shown in Fig. 2, the replacement of the pion cloud by the kaon one brings about a sizable decrease of the strange electric form factor almost by a factor of three.

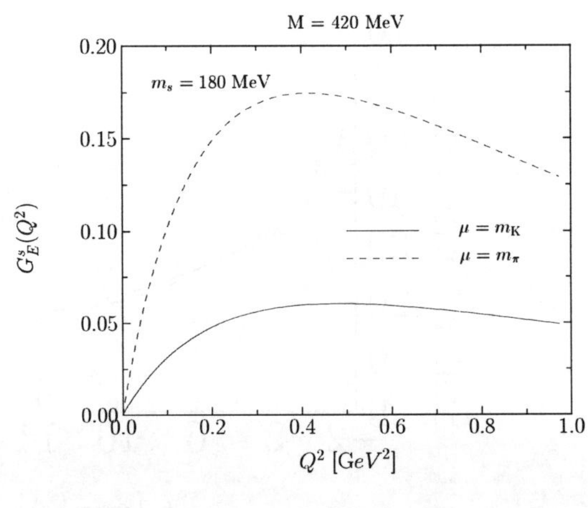

Figure 2: The strange electric form factor G_E^s as functions of Q^2: The solid curve corresponds to the $\mu = m_K$, while dashed curve draws $\mu = m_\pi$. The constituent quark mass M and m_s are 420 MeV and 180 MeV, respectively.

Such a drastic reduction of the strange electric form factor can be easily under-

[a] Ref.[26] contains a numerical mistake in the strange magnetic form factor. Correct and improved results will appear elsewhere.

stood explicitly by evaluating the strange electric radii. In any hedgehog model the strange radii depend on the inverse of the meson mass μ which suppresses the tail of the profile, *i.e.* $\langle r^2\rangle_s \sim 1/\mu$. From such a behavior of the $\langle r^2\rangle_s$, we can derive the relation

$$\frac{\langle r^2\rangle_s^{\rm Sachs}\big|_{\mu=m_\pi}}{\langle r^2\rangle_s^{\rm Sachs}\big|_{\mu=m_{\rm K}}} = \frac{m_{\rm K}}{m_\pi} \simeq 3.5. \tag{5}$$

Eq.(5) explains the decrease of the $\langle r^2\rangle_s$ with $\mu = m_{\rm K}$.

The SU(3) symmetry with the strange quark mass makes it possible to extract the strange magnetic moment from the experimental data of the octet baryon magnetic moments [21,27]. In Table I we list the recent results by these fitting procedures.

Finally, We want shortly to present a recent calculation of the tensor charges of the nucleon [28]. The tensor charge δq is related to the transverse quark distribution $h_1(x)$. However, $h_1(x)$ is not measurable in inclusive deep-inelastic scattering. That is the reason why it has not been extensively studied for long. In fact, Ralston and Soper [29] proposed the $h_1(x)$ which can be measured in polarized Drell-Yan processes almost 20 years ago. Recently, It was also suggested that the $h_1(x)$ can be measured in other exclusive hard reactions [30,31,32].

Jaffe and Ji demonstrated that the first moment of the $h_1(x)$ is related to the tensor charge of the nucleon:

$$\int_1^0 dx\,\left(h_1(x) - \bar{h}_1(x)\right) = \delta q, \tag{6}$$

where $\bar{h}_1(x)$ is an antiquark transversity distribution. Note that the tensor charges depend on the renormalization scale, though their dependence on it is very weak. The normalization point pertinent to the χQSM is not determined from the first principle. However, as mentioned in section I, the renormalization point in the χQSM has be chosen by $\rho^{-1} \simeq 600$ MeV, but there may be a factor of order unity.

In Tables II-III the results of the tensor charges are summarized. As shown in Table I the rotational $1/N_c$ corrections are of great significance numerically, while the m_s corrections are relatively small. In contrast to the axial charges, the tensor ones in the χQSM are closer to their values in the nonrelativistic constituent quark model and in particular the strangeness contribution to the tensor charge δs is compatible with zero, while that to the axial ones Δs in the χQSM is negative and distinctive from zero [33]. This difference between the tensor charges and the axial ones can be explained by calculating them in the gradient expansion and scrutinizing their dependence on the size of the soliton [34].

Our aim in this talk has been to review the recent investigation on strangeness in the nucleon in the SU(3) chiral quark-soliton model(χQSM). The rotational $1/N_c$ and linear m_s corrections were taken into account. The only parameter we have in the model is the constituent quark mass M which is fixed to 420 MeV by the mass splitting of the SU(3) baryon octet and decuplet. The results for the scalar form factors are in remarkable agreement with empirical data. The strange vector form factors were also discussed. The effect of kaon cloud turns out to be very large. By taking into account this kaonic effect, the strange charge radius is reduced to be almost 3 times smaller than that without the kaon cloud. It was also shown that

Table 1: Magnetic moments and their flavor decomposition for 4 fits described in Ref.[27], for model calculations (χQSM) and for fit 1 of Ref.[21] (HPM)

fit/model	$\mu^{(u)}$	$\mu^{(s)}$
(1)	4.48	0.25
(2)	3.92	0.25
(3)	4.00	0.37
(4)	4.18	0.41
χQSM	3.02	0.03
HPM	4.08	0.33

Table 2: Tensor charges $g_T^{(0)}$, $g_T^{(3)}$ and $g_T^{(8)}$ with the constituent quark mass $M = 420$ MeV. The current quark mass m_s is chosen as $m_s = 180$ MeV. The final model predictions are given by $\mathcal{O}(\Omega^1, m_s^1)$.

	$\mathcal{O}(\Omega^0, m_s^0)$	$\mathcal{O}(\Omega^1, m_s^0)$	$\mathcal{O}(\Omega^1, m_s^1)$
$g_T^{(0)}$	0	0.69	0.70
$g_T^{(3)}$	0.79	1.48	1.54
$g_T^{(8)}$	0.09	0.48	0.42

Table 3: Each flavor contribution to the tensor charges as varying the constituent quark mass M. The current quark mass m_s is chosen as $m_s = 180$ MeV.

M [MeV]	δu	δd	δs
420	1.12	-0.42	-0.008

by using the octet magnetic moments we were able to extract the strange magnetic moment.

The tensor charges of the nucleon are presented within the same framework. The χQSM predicts the number of the transversely polarized strange quarks δs in the transversely polarized nucleon compatible with zero, whereas it yields the negative nonzero number of the polarized strange quarks Δs in the longitudinally polarized nucleon, which is consistent with the corresponding experimental value.

Acknowledgments

The works presented in this talk were done in collaboration with A. Blotz, M.V. Polyakov, M. Praszałowicz, C. Schneider, T. Watabe and K. Goeke. The work is supported in part by the research fund of Pusan National University.

References

1. T.P. Cheng, *Phys. Rev.* **D13** (1976) 2161.
2. J. Gasser, H. Leuwyler, and M.E. Sainio, *Phys. Lett.* **B253** (1991) 252; *Phys. Lett.* **253B** (1991) 260.
3. B. Adeva *et al.*, *Phys. Lett.* **302B** (1993) 533.
4. D. Adams *et al.*, *Phys. Lett.* **329B** (1994) 399.
5. P.L. Anthony *et al.*, *Phys. Rev. Lett.* **71** (1993) 959.

6. K. Abe *et al.*, *Phys. Rev. Lett.* **75** (1995) 25.
7. D.B. Kaplan and A. Manohar, *Nucl. Phys.* **B310** (1988) 527.
8. R.L. Jaffe, *Phys. Lett.* **229B** (1989) 275.
9. N.W. Park, J. Schechter and H. Weigel, *Phys. Rev.* **D43** (1991) 869.
10. N.W. Park and H. Weigel, *Nucl. Phys.* **A541** (1992) 453.
11. W. Koepf, E.M. Henley and S.J. Pollock, *Phys. Lett.* **288B** (1992) 11.
12. T.D. Cohen, H.Forkel and M. Nielsen, *Phys. Lett.* **B316** (1993) 1.
13. S.T. Hong and B.-Y. Park, *Nucl. Phys.* **A561**, 525 (1993).
14. M.J. Musolf and M. Burkardt, *Z. Phys.* **C61** (1994) 433.
15. H. Forkel, M. Nielsen, X. Jin and T.D. Cohen, *Phys. Rev.* **C50** (1994) 3108.
16. D.B. Leinweber, *Phys.Rev.* **D53** (1996) 5115.
17. H.W. Hammer, Ulf-G. Meißner and D. Drechsel, *Phys.Lett.* **B367** (1996) 323.
18. M.J. Musolf and H. Ito, DOE-ER-40561-245 [`nucl-th/9607021`], (1996).
19. H.-C. Kim, T. Watabe and K. Goeke, *Nucl. Phys* **A616** (1997) 606.
20. Ulf-G. Meissner, V. Mull, J. Speth and J.W. van Orden, *Phys. Lett.* **B408** (1997) 381.
21. S.-T. Hong, B.-Y. Park and D.-P. Min, SNUTP-97-032 [`nucl-th/9706008`], (1997).
22. M. Mueller *et al.*, *Phys. Rev. Lett.* **78** (1997) 3824.
23. HAPPEX Collaboration (K.A. Aniol *et al.*, *Phys. Rev. Lett.* **82** (1999) 1096.
24. C.V. Christov, A. Blotz, H.-C. Kim,P. Pobylitsa, T. Watabe, Th. Meissner, E. Ruiz Arriola and K. Goeke, *Prog. Part. Nucl. Phys.* **37** (1996) 91.
25. H.-C. Kim, A. Blotz, C. Schneider, and K. Goeke, *Nucl. Phys.* **A596** (1996) 415.
26. H.-C. Kim, T. Watabe, and K. Goeke, *Nucl. Phys.* **A616** (1997) 606.
27. H.-C. Kim, M.V. Polyakov, M. Praszalowicz and K. Goeke, *Phys. Rev.***D58** (1998) 114027.
28. H.-C. Kim, M.V. Polyakov, and K. Goeke, *Phys. Lett.* **387B** (1996) 577.
29. J. Ralston and D. Soper, *Nucl. Phys.* **B152**, 109 (1979).
30. R.L. Jaffe and X. Ji, *Phys. Rev. Lett.* **71**, 2547 (1993).
31. J. Collins, *Nucl. Phys.* **B394** (1993) 169.
32. C. Bourrely and J. Soffer, *Nucl. Phys.* **B243**, 329 (1994); *Nucl. Phys.* **B445**, 341 (1995).
33. A. Blotz, M. Praszałowicz, and K. Goeke, *Phys. Rev.* **D53** (1996) 485.
34. H.-C. Kim, M.V. Polyakov, and K. Goeke, *Phys. Rev.* **D53** (1996) R4715.

QUARK RECOMBINATION MODEL FOR POLARIZATIONS IN INCLUSIVE HYPERON PRODUCTIONS AT HIGH ENERGY

N. NAKAJIMA, K. SUZUKI, H. TOKI

Research Center for Nuclear Physics (RCNP), Osaka University, Mihogaoka 10-1, Ibaraki, Osaka 567-0047, Japan

K.-I. KUBO

Department of Physics, Tokyo Metropolitan University, Hachiohji, Tokyo 192-0397, Japan

We investigate the transverse polarization of hyperons produced by the photon induced reactions using the quark recombination model. This model reproduces polarizations of hadrons produced by the hadron-hadron collisions and accounts for the origin of the empirical rule by DeGrand and Miettinen. We find significant polarizations in the hyperon photoproduction by applying this model to the $\gamma N \to \Lambda\ (\Sigma^0) + X$ reaction.

1 Introduction

Significant transverse polarizations and analyzing powers of inclusively produced hadrons Λ in unpolarized hadron-hadron collisions are observed at high Feynman $x_F(= 2p_L/\sqrt{s}$, p_L is the longitudinal momentum of the observed particle) and low transverse momentum p_T region[1] against predictions of the perturbative QCD[2]. These data indicate some non-perturbative mechanism being intact for the production of the polarization. Several models are proposed to account for these phenomena[3,4,5,6,7]. It is generally expected that the spin polarization is generated by some soft processes in the initial state and/or final state thorough the hadron-hadron collision.

Yamamoto *et al.* have constructed the quark recombination model to explain the generation of the polarization[4,5]. In this model, a final state hadron is produced by the simple recombination of relativistic quarks and/or diquarks, and polarizations are generated by the scalar confining force through the hadronization process in purely non-perturbative way. It was shown that this model reproduces the empirical rule of DeGrand and Miettinen(DM)[3], and provides the polarizations in agreement with the experimental data. In the framework of this model, we naturally expect non-vanishing polarizations even in the hyperon photoproduction. Here, we investigate polarizations of the hyperons in the photon induced reaction using the quark recombination (QRC) model [3,4].

2 Quark Recombination Model

Let us briefly introduce the quark recombination model[4]. It is assumed that a fast valence quark (or diquark) directly coming from the projectile hadron picks up slow quarks created by the hadronization to form the finale state hadron. The particle production at high x_F is dominated by this direct process, while the middle and small x_F hadron production are described by the standard fragmentation, in which all the quarks are randomly created by the string breaking. Hence, the produced hadron at the high x_F contains some information on the valence quark structure of the beam hadron. In fact, the polarization of the hyperon in the final state reflects the spin structure of the beam hadron (see assumption *[4]*).

Our model is based on the following assumptions;

1. The produced hadron is formed by direct recombination process, since significant polarizations are observed in the high x_F region.

2. Each parton which participates in this reaction has the intrinsic transverse momentum distribution.

3. Quark and diquark are recombined by the scalar confinement interaction in the hadronization process.

4. SU(6) spin-flavor symmetry is assumed for the initial and final state hadron structure.

Now we concentrate on the Λ hyperon production in the proton-proton collision. In this case, the SU(6) spin-flavor symmetry tells us that the valence $(ud)^0$-diquark from the beam proton picks up a slow s-quark in order to form the final state Λ. The spin of the Λ is determined by the s-quark spin. Thus, the polarization of the Λ, $P_N(\Lambda)$, is defined as

$$P_N(\Lambda) \equiv \frac{\sigma(\Lambda\uparrow) - \sigma(\Lambda\downarrow)}{\sigma(\Lambda)} = \frac{\sigma(s\uparrow) - \sigma(s\downarrow)}{\sigma(\Lambda)} \tag{1}$$

where the spin direction is fixed by $p_{beam} \times p_\Lambda$.

The Λ production probability in the Infinite Momentum Frame (IMF) is given by

$$S_{p\to\Lambda} = \int [dx_i dy_i dz_i/x_i] G_\Lambda(x_3, x_4, y_3, y_4, z_3, z_4) \times |M(x_i, y_i, z_i)|^2 f_s(x_2, y_2, z_2) G_{(ud)^0/p}(x_1, y_1, z_1) \Delta^3 \Delta^4 \tag{2}$$

where we choose the x-axis as the beam direction and z as the orientation of the Λ spin in the final state, Δ^3 and Δ^4 express the delta functions which correspond to energy-momentum conservation in this process. $G_{(ud)^0/p}$ is the momentum distribution of the (ud)-diquark in the projectile proton. We assume the following form;

$$G_{(ud)^0/p}(x_1, y_1, z_1) = q_{(ud)^0/p}(x_1)\, e^{-(y_1^2 + z_1^2)} \tag{3}$$

where the longitudinal momentum distribution $q_{(ud)^0/p}(x_1)$ is the (ud)-diquark distribution function in the proton taken from Ref. [8]. On the other hand, the transverse y and z momentum distributions are assumed to be the Gaussian form with the average transverse momentum being about 400 MeV. f_s is the momentum distribution of the piked-up s-quark;

$$f_s(x_2, y_2, z_2) = \theta(x_1 - x_2) e^{-(y_2^2 + z_2^2)}. \tag{4}$$

where we take the Gaussian transverse momentum distribution. We assume that the picked up s-quark is slower than the (ud)-diquark, which is expressed by θ-function.

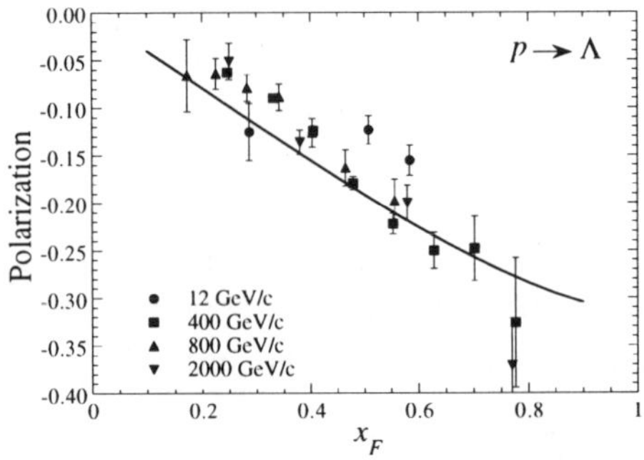

Figure 1. Λ polarization in the pp collision at $p_T = 1\text{GeV}/c$ with the experimental data.

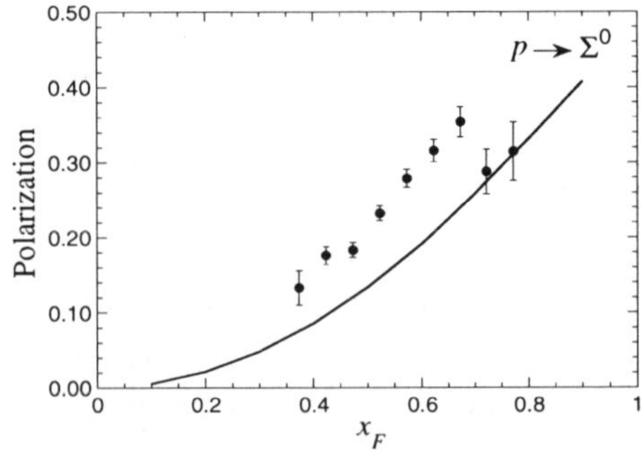

Figure 2. Σ polarization in the pp collision at $p_T = 1\text{GeV}/c$ with the experimental data.

G_Λ is the light-cone wave function of the final state Λ;

$$G_\Lambda(x_3, x_4, y_3, y_4, z_3, z_4) = A \exp\left[-\frac{1}{8\beta^2}\left[\frac{k_t^2 + m_{ud}^2}{x_3} + \frac{k_t^2 + m_s^2}{x_4}\right]\right]. \tag{5}$$

M is the elementary hadronization amplitude of the (ud)-diquark and s-quark to produce the Λ hyperon. We shall calculate them explicitly using the scalar confinement force, and find that the non-trivial spin dependence arises from the interference between the leading and higher order amplitudes.

We finally arrive at the following expression for the polarization,

$$P_N(\Lambda) = R_0 \frac{\int [dx_i dy_i dz_i / x_i] G_\Lambda^2 \sigma_{dep.} f_s G_{(ud)^0/p} \Delta^3 \Delta^4}{\int [dx_i dy_i dz_i / x_i] G_\Lambda^2 \sigma_{ind.} f_s G_{(ud)^0/p} \Delta^3 \Delta^4} \tag{6}$$

where the spin independent cross section is

$$\sigma_{ind.} = (x_F x_4 x_2)\left[\left(\frac{x_F x_4 + x_2}{x_F x_4 x_2} m_2\right)^2 + \left(\frac{x_F x_4 y_2 - x_2 y_4}{x_F x_4 x_2} \bar{p}_t\right)^2\right] \tag{7}$$

and the spin dependent one

$$\sigma_{dep.} = -(x_F x_4 x_2)\left(\frac{x_F x_4 y_2 - x_2 y_4}{x_F x_4 x_2} \bar{p}_t\right). \tag{8}$$

Note that, if we took the vector interaction instead of the scalar, the spin dependent part of the cross section would vanish in the IMF. R_0 in Eq. (6) is a free parameter and will be fixed to reproduce the experimental data of the Λ polarization.

Similarly, we can obtain the formula for the polarization of Σ^0 production. Since the Σ hyperon has the isospin 1, Σ^0 hyperon is formed by the recombination of a fast spin-1 ud-diquark and a slow s-quark. We need another parameter R_1 to be fixed by the Σ polarization data. From experimental data for Λ and Σ, we determine $R_0 = -12$ GeV and $R_1 = 50$ GeV. Our results are shown in Fig. 1 and 2 for the Λ and Σ polarizations, which are consistent with experiments. We can calculate the polarization of other hadrons, e.g. Ξ, which is also in good agreement with experiments.

3 Photon induced hyperon production

In our model, polarizations are generated by the non-perturbative hadronization process. Therefore, even in the photon induced reaction, we expect finite polarizations of the produced hyperons, because the hadronization process itself is independent of the beam. Let us consider the unpolarized γ-hadron reaction as schematically shown in Fig. 3. It is known that the real photon has the hadronic (vector meson) structure, since the real photon has enough time to turn into a $\bar{q}q$ system. In fact, the quark distribution of the real photon observed by the deep inelastic lepton-photon scattering contains the substantial hadronic components. We show in Fig. 4 the parametrization of the quark distribution of the real photon taken from ref. [9].

We apply our model to the photoproduction of hyperons. we replace the diquark distribution of the proton in the previous formula Eq. (6) with the quark distribution of the photon $G_{q/\gamma}$. In this case, the u(or d or s)-quark picks up a diquark to form the Λ or Σ^0 hyperons. Notice that both spin-0 and spin-1 diquark recombination processes contribute to this reaction. We get the following expression for the Λ polarization in the photon induced reaction,

$$P_N(\Lambda) = \frac{\sum_l R_l \int [dx_i dy_i dz_i/x_i] G_\Lambda^2 \sigma_{dep.,l} f_{(qq)^l} G_{q/\gamma} \Delta^3 \Delta^4}{\sum_l \int [dx_i dy_i dz_i/x_i] G_\Lambda^2 \sigma_{ind.,l} f_{(qq)^l} G_{q/\gamma} \Delta^3 \Delta^4} \tag{9}$$

where l runs over possible combinations of quarks and diquarks.

We show in Fig. 5 and 6 our results for the Λ and Σ polarizations in the γN reaction. The magnitude of the polarization is similar with that of the proton-proton collision. These theoretical predictions will be clarified by future experiments such as the PEARL facility.

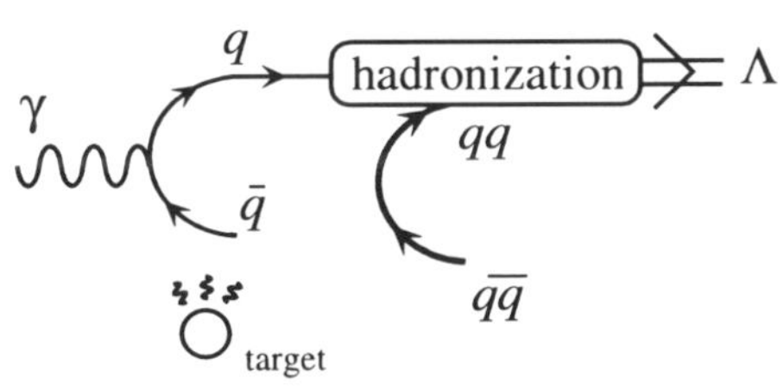

Figure 3. Hyperon production by the γ-hadron reaction

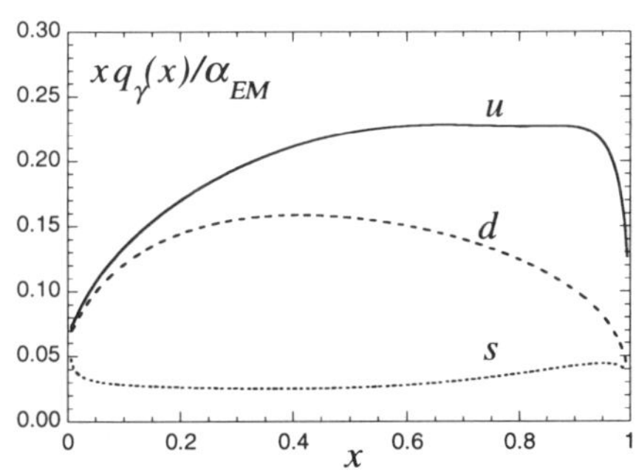

Figure 4. Photon structure function taken from ref. [9]

4 Summary

In conclusion, we have studied the spin polarizations of the hyperon photoproduction based on the quark recombination model. Our model may be applicable for the process with the photon beam , whose energy is larger than about 5GeV, where the quark degrees of freedom are essential rather than the hadronic components. We have reproduced various polarizations in the inclusive hyperon production in the hadron-hadron collisions with the model parameters being fixed to reproduce the absolute magnitudes of the polarizations of Λ and Σ. With these parameters we have found that about 20 $\sim$ 30% negative polarizations of Λ in the photon induced reaction, which will be tested by future experiments. It is also possible to extend the present model to the e^+e^- annihilation process. However, the polarization almost vanishes in the $e^+e^- \to \Lambda + X$ case, because the intrinsic transverse momentum distribution of the initial quark can be neglected in this case.

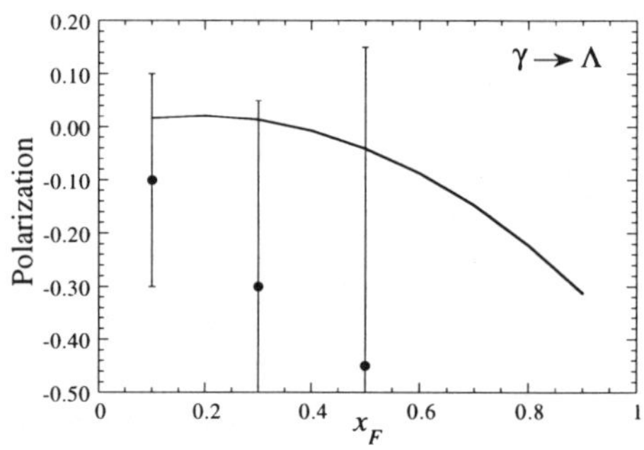

Figure 5. Λ Polarization in the γN reaction. Experimental data are taken from ref. [10].

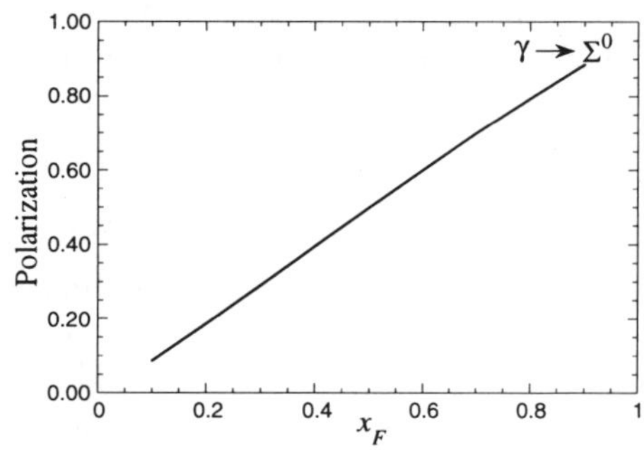

Figure 6. Σ Polarization in the γN reaction.

References

1. G. Bunec et al., *Phys. Rev. Lett.* **36**, 1113 (1976)
   Recent experimental references are found in ref. [4].
2. G.L. Kave, J. Pumplin and W.Repko, *Phys. Rev. Lett.* **41**, 1689 (1978)
3. T.A. DeGrand and H.I. Miettinen, *Phys. Rev.* D **23**, 1227 (1981)
4. Y. Yamamoto, K.-I. Kubo and H. Toki, *Prog. Theor. Phys.* **98**, 95 (1997)
   N. Nakajima, K. Suzuki, H. Toki and K.-I. Kubo, to be published.
5. K. Suzuki, N. Nakajima, H. Toki, and K.-I. Kubo, preprint (1998)
6. B. Andersson, G, Gustafson and G. Ingelman, *Phys. Lett.* B **85**, 417 (1979)
   J. Soffer and N,A Trnqvist, *Phys. Rev. Lett.* **68**, 907 (1992)
   L. Zuo-tang and C. Boros, *Phys. Rev. Lett.* **79**, 3608 (1997)
7. C. Boros, L. Zuo-tang and M. Ta-chung, *Phys. Rev.* D **51**, 4867 (1995)
   M. Anselmino, M. Boglione and F. Murigia, *Phys. Lett.* B **362**, 164 (1995)

S.M. Troshi and N.E. Tyurin, *Phys. Rev.* D **54**, 838 (1996)
X. Artru, J.Czyzewski and H. Yabuki, *Z. Phys.* C **73**, 527 (1997)
J. Qiu and G. Sterman, hep-ph/9806356
M. Anselmino and F.Murgia, hep-ph/9808426

8. S. Ekelin and S.Fredriksson, *Phys. Lett.* **95** 373 (1985)
9. M. Glück, E. Reya and A. Vogt, *Phys. Rev.* D **46**, 3986 (1992)
10. K. Abe, et al., *Phys. Rev.* D **29**, 1877 (1984)

KNY COUPLING CONSTANTS AND FORM FACTORS FROM THE CHIRAL BAG MODEL

M. T. JEONG
Department of Physics, Dongshin University, Naju 520-714, Korea
E-mail : mjeong@dsweb.dongshinu.ac.kr

Il-T. CHEON
Department of Physics, Yonsei University, Seoul 120-749, Korea
E-mail : itcheon@phya.yonsei.ac.kr

The form factors and coupling constants for $KN\Lambda$ and $KN\Sigma$ interactions have been calculated in the framework of the Chiral Bag Model with vector mesons. Taking into account vector meson (ρ, ω, K^*) field effects, we find $-3.88 \leq g_{KN\Lambda} \leq -3.67$ and $1.15 \leq g_{KN\Sigma} \leq 1.24$, where the quark-meson coupling constants are determined by fitting the renormalized πNN coupling constant, $[g_{\pi NN}(0)]^2/4\pi = 14.3$. It is shown that vector mesons make significant contributions to the coupling constants $g_{KN\Lambda}$ and $g_{KN\Sigma}$. Our values are existing within the experimental limits compared to the phenomenological values extracted from the kaon photoproduction experiments.

1 Introduction

The quantum chromodynamics(QCD) is believed to be the basic theory of the strong interaction. The MIT bag model[1], soliton model[2], and skyrme model[3] are amongst the best known phenomenological models of QCD. The chiral bag models(CBM)[4,5] which have been studied by many groups are based on the σ model. Araki[6] calculated the values of g_A, $g_{\pi NN}$ and μ_N in the framework of the nonlinear chiral soliton model. As a result, they showed that it gave too a large g_A and too small nucleon magnetic moments, and the introduction of the $\omega-$quark interaction was able to give an overall agreement of g_A, $g_{\pi NN}$ and μ_N. The CBM with vector mesons(CBMVM) improves remarkably the theoretical curves of nucleon form factors as well as the values of nucleon magnetic moments and nucleon e.m. radii[7]. Also, Theberge et al.[5] calculated the magnetic moments for all members of the $J^P = (1/2)^+$ octet baryons $(n, p, \Sigma^{\pm}, \Lambda, \Xi^{-,0})$ including the strange baryons in the framework of the CBM, and they showed excellent agreement with data within the 10% level. Therefore, one can make use of the CBMVM to analyze the hyperons (Σ, Λ).

The most fundamental coupling constants in the kaon-nucleon physics are the $KN\Lambda$ and $KN\Sigma$ coupling constants. The extraction of precise values for $g_{KN\Lambda}$ and $g_{KN\Sigma}$ is not yet allowed due to the limited set of data, while the πNN coupling constant $g_{\Pi NN}$ is determined accuratively through either the nucleon-nucleon scattering or the pion-nucleon scattering experiment. Within a phenomenological approach, Adelseck et al.[8,9] extracted the coupling constants by performing a least-squares search similar to that of Thom's[10]. However, these results turned out to have large uncertainties, $-1.29 \geq g_{KN\Lambda}/\sqrt{4\pi} \geq -4.17$, due to the simultaneous determination of many other unknown coupling constants. Therefore, it is necessary to explore theoretical predictions because of uncertainties and difficulties in extracting the precise values for these coupling constants from the experiments.

The aim of this paper is to calculate the coupling constants $g_{KN\Lambda}$ and $g_{KN\Sigma}$ in the framework of the CBMVM. We analyze the hadronic πNN coupling constant at first, and then, we calculate $g_{KN\Lambda}$ and $g_{KN\Sigma}$ using extension from SU(2) CBMVM to the SU(3) and taking into account the vector mesons(ρ, ω, K^*) as well as the pseudoscalar mesons(π, K). The quark-meson coupling constants are determined by fitting the renormalized πNN coupling constant. Our predictions are compared with the values extracted from phenomenological analyses and other theoretical QCD calculations. Also, form factors are suggested for πNN, $\pi N\Delta$, $KN\Lambda$ and $KN\Sigma$.

2 Interaction Lagrangians with Vector Mesons

The Lagrangian containing vector mesons can be derived by a local gauge transformation of the CBM Lagrangian[11] as following,

$$\begin{aligned}
\mathcal{L} &= \bar{q}i\gamma^\mu\left(\partial_\mu - \frac{i}{2}g_v\vec{\tau}\cdot\vec{\rho}_\mu - \frac{i}{2}g_v\omega_\mu\right)q\Theta_V - B\Theta_V - \frac{1}{2}\bar{q}q\Delta_S \\
&+ \bar{q}\gamma^\mu\frac{1}{1+g_\pi^2\pi^2}[g_\pi\gamma_5\vec{\tau}\cdot D_\mu\vec{\pi} - g_\pi^2\vec{\tau}\cdot(\vec{\pi}\times D_\mu\vec{\pi})]q\Theta_V \\
&+ \frac{1}{2}\frac{1}{(1+g_\pi^2\pi^2)^2}D^\mu\vec{\pi}\cdot D_\mu\vec{\pi} - \frac{1}{2}\frac{1}{(1+g_\pi^2\pi^2)}m_\pi^2\vec{\pi}^2 - \frac{1}{4}\vec{\rho}^{\mu\nu}\cdot\vec{\rho}_{\mu\nu} \\
&+ \frac{m_\rho^2}{2}\vec{\rho}^\mu\cdot\vec{\rho}_\mu - \frac{1}{4}\omega^{\mu\nu}\omega_{\mu\nu} + \frac{m_\omega^2}{2}\omega^\mu\omega_\mu + \frac{3g_\pi g_v^2}{4\pi^2}\varepsilon^{\mu\nu\alpha\beta}\partial_\mu\omega_\nu\vec{\rho}_\alpha\cdot\partial_\beta\vec{\pi}, \quad (1)
\end{aligned}$$

where

$$\begin{aligned}
\vec{\rho}_{\mu\nu} &= \partial_\mu\vec{\rho}_\nu - \partial_\nu\vec{\rho}_\mu + g_v\vec{\rho}_\mu\times\vec{\rho}_\nu, \\
\omega_{\mu\nu} &= \partial_\mu\omega_\nu - \partial_\nu\omega_\mu, \\
D_\mu\vec{\pi} &= \partial_\mu\vec{\pi} + g_v\vec{\rho}_\mu\times\vec{\pi}. \quad (2)
\end{aligned}$$

B is the vacuum energy constant, D_μ is the covariant derivative, $g_\pi = 1/2f_\pi$ with the pion decay constant f_π, and $q(x)$ and $\pi(x)$ are quark and pion

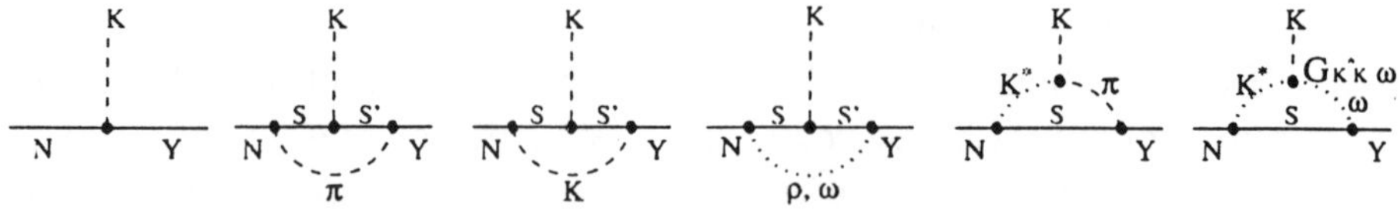

(1a) (1b) (1c) (1d) (1e) (1f)

Fig. 3. Feynman diagrams for the $KN\Lambda$ and $KN\Sigma$ form factors.

fields, respectively. Θ_V and Δ_S denote the volume and surface δ functions. With this effective Lagrangian of Eq(1), one can derive the interactions of the pion-quark and vector meson-quark as

$$\mathcal{L}_{\pi qq} = g_\pi \int d^3x \bar{q}(x)\gamma^\mu\gamma_5\vec{\tau}\cdot D_\mu\vec{\pi}(x)q(x)\Theta_V, \quad (3)$$

$$\mathcal{L}_{vqq} = g_v \int d^3x \bar{q}(x)\gamma^\mu[\vec{\tau}\cdot\vec{\rho}_\mu(x)+\omega_\mu(x)]q(x)\Theta_V. \quad (4)$$

The gauge coupling constant g_v in Eq.(4) can be evaluated by the $\rho\to 2\pi$ or $\omega\to\pi\gamma$ decay processes.

Extending from the basic doublet (u,d) to a triplet (u,d,s) and substituting the λ_i of $SU(3)$ generators for the τ_i of $SU(2)$, one can derive the Kqq and K^*qq interactions as

$$\mathcal{L}_{Kqq} = g_k \int d^3x \bar{q}(x)\gamma^\mu\gamma_5\vec{\lambda}\cdot D_\mu\vec{K}(x)q(x)\Theta_V, \quad (5)$$

$$\mathcal{L}_{K^*qq} = g_v^* \int d^3x \bar{q}(x)\gamma^\mu\vec{\lambda}\cdot\vec{K}^*_\mu(x)q(x)\Theta_V. \quad (6)$$

3 The KNY Form Factors

The $KN\Sigma$ and $KN\Lambda$ form factors can be obtained by calculating the Feynman diagrams shown in Fig.1 with Lagrangians (3), (4), (5), (6) and the Wess-Zumino term. The form factor $g_{KNY}(k_k^r)$ associated with each diagram is given as

$$\begin{aligned}\frac{g_{KNY}(k_k^r)}{2M} &= A g_k U(k_k^r)\\ &+ \frac{g_k}{144\pi^2}\sum_{SS'}\Big\{g_\pi^2 U(k_k^r)\int dk_\pi \frac{B_{SS'}k_\pi^4 U^2(k_\pi)}{\omega_\pi(\omega_\pi+\omega_{NS})(\omega_\pi+\omega_K^r+\omega_{NS'})}\\ &+ g_k^2 U(k_k^r)\int dk_k \frac{C_{SS'}k_K^4 U^2(k_k)}{\omega_k(\omega_k+\omega_{NS})(\omega_k+\omega_k^r+\omega_{NS'})}\Big\}\end{aligned}$$

$$
\begin{aligned}
&+ \frac{g_k}{144\pi^2}\sum_{SS'}\left\{ g_v^2 \int dk_v k_v^2 U(k_k^r)\right.\\
&\quad \times\left(\frac{D_{SS'}S_1^2(k_v)}{\omega_v(\omega_v+\omega_{NS})(\omega_v+\omega_k^r+\omega_{NS'})}+\frac{E_{SS'}S_0^2(k_v)}{\omega_v^2(\omega_v+\omega_k^r)}\right)\\
&+ g_\pi g_v \int dk_\pi (\vec{k_\pi}\cdot\vec{k_v^*})U(k_\pi)\\
&\quad \times\left(\frac{F_S S_1(k_v^*)}{(\omega_\pi+\omega_v^*)(\omega_\pi+\omega_{NS})(\omega_v^*+\omega_{NS})}+\frac{G_S S_0(k_v^*)}{\omega_\pi^2\omega_v^{*2}}\right)\\
&+ \frac{9g_v^2 g_v^{*2}}{4\pi^2}\int dk_v k_v \frac{(\hat{k_k}\cdot\vec{k_v}\times\vec{k_v^*})(\omega_v+\omega_v^*+\omega_{NS})}{\omega_v\omega_v^*(\omega_v+\omega_{NS})(\omega_v^*+\omega_{NS})(\omega_v+\omega_v^*)}\\
&\quad \left.\times[F_S S_1(k_v^*)S_0(k_v)+G_S S_1(k_v)S_0(k_v^*)]\right\}, \qquad (7)
\end{aligned}
$$

where k_k^r and ω_k^r are the real meson momentum and energy, k_ϕ and ω_ϕ are the virtual meson momentum and energy, $\vec{k_v}' = \vec{k_v} + \vec{k_\phi^r}$ and $\omega_v' = \sqrt{m_v^2 + k_v'^2}$, $\omega_{NS} = m_S - m_N$, and $g_\phi = 1/(2f_\phi)$ with the experimental value of the pion decay constant $f_\pi = 93.3MeV$ and the kaon decay constant $f_k = 114MeV$. The suffixes S and S' denote N, Λ and Σ having strangeness 0 or -1 for the octet with $J^P = 1/2^+$, and Δ and Σ^* having strangeness 0 or -1 for the decuplet with $J^P = 3/2^+$. Vertex functions in Eq.[7] are introduced in ref.[11], and coefficients in Eq.(7) can be calculated by substituting the SU(3) wave functions of three quarks in ref.[12]. The hadronic coupling constant, $g_v^* = g_v$, can be inferred from the SU(3) symmetry assumption.

The πNN coupling constant is determined by the values of the πNN form factors at $k_\pi^r = 0$. The result of numerical calculation for Eq.(7) at $k_\pi^r = 0$ is expressed by

$$
\frac{[g_{\Pi NN}]^2}{4\pi} = g_{\pi q}^2\left[4.154 + 0.630g_{\pi q}^2 + 0.00641g_v^2 + 0.00933g_{\pi q}g_v + 0.00044g_v^4\right]^2
$$

The numerical values of each term have been calculated with $R = 0.90fm$. The renormalized πNN coupling constant $[g_{\Pi NN}(0)]^2/4\pi = 14.3$ can be obtained by substituting the vqq coupling constant $g_v = 5.21$ and the πqq dimensionless coupling constant $g_{\pi q} = m_{\pi^+}/(2f_\pi)$ with the experimental value of the pion decay constant $f_\pi = 93MeV$. Calculating the Feynman diagrams of Figs.(1a), (1a $\sim$ 1b) and (1a $\sim$ 1e), we obtain the πNN coupling constant $[g_{\Pi NN}(0)]^2/4\pi = 9.70$, 11.4 and 14.3, and the $\pi N\Delta$ coupling constant $G_3 = g_{\Pi N\Delta}(0)/2M = 1.41/m_\pi$, $1.65/m_\pi$ and $2.01/m_\pi$, respectively. Here, the last value, $G_3 = 2.01/m_\pi$, reproduces the $\Delta\pi N$ decay width $\Gamma = 122MeV$.

Table 1: The $KN\Lambda$ and $KN\Sigma$ coupling constants . Set I and II are the results extracted from the analyses of the kaon photoproduction. III is the Skyrme Model predictions, and IV is a QCD sum rule predictions. V is the present calculations from the SU(3) CBMVM at radius $R = 0.90\ fm$.

	I[8]	II[13]	III[14]	IV[15]	V(0.90fm)
$g_{KN\Lambda}/\sqrt{4\pi}$	-4.17 ± 0.75	-3.16 ± 0.01	3.73	-2.76	-3.77
$g_{KN\Sigma}/\sqrt{4\pi}$	$+1.18 \pm 0.66$	$+0.91 \pm 0.10$	1.82	$+0.44$	$+1.19$

Substituting the $f_\pi(f_k) = 93(114)\ MeV$, $g_v = g_v^* = 5.21$ and $R = 0.90 fm$ into Eq.(7), we obtain $g_{KN\Lambda}/\sqrt{4\pi} = -3.77$ and $g_{KN\Sigma}\sqrt{4\pi} = +1.19$ at $k_k^r = 0$. The values extracted from the kaon photoproduction[8,13] are listed in Table 1 with our value and other theoretical predictions. Our values for $g_{KN\Lambda}$ and $g_{KN\Sigma}$ are existing within the experimental limits of the phenomenolgical values extracted from the kaon photoproduction of ref.[8]. It should be stressed that our results support the values extracted from the phenomenologocal analyses, although they have large uncertainties due to the simultaneous determinations of many unknown coupling constants. In Fig.2, we show the numerical results for the hadronic form factors $g_{\Pi NN}(k_\pi)$, $g_{\Pi N\Delta}(k_\pi)$, $g_{KN\Lambda}(k_k)$ and $g_{KN\Sigma}(k_k)$, where we use the bag radius $R = 0.90\ fm$. We can see that the $KN\Sigma$ form factor falls down considerably faster than the other form factors.

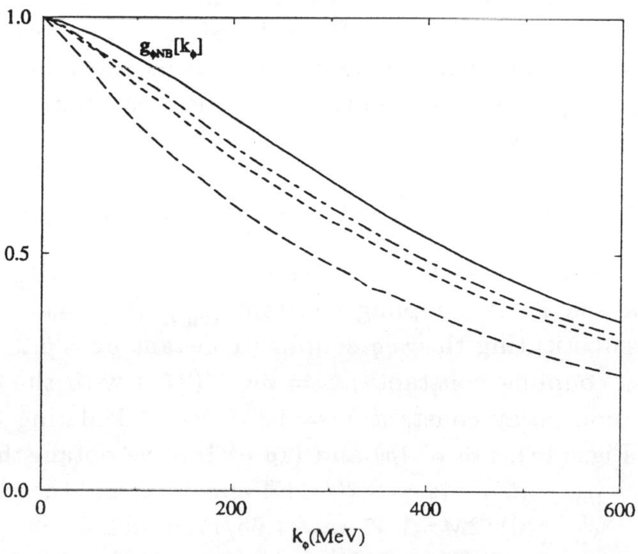

Fig.2. The solid, short-dashed, dot-dashed and long-dashed curves represent πNN, $\pi N\Delta$, $KN\Lambda$ and $KN\Sigma$ form factors, respectively.

4 Summary

Our calculations based on the SU(3) chiral bag model with vector mesons contain two parameters, i.e. the bag radius R and the quark-vector meson coupling constant g_v. However, g_v is determined by the renormalized πNN coupling constant $g^2_{\Pi NN}/(4\pi) = 14.3$ through the relation given in Eq.(7). Accordingly, we have the bag radius R as only one parameter. In the range of $0.80 \leq R \leq 1.00\ fm$, we obtain the value of the $KN\Lambda$ and $KN\Sigma$ coupling constants as follows

$$-3.88 \leq g_{KN\Lambda} \leq -3.67, \quad +1.24 \geq g_{KN\Sigma} \geq +1.15.$$

These values are considerably close to those phenomenological extracted from the kaon photoproduction, while other theoretical predictions fall short of the experimental limits. The reason of these enhancements is actually the point that we take into account meson $(\pi, K, \rho, \omega, K^*)$ field effects. It is again stressed that our results support the values obtained from the phenomenological analyses.

References

1. DeGrand, R. L. Jaffe, K. Johonson, and J. Kiskis, Phys. Rev. **D12**, 2060 (1975).
2. R. Friedberg and T. D. Lee, Phys. Rev. **D15**, 1694 (1977).
3. Adkins, C. R. Nappi and E. Witten, Nucl. Phys. **B228**, 552 (1983).
4. G. Kalbermann and J. M. Eisenberg, Phys. Rev. **D28**, 7 1(1983).
5. S. Theberge and A. W. Thomas, Nucl. Phys. **A393**, 252 (1983).
6. M. Araki, Phys. Rev. **C35**, 1954(1987).
7. M. T. Jeong and Il-T. Cheon, Phys. Rev. **D43**, 372 5(1991).
8. R. A. Adelseck and B. Saghai, Phys. Rev. **C42**, 108 (1990).
9. R. A. Adelseck, C. Benhold and L. E. Wright, Phys. Rev. **C32**, 1681 (1985).
10. H. Thom, Phys. Rev. **25**, 151 (1966).
11. M. T. Jeong and Il-T. Cheon, Eur. Phys. J. A 4, 357 (1999).
12. J. J. J. Kokkedee, **The Quark Model** (W. A. Benjamin Inc., New York, 1969).
13. J.-C. David et al., Phys. Rev. **C53**, 2613 (1996).
14. C. Gobbi, D. O. Riska and N. N. Scoccola, Nucl. Phys. **A554**, 671 (1992).
15. S. Choe, M. K. Cheoun and S. H. Lee, Phys. Rev. **53**, 1363 (1996).

WEAK PRODUCTION OF Λ IN THE PROTON-NUCLEON SCATTERING

T. INOUE, K. SASAKI AND M. OKA

Department of Physics, Tokyo Institute of Technology, Meguro, Tokyo 152 Japan

The $pn \rightarrow p\Lambda$ weak scattering is studied theoretically. The weak transition is described by the one meson exchange mechanism and the direct quark-quark interaction mechanism. The two baryon scattering wave functions are calculated in quark cluster model. The total and partial lambda production cross section and it's energy dependence are predicted.

1 Introduction and Formalism

The non-leptonic weak processes are studied for many years. It is know that the strong interaction at low energy is essential to understand this kind of process, and that comprehensive understanding is not achieved. The baryon-baryon weak interaction is a one of the non-leptonic weak processes and expected to be a new source of infomation of the strong interaction effect to the weak process. In this paper, we study this hyperon producing scattering, $pn \rightarrow p\Lambda$ from the fundamental theories.

We calculate the T matrix of the $pn \rightarrow p\Lambda$ scattering and obtain the crossection. We employ the distorted wave Born approximation, *i.e.*the T matrix is calculated by the convolution of the weak transition potential, V_W, and the two baryon scattering wave functions, $\chi^{(\pm)}$. As the weak transition potential contains nonlocal terms, a two-fold integral is required.

$$T^{DWBA} = \int d^3\vec{r} \int d^3\vec{r}' \chi^{(-)\dagger}(\vec{r}') V_W(\vec{r}', \vec{r}) \chi^{(+)}(\vec{r})$$

We consider only relative S wave state for the final $p\Lambda$. Even then, the initial and final interactions induces $\Lambda N \leftrightarrow \Sigma N$ coupling and $S \leftrightarrow D$ coupling. Therefore we consider the intermediate ΛN and ΣN state with relative S and D waves and the NN states with S, P and D waves.

Our weak $NN \rightarrow \Lambda N$ and $NN \rightarrow \Sigma N$ transition potential, V_W, is constructed by two different types of mechanism, hybrid model. The one is the one-meson-exchange mechanism. We consider the one π exchange and one K exchange. We take form factors into account. These mechanisms are for the larger distance. Another mechanism is based on constituent quark picture of baryon. In this mechanism, the weak interaction between two constituent quarks in the initial state, produce the strange constituent quark in the final state. The effect of this interaction was estimated by using effective weak Hamiltonian, $H_{eff}^{\Delta S=1}$, and quark cluster model [1]. This part is for the shorter distance.

We obtain the wave function of the scattering NN and YN system, $\chi^{(\pm)}$, in the quark cluster model. The Schrödinger equation for the relative motion (RGM

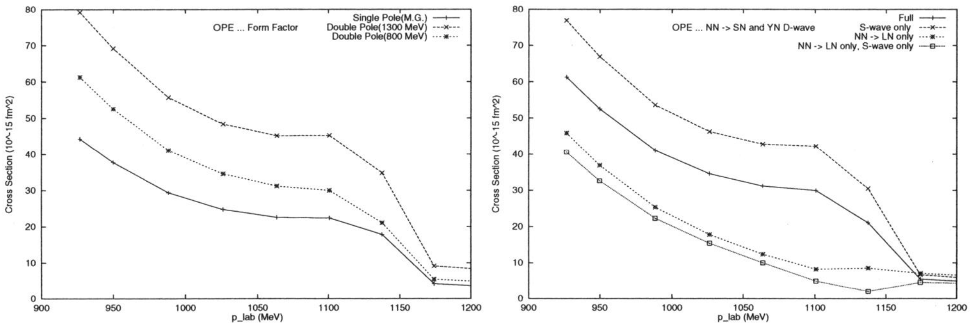

Figure 1. The Λ production cross section in OPE.

equation),

$$\int (H_{RGM}(\vec{r}', \vec{r}) - EN_{RGM}(\vec{r}', \vec{r}))\chi(\vec{r})d\vec{r} = 0 \tag{1}$$

determines the wave function $\chi^{(\pm)}$. There are one gluon exchange induced quark-quark force and one meson exchange induced cluster-cluster force in the Hamiltonian. We tune the meson exchange force so that the resulting NN wave function well reproduce the experimental data of phase shift. Therefor our NN wave function is phenomenological. On the other hand, the YN wave functions is calculated with the same meson exchange force as NN . Therefore, our YN wave functions are only theoretical.

2 Result and Discussion

Fig 1 left shows the Λ production cross section when we employ only the one-pion-exchange (OPE) induced weak transition potential. The horizontal axis is the laboratory momentum of proton. We test three different form factors for the OPE potential. First, we test the following "double pole" form factor [2],

$$V_{OPE}(\vec{q}) \to V_{OPE}(\vec{q}) \left(\frac{\Lambda_\pi^2 - m_\pi^2}{\Lambda_\pi^2 + \vec{q}^2} \right)^2 \tag{2}$$

with the cut off parameter $\Lambda_\pi = 1300$ MeV according to the Bonn NN force. The result is plotted in dashed line in Fig 1 left. Second, we take the single pole form factor,

$$V_{OPE}(\vec{q}) \to V_{OPE}(\vec{q}) \left(\frac{\Lambda_\pi^{SP^2} - m^2}{\Lambda^2 + \vec{q}^2} \right) \tag{3}$$

with the cut off parameter $\Lambda_\pi^{SP} = 630$ MeV [3]. The result is given by full line in Fig 1 left. One see the strong sensitivity on the choice of the form factor. This situation can be understood by considering the large momentum transfer involved.

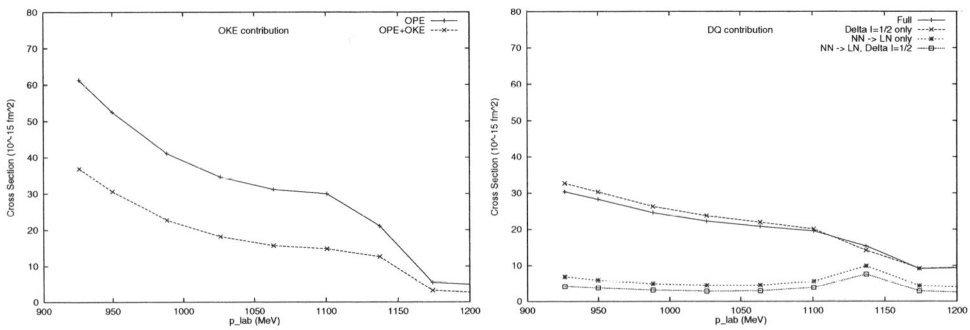

Figure 2. The effect of OKE and DQ respectively.

We insist that choosing an appropriate form factor is crucial for this process. It is reported that the DP form factor with Bonn cut off parameter Λ =1300 MeV is not suitable for the intermediate energy process $NN \rightarrow NN\pi$ but smaller cut off parameter is phenomenologically preferred [4]. Besides this, our previous study of the non-mesonic decay of hypernuclei suggests that a softer form factor is necessary so that unrealistically strong pion-exchange is suppressed [5]. Accordingly we employ a soft cut off $\Lambda = 800 MeV$. The result, given as the dotted line Fig 1 left, places between the above two choices. In the following, we use this form factor for the one pion exchange potential. The production cross section starts at about $6 \times 10^{-14}\text{fm}^2$ and decrease to about $4 \times 10^{-14}\text{fm}^2$ as the proton laboratory momentum increase from the Λ production threshold to the Σ production threshold.

The effect of the $NN \rightarrow \Sigma N$ weak transition and the YN D wave contribution in OPE-only calculation are shown in Fig 1 right. The upper two lines are the results when we consider the ΣN intermediate states that follows that $NN \rightarrow \Sigma N$ weak transition. The lower two lines are the results when we omit the ΣN intermediate state. In our calculation, $NN \rightarrow \Sigma N$ weak transition enhances the cross section much especially at large momentum. The YN D wave contribution is also significant when we take $NN \rightarrow \Sigma N$ weak transition into account. Its contribution is destructive to the S wave one.

Fig 2 left shows the effect of the one kaon exchange(OKE) mechanism. The double pole form factor with Bonn cut off parameter is taken into account. This figure shows that the one kaon exchange potential cancel one pion exchange one much. This tendency is well known in the study of non-mesonic decay of hypernuclei, and this cancelation are considered as important role of one kaon exchange potential. The production cross section predicted in OPE and OKE combined mechanism start at about $4 \times 10^{-14}\text{fm}^2$ and decree to about $2 \times 10^{-14}\text{fm}^2$.

We turn to the direct quark mechanism. If we take only direct quark mechanism, the production cross section becomes that shown in Fig 2 right. The upper two lines are the result when we consider the $NN \rightarrow \Sigma N$ weak transition. The lower two lines are the result when we omit the $NN \rightarrow \Sigma N$ weak transition. In direct quark

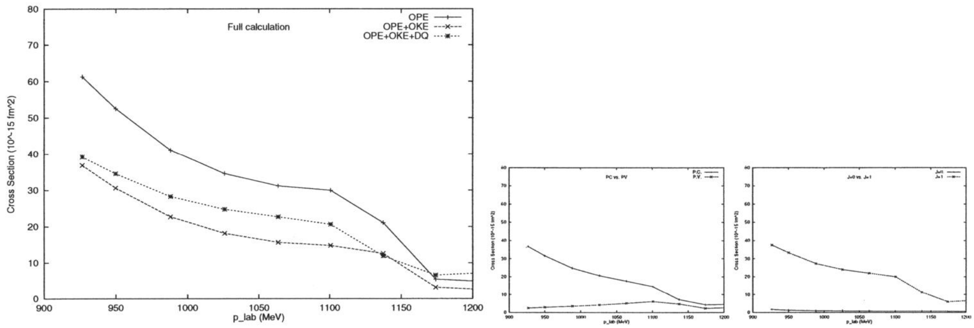

Figure 3. The Λ production cross section in OPE+OKE+DQ.

mechanism, $NN \rightarrow \Sigma N$ transition has very large contribution at all momentum region, from Λ production threshold to Σ production threshold. The cross section is a little smaller than meson exchange one, start at about $3 \times 10^{-14}\mathrm{fm}^2$ and decree to about $2 \times 10^{-14}\mathrm{fm}^2$.

Finally we go to the full of our approach, OPE+OKE+DQ mechanism. The production cross section in our full combined mechanism becomes that show in Fig 3 left. This figure shows that DQ has constructive contribution to meson exchange one. The production cross section increase about $1 \times 10^{-14}\mathrm{fm}^2$ at around 1050 MeV. We study the property of this production process. Fig 3 center shows the parity conserving (PC) and parity violating (PV) partial cross sections. One wee the very strong PC dominance especially at low momentum. Fig 3 right shows the J=0 and J=1 partial cross sections. One see very strong J=1 dominance and that J=0 transition is negligible.

3 Conclusion

We study Λ producing p-n weak scattering in DWBA. The weak transition potential is constructed by one pion exchange mechanisms, one kaon exchange mechanisms and direct quark mechanisms. The distorted waves are calculated in quark cluster model.

First we study the process in the one pion exchange mechanism. We find that the cross section is sensitive to the form factor of pion vertex, and chose the rather reasonable one. Then we add the one kaon exchange. The kaon exchange reduce cross section much, as know at non-mesonic decay of hypernuclei. Finally, we add the direct quark mechanism. We find sizable constructive contribution. In both meson exchange and direct quark exchange, the $NN \rightarrow \Sigma N$ weak transition has very significant role. Final result of the cross-section is about $2 \sim 4 \times 10^{-14}\mathrm{fm}^2$. The process is dominated by J=1 transition and parity conserving transition.

We hope that we can compare the result to experimental data in near future.

References

1. T.Inoue, S.Takeuchi and M.Oka, Nucl. Phys. **A597**(1996)563
2. A.Parreno, A.Ramos and C.Bennhold, Phys. Rev. **C56**(1997)339
3. B.H.J. Mckellar and B.F. Gibson, Phys. Rev. **C30**(1984)322
4. T.S.Lee and A. Matsuyama, Phys. Rev. **C36**(1987)1459
5. K.Sasaki, T.Inoue and M.Oka, nucl-th/9906036

MEASUREMENT OF SPIN-ORBIT SPLITTING IN ${}^{13}_{\Lambda}$C HYPERNUCLEUS

A. Sakaguchi[a], S. Ajimura[a], R.E. Chrien[b], P.M. Eugenio[c], G. Franklin[c], J. Franz[d], T. Fukuda[e], L. Gan[f], H. Hayakawa[a], P. Khaustov[c], T. Kishimoto[a], H. Kohri[a], K. Matsuoka[a], M. May[b], S. Minami[a], T. Mori[a], K. Morikubo[a], J. Nakano[g], H. Noumi[e], H. Outa[e], K. Paschke[c], P. Pile[b], B. Quinn[c], A. Rusek[b], E. Saji[a], R. Sawafta[h], R. Sutter[b], Y. Shimizu[a], M. Sumihama[a], T. Tamagawa[g], H. Tamura[i], K. Tanida[g], L. Tang[f] and L. Yuan[f]

(BNL-AGS E929 Collaboration)

[a] *Department of Physics, Osaka University, Toyonaka, Osaka 560-0043, Japan*
[b] *Brookhaven National Laboratory, Upton, NY 11973, USA*
[c] *Department of Physics, Carnegie Mellon University, Pittsburgh, PA 15213, USA*
[d] *University of Freiburg, D79104 Freiburg, Germany*
[e] *High Energy Accelerator Research Organization, Tsukuba 305-0801, Japan*
[f] *Department of Physics, Hampton University, Hampton, VA 23668, USA*
[g] *Department of Physics, University of Tokyo, Tokyo 113-0033, Japan*
[h] *Department of Physics, North Carolina A&T State University, Greensboro, NC 27411, USA*
[i] *Department of Physics, Tohoku University, Sendai 980-8578, Japan*

An experiment, E929, has been carried out at BNL/AGS to estimate the spin-orbit splitting in the ${}^{13}_{\Lambda}$C hypernucleus by measuring $1\hbar\omega$ γ-ray transitions. About 250 events of $1\hbar\omega$ γ-ray transitions coincidence with the ^{13}C(K$^-$,π^-) reaction were obtained by analyzing about 50 % of the experimental data. Even though the energy calibration of the γ-ray detector is still primitive, overall energy resolution of about 550 keV at 11 MeV is obtained, and it corresponds to the sensitivity roughly down to 150 keV in estimating the spin-orbit splitting. Preliminary γ-ray energy spectra are shown.

1 Introduction

Spectroscopic study on the Λ hypernuclei has been done for many years. One of the interests in the study was the property of the Λ-nucleus interaction. How is it similar with and how is it different from the well known nucleon-nucleus interaction ? The single independent particle motion of Λ in the mean field of the nucleus was clearly demonstrated by the associated production reaction; i.e., the (π^+,K$^+$) reaction.[1,2] The simple shell structure of the heavy Λ hypernuclei viewed by the (π^+,K$^+$) reaction is telling the smallness of the Λ-nucleon residual interaction, which makes the escape and spreading widths wider for the particle unbound Λ shell-model states, and it is also telling smaller spin dependent interaction on the Λ particle in the nuclear medium than that on the nucleon. Experimentally, the smallness of the spin dependent interaction made the detailed study of the Λ-nucleus interaction difficult. Various attempts to determine the spin dependent interaction were not so successful and results were unclear with a few exceptions. On the other hand, theoretical studies on the Λ-nucleus interaction, based on the SU(3) flavour symmetry assumption with the microscopic nucleon-nucleon interaction, had a great progress in the recent years. Spin doublets in the Λ hypernuclei were discussed with the spin-spin, spin-orbit and tensor interactions, and effect of the Λ-Σ coupling was estimated.[3] To catch up with the theoretical progresses, new experiments have been proposed. These new types of experiments necessarily attempt to be high resolution; i.e., sub-MeV resolution at least and a few keV resolution in the best case.

In this paper, we report preliminary results of the NaI(Tl) γ-ray detector experiment, AGS/E929, for the spin-orbit splitting measurement in the $^{13}_{\Lambda}$C hypernucleus attempting the sub-100keV sensitivity.

2 Basic Ideas of the Experiment

The measurement of the Λ $1\hbar\omega$ γ-ray transition by NaI(Tl) detectors was reported by May *et al.* at the first time for the $^{13}_{\Lambda}$C hypernucleus.[4] Among many excited states of the $^{13}_{\Lambda}$C hypernucleus, there are two excited states for which $p_{1/2\Lambda}$ and $p_{3/2\Lambda}$ couple to the $^{12}C_{g.s.}$ core, respectively. The γ-ray transition observed, E_γ=10.95±0.1±0.2 MeV, was the transition from the excited states to the ground state of the hypernucleus. Due to the limited yield of the decay γ-rays (about 15 events), most of the γ-rays corresponded to the $p_{1/2\Lambda}\rightarrow s_{1/2\Lambda}$ transition. An experiment with higher statistics makes it possible to measure both $p_{1/2\Lambda}\rightarrow s_{1/2\Lambda}$ and $p_{3/2\Lambda}\rightarrow s_{1/2\Lambda}$ transitions, and we can estimate the spin-orbit splitting between the $p_{1/2\Lambda}$ and the $p_{3/2\Lambda}$ states if a selective measurement of the two transitions is possible. A sub-100keV sensitivity is expected by γ-ray detectors with a 300 keV energy resolution at 11 MeV.

The selective excitation of the two states in $^{13}_{\Lambda}$C hypernucleus is possible by using the difference of the angular dependences of the differential cross section of the (K^-,π^-) reaction. Fig.1 shows the differential cross section of the $^{13}C(K^-,\pi^-)^{13}_{\Lambda}C$ reaction as a function of the π^- scattering angle at E_{K^-}=0.9GeV/c by a DWIA calculation.[5] Due to the small momentum transfer in the (K^-,π^-) reaction, the differential cross section of the $p_{1/2\Lambda}$ population (momentum transfer in the reaction ΔL=0) is forward peaking, and that of the $p_{3/2\Lambda}$ population (ΔL=2) is large at larger angle. By using the difference of the angular distribution of the cross sections, we can selectively populate either of the excited state by selecting the angular region of the scattered π^-. We wish to emphasize that this method of the spin-orbit splitting estimation is almost free from systematic errors because the π^- detection with the spectrometer system and the detection of γ-rays with the γ-ray detectors are almost independent. This is not the case for the experiment only with a spectrometer, since the momentum and scattering angle measured with a spectrometer usually have an instrumental correlation. A small residual correlation between the π^- scattering angle and the γ-ray energy comes from the Doppler-shift, but the effect is small and correctable.

More details of the ideas and design principle of the experiment are described in the experimental proposal.[6]

3 Experimental Setup

The experiment has been carried out at the D6 beamline of the AGS experimental facility in Brookhaven National Laboratory in June/July 1998. The beamline pro-

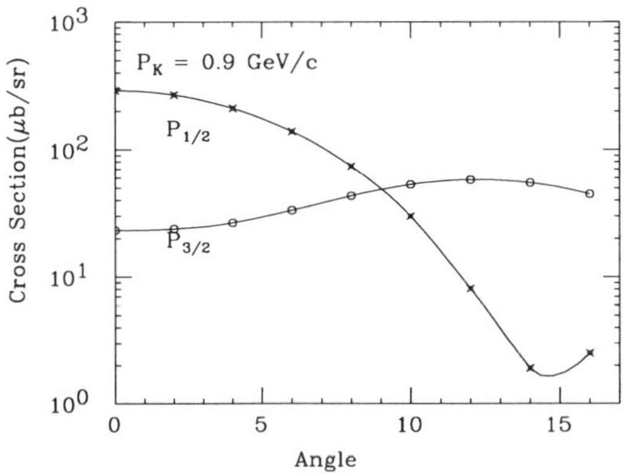

Figure 1: Differential cross section of the $^{13}C(K^-,\pi^-)^{13}_{\Lambda}C$ reaction as a function of the π^- scattering angle by a DWIA calculation.

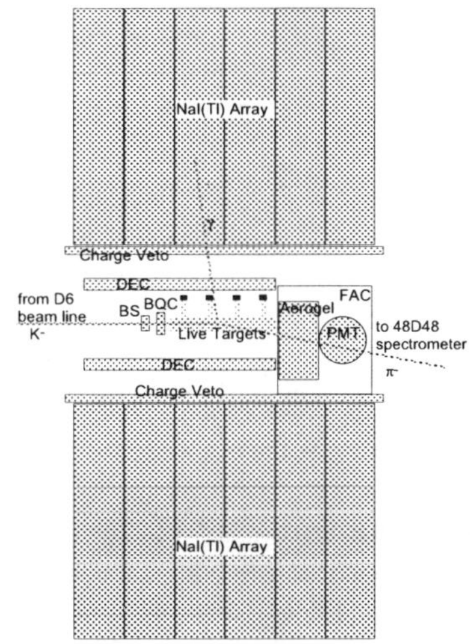

Figure 2: A schematic view of the detector setup around the target.

vided good quality of intense K^- beam, and products of the (K^-,π^-) reaction were measured with the large acceptance spectrometer system, 48D48. Details of the D6 beamline and the 48D48 spectrometer can be found elsewhere.[7]

The nominal beamline momentum was set to 930 MeV/c which was optimized for the yield of the (K^-,π^-) reaction, typical beam intensity was 10^5 K^-/spill (1 spill $\sim$ 4 sec), and 21×10^9 K^- beam bombarded on the target in total. The K^-/π^- ratio was better than 1 by the electrostatic mass separator in the beamline, and clear identification of the K^- beam particles was possible by the beam aerogel Čerenkov detector and the beamline TOF information in the trigger-level and in the off-line analysis, respectively. The π^- particles from the reaction were analyzed with the 48D48 spectrometer system. The aerogel Čerenkov detector and the TOF system in the spectrometer made the clear identification of π^- particles feasible. Typical missing mass resolution of the spectrometer system was 10 MeV/c^2 including the contribution from the beamline momentum analysis.

In addition to the standard detector setup of the D6 beamline and the 48D48 spectrometer system, new detectors were installed in the target region. Fig.2 shows a schematic view of the detector setup around the target. Our ^{13}C target consisted of 4 sets of ^{13}C benzene based liquid scintillator contained in glass cells (Live Targets). The scintillation light was read out by PM tubes attached to both ends of each glass cell through light guides. The information was used to positively identify the formation of the hypernuclei as described in Sec.4. In the upstream of the target, the beam defining plastic scintillation counter (BS) and the total reflection type quartz Čerenkov detector (BQC) were placed. In the downstream of the target, the aerogel Čerenkov detector (FAC) was placed. Since the most strong background

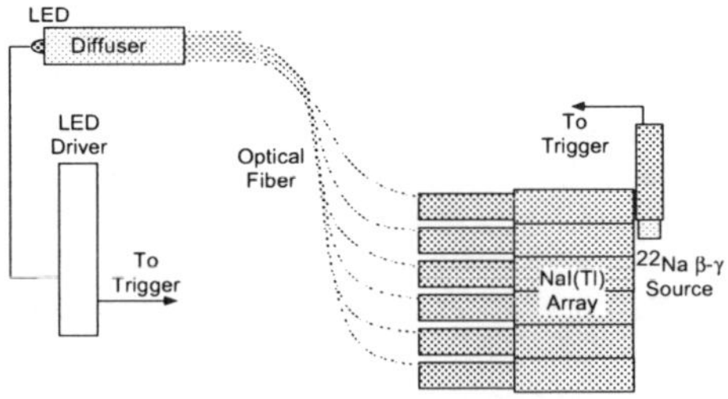

Figure 3: A schematic view of the energy calibration system for the NaI(Tl) detectors. See text for more detail.

in the measurement of the (K$^-$,π^-) reaction was the K$^-$ decay in-flight, BQC and FAC information was used to minimize the decay region of K$^-$. The decay particle counters (DEC) were placed at the top and the bottom of the target, and detected charged particles from the target cell.

To measure the γ-rays coincidence with the (K$^-$,π^-) reaction, 6×6 arrays of NaI(Tl) detectors were placed above and below the target with charge veto plastic scintillation detector (Charge Veto). Each NaI(Tl) module had a 2.5"×2.5"×12" size. The distance from the target to the NaI(Tl) detector arrays was about 10 cm. For the energy calibration, the LED light signal driven by a pulser sent to each NaI(Tl) module through an optical fiber. The LED was pulsed about 5 times in a beam spill, and could be used for a monitoring of the detector energy gain. To make a continuous monitoring for the absolute energy calibration, a ^{22}Na γ-ray source was attached to the top NaI(Tl) array, and the 1.275 MeV γ-ray was monitored as a part of the data acquisition stream. Fig.3 shows a schematic view of the energy calibration system for the NaI(Tl) detectors. Absolute energy calibration was done for all NaI(Tl) modules by the ^{22}Na γ-ray source and γ-rays from the ^{58}Ni(n,γ) reaction (Eγ=8.54MeV and 9.00 MeV) about in every week and every month, respectively.

4 Data Reduction and Results

The recorded experimental data contained the (K$^-$,π^-) events, large amount of background events from the K$^-$ decay in-flight between BQC and FAC, and mis-identified K^- elastic scattering events. The K$^-$ elastic scattering events were removed by the mass-squared cut calculated from the momentum and TOF information in the 48D48 spectrometer. The energy deposit in the Live Targets for the background K$^-$ decay in-flight events is essentially comparable with 1 MIP energy deposit. On the other hand, if the ^{13}C(K$^-$,π^-) reaction creates a Λ hypernucleus and it decays with non-mesonic process (proton-stimulated), the energy deposit in the Live Target at the reaction vertex is much higher than that of 1 MIP due to additional energy deposit by the decay proton. By selecting large energy deposit in one of the Live Target, we could efficiently select the formation of hypernuclei (efficiency $\sim$ 30%) and reject the K$^-$ decay in-flight background (reduction $\sim$ a few 0.1% with a tight cut). DEC information was also useful to detect decay pro-

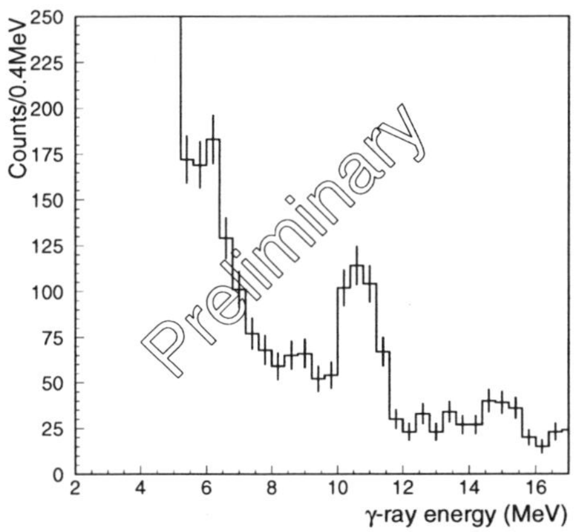

Figure 4: Energy spectrum of γ-ray coincidence with the (K$^-$,π^-) reaction. The peak around 11 MeV corresponds to the $p_\Lambda \rightarrow s_\Lambda$ $1\hbar\omega$ transition.

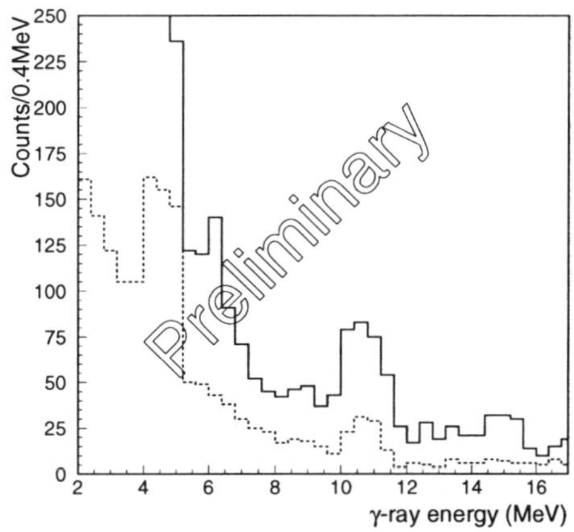

Figure 5: Same histograms as Fig.4 with cuts on the π^- scattering angle; $2° \leq \theta_{\pi^-} \leq 9°$ (solid line) and $9° \leq \theta_{\pi^-} \leq 16°$ (dashed line).

tons from the non-mesonic decays of hypernuclei, and improved the efficiency of the formation tagging of the hypernuclei.

After the reduction of the background, the missing mass was calculated by assuming the ^{13}C(K$^-$,π^-) kinematics. Since the spectrometer analysis was still preliminary and absolute momentum calibration of the spectrometer was not final, a loose cut on the excitation energy was set; i.e., the cut $0 \leq E_X \leq 50$ MeV was used. For the events selected by the excitation energy, the γ-ray energies measured by the NaI(Tl) arrays were accumulated as shown in Fig.4 for 50 % of total data. The peak around 11 MeV corresponds to the reported $1\hbar\omega$ γ-ray transition in the $^{13}_{\Lambda}$C hypernucleus. The γ-rays with energy around 11 MeV deposit energy in NaI(Tl) by several steps of interaction and has transverse spread of energy deposit which is larger than the size of our NaI(Tl) modules. The energy deposits in a 3×3 array of the NaI(Tl) modules were summed up to achieve a larger full-energy peak efficiency for the 11 MeV γ-rays. The yield of the γ-rays in the peak around 11 MeV was about 250 counts with 50 % of total data, which was almost consistent with the yield estimation in the experimental proposal. The energy calibration of the NaI(Tl) detector was still preliminary; i.e., only the weekly measured energy calibration data with the ^{22}Na calibration source were used to correct the gain drift of the NaI(Tl) detectors. The energy resolution of the NaI(Tl) detector arrays including the imperfection of the energy calibration was estimated with the 4.44 MeV γ-rays from the ^{12}C nucleus. The quasi-free (K$^-$,π^-) reaction, in which the Λ particles escape from the hypernuclei, leave the residual nuclei in their excited states. The γ-rays from the quasi-free events could be accumulated from the same dataset used for the hypernuclear γ-ray measurement. By a simple extrapolation of the γ-ray energy resolution from 4.44 MeV to 11 MeV, the resolution at 11 MeV was estimated to be 550 keV. This energy resolution roughly corresponds to 150 keV sensitivity in the spin-orbit splitting determination.

Fig.5 shows also the γ-ray energy spectra with different cuts on the π^- scattering angle; i.e., the forward angle $2^\circ \leq \theta_{\pi^-} \leq 9^\circ$ (solid line) and the backward angle $9^\circ \leq \theta_{\pi^-} \leq 16^\circ$ (dashed line) cuts. If our basic idea described in Sec.2 works well, the difference of the peak positions of the two histograms in Fig.5 tells the spin-orbit splitting for the $p_{1/2\Lambda}$ and $p_{3/2\Lambda}$ doublets in the ${}^{13}_{\Lambda}$C hypernucleus. Since the analysis is still preliminary and still in progress; e.g., the energy calibration of the γ-ray detectors, spectrometer analysis, effect of Doppler-shift, etc., we postpone the numerical evaluation of the spin-orbit splitting while it looks very small.

5 Summary

We carried out a high statistics γ-ray measurement coincidence with the ^{13}C(K$^-$,π^-) reaction to estimate the spin-orbit splitting between $p_{1/2\Lambda}$ and $p_{3/2\Lambda}$ states in the ${}^{13}_{\Lambda}$C hypernucleus. About 250 events of $p_\Lambda \rightarrow s_\Lambda$ γ-ray transition were obtained after analyzing about 50 % of the total data. We expect about 30 times improvement in statistics compared with the previous experiment after analyzing the full data, and it will make precise determination of the Λ spin-orbit splitting in sub-100keV level feasible. Currently obtained energy resolution for 11 MeV γ-ray was estimated to be 550 keV, and it means we already have the sensitivity of the spin-orbit splitting determination down to 150 keV. We expect an improvement of the energy resolution of the NaI(Tl) detectors down to 300 keV with more careful energy calibration of the γ-ray detectors.

Acknowledgments

We thank the staff of the Brookhaven AGS for their support in running the experiment. We also thank the technical staff of Brookhaven Medium Energy Group for their support in preparing a part of the experimental equipment. We thank T. Motoba for useful discussions during the preparation of the experimental proposal. We are also grateful for the support given by the Ministry of Education, Science and Culture, Japan, the U.S. Department of Energy, and the National Science Foundation.

References

1. R.E. Chrien, BNL (π^+,K$^+$) Collaboration, *Nucl. Phys.* A **478**, 705c (1998); P.H. Pile *et al.*, *Phys. Rev. Lett.* **66**, 2585 (1991).
2. T. Hasegawa *et al.*, *Phys. Rev.* C **53**, 1210 (1996).
3. See proceedings of talks on theoretical studies in this workshop.
4. M. May *et al.*, *Phys. Rev. Lett.* **78**, 4343 (1997).
5. T. Motoba, private communication.
6. T. Kishimoto, *et al.*, "Spin-orbit splitting of single Λ studied by the ^{13}C(K$^-$,$\pi^-\gamma$) reaction", BNL/AGS proposal E929.
7. P.D. Barnes, the BNL H Particle Collaboration E-813, *Nucl. Phys.* A **547**, 3c (1992).

ENERGY-INDEPENDENT AND HERMITIAN ΛN EFFECTIVE INTERACTION AND PROPERTIES OF Λ IN HYPERNUCLEI

S. FUJII

Department of Physics, Kyushu University, Fukuoka 812-8581, Japan
E-mail: fuji2scp@mbox.nc.kyushu-u.ac.jp

R. OKAMOTO, K. SUZUKI

Department of Physics, Kyushu Institute of Technology, Kitakyushu 804-8550, Japan
E-mail: okamoto@mns.kyutech.ac.jp, suzuki@mns.kyutech.ac.jp

A lambda-nucleon (ΛN) effective interaction and the single-particle potential of Λ in Λ hypernuclei are derived in the framework of the unitary-model-operator approach (UMOA), starting from a bare hyperon-nucleon (YN) interaction. The UMOA is applied to $^{17}_{\Lambda}$O and $^{41}_{\Lambda}$Ca. The single-particle energies of Λ are calculated for the $0s_{1/2}$, $0p_{3/2}$, $0p_{1/2}$ (for both $^{17}_{\Lambda}$O and $^{41}_{\Lambda}$Ca), $0d_{5/2}$, $0d_{3/2}$ and $1s_{1/2}$ (for only $^{41}_{\Lambda}$Ca) states, employing various YN potentials in free space constructed by the Nijmegen and the Jülich groups.

1 Introduction

There have been a number of attempts to study nuclear many-body problem. The unitary-model-operator approach (UMOA) [1] is one of many-body theories to describe a many-body system in terms of an effective interaction. The effective interaction is introduced as an interaction which acts in a model space and yields some of the same eigenvalues of the original Hamiltonian. The effective interaction in the UMOA derived from a bare interaction has some desirable properties for describing a many-body system: (a) It is energy-independent. (b) It is Hermitian. (c) It has the property of decoupling between a model space and its complement. This method has been developed and applied to the description of the ground-state properties of ordinary nuclei ^{16}O [1] and ^{40}Ca [2], using various realistic nucleon-nucleon (NN) interactions. Some improvements have been attained in comparison with other microscopic calculations.

In the previous works, we extended the formulation of the UMOA to the description of the structure of Λ hypernuclei and made a calculation of the single-particle energies of Λ in $^{17}_{\Lambda}$O [3,4], employing several hyperon-nucleon (YN) interactions. We confirmed the importance of the ΣN-ΛN coupling, as has been discussed in many other works. We also made clear the importance of the careful treatment of the Pauli-blocking effect for nucleons in $^{17}_{\Lambda}$O. However, the calculated single-particle energies of Λ and the matrix elements of the ΛN effective interaction were scattered dependently on the YN interactions employed. This is mainly because of the uncertainty of the YN interaction given in the present day. Therefore, it is necessary to examine the dependence of calculated results on various YN interactions. This kind of study could make us understand the structure of hypernuclei itself and also help us to determine the YN interaction. In this study, we apply the UMOA not only to $^{17}_{\Lambda}$O but also to $^{41}_{\Lambda}$Ca, using various YN interactions in free space, namely, the Nijmegen soft-core (NSC) [5], NSC97 models a-f [6], the Jülich $\tilde{\text{A}}$ (J$\tilde{\text{A}}$) and $\tilde{\text{B}}$ (J$\tilde{\text{B}}$) [7].

2 Formulation of the UMOA for Deriving the ΛN Effective Interaction

Let us outline the formulation of the UMOA to obtain the ΛN effective interaction and the single-particle potential of Λ. We first consider a two-body Hamiltonian of a ΣN-ΛN coupled system given by

$$H_{YN} = h_{YN} + \Delta m + v_{YN}, \tag{1}$$

where

$$h_{YN} = \begin{pmatrix} t_\Lambda + u_\Lambda + t_N + u_N & 0 \\ 0 & t_\Sigma + u_\Sigma + t_N + u_N \end{pmatrix}, \tag{2}$$

$$\Delta m = \begin{pmatrix} 0 & 0 \\ 0 & m_\Sigma - m_\Lambda \end{pmatrix} \tag{3}$$

and

$$v_{YN} = \begin{pmatrix} v_{\Lambda N-\Lambda N} & v_{\Lambda N-\Sigma N} \\ v_{\Sigma N-\Lambda N} & v_{\Sigma N-\Sigma N} \end{pmatrix}. \tag{4}$$

The terms t_k and u_k for $k = \Lambda, \Sigma$ and N are the kinetic and potential energies of Λ, Σ and N, respectively, and v_{YN} is the bare two-body YN interaction including the ΛN and ΣN channels. The terms m_Λ and m_Σ are the rest masses of Λ and Σ, respectively. The Hamiltonian H_{YN} describes a subsystem of interacting two particles ΛN or ΣN in the one-body potentials u_Λ, u_Σ and u_N.

In order to derive a ΛN effective interaction we apply a general theory of constructing a Hermitian effective interaction [1]. We introduce a unitarily transformed YN interaction given by

$$\tilde{v}_{YN} = e^{-S_{YN}}(h_{YN} + \Delta m + v_{YN})e^{S_{YN}} - (h_{YN} + \Delta m), \tag{5}$$

where S_{YN} is the correlation operator defined in a space of two-particle states of ΛN and ΣN. Let $P_{\Lambda N}$ be the projection operator onto the model space consisting of low-momentum ΛN states. The complement of the $P_{\Lambda N}$ space includes a space of high-momentum ΛN states referred to as the $Q_{\Lambda N}$ space and a space of all the ΣN states referred to as the $Q_{\Sigma N}$ space. In the formulation of the UMOA the correlation operator S_{YN} is determined by solving the equation of decoupling for $\tilde{v}_{YN}$ between the model space and its complement as

$$(Q_{\Lambda N} + Q_{\Sigma N})\tilde{v}_{YN}P_{\Lambda N} = P_{\Lambda N}\tilde{v}_{YN}(Q_{\Lambda N} + Q_{\Sigma N}) = 0. \tag{6}$$

If we assume that the second term $h_{YN} + \Delta m$ in Eq. (5) is decoupled as

$$(Q_{\Lambda N} + Q_{\Sigma N})(h_{YN} + \Delta m)P_{\Lambda N} = P_{\Lambda N}(h_{YN} + \Delta m)(Q_{\Lambda N} + Q_{\Sigma N}) = 0, \tag{7}$$

Eq. (6) becomes

$$(Q_{\Lambda N} + Q_{\Sigma N})e^{-S_{YN}}H_{YN}e^{S_{YN}}P_{\Lambda N} = 0. \tag{8}$$

The solution for S_{YN} to this equation has been known [1,4]. With the solution S_{YN}, the ΛN effective interaction $\tilde{v}_{\Lambda N}$ acting in the $P_{\Lambda N}$ space is given by

$$\tilde{v}_{\Lambda N} = P_{\Lambda N} \tilde{v}_{YN} P_{\Lambda N}. \tag{9}$$

The single-particle potential u_Λ in Eq. (2) is introduced as a self-consistent potential of Λ given by

$$\langle \alpha_\Lambda | u_\Lambda | \alpha'_\Lambda \rangle = \sum_{\xi_N:\text{occupied}} \langle \alpha_\Lambda \xi_N | \tilde{v}_{\Lambda N} | \alpha'_\Lambda \xi_N \rangle, \tag{10}$$

where $|\alpha_\Lambda\rangle$ ($|\alpha'_\Lambda\rangle$) and $|\xi_N\rangle$ are the single-particle states of Λ and N, respectively.

Equations (1)-(10) give a set of self-consistent equations. Assuming a starting value for u_Λ in Eq. (2), these equations can be solved in an iterative manner. Therefore, the ΛN effective interaction $\tilde{v}_{\Lambda N}$ and the Λ single-particle potential u_Λ are determined simultaneously. In an actual calculation, however, we must employ some approximation procedures. For instance, when we solve the set of self-consistent equations for a closed-shell nucleus plus one Λ system, the Pauli-blocking effect for nucleons should be considered. In the usual G-matrix calculation, the angle-average technique by Wong [8] and Sauer [9] is often used for the treatment of the Pauli-blocking effect. We also employ the angle-average technique basically. We have examined that the careful treatment of the Pauli-blocking effect was necessary to obtain a reliable result. The other detailed formulation and the calculation procedure are given in Ref. 4.

3 Results of Numerical Calculations of $^{17}_{\Lambda}$O and $^{41}_{\Lambda}$Ca

We made calculations employing the Jülich Ã (JÃ), $\tilde{\text{B}}$ (J$\tilde{\text{B}}$) [7], Nijmegen soft-core (NSC) [5] and NSC97 models a-f [6] potentials for the YN interaction in free space, and the quantum numbers for relative angular momentum $l \leq 4$ were considered for each potential. We used the harmonic oscillator basis states with a common

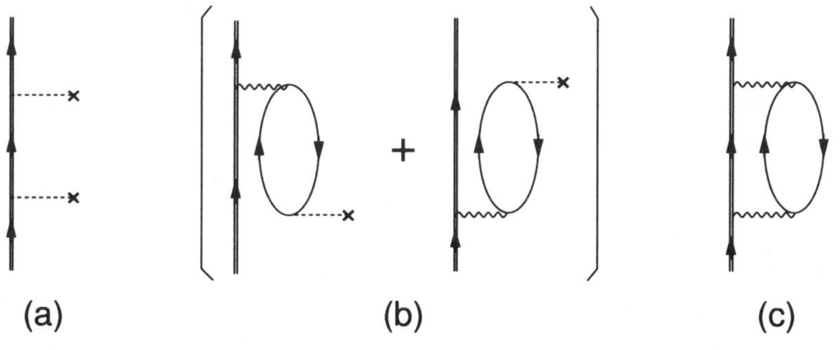

Figure 1: Calculated diagrams in the second order for the Λ single-particle energies. The ΛN effective interaction $\tilde{v}_{\Lambda N}$ is represented by the wavy-line vertex, and the $\times$ vertex with the dashed line represents the non-diagonal part of the one-body Hamiltonian $h_k = t_k + u_k$ for $k = \Lambda$ and N. The propagations of N and Λ are represented by the single and double external lines, respectively.

Table 1: Contributions to the single-particle energies of Λ in ${}^{17}_{\Lambda}$O for the NSC97f potential. Here KE and PE stand for the kinetic energy and the first-order potential energy, respectively. Rows (a), (b) and (c) are the contributions of diagrams (a), (b) and (c) in Figure 1, respectively. All entries are in MeV.

	$0s_{1/2}$	$0p_{3/2}$	$0p_{1/2}$
KE	10.50	17.50	17.50
PE	-23.13	-17.20	-16.09
(a)	-1.84	-2.35	-2.42
(b)	-1.20	-0.37	-0.13
(c)	-0.55	-0.98	-1.25
Total	-16.22	-3.40	-2.40

Table 2: Results of ${}^{41}_{\Lambda}$Ca. The notation is the same as in Table 1.

	$0s_{1/2}$	$0p_{3/2}$	$0p_{1/2}$	$0d_{5/2}$	$0d_{3/2}$	$1s_{1/2}$
KE	9.00	15.00	15.00	21.00	21.00	21.00
PE	-27.88	-22.67	-21.68	-18.03	-16.57	-16.26
(a)	-2.21	-3.25	-3.31	-3.74	-3.88	-3.01
(b)	-2.25	-1.45	-1.11	-0.50	-0.03	0.45
(c)	-0.60	-1.03	-1.29	-1.33	-1.76	-1.73
Total	-23.94	-13.41	-12.39	-2.61	-1.25	0.44

$\hbar\Omega = 14$MeV for ${}^{17}_{\Lambda}$O and $\hbar\Omega = 12$MeV for ${}^{41}_{\Lambda}$Ca for Λ, Σ and N. The dependence of the single-particle energy on the value $\hbar\Omega$ has been already discussed [10] for ^{16}O, and we confirmed that the dependence was quite small around $\hbar\Omega = 14$MeV if perturbation corrections were included. In the present study we calculated the diagrams for the single-particle energy of Λ up to second order as shown in Figure 1. As for the nucleon single-particle potential u_N, we used the fixed data [1] calculated using the Paris [11] potential, and we have assumed that $u_\Sigma = 0$.

We display the single-particle energies of Λ for ${}^{17}_{\Lambda}$O and ${}^{41}_{\Lambda}$Ca in Table 1 and Table 2, respectively, with the contributions of the perturbation corrections given in Figure 1 for the NSC97f potential. Correction (a) in Figure 1 is induced by the presence of non-diagonal terms in the one-body Hamiltonian $h_\Lambda = t_\Lambda + u_\Lambda$. This effect should be treated in the Hartree-Fock calculation. However, in the present calculation, we treat this effect perturbatively. These non-diagonal terms work to change the single-particle wave function of Λ. Therefore, this procedure is indispensable in this approach since we have taken a fixed value of $\hbar\Omega$ commonly for Λ and N. This correction (a) contributes most attractively to the single-particle energies among all the corrections in Figure 1. One finds that the magnitude of every correction is considerably small in comparison with the first-order potential energies. Almost all the perturbation corrections contribute, however, attractively to the single-particle energies of Λ. Therefore, the sum of the perturbation corrections for every state has sizable contribution attractively. It should be noted that the values of the KE plus PE in the $0p$ states for ${}^{17}_{\Lambda}$O and in the $0d$ and $1s$ states for ${}^{41}_{\Lambda}$Ca are positive (unbound). Therefore, the perturbation corrections play a decisive role when we argue that the Λ particles in the $0p$ states for ${}^{17}_{\Lambda}$O and in the $0d$ and $1s$ states for ${}^{41}_{\Lambda}$Ca are bound or not. We note here that the sums of the perturbation

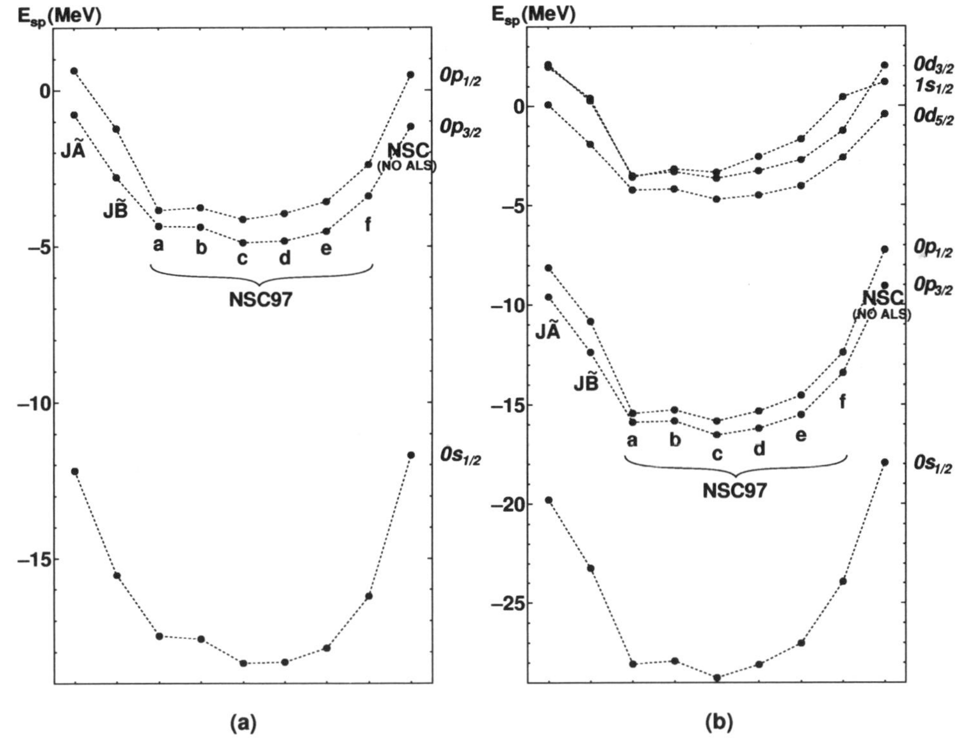

Figure 2: The single-particle energies of Λ $E_{\rm sp}$ in ${}^{17}_{\Lambda}$O (a) and ${}^{41}_{\Lambda}$Ca (b) including the perturbation corrections up to second order. All entries are in MeV.

corrections for all the states do not change the mutual spacings of the single-particle states significantly. This trend is very different from the situation observed in ordinary nuclei such as ^{16}O and ^{40}Ca, in which the correction terms contribute to about 40% of the calculated spin-orbit splitting of the $0p$ or $0d$ states [1,2]. The role of many-body correlations, such as the core polarization, does not seem to be important in the spin-orbit splitting of Λ.

We present, in Figure 2, the single-particle energies of Λ which include the perturbation corrections up to second order as shown in Figure 1 for the various YN potentials. We note that there are considerable differences among the calculated results for the YN interactions employed. In Figure 2(a) the $0p_{1/2}$ state in ${}^{17}_{\Lambda}$O is unbound for the J$\tilde{\rm A}$ and NSC. The results for the NSC97 models are more binding, on the whole, than those for the other interactions. Furthermore, the magnitude of the spin-orbit splitting of the $0p$ states for each of the NSC97 models is much smaller than the old NSC. This feature comes from the fact that the strength of the spin-orbit terms in the NSC97 models is weaker than that in the old NSC, and the antisymmetric spin-orbit terms are included in the NSC97 models. It is noted here that, in Figure 2, we used the computer program for the old NSC in which the antisymmetric spin-orbit term is not included. If we use the computer program for the NSC given recently, the spin-orbit splitting of the $0p$ state in ${}^{17}_{\Lambda}$O becomes 1.33MeV. This value is smaller than the value 1.66 MeV shown in Figure 1(a). Similar tendency can be found for ${}^{41}_{\Lambda}$Ca in Figure 2(b).

4 Summary

In this work we applied the UMOA to ${}^{17}_{\Lambda}$O and ${}^{41}_{\Lambda}$Ca, employing various YN interactions. We confirmed that there were considerable differences among our results for the YN interactions employed. This is mainly because of the uncertainty in the YN interaction.

Some problems have still remained in the application of the UMOA to Λ hypernuclei. In the present calculation, single-particle energies of Σ in intermediate states are neglected. However, the single-particle spectrum of Σ has a possibility of giving rise to some contributions to the ΛN effective interaction. Although properties of single-particle energies of Σ have not always been made clear, we should investigate how much the spectrum of Σ affects in the determination of the ΛN effective interaction.

Another refinement to be made is that the unitarily transformed Hamiltonian [1] contains three-or-more-body effective interactions. The effect of these many-body interactions has been evaluated in the previous works of calculating nuclear properties, and we have concluded that the contributions of these many-body terms are much smaller than those of two-body effective interactions but sizable in some cases. We therefore should estimate the effects of many-body interactions also in hypernuclei. We shall report our results on this point in a forthcoming paper.

References

1. K. Suzuki, R. Okamoto and H. Kumagai, *Phys. Rev.* C **36**, 804 (1987); K. Suzuki and R. Okamoto, *Prog. Theor. Phys.* **92**, 1045 (1994).
2. H. Kumagai, K. Suzuki and R. Okamoto, *Prog. Theor. Phys.* **97**, 1023 (1997).
3. S. Fujii, R. Okamoto and K. Suzuki, *Prog. Theor. Phys.* **99**, 151 (1998).
4. S. Fujii, R. Okamoto and K. Suzuki, *Nucl. Phys.* A **651**, 411 (1999).
5. P. M. M. Maessen, Th. A. Rijken and J. J. de Swart, *Phys. Rev.* C **40**, 2226 (1989).
6. Th. A. Rijken, V. G. J. Stoks and Y. Yamamoto, *Phys. Rev.* C **59**, 21 (1999).
7. A. Reuber, K. Holinde and J. Speth, *Nucl. Phys.* A **570**, 543 (1994).
8. C. W. Wong, *Nucl. Phys.* A **91**, 399 (1967).
9. P. U. Sauer, *Nucl. Phys.* A **150**, 467 (1970).
10. K. Suzuki and R. Okamoto, *Prog. Theor. Phys.* **76**, 127 (1986).
11. M. Lacombe, B. Loiseau, J. M. Richard, R. Vinh Mau, J. Côté, P. Pirès and R. de Tourreil, *Phys. Rev.* C **21**, 861 (1980).

IMPLICATION OF THE SU_6 QUARK MODEL YN POTENTIAL FOR Λ AND Σ HYPERNUCLEAR STATES

M. KOHNO, Y. FUJIWARA*, T. FUJITA*, C. NAKAMOTO** AND Y. SUZUKI***

Physics Division, Kyushu Dental College, Kitakyushu 803-8580, Japan
**Department of Physics, Kyoto University, Kyoto 606-8502, Japan*
***Suzuka National College of Technology, Suzuka 510-0294, Japan*
****Department of Physics, Niigata University, Niigata 950-2102, Japan*

Using the SU_6 quark-model baryon-baryon interaction which was recently developed by the Kyoto-Niigata group, we calculate NN, ΛN and ΣN G-matrices in ordinary nuclear matter. First, the Λ and Σ single-particle potentials in nuclear medium are shown. The Σ potential turns out to be repulsive. Next, we discuss the strength of the single-particle spin-orbit potential S_B, following the Scheerbaum's prescription. The S_Λ becomes small because of the cancellation between spin-orbit and anti-symmetric spin-orbit components. The short-range correlation is found to further reduce S_Λ.

1 Introduction

The SU_6 quark-model provides a unified framework to describe baryon-baryon interactions including hyperons. Recently the Kyoto-Niigata group [1,2] has developed a modern quark-model baryon-baryon interaction, which reproduces essential features of the NN and YN scattering data below 300 MeV. In this model, the quark-quark interaction in a framework of $(3q)$-$(3q)$ resonating-group method is composed of a phenomenological quark-confining potential and the Fermi-Breit interaction with a minimal effective meson-exchange potential of scalar and pseudoscalar nonets directly coupled to quarks. Since the experimental information is still scarce in the strangeness sector, it is valuable to examine the implication of such interactions guided by the quark picture of baryons on the hypernuclear physics.

In this report, we present G-matrix calculations for the NN, ΛN and ΣN interactions in ordinary nuclear matter in the lowest order Brueckner theory, employing the quark-model interaction called FSS [2]. First we discuss the hyperon single-particle potential calculated in nuclear matter. Although the Λ potential has been established experimentaly, the Σ potential is basically unknown. Even the sign of the Σ potential has been controversial [3]. Thus it is interesting to discuss the theoretical prediction of the Σ potential using the quark-model hyperon-nucleon interaction. We find that the FSS gives a repulsive Σ single-particle potential.

Next, we examine in detail single-particle spin-orbit strengths of N, Λ and Σ based on the calculated G-matrices, extending the Scheerbaum's formulation [5] to hyperons. One of the salient features of the quark-model description is that the antisymmetric LS force ($LS^{(-)}$ force) originating from the Fermi-Breit LS interaction is considerably strong in the isospin $I = 1/2$ channel [4], which leads to very small spin-orbit splitting for the Λ single-particle states. In view of recent experimental progress, a quantitative analysis of single-particle ℓs potentials is interesting. Our calculation shows that the short-range correlation further reduces the S_Λ to be less than $\frac{1}{10}S_N$.

2 G–matrices of the quark model potential

Since the quark-model interaction is defined in a form of RGM kernel, we first define partial wave Born amplitudes in momentum space by numerical angle-integration. Then it is straightforward to apply them to solve Bethe-Goldstone equations in nuclear matter. Fig. 1 shows calculated saturation curves for ordinary nuclear matter with the QTQ prescription as well as the continuous choice for intermediate spectra. The saturation point of the quark-model, FSS, is seen not to deviate from the Coester band, which indicates that the FSS has similar saturation properties with other realistic meson-exchange forces, the Paris potential [6] and the Bonn potential [7]. The behaviour of the calculated energy at higher densities suggests that the short-range repulsion of the NN interaction in the FSS is not strong enough at higher energies.

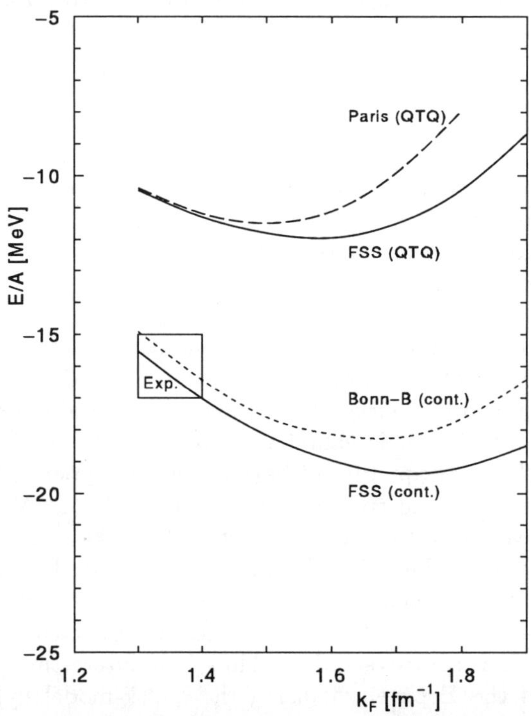

Figure 1. Nuclear matter saturation curves obtained for the quark-model potential FSS, together with the results with the Paris potential and the Bonn potential.

Figs. 2 and 3 show the momentum dependence of the Λ and Σ single particle potentials in nuclear matter obtained from quark-model G-matrices with the continuous prescription for intermediate spectra. For comparison, the results by Schulze *et al.* [8] with the Nijmegen soft core (NSC) YN force [9] are also shown. Partial wave contributions of the single-particle potential $U_B(k=0)$ for the Λ and

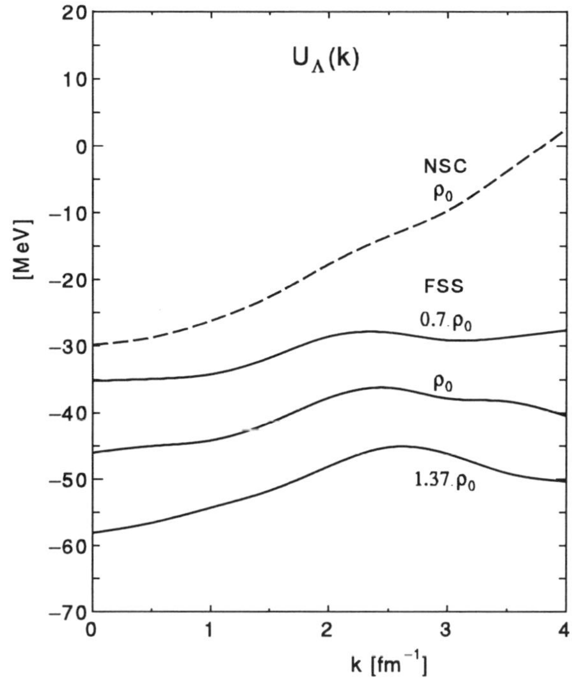

Figure 2. The Λ single-particle potential $U_\Lambda(k)$ in nuclear matter, using the quark-model potential FSS, for three densities of $\rho = 0.7\rho_0, \rho_0, 1.37\rho_0$, where the normal density ρ_0 corresponds to $k_F = 1.35$ fm^{-1}. The dashed curve is the result by Schulze *et al.* [7] with the Nijmegen soft-core YN potential [8].

the Σ in nuclear matter at $k_F = 1.35$ are tabulated in Table 1.

Table 1. Λ and Σ single-particle potentials in nuclear matter calculated from quark-model G-matrices with the continuous prescription for intermediate spectra at $k_F = 1.35$ fm^{-1}.

	$U_\Lambda(0)$ [MeV]		$U_\Sigma(0)$ [MeV]			
$k_F = 1.35$ fm^{-1}	Schulze *et al.* NSC	ours FSS	Schulze *et al.* NSC		ours FSS	
I	1/2	1/2	1/2	3/2	1/2	3/2
$^1S_0$	−15.3	−20.0	6.7	−12.0	6.2	−8.9
$^3S_1 + {^3D_1}$	−13.0	−21.2	−14.9	6.7	−20.3	48.9
$^1P_1$	1.6	−0.3	1.8	−2.7	1.7	−4.3
$^3P_0$	0.2	0.6	2.6	−2.0	3.0	−2.3
$^3P_1$	2.0	0.5	−5.3	6.6	−8.7	8.4
$^3P_2 + {^3F_2}$	−4.0	−3.7	−0.5	−1.9	−1.3	−1.2
subtotal			−9.8	−5.5	−21.0	41.4
total	−29.8	−46.0	−15.3		+20.4	

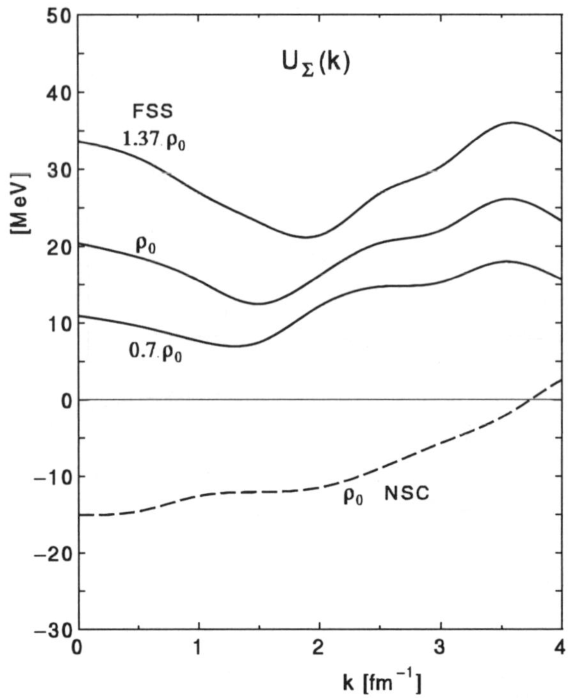

Figure 3. The Σ single-particle potential $U_\Sigma(k)$ in nuclear matter, using the quark-model potential FSS, for three densities of $\rho = 0.7\rho_0, \rho_0, 1.37\rho_0$, where the normal density ρ_0 corresponds to $k_F = 1.35$ fm^{-1}. The dashed curve is the result by Schulze *et al.* [7] with the Nigmegen soft-core YN potential [8].

The FSS is seen to produce a larger potential depth $U_\Lambda(0)$ than the NSC. It is to be remarked that the FSS predicts a positive $U_\Sigma(0)$, reflecting the characteristic repulsion in the ${}^3S_1 + {}^3D_1$ channel of $I = 3/2$.

3 Single-particle spin-orbit strength for hypernuclear states

The single-particle spin-orbit potential is well described by the Thomas form of

$$U_B(r) = -\frac{\pi}{2} S_B \frac{1}{r} \frac{d\rho(r)}{dr} \boldsymbol{\ell} \cdot \boldsymbol{\sigma},$$

where $\rho(r)$ is the total density of nucleons. Scheerbaum [5] derived expressions for the strength factor S_B for the nucleon arising from the effective two-body spin-orbit interaction. We apply his formulation for a spin-saturated core of nucleons to hypernuclear single-particle energies and obtain the following expression,

$$S_B(k) = -(1 + \delta_{B,N}) \frac{3\pi}{4{k_F}^3} \xi (1+\xi)^2 \sum_{I,J} \frac{2I+1}{2I_B+1} (2J+1) \frac{1}{2\pi^2} \int_0^{q_{max}} dq \; W(k,q)$$

$$\times \left\{ (J+2) G^J_{1J+1,1J+1}(q,q;K) + G^J_{1J,1J}(q,q;K) \right.$$
$$\left. -(J-1) G^J_{1J-1,1J-1}(q,q;K) - 2\sqrt{J(J+1)}\; G^J_{1J,0J}(q,q;K) \right\},$$

where k denotes the momentum of the particle B considered, q the relative momentum, K the total momentum, ξ a ratio M_N/M_B, $W(k,q)$ a phase-space factor and $G^J_{S\ell,S'\ell}(q,q;K)$ stands for a diagonal G-matrix element with the total angular momentum J. The last term $G^J_{1J,0J}$ corresponds to the $LS^{(-)}$ contribution which is absent in the NN case. Detailed accounts of these expressions will be given elsewhere.

Table 2 shows the values of $S_B(k=0)$ for N, Λ and Σ, which are calculated from the FSS G-matrices with the continuous choice for intermediate spectra. Each contribution from the LS and the $LS^{(-)}$ components in even and odd as well as $I=1/2$ and $I=3/2$ channels is shown in Table 3. It is apparent that in the case of S_Λ the $LS^{(-)}$ contribution almost cancels the LS one, which makes the ratio S_Λ/S_N to be less than 1/10. It is noted that in the quark-model description Born amplitudes already give a comparable order of magnitude of S_B, namely $S_\Lambda/S_N \sim 1/5$. For the Σ, the contribution from the $LS^{(-)}$ force has an opposite sign to the Λ, and the ratio S_Σ/S_N turns out to be about 1/2.

Table 2. Single-particle spin-orbit strength S_B for N, Λ and Σ.

	S_B from quark-model G-matrices			[in MeV·fm^5]
k_F [fm^{-1}]	1.20	1.35		1.50
N	−42.1	−41.0	(1)	−39.8
Λ	−2.7	−3.5	$(\sim \frac{1}{12})$	−4.1
Σ	−21.9	−22.3	$(\sim \frac{1}{2})$	−22.2

Table 3. Details of $S_\Lambda = -3.5$ MeV·fm^5 and $S_\Sigma = -22.3$ MeV·fm^5 at $k_F = 1.35$ fm^{-1}. Entries are in MeV·fm^5.

		$I=1/2$		$I=3/2$	
		odd	even	odd	even
S_Λ	LS	−17.2	0.6	—	—
	ALS	12.8	0.3	—	—
S_Σ	LS	2.9	0.1	−12.2	−1.6
	ALS	−10.9	−0.6	0.1	−0.0

4 Summary

We calculated NN, ΛN and ΣN G-matrices in ordinary nuclear matter by solving the Bethe-Goldstone equations for the exchange kernel of our recent quark-model interaction called FSS [2]. We showed that the FSS predicts a repulsive Σ

single-particle potential, the strength of which is about 20 MeV. This result seems to be consistent with the conclusion presented by Dabrowski [10] in this workshop, which is based on the analysis of recent (π^-, $K^\pm$) experiments [11] at BNL. The future experiments is expected to settle the problem of the Σ single-particle potential.

We also presented a detailed analysis of the strength S_B of single-particle spin-orbit potentials of N, Λ and Σ in nuclear medium, for which the antisymmetric LS force originating from the Fermi-Breit spin-orbit interaction between quarks plays a characteristic role. The relative ratio S_Λ/S_N is already small, $\sim 1/5$, in the Born approximation, for which the cancellation between LS and antisymmetric LS contributions is essential. The short-range correlation is found to further reduce the ratio to be less than 1/10. This small spin-orbit potential for the Λ might be confirmed in the future experiments. On the other hand, the short-range correlation does not change much the ratio of S_Σ/S_N, which is about 1/2.

References

1. C. Nakamoto, Y. Suzuki and Y. Fujiwara, Phys. Lett. **B318** (1993) 587.
2. Y. Fujiwara, C. Nakamoto and Y. Suzuki, Phys. Rev. Lett. 76 (1996) 2242; Phys. Rev. **C54** (1996) 2180.
3. C. J. Batty, E. Friedman and A. Gal, Phys. Lett. **B335** (1994) 273.
4. O. Morimatsu, S. Ohta, K. Shimizu and K. Yazaki, Nucl. Phys. **A420** (1984) 573.
5. R. R. Scheerbaum, Nucl. Phys. **A257** (1976) 77.
6. M. Lacombe, B. Loiseau, J. Richard, R.Vinh Mau, J. Côté, P. Pirès and R. de Tourreil, Phys. Rev. **C21** (1980) 861.
7. R. Machleidt, Adv. Nucl. Phys. **19** (1989) 189.
8. H.-J. Schulze, M. Baldo, U. Lombardo, J. Cugnon and A. Lejeune, Phys. Rev. **C57** (1998) 704.
9. P. Maessen, Th. Rijken and J. de Swart, Phys. Rev. **C40** (1989) 2226.
10. J. Dabrowski, the talk given in this workshop SNP'99.
11. R. Sawafta, Nucl. Phys. **A585** (1995) 103c; **A639** (1998) 103c.

LASER-ELECTRON-PHOTON PROJECT AT SPring-8; WHERE DO WE STAND AND WHERE DO WE GO ?

Mamoru Fujiwara[a,b] and Hidetoshi Akimune[b]
[a)] *Research Center for Nuclear Physics, Osaka University, Mihogaoka 10-1 Ibaraki, Osaka, 567, Japan*
and
[b)] *Advanced Science Research Center, Japan Atomic Energy Research Institute, Tokai Ibaraki, 319-1195 Japan*
E-mail: fujiwara@rcnp.osaka-u.ac.jp

A new facility for GeV photon beams in the energy range of E_γ = 1.5~3.5 GeV is constructed at SPring-8. The GeV Photons are produced via Compton backscattering, and are used to study the sub-nucleonic structure of the nucleon. The main subject of this project is the search for the quark behavior in the nucleon - the building blocks of nucleons. Fundamental role of quark in nucleon or in a nucleus will be studied as the current topics in physics to reveal the underlying basic principle. This Spring-8 project called Laser-Electron-Photon (LEP) Project is expected to offer an unique opportunity to study the nucleon in view of the quark structure. Using the Compton backscattered photons induced by a short-wave-length laser light, the polarized GeV photons with a high emmitance are generated. This special feature makes real-photon experiments complementary with electron-scattering experiments.

1 Introduction

Since the discovery of the nucleus by Rutherford in 1911 [1], almost one century has been devoted for understanding atomic nuclei. We now recognize that a nucleus consists of nucleons (protons and neutrons), and a nucleon consists of sub-nucleonic matter, quarks and gluons. One of the current subjects towards the next 21 century is to understand the structure of the nucleon in terms of PQCD (Perturbed Quantum Chromo-Dynamics).

In order to study the nucleon structure, a new facility called the Laser-Electron-Photon (LEP) Project is under construction at SPring-8 [2], where a GeV photon beam in the energy range of E_γ=1.5~3.5 GeV becomes available. The GeV photons are produced via Compton backscattering between laser light and the electron beam at 8 GeV. This project will offer an unique opportunity to study the quark system in a nucleon or in a nucleus by means of real polarized photons with the energy of 1.5~3.5 GeV [3,4].

2 GeV Photon beam at SPring-8

Fig. 1 shows a schematic diagram of the Compton backscattering process to obtain high energy photons. The energy of the Compton backscattered photons is obtained as a function of the scattering angle θ;

$$\omega = \frac{4\omega_0 E_e^2}{m^2 + 4\omega_0 E_e}\left(\frac{1}{1+\left(\frac{E_e^2}{m^2+4\omega_0 E_e}\right)\theta^2}\right). \tag{1}$$

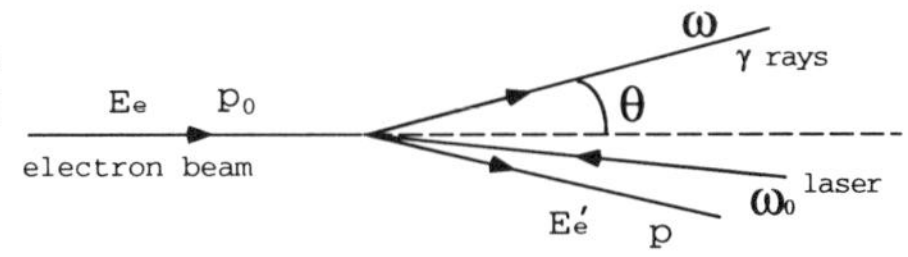

Figure 1: Backscattered Compton Process via the collision between high energy electron and laser light.

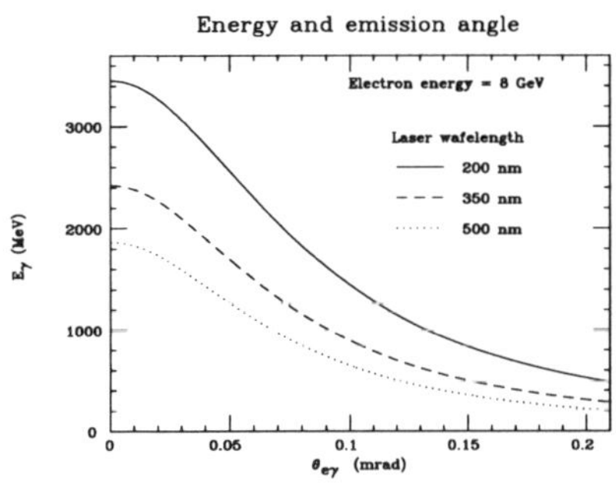

Figure 2: Relation between Compton backscattered photons and its scattering angle. The γ ray energies are plotted for laser light with the wave lengths of 200 nm, 350 nm and 500 nm. Higher the electron energy is, much more the scattered photons concentrate to forward direction(*i.e.* the direction of incident electron momentum. When the angle is large, the energy of photons slowly changes and it becomes independent on incident electron beam.

The energy of the scattered photon is proportional to E_e^2 and to the laser photon energy ω_0, respectively.

Fig. 2 shows that a short–wave laser light and a high–energy electron beam are essential to produce the high–energy photon beam. Using 351.1 nm (ω_0=3.53 eV) argon laser, we can get photons with a maximum energy of 2.49 GeV from 8 GeV electrons. On the other hand, when the electron energy is low, the available photon energy becomes much lower since the photon energy is proportional to E_e^2.

An great advantage using the Compton backscattered photons for an experiment is the polarization; a polarized photon beam is easily obtained using a linear– or circular–polarized laser–light.

We estimate the limit of the photon beam intensity to be $\sim 10^7$/sec. at SPring-8. This upper limit comes from the technical restriction for the loss rate of the electron beam intensity in the 8 GeV electron storage ring at SPring-8. When we have 10^7 Compton scattering per second, the loss rate of the stored electron beam becomes 10% over the period of 10 hours.

Angular dependences of the scattered photon beam are easily read in Eq. (1). All scattered photons concentrate within the narrow angular cone defined by the width of $\Gamma=\sqrt{(m^2+4\omega_0 E_e)/E_e^2}$ as shown in Fig. 2. Scattered photons in the energy range between 1.5 and 3.5 GeV concentrate within $\Gamma \approx m/E_e = 0.063$ mrad, irrespective of the wave-length of laser–light. Even if we install the target for an experiment at the point of 100 m downstream from the photon-electron collision region, the target size required could be less than 1 cm in diameter.

Since the photon beam concentrates in the extremely forward direction, the photon energy is not determined by means of a slit method in the case of the laser Compton backscattering with a high energy electron. In order to get information on the photon energy with an appreciable resolution, tagging of scattered electron energies is only a possible way.

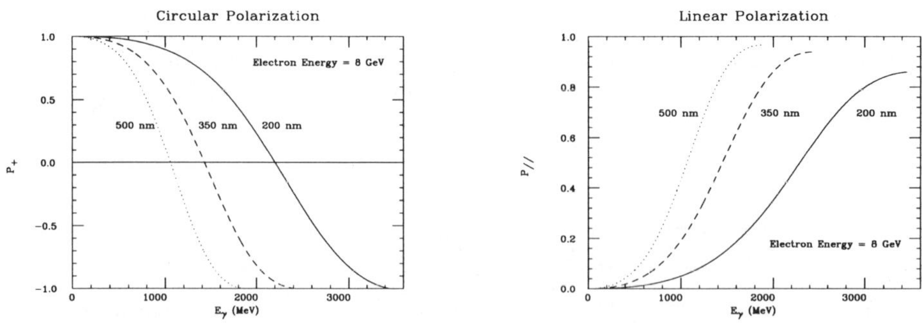

Figure 3: Polarization of the backscattered photons induced by the 100% circular polarized laser light (left panel) and linear polarization of the backscattered photons induced by the 100% linear polarized laser light (right panel).

Polarization is the unique feature of the Compton backscattered γ-rays. Fig. 3 shows the polarization property as a function of the γ energy.

3 Physics with GeV photons

3.1 $s\bar{s}$ component of nucleon

The constituent quark model predicts that the ratio (μ_n/μ_p) of the neutron and proton magnetic moments is -2/3, which agrees with the experimental value -0.685. This supports the basic idea of "quark". However, the recent experiments from the lepton deep inelastic scattering address a serious question; the magnetic moments from constituent quarks only contribute 10%. The strange quarks give 10%~20% contributions [5,6].

The ϕ meson with the mass of 1020 MeV has the pure $s\bar{s}$ wave function. Thus, there is a possibility of pinning down $s\bar{s}$ components in a nucleon by observing the interference effect in the ϕ meson production processes. Polarization

Fig. 4 shows the important three cases of the ϕ meson production processes with γ rays in the energy region of 1~3.5 GeV. One process is the pomeron exchange process, where a photon changes to a vector meson and is scattered from the multi-gluon exchange process (shown in Fig. 4c). Since the pomeron has the same quantum number as vacuum, information on the nucleon spin is not carried out.

Fig. 4a) and b) show the knockout process, where the $s\bar{s}$ and uud component is knocked out by a high energy photon. In this case, the knocked $s\bar{s}$ and uud quarks are the original ingredient of the nucleon. The $s\bar{s}$ pair carries information on the spin direction of nucleon together with the spin information of uud quarks. The $s\bar{s}$ quark pair and the uud quarks have to make the total nucleon spin 1/2 and the magnetic moments -0.685.

If the contribution of the knockout process in Fig. 4c) is experimentally checked,

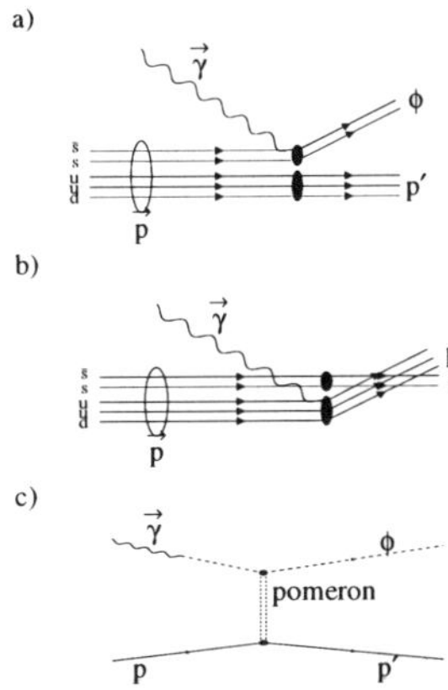

Figure 4: Three photo-production processes of ϕ meson. a) Knock-out process of $s\bar{s}$ pair by photons. b) Knock-out process of the uud component. c) Vector dominance process (VMD process).

we can estimate the $s\bar{s}$ component in a nucleon. But, the situation is not easy. In the high energy region, the pomeron exchange process dominates. In the low energy region above the threshold of the ϕ production, the $s\bar{s}$ knockout process can compete the pomeron exchange process. Thus, the interference effect between the two processes in the energy region of 2~3.5 GeV becomes one of the most promising way to extract the information on the $s\bar{s}$ component.

At SPring-8, we will measure the interference effect between the pomeron exchange and knock-out process. Titov *et al.* [7] conclude that it is only possible to observe this interference effect in the photon experiment at 2~3.5 GeV, and the asymmetry observable is quite sensitive to the $s\bar{s}$ component.

3.2 Gerasimov–Drell–Hearn sum rule

The Gerasimov–Drell–Hearn (GDH) sum rule relates to the total photo–absorption cross sections by a polarized nucleon with the anomalous magnetic moment. A simple relation between the integrated photo–absorption cross section value and the magnetic moment of a nucleon is deduced by assuming fundamental principles such as the Lorentz and gauge invariance, causality, and unitarity [8,9,10]. The anomalous magnetic moment of the nucleon (κ^2) has a relation to the difference between the total cross sections with the nucleon spin parallel ($\sigma_{3/2}$) and anti-parallel ($\sigma_{1/2}$) to the circular polarized photons, integrated as a function of photon energy (ν).

$$\int_0^\infty \frac{\sigma_{3/2} - \sigma_{1/2}}{\nu} d\nu = \frac{2\pi^2\alpha}{m^2}\kappa^2, \tag{2}$$

where m stands for nucleon mass. This sum rule has never been tested because it requires measurements of absolute values of cross sections with a polarized photon beam on a polarized target. There are the approved experimental projects at ELSA ($0.5 \leq E_\gamma \leq 3$ GeV) and at Mainz ($140 \leq E_\gamma \leq 800$ MeV) [11]. However, at SPring-8, much precise measurements of the high energy contribution of the DGH sum rule could be possible due to a high energy and high polarized photon beam.

The total photo-absorption cross sections at the energy range of 1.5~3.5 GeV will provide a means of checking the QCD model[10,12].

The Compton scattering experiment from proton is complementary to the GDH experiment. Gell-Mann, Goldberger, and Thirring describe the Compton scattering between the polarized photon and polarized proton[13] and point out that the Compton scattering relates to the proton magnetic moment and the charge distribution of proton. Thus, if we can measure the Compton scattering at 1.5~3.5 GeV, the experiment will provide detailed information on the microscopic structure of proton.

3.3 Photo–disintegration of deuteron

Photo–disintegration of a deuteron is a simplest reaction in which a deuteron is broken up into a proton and a neutron by photo–absorption. Deuteron is essentially a two–nucleon system bound by the strong force. The study of the deuteron system has a long history. But, valuable information is still missing in the region above 1 GeV. One may see an interplay between the nucleon and quark aspects from the photo–breakup experiment on deuteron.

A transition from a meson picture to a quark picture will appear in the region between 1 and 3 GeV in order to understand the various observables associating to deuteron. One may distinguish a marked difference between the nucleon- and quark- pictures via the polarization measurements with a polarized beam.

SPring–8 is an unique facility which provide the feasibility of the above–mentioned experiments with a high quality γ beam of the energy range between 1.5 and 3 GeV.

3.4 Physics with vector mesons

Vector mesons of $\rho(770)$, $\omega(782)$ and $\phi(1020)$ couple strongly to photons in the nuclear medium. The vector dominance model predicts that the vector mesons are created with the same helicity with that of incident photons. The validity of the vector dominance model has been checked for $\rho(770)$ only within the accuracy of 10%. This validity should be examined for $\omega(782)$ and $\phi(1020)$ with a high accuracy at high energies.

Theorists predict a decrease of the mass of vector mesons in the nuclear medium. This mass reduction is due to the partial restoration of the chiral symmetry breaking in nuclear environments with a high nucleon–density[14,15]. It is still not known whether the vector mesons, $\rho(770)$ and $\omega(782)$, become heavier or lighter in dependence on the nuclear matter density and on the nuclear temperature[16,17].

Photo production of vector mesons is possible at SPring-8. The mass of the produced vector meson is measured by reconstructing the invariant mass of the decay particles. At present, such mass measurements are tried via the hadron-hadron and heavy ion collisions. But, in the case of hadron-hadron and heavy ion collisions, the impact parameter is not well defined. Thus, it would be very difficult to experimentally determine the nuclear matter density at the position of the vector meson production. If the density dependence of the mass change is very small, the precise mass-change measurements would be required and are generally difficult. In the photo-production, the advantage is that the density is those obtained from the electron scattering, and that the incident channel and outgoing channel are theoretically predictable if we measure the lepton-pair from the decay of the vector meson.

Acknowledgment

We thank Dr. Nakano, Dr. Hotta, Y. Ohashi, S. Daté, H. Ohkuma, and N. Kimagai for their collaboration in constructing the LEP facility at the BL33B2 course at SPring-8. We are grateful to acknowledge the support of the Japan Society for Promotion of Science (JSPS) under the Japan-Korea Cooperative Science Program.

References

1. E. Rutherford, Philos. Mag. **21**, 669 (1911).
2. H. Kamitsubo, in Proceedings of FRONTIER96, edited by H. Toki, T. Kishimoto and M. Fujiwara, World Scientific, March 7-9, 1996, pp. 147.
3. M. Fujiwara, T. Hotta, T. Kinashi, K. Takanashi, T. Nakano, Y. Ohashi, S. Daté, H. Okuma, and N. Kumagai, Acta Physica Polonica B **29**, 141 (1998).
4. T. Nakano, H. Ejiri, M. Fujiwara, T. Hotta, K. Takanashi, H. Toki, S. Hasegawa, T. Iwata, K. Okamoto, T. Murakami, J. Tamii, K. Imai, K. Maeda, K. Maruyama, S. Daté, M.M. Obuti, Y. Ohashi, and H. Ohkuma, Nucl. Phys. A **629**, 559c (1998).
5. EMC Collaboration, J. Ashman et al., Phys. Lett. **B 206**, 364 (1988).
6. SMC, D. Adams et al., Phys. Lett. **B 329**, 399 (1994).
7. A.I. Titov, Y. Oh and S.N. Yang, Phys. Rev. Lett. **79**, 1634 (1997).
8. S.B. Gerasimov, Sov. J. Nucl. Phys. **2**, 430 (1966).
9. S.D. Drell and A.C. Hearn, Phys. Rev. Lett. **16**, 908 (1966).
10. D. Drechsel, Prog. Part. Nucl. Phys. **34**, 181 (1995).
11. G. Anton, Prog. Part. Nucl. Phys. **34**, 173 (1995).
12. A. M. Sandorfi, in Proceedings of FRONTIER96, edited by H. Toki, T. Kishimoto and M. Fujiwara, World Scientific, March 7-9, 1996, pp. 17.
13. M. Gell-Mann, M. Goldberger, and W. Thirring, Phys. Rev. **96**, 1612 (1954).
14. T. Hatsuda and T. Kunihiro, Phys. Rev. Lett. **55**, 158 (1985).
15. G.E. Brown, Nucl. Phys. **A 488**, 689 (1988).
16. T. Hatsuda et al., Nucl. Phys. **B 394**, 221 (1993).
17. T. Hatsuda and S.H. Lee, Phys. Rev. Lett. **46**, R34 (1992).

LONGITUDINAL AND TRANSVERSE CROSS SECTIONS IN THE $^1H(E,E'K^+)\Lambda$ REACTION.

W. KIM

Kyungpook National University, Taegu, S. Korea

The $^1H(e,e'K^+)\Lambda$ reaction was studied as a function of the squared four-momentum-transfer, Q^2, and the virtual photon polarization, ε. For each of four Q^2 settings, from 0.5 to 2.0 $(GeV/c)^2$, the longitudinal and transverse virtual photon cross sections were extracted in measurements at three virtual photon polarizations. The σ_L/σ_T ratio is found to have a Q^2 dependence that differs from existing theoretical predictions.

The electromagnetic production of kaons is important for further progress in many fields and subfields of physics such as hypernuclear production and spectroscopy, and some aspects of QCD model building. The $(e,e'K)$ reaction is important for an improved understanding of the basic coupling constants needed in various nucleon/meson and quark models[1]). The use of virtual photons as opposed to real photons offers access to a much richer body of information because both the virtual photon mass and polarization can be varied independently.

We report the first results of a study of associated kaon electroproduction on hydrogen using the medium-energy, high-intensity, electron beam at the Jefferson Laboratory. The excellent beam qualities of the CEBAF accelerator combined with precision magnetic spectrometers allowed a detailed investigation of the reaction with high systematic and statistical precision, whereas the only prior measurement[3] provided virtually no constraints for the available theories.

Experiment E93-018 was among the first experiments performed in experimental HALL C at Jefferson Lab utilizing the 100% duty factor electron beam. The beam energy was varied between 2.445 and 4.045 GeV with a beam current between 10 and 30 μA. The beam current was measured with two resonant cavities and a parametric current transformer. Two-inch thick tungsten collimators with large octagonal appertures were used to define the solid angle of each spectrometer. The target was a 4.36 ± 0.01 cm liquid-hydrogen cell with a proton abundancy of 99.8%. The background from the beam interaction with the aluminum walls of the target was measured separately using a replica of the empty target cell. The scattered electrons were detected in the High Momentum Spectrometer in coincidence with the kaons detected in the Short Orbit Spectrometer. The detector stack in each spectrometer had four segmented-plastic-scintillator arrays used for trigger formation, time of flight, and energy loss information. Two six-plane multiwire planar drift chambers were used to perform track reconstruction in each spectrometer. Gas threshold Čerenkov counters and segmented lead-glass shower counter arrays provided the signals needed for electron identification.

Two peaks, identified clearly in Fig.1, correspond to the masses of Λ (1115) and Σ^0 (1192), respectively. Because of kinematical restrictions in this experiment, the production of higher mass hyperons is forbidden. Subtraction of random coincidence events and target wall contributions yields a clean missing mass spectrum with virtually no events below the Λ threshold. For the purpose of this work, all events with a missing mass value between 1.1 and 1.175 GeV were accepted as valid $e + p \to e' + K^+ + \Lambda$ coincidence events and used in subsequent calculations.

In the one-photon exchange approximation, the cross-section for the scattering of an electron from a hydrogen target, where none of the initial or final spins are detected, can be reduced, in parallel kinematics (appropriate for our experimental setup), to:

$$\frac{d^5\sigma}{dE_{e'}\, d\Omega_{e'}\, d\Omega_{K^+}} = \Gamma\Big(\frac{d\sigma}{d\Omega_{K^+}}\Big)^{CM} = \Gamma(\sigma_T + \varepsilon\ \sigma_L) \tag{1}$$

where Γ is the virtual photon flux, σ_T is the unpolarized (transverse) cross section, σ_L is the longitudinal cross section, $\Omega_{e'}$ is the scattered electron solid angle, Ω_{K^+} is the kaon solid angle, $E_{e'}$ is the scattered electron energy, ε is the virtual photon polarization.

The quantities σ_T and σ_L completely describe the parallel kinematics cross section behavior. Generally they are functions of the kinematic variables Q^2, the squared four-momentum of the virtual photon, the total energy of the photon-proton system W, and $t = (q - K)^2$, the squared four-momentum-transfer between the photon and the kaon. Here q and K denote the four-momenta of the virtual photon and kaon, respectively.

The separation of σ_T and σ_L is possible because of the so-called Rosenbluth (L/T separation) technique. This separation is achieved by keeping Q^2, W, and t constant (so σ_T and σ_L are kept constant), averaging over some (small) region around $\theta = 0$ (θ here is the polar angle between the virtual photon and the kaon directions), and repeating the above steps for different values of ε (by varying the electron kinematics). By following the above procedure at each Q^2 value, three different ε settings were probed in experiment E93-018.

For each kinematic setting, the sample of coincident $e+p \to e' + K^+ + \Lambda$ events was used to compute the laboratory cross section. Then from Eq. 1, the center-of-mass cross section $(\frac{d\sigma}{d\Omega_{K^+}})^{CM}$ was computed. The uncertainties in these kaon electroproduction cross sections are at the same level or smaller than those in the existing world data set[4,5,6]. In order to facilitate comparison of data at different values of W, all data were scaled to W=2.15 GeV as outlined in Ref. [6]. For the cases where more than one ε setting was measured only the highest value of the cross-section is quoted for comparison with older data sets.

In Fig.2, the calculated values of R from this experiment are compared with the only other existing data[3]; the two sets of measurements agree reasonably well, although in order to properly show the uncertainties of the previous measurement the scale in R had to be extended to unphysical negative values.

Two theoretical predictions for σ_L/σ_T, by Williams, Ji and Cotanch[2] (hereafter referred to as WJC), and the Saclay-Lyon (SL) model described in [7] are included in Fig.2. Both calculations use phenomenological fits of previous world data sets to extract coupling constants and relevant form factors. The estimated uncertainty in these quantities is indicated by the height of the shaded areas. The present measurements agree generally in magnitude with these model calculations, at least for the lower Q^2 settings. Both theoretical models predict that the ratio R increases with momentum transfer whereas our data indicate a flat behavior over this range of Q^2 for the given kinematic settings. This discrepancy may be partially explained by a difference in the true kaon electromagnetic form factor and its parameterization, as used in the models. In this kinematic range for Λ-production the data support the existence of a large longitudinal component to the kaon electroproduction cross section which is expected to be dominated by kaon exchange. In general, the L/T current separation is sensitive to the prescription for establishing current conservation in electroproduction models, especially the combination of the non-gauge invariant terms involving the electric proton and kaon form-factors. The precision of the present data serves to refine the existing kaon electroproduction models and to constrain future models.

The first results from measurements of elementary kaon electroproduction cross sections performed during experiment E93-018 are reported. The center-of-mass cross sections reported here for the reaction $e+p \rightarrow e' + K^+ + \Lambda$ are in good agreement with the existing world data set. The present data demonstrate the ability to separate the longitudinal and transverse cross-sections with a precision significantly greater ($\sim$ three times) than the only prior measurement. The data indicate a strong contribution from the longitudinal part of the cross section, however this contribution is much less than model predictions for the $Q^2 = 2.0$ point. Current calculations indicate that R increases with Q^2 whereas our data support a flat behavior in this kinematic range. The precision of the separated cross-sections should serve to constrain and refine theoretical models of associated strangeness production. Considering the considerable strength observed in the longitudinal part of the cross-section even for the lowest Q^2 measured and in view of the difference in R between data and model calculations at our highest Q^2 setting, precision L/T measurements extending the Q^2 range to both lower and higher values are called for.

We acknowledge the support of the staff of the Accelerator Division of Thomas Jefferson National Accelerator Facility. This work was supported in part by the U.S. Department of Energy and the Korean Science and Engineering Foundation.

References

1. R. A. Schumacher, Nucl. Phys. A **585** 63c, 1995.
2. R. A. Williams, C. R. Ji and S. R. Cotanch, Phys. Rev. C **46**, 1617 (1992).
3. C.J. Bebek *et al.*, Phys. Rev. D **15**, 3082 (1977).
4. C. N. Brown *et al.*, Phys. Rev. Lett. **28**, 1086 (1972).
5. P. Brauel *et al.*, Z. Phys. C **3**, 101 (1979).
6. C.J. Bebek *et al.*, Phys. Rev. D **17**, 1693 (1978).
7. J. C. David *et al.*, Phys. Rev. C **53**, 2613 (1996), and B. Saghai, private communication.

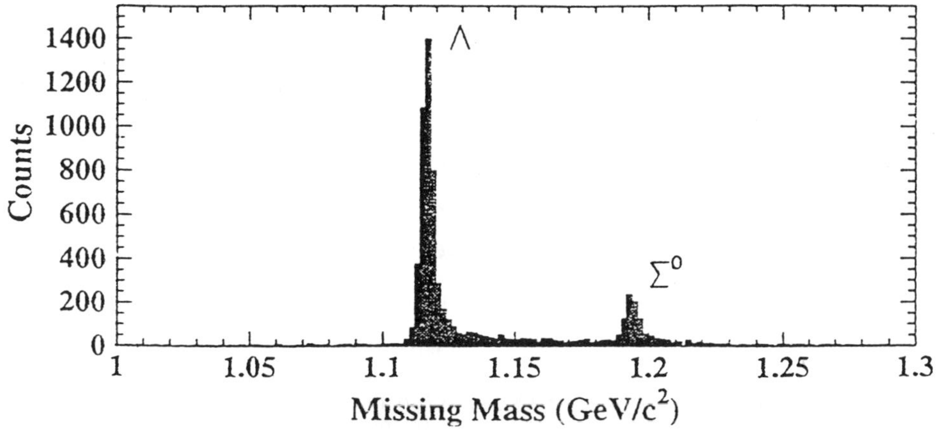

Figure 1: Missing mass spectrum for the $^1H(e, e'K^+)$ reaction after subtracting random coincidences and target wall contributions. Because no radiative corrections were performed here, radiative tails are seen projecting towards higher missing mass values.

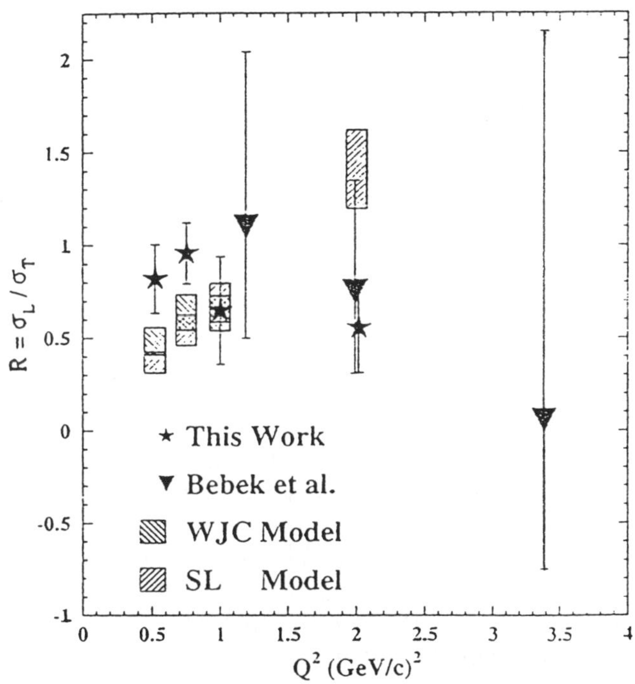

Figure 2: The ratio $R \equiv \sigma_L/\sigma_T$ as extracted from these data and and previous data. This is the world's complete set of data for the $^1H(e, e'K^+)\Lambda$ reaction. The calculations are as outlined in this paper. At $Q^2 = 2.0\ (GeV/c)^2$, the WJC prediction for R is ~ 3.5 (off scale).

THE KAON PHOTO PRODUCTION NEAR THRESHOLD

MYUNG KI CHEOUN, B.S.HAN, K.S.KIM, AND IL-TONG CHEON
Department of Physics, Yonsei University, Seoul, 120-749, Korea
E-mail:cheoun@phya.yonsei.ac.kr

B.G.YU
Hankuk Hangkong University, Koyang 200-1, Korea

The kaon photo-production is investigated near the kaon threshold energy region. In this energy region most of the coupling constants necessary to describe this reaction can be determined from the experimental results and theoretical estimation, so that the cross sections can be predicted with the leading coupling constant, $g_{KN\Lambda}$. We show that the E_0^+ amplitude for $p(\gamma, K^+)\Lambda$ process at the kaon threshold is presented only by the static properties of the related particles. The predicted total cross sections indicate that the future experiments of $p(\gamma, K^+)\Lambda$ near threshold could be very sensitive and meaningful tests for the $KN\Lambda$ coupling constant.

1 INTRODUCTION

The kaon-nuclear physics has many interesting features in the intermediate energy nuclear physics. Compared to the pion physics which are described by u and d quarks it has an additional degree of freedom, s quark and as a result yields many exotic states in nuclear physics [1] by hadronic or electro-magnetic processes. But in the intermediate energy physics one lives under so called the **infrared slavery**. Although Quantum Chromo Dynamics (QCD) is accepted as the correct theory for the strong interactions it has some difficulties from this non-perturbative region, consequently brings one to treat the hadrons such as the nucleon or the mesons as basic particles i.e. Quantum Hadron Dynamics (QHD) with some undetermined coupling constants. In principle, of course, it is possible to calculate these coupling constants using the calculations based on QCD model, for example, chiral bag model [3] or QCD sum rule [2].

On the other hand these coupling constants can be also determined from the kaon production by the electro-magnetic probes. Many papers [5,6] already discussed this problem. Adelseck et al. [5] tried to determine the related coupling constants phenomenologically by a least-square fit to the available data in this process like Thom's method 30 years ago and deduced some values for them. But the leading coupling constant $g_{KN\Lambda}$ turned out to be too divergent in contrast with the case of πNN coupling constant from the photopion production and was not consistent with the SU(3) prediction value $3.0 \leq |g_{KN\Lambda}/\sqrt{4\pi}| \leq 4.4$.

Given the above situations, it is natural to approach this reaction near threshold because one needs fewer coupling constants than the region beyond threshold. For instance, in case of the photo pion production at threshold, the Low Energy Theorem (LET) demonstrates that the experimental results of electro- and photo- pion production near threshold can be well described by only the macroscopic properties of related particles. It means that one can determine $g_{\pi NN}$ coupling constant from the photopion production experiments near threshold. However, it is well known that there are some obstacles in order to apply **the LET's concept or techniques** to the photokaon process (see ref. [4]). Our scheme against such difficulties is presented at this talk with the explanation of the deviations from the pseudo-scalar (PS) and pseudo-vector (PV) coupling schemes for the Born terms.

2 TRANSITION MATRIX ELEMENTS

The general hadronic transition current matrix elements in the $p(\gamma, K^+)\Lambda$ processes are given by

$$M^\mu = \frac{eg}{(M_i + M_f)} \Sigma_{i=1,8} \bar{u}(P_f, S_f) Q_i^\mu A_i(\nu, \nu_1) u(P_i, S_i), \tag{1}$$

with dimensionless kinematical variables

$$\nu = \frac{(P_i + P_f) \cdot k}{2M^2} \ , \ \nu_1 = \frac{q \cdot k}{2M^2} \ , \tag{2}$$

where k and q are the four momenta of the initial photon and the outgoing kaon respectively. The $u_{i,f}$ are the Dirac spinors of the initial and the final hardrons and $KN\Lambda$ coupling constant is designated as g. We assume the PS and PV coupling constants on the $KN\Lambda$ interaction have the relation $g^{PS}/M_i + M_f = g^{PV}/m$, where m is the relevant meson mass and $M_{i,f}$ are the initial and final hadron masses, at the kaon threshold. Here Q_i is the operator constructed by the various symmetries. The Ball amplitudes A_i are, in principle, model dependent.

In this talk, we skip the detailed derivation from our model (see ref. [4]). Instead, we simply explain our strategy and the results for this reaction. For the extended Born terms, which are the Born terms added by vector mesons' contributions, we use well known PV Feymann digrams. The coupling constants of the relevant particles are calculated from effective lagrangian theory whose basic coupling constants are determined to satisfy the available experimental data.

For the resonances, we use the following scheme. The N1(1470) resonance contribution to the E_0^+ amplitude is assumed to be negligible because it is formed far below than the kaon threshold energy and its parity differs from the $K\Lambda$ system and its weak coupling constants to the $K\Lambda$ channel. For the N4(1650) resonance we exploit the Breit-Wigner form with experimental branching ratio of $\gamma\ {}^1H$ and $K^+\Lambda$ states. The unitarity condition factor attached to this resonance is determined with analyses of the total cross section's dependence on this factor. The contributions from the hyperon resonances are discussed below.

The E_0^+ amplitude at threshold can be expressed as

$$E_0^+|_{thr.} = -\frac{e}{(M + M_f)} \left[\frac{M}{4\pi W}\sqrt{\frac{E_i + M}{2M}}\right] \ (A_1 + \frac{1}{2}(\mu + \delta) A_5) \ . \tag{3}$$

Here E_i is the energy of the initial nucleon and

$$A_1 + \frac{\delta + \mu}{2} A_5 = B^S \pm B^V \ . \tag{4}$$

Here + is for $p(\gamma, K^+)\Lambda$ and - for $n(\gamma, K^0)\Lambda$ process, respectively. $\mu = m/M$ and $\delta = (M_f - M_i)/M$.

2.1 BORN TERMS

With the simple expansions of the kinematical variables at the meson threshold, one can obtain the following results for $p(\gamma, K^+)\Lambda$ through expansion of A_i amplitudes with keeping up to the order $O(\mu^2, \delta^2, \mu\delta) = \mathrm{O}(2)$.

For $p(\gamma, K^+)\Lambda$ process,

$$B^S + B^V|_{PV} = -\ g[(1 - \frac{1}{2}(\mu+\delta) + \frac{1}{4}(\delta+\mu)\ [(\delta+\mu) + \mu(\kappa_p + \kappa_\Lambda + \frac{\kappa_T}{\alpha})\ ]) + O(3)], \quad (5)$$

and for $n(\gamma, K^0)\Lambda$ process,

$$B^S - B^V|_{PV} = g\ [\frac{(\delta+\mu)\mu}{4}(\kappa_n + \kappa_\Lambda - \frac{\kappa_T}{\alpha}) + O(3)], \quad (6)$$

where α is designed for the ratio of $g_{K\Lambda N}$ and $g_{K\Sigma N}$ [4]. It should be noted that the contributions from the (transition) magnetic moments appear firstly in the O(2) order and they are the only term contributing for the $n(\gamma, K^0)\Lambda$ process . In specific, the contribution from the u-channel of the hyperon resonance, $\Lambda(1405)$ is also expected to show up in the O(2) order multiplied by the transition magnetic moment. The comparison with the $p(\gamma, \pi^0)p$ process, which has the same isospin structure with the $p(\gamma, K^+)\Lambda$ process, shows that the 1 st term peculiar to this reaction in eq.(6) originated from the absence of the Dirac coupling in the u-channel. The same calculations as the above PV coupling scheme lead one to have the following analytical results for the PS coupling scheme,

$$B^S + B^V|_{PS} = -\ (1 - \frac{1}{2}(\mu+\delta) + \frac{1}{4}(\delta+\mu)\ [(\delta+\mu) + (\delta+\mu-2)(\kappa_p + \kappa_\Lambda + \frac{\kappa_T}{\alpha})\ ]) + O(3), \quad (7)$$

for $p(\gamma, K^+)\Lambda$ process,

$$B^S - B^V|_{PS} = \frac{(\delta+\mu)(\delta+\mu-2)}{4}(\kappa_n + \kappa_\Lambda - \frac{\kappa_T}{\alpha}) + O(3), \quad (8)$$

for $n(\gamma, K^0)\Lambda$ process near threshold respectively. In case of K^+ production the PV coupling scheme shows larger prediction than the PS coupling, but it depends on $g_{KN\Lambda}$. However the deviations between them are drastic in case of K^0 production just because of the appearance of $\kappa = \kappa_n + \kappa_\Lambda + \kappa_\alpha/\alpha$ term at O(1) order compared to O(2) order in case of K^+ production. It is remarkable that the magnetic moments parts appears at O(1) order compared to the appearance of O(2) order in the PV theory and as a result the contribution from the hyperon resonance, $\Lambda(1405)$, would be quite large compared to the PV case. The whole discussions on the difference of both PV and PS schemes are detailed in ref. [4].

2.2 THE VECTOR MESON'S POLES CONTRIBUTION

Here we consider the additional contributions to E_0^+ amplitude from the t-channel of the vector mesons ($K1$ and K^*). For the simple interpretation of our analytical results we are not considering the widths of those vector mesons. First, the K^* meson ($892.1\ MeV, J^\pi = 1^-$) has anomaly coupling with the incident photon, similarly to the ρ, ω mesons in photopion productions. The result of K^* at kaon threshold is as follows ;

$$A_1^{K^*} + \frac{(\delta+\mu)}{2} A_5^{K^*}|_{thr.} = \frac{1}{2M} \frac{G_{K^*K^+\gamma}}{M} [(G^V + G^T) \frac{M^3(\delta+\mu)^2 \mu}{1+\mu+\delta}] \,/\, (t - m_{K^*}^2) \,. \quad (9)$$

The $G_{K^*K^+\gamma}$ coupling constant is extracted from the decay width of $K^* \to K^+\gamma$. Our value deduced that way is 0.8341 e smaller than that of the Thom's because we used 50 keV for this decay width instead of the estimated value 86 KeV of Thom. The hadronic coupling constants, $G_{PK^*\Lambda}^V = G^V$ and $G_{PK^*\Lambda}^T = G^T$ are inferred from the SU(3) symmetry assumption i.e. the relations of the coupling constants in the pseudo-scalar mesons are also hold in the vector mesons coupling to the hadrons ;

$$G_{K^*\Lambda N}^{V,T} = -(\frac{1}{\sqrt{3}})(3 - 2\alpha_{E,M}) G_{\rho NN}^{V,T} \,, \quad (10)$$

where we used 0 and 3/4 for $\alpha_{E,M}$ similarly to the KNΛ case.

On the other hand the $K1$ vector meson ($1270\ MeV, J^\pi = 1^+$) is said to play important roles in kaon photo-production.

At the kaon threshold it contributes to the E_0^+ amplitude as the following A_1 form ;

$$A_1^{K1}|_{thr.} = \frac{G_{K1K^+\gamma}}{M} k_0 q_0 [G_{K1K^+P}^V + G_{K1K^+P}^T \frac{M_\Lambda - M}{M_\Lambda + M}] \,/\, (t - m_{K1}^2) \,. \quad (11)$$

Here k_0 is the photonic energy at threshold. We see the tensor parts affect little to the E_0^+ amplitude. The photonic coupling constant of the K1 axial vector meson, $G_{K1K^+\gamma}$ can be estimated through the vector meson dominance model.

For the hadronic coupling constants one does not have any informations from the both experimental and theoretical sides to our knowledge. Here we assumed that the ratio of the hadronic coupling constants between the vector mesons and the pseudo-scalar mesons, K^* (1^-) and $\pi(0^-)$, are also hold in the ratio between the axial vector meson and the scalar mesons in the following way ;

$$\frac{G_{K1P\Lambda}^V}{g_{\sigma NN}} = \frac{G_{K^*P\Lambda}^V}{g_{\pi NN}} \quad (12)$$

The coupling constants determined this way and the effects of K^* and $K1$ axial vector meson to E_0^+ amplitude near threshold are also tabulated and compared with others obtained from the fitting procedures in ref.[4].

2.3 RESONANCES

For the contribution of the N4 resonance we use the following Breit-Wigner form ref.[4] ;

$$E_0^+(N4) = [2q_0k_0]^{-1/2}\frac{M_0\Gamma\sqrt{x_\gamma x_K}}{M_0^2 - W^2 - i\Gamma M_0}e^{i\theta} \tag{13}$$

, where k_0, q_0, M_0 and Γ are the photon, the kaon momentum at resonance, the mass and width of the N4 resonance respectively while the total invariant energy is given as W. The x_γ and x_K mean the branching ratio into γ^1H and $K^+\Lambda$ states from N4 resonance. We use 7.8 $\cdot 10^{-3}$ for the value of $\sqrt{x_\gamma x_k}$. The background resonance is taken into account by the $e^{i\theta}$ factor in eq.(15). We use the value of 108.3 degree in Workman [4].

The hyperon resonance through the u-channel is already shown to be small near threshold by the same argument of the u-channel of the Λ particle in case of the PV coupling. However the PS scheme shows completely different features, this dominates the E_0^+ amplitude near threshold.

3 TOTAL AMPLITUDES AND TOTAL CROSS SECTIONS

To predict the future experiments we plot in figure 1 the results of the total cross section of this $p(\gamma, K^+)\Lambda$ process near threshold with the various values of $g_{KN\Lambda}/\sqrt{4\pi}$. The total cross section near threshold is given by

$$\sigma_T = 4\pi\frac{q}{k}|E_0^+(Extended\ Born) + E_0^+(N4)|^2. \tag{14}$$

The results of the total cross section strongly depend on the $g_{KN\Lambda}/\sqrt{4\pi}$ as expected and show a very interesting feature. In specific, recent data from ELSA [7] show two interesting data points around this energy region. A deduction of $g_{KN\Lambda}/\sqrt{4\pi}$ in this model from these data give values, - 0.8 $\sim$ - 1.2, whose values are smaller than the SU(3) predicted. Since the data around here show strong enhancements near threshold, we need more refined model taking the full resonances in theoretical side and more data in experimental side in order to make a final conclusion for this leading coupling constant.

4 CONCLUSIONS

As a summary, the E_0^+ amplitude near photo kaon threshold can be depicted by only $g_{KN\Lambda}/\sqrt{4\pi}$ with som static properties of the related particles. Therefore one can determine the exact $g_{KN\Lambda}$ coupling constant by the experiment near threshold if one knows exactly the N4 resonance. Even if one has some uncertainty explained above from the N4 resonance the future experiments would play an important role in determining this coupling constant as shown in figure 1.

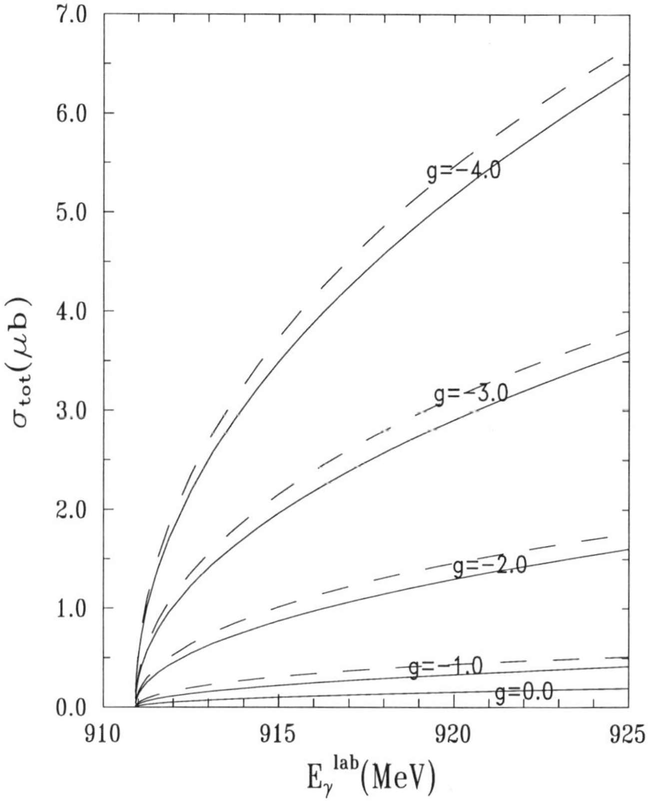

Figure 1. The total cross section of the $p(\gamma, K^+)\Lambda$ process as a function of E_γ^{lab}. The coupling constants $g = g_{KN\Lambda}/\sqrt{4\pi}$ from 0.0 to - 4.0 are indicated respectively. The region between solid and dashed lines stands for the uncertainty from the N4 resonance.

References

1. "Hyper Nuclear and Strange Particle Physics", edited by T.Hukuda, et al., Nucl.Phys.**A547**,(1992).
2. S.H.Choi, M.K.Cheoun and S.H.Lee, Phys. Rev.**C53**, 1363(1996).
3. Zh. Li, Phys. Rev. **C 52**, 1648(1995) and M.T. Jung at this Proceedings.
4. Myung Ki Cheoun, B.S.Han, B.G.Yu, and Il-Tong Cheon, Phys. Rev. **C54**, 1811(1996). The detailed references are herein.
5. R.A.Adelseck and B.Saghai, Phys. Rev. **C45**, 2030(1992).
6. J.-C.David, C.Fayard, G.H.Lamot, and B.Saghai, Phys. Rev. **C53**, 2613(1996).
7. M.Q.Tran, *et. al*, Phys.Lett **B445**, 20(1998).

KAON PHOTOPRODUCTION ON THE NUCLEON UP TO $E_{\gamma}^{\rm LAB} = 2.1$GEV

B. S. HAN, MYUNG KI CHEOUN, AND IL-TONG CHEON
Department of Physics, Yonsei University, Seoul 120 – 749, Korea

BYUNG GEEL YU
Department of General Studies, Hankuk Aviation University, Koyang 412-791, Korea

Using an isobaric model based on effective chiral Lagrangians and taking into account the crossing symmetry, the kaon photoproduction ($\gamma + p \to K^+ + \Lambda$) is analyzed from its threshold to $E_{\gamma}^{\rm Lab}$=2.1 GeV energy region. Here we use the *pseudovector* and *pseudoscalar* coupling schemes. In this calculation, we consider all the possible resonances such as a vector meson($K^*(890)$), an axial vector meson($K_1(1270)$), nucleon resonances($J \leq 5/2$), and hyperon resonances($J \leq 3/2$). Considering the broken SU(3) symmetry and experimental values together with the nonrelativistic quark model predictions for the strong and electromagnetic decay widths of the resonances, we determine the unknown parameters by least χ^2 fit. The strong form factors are also introduced to the transition amplitude of the reaction in a gauge invariant manner by Haberzettl's gauge prescription and Ohta's one, and we examine the effects of such form factors and the difference between the two gauge prescriptions. Through this analysis all the available data are reproduced well and it is shown that an introduction of strong form factors to this reaction improves the χ^2 and makes negligible difference between the pseudovector and pseudoscalar coupling schemes. Without the form factor, pseudoscalar coupling scheme of kaon to baryon is superior to pseudovector one.

Due to the advance of high energy and high duty cycle electron accelerators and detectors at the Thomas Jefferson National Accelerator Facility(TJNAF), ELectron Stretcher Accelerator (ELSA) and European Synchrotron Radiation Facility(USRF), the high-current and polarized beams in the energy domain of a few GeV's are provided. Consequently, the sufficient and high-precision experimental data of kaon photoproduction are available now or will be in the near future. Therefore, it is expected that the more comprehensive understanding for the kaon photoproduction will be possible.

The most widely used method to investigate the kaon photoproduction is an isobaric model between threshold and roughly $E_{\gamma}^{\rm Lab} = 2.1$ GeV. Until now many works using isobaric approaches have been published[1,2,3,4,5,6,7].

All these formalisms use the Feynman diagrammatic techniques where vertex functions are obtained from the relevant effective Lagrangians. As is well known, there are two coupling types of kaon and baryons with $J = \frac{1}{2}$ in strong vertex. One is a pseudoscalar(PS) coupling and the other is a pseudovector(PV) coupling

Through the chiral rotation[8], the two coupling schemes are related to each other, in the tree approximation the difference between the two schemes are inevitable.

Nevertheless, until now, most calculations base on isobaric model using effective Lagrangian were carried out by PS coupling scheme. Although some works[9,10,11] were done by using of PV coupling they are restrictive: In Ref. [9], Bennhold *et al.* tried to fit phenomenologically the available data for $\gamma + p \to K^+ + \Lambda$ and obtained very different sets of coupling constants for the two coupling schemes.

However, in their paper any nucleon or hyperon resonance was not considered and the values of $g_{KN\Lambda}$ and $g_{KN\Sigma}$ didn't satisfy the SU(3) symmetry constraints. Feuster *et al.* in Ref. [11], by using the unitary model, investigated photon- and meson-induced reactions. However, they didn't consider contribution of the spin $\frac{5}{2}$ particle which played an important role to reproduce the polarization data in the Ref. [5,12]. They didn't also take into account the contributions from hyperon resonances. As pointed out in Ref. [6] $\Lambda(1405)$ gave a dominant contribution to the radiative kaon capture which is related to kaon photoproduction by crossing symmetry[13,14,15,16,17]. Therefore, including more resonances, the more extensive study using the PV-coupling scheme and scrupulous comparison for both coupling schemes are needed.

In this report, we analyze the kaon photoproduction ($\gamma + p \to K^+ + \Lambda$) up to $E_\gamma^{\rm Lab} \simeq 2.1$GeV. Here we use PV and PS coupling schemes and form factors and investigate the difference between the two coupling schemes and effects of the form factors.

The hadronic form factors are introduced in the kaon photoproduction in order to consider the effects of the internal structures of hadrons and improve the divergent behavior of the kaon photoproduction amplitudes in an isobaric model as the incident photon energy increases.

On the other hand, it is well known that the inclusion of form factors at the hadronic vertices gives rise to gauge violation of the amplitude. To restore the gauge invariance we use the Ohta's prescription[19] and Haberzettl's[20]. The difference between the two gauge prescriptions is that for the Ohta's prescription the form factor effects disappear in the A_2 amplitude of Born term and for Haberzettl's one, A_2 is multiplied by the form factor, are multiplied $(a_1 f(M_\Lambda^2, s, m_K^2) + (1 - a_1) f(M_\Lambda^2, M^2, t))$[21] where we use the free parameter $a_1 = 1/2$. In this calculation, we use the phenomenological form factors used in Ref. [18]:

In our fitting procedure, we consider the 26 intermediate particles listed in Table 1. Since most of the coupling constants are unknown they are determined by fitting the total cross section, differential cross section, Λ-polarization asymmetry (P), target-polarization asymmetry (T), and radiative kaon capture ratio to their experimental data. In fact the parameters to be determined are product of electromagnetic and strong coupling constants and defined in Ref. [22].

According to Adelseck *et al.*'s analysis of the existing differential cross section data for the kaon photoproduction up to $E_\gamma^{\rm Lab}$ =1.4 GeV, they obtained some simple models, which was consistent with the broken $SU(3)$ symmetry, by the excluding the Orsay's 22 data[23] which reveal the internal inconsistency[4]. On the same lines, we use the same data set as used in Ref. [4] and the data in Ref. [24], then we include 251 the experimental data: 197 differential cross section, 21 total cross section, 30 Λ-polarization asymmetry, 3 target-polarization asymmetry, and 1 radiative kaon capture ratio data. Our fitting procedure is fulfilled by using CERN library package MINUIT[25].[a]

In our fitting procedure we fix the leading coupling constants as the $SU(3)$ symmetry predicting values, $\frac{g_{KN\Lambda}}{\sqrt{4\pi}} = -3.74$ and $\frac{g_{KN\Sigma}}{\sqrt{4\pi}} = 1.09$, which is conformed

[a] note added: After our analysis was finished, we know that new data for $\gamma + p \to K^+ + \Lambda$ from ELSA. Therefore we didn't include the new data in our calculation.

Table 1. List of the particles considered in our study.

particles	$L_{2I(I)\,2J}$	J^P	mass (MeV)	Γ(MeV)	particles	$L_{2I(I)\,2J}$	J^P	mass (MeV)	Γ(MeV)
p		$\frac{1}{2}^+$	938.28		N8	D_{15}	$\frac{5}{2}^-$	1675	150
Λ		$\frac{1}{2}^+$	1115.6		N9	F_{15}	$\frac{5}{2}^+$	1680	130
Σ		$\frac{1}{2}^+$	1192.46		Λ1	S_{01}	$\frac{1}{2}^-$	1405	50
K^+		0^-	493.667		Λ2	P_{01}	$\frac{1}{2}^+$	1600	150
K^*		1^-	892	49.8	Λ3	S_{01}	$\frac{1}{2}^-$	1670	35
K_1		1^+	1270	90	Λ4	S_{01}	$\frac{1}{2}^-$	1800	300
N1	P_{11}	$\frac{1}{2}^+$	1440	350	Λ5	P_{01}	$\frac{1}{2}^+$	1810	150
N2	S_{11}	$\frac{1}{2}^-$	1535	150	Σ1	P_{11}	$\frac{1}{2}^+$	1660	100
N3	S_{11}	$\frac{1}{2}^-$	1650	150	Σ2	S_{11}	$\frac{1}{2}^-$	1750	90
N4	P_{11}	$\frac{1}{2}^+$	1710	100	Λ6	D_{03}	$\frac{3}{2}^-$	1520	15.6
N5	D_{13}	$\frac{3}{2}^-$	1520	120	Λ7	D_{03}	$\frac{3}{2}^-$	1690	60
N6	D_{13}	$\frac{3}{2}^-$	1700	100	Σ3	D_{13}	$\frac{3}{2}^-$	1670	60
N7	P_{13}	$\frac{3}{2}^+$	1720	150	Λ8	P_{03}	$\frac{3}{2}^+$	1890	100

by the fact that the variation of the coupling constants in the ranges given by about 20% broken SU(3) symmetry gives nearly same χ^2

We fulfill the least χ^2 fit using Haberzettl's gauge prescription for PV and PS coupling schemes. The resulting parameters are given in 2nd and 3rd columns in Table 2 (Habe(PV), Habe(PS)). We also obtain same particle set with different parameter values listed in Table 2 (Ohta(PV), Ohta(PS)). To investigate the form factor effect we also do fitting procedure without form factors using same particles obtained by the cases with form factors (NoformPV, NoformPS). To obtain the particle set, we investigate the contribution of each article to the χ^2 value and exclude the particles playing minor roles in this calculation for more simpler model.

As results, we present the χ^2 per particle for each physical observables in Table 3. The results in Table 3 show that except for the Noform(PV), the others give good χ^2. The results in Table 3 show that an introduction of the form factors gives more improved results than the case without them. Although the form factors improve the total χ^2, the improvement is significant in the total and differential cross sections but not in the polarization observables. In the target polarization asymmetry, Noform(PS) is better than Habe's and Ohta's. The results for target polarization asymmetry (T) show that PS scheme is much better than PV one (see Table 3). However, since there exit only 3 data points available for T, this result is not decisive.

Without the form factors, the PS scheme is superior to PV one, but an introduction of the form factor gives the negligible difference between the two coupling schemes(see Table 3), which can be explained by following facts: *The difference between the two coupling schemes is shown up only in the spin-1/2 particles' contributions.* In our results given in Table 2, the masses of spin-1/2 particles having large coupling constants are located below the threshold c.m total energy and the

Table 2. The parameters in the 2nd and the 3rd columns are obtained by using the Haberzettl's gauge prescription, 4th and 5th columns by the Ohta's prescription and the last two columns by fit without strong form factors. The parameters are defined in Ref. [22].

parameters	Habe(PV)	Habe(PS)	Ohta(PV)	Ohta(PS)	Noform(PS)	Noform(PV)
$g_{KN\Lambda}/\sqrt{4\pi}$	-3.74	-3.74	-3.74	-3.74	-3.74	-3.74
$g_{KN\Sigma}/\sqrt{4\pi}$	1.09	1.09	1.09	1.09	1.09	1.09
$G_V^{K^*}/4\pi$	-0.16	-0.22	0.37	0.37	-0.34	-0.19
$G_T^{K^*}/4\pi$	-0.83	-0.83	-0.83	-0.83	-0.03	-0.33
$G_V^{K_1}/4\pi$	-0.0042	-0.01	0.084	0.082	0.26	0.54
$G_T^{K_1}/4\pi$	-0.305	-0.39	-3.41	-3.48	-0.56	-0.71
$G_{N1}/\sqrt{4\pi}$	-3.78	-3.11	-4.11	-3.09	0.61	4.40
$G_{N2}/\sqrt{4\pi}$	0.104	0.19	0.12	0.18	0.03	-0.23
$G_{N3}/\sqrt{4\pi}$	-0.022	-0.015	0.0057	0.014	0.02	0.05
$G_{N4}/\sqrt{4\pi}$	0.179	0.22	0.0853	0.091	-0.05	-0.13
$G_{\Lambda1}/\sqrt{4\pi}$	-1.51	-1.41	-0.26	-0.27	-0.31	-0.38
$G_{\Lambda3}/\sqrt{4\pi}$	0.008	0.008	0.008	0.008	0.008	0.008
$G_{\Lambda4}/\sqrt{4\pi}$	0.023	0.023	0.023	0.023	0.023	0.023
$G_{\Sigma2}/\sqrt{4\pi}$	0.014	0.014	0.014	0.014	0.014	0.014
G_{N6}^1	32.154	23.8	23.62	20.84	3.586	7.634
G_{N6}^2	42.965	37.9	30.88	29.78	1.484	-4.201
G_{N7}^1	-0.076	0.36	0.70	0.66	-0.353	-0.468
G_{N7}^2	1.00	1.33	2.17	2.42	-0.927	1.585
G_{N8}^1	-40.85	-35.27	-36.49	-38.92	9.699	2.297
G_{N8}^2	34.92	10.95	18.11	4.69	-28.114	-6.864
Λ_1^2	0.50	0.50	0.80	0.80		
$\Lambda_{1/2}^2$	0.60	0.88	0.53	0.76		
$\Lambda_{3/2}^2$	0.79	0.89	1.13	1.15		
$\Lambda_{5/2}^2$	0.75	0.79	0.75	0.79		
χ^2/N	1.0	1.04	1.24	1.26	1.3	2.72

hadronic form factors rapidly decrease as the momentum square of the exchanged particle goes away from its mass shell. Consequently, the difference between the two coupling scheme are negligible.

For the lack of space, we present only the total cross section in Fig 1. The experimental data are almost reproduced well by Habe(PV) up to 2.1 GeV. The AS1 and Noform(PV) overpredict above 1.5 GeV and Noform(PS) and Ohta(PV) show larger values than experimental data in the energy higher than 1.7 GeV. From the analysis of the total cross section, we can conclude that the finite size of internal structure of hadron should be considered in the isobaric model based on effective Lagrangians.

As a conclusion it is found that the experimental data of the kaon photoproduction are well reproduced in our model space. The hadronic form factors yield

Table 3. The partial χ^2 per data: σ_{tot}, $d\sigma/d\Omega$, P, and T are the χ^2 values of the total and differential cross sections, Λ-polarization asymmetry and a target polarization asymmetry, respectively. The last column is the total χ^2 per particle.

$\chi^2 \rightarrow$	σ_{tot}	$d\sigma/d\Omega$	P	T	χ^2_{tot}
Habe(PV)	0.975	0.957	0.993	4.36	1.0
Habe(PS)	1.01	0.99	1.11	3.16	1.03
Ohta(PV)	1.22	1.07	1.055	8.13	1.24
Ohta(PS)	1.27	1.08	1.008	3.97	1.26
Noform(PV)	5.66	1.7	4.19	9.39	2.72
Noform(PS)	2.3	1.2	1.17	0.3	1.3

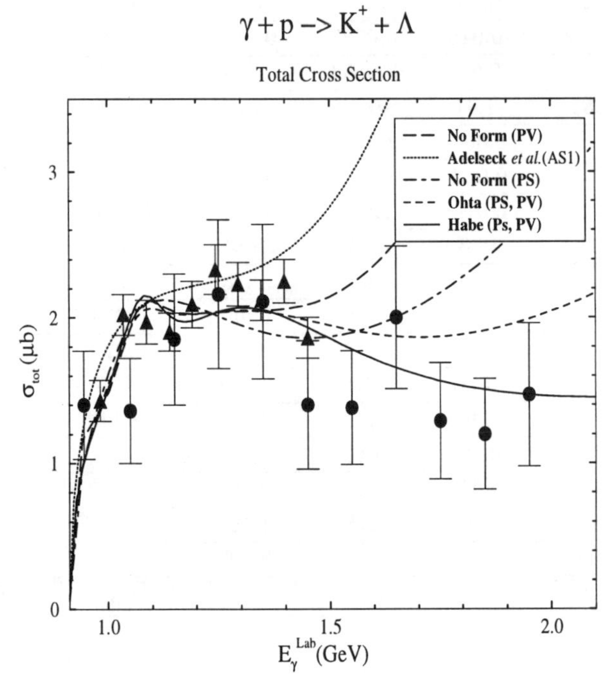

Figure 1. The total cross sections up to energy E_γ=2.1 GeV.

more improved results than those of the case without them and diminish the difference between the PV and PS coupling schemes. The form factor, used in our calculation, is a function of momentum square of the exchanged particle. Although the study about the hadronic form factor, which is attached to the vertex with off shell meson leg, has been done well, the information for the form factor at vertex with the off mass shell nucleon(or hyperon) leg, is scarcely known. Therefore, it is necessary to investigate the form factor through the more elaborate phenomenological and microscopic studies. As mention in other works[5,12], our analysis for the polarization observable shows the striking model dependence, so that the forthcoming data of TJNAF, ESRF, and ELSA, are expected to sort out the correct one from the existing models. Since our model can be applied directly to the reaction, $\gamma + p \rightarrow K^0 + \Lambda$, future experiments on this reaction will verify the validity of our model.

References

1. H. Thom, Phys. Rev. **151**, 1322 (1966).
2. F. M. Renard and Y. Renard, Nucl. Phys. **B25**, 490 (1971); Y.Renard, *ibid.*, **B40**, 499(1972).
3. R.A.Adelseck, C.Bennhold and L.E.Wright, Phys. Rev. C **32**,1681(1985).
4. R. A. Adelseck and B. Saghai, Phys. Rev. C **42**, 108 (1990).
5. J.-C.David, C.Fayard, G.H.Lamot and B.Saghai, Phys. Rev. C **53**, 2613 (1996).
6. R.A.Williams, C.-R. Ji and S. R. Cotanch, Phys. Rev. **C46**, 1617 (1992) and S. R. Cotanch, R. A. Williams and C.-R. Ji, Phys. Scripta **48**, 217(1993).
7. T. Mart, C. Bennhold and C.E. Hyde-Wright, Phys. Rev. **C51**, 1074 (1995).
8. S. Weinberg, Phys. Rev. Lett. **18** (1967) 188.
9. C. Bennhold and L. E. Wright, Phys. Rev. **C36**, 438 (1987).
10. M. K. Cheoun, B. S. Han, B. G. Yu, and Il-Tong Cheon, Phys. Rev. C **54**, 1811 (1996).
11. T. Feuster and U. Mosel, Los Alamos eprint nucl-th/9803057.
12. B.Saghai and F.Tabakin, Phys. Rev. C **53**, 66(1996).
13. K. Johnson and E.C.G Sudarshan, Ann. Phys. **13**, 126 (1961).
14. G. Velo and D. Zwanziger, Phys. Rev. **186**, 1337 (1969).
15. H. Burkhardt, J. Lowe, and A. A. Rosenthal, Nucl. Phys. **A440**, 653 (1985).
16. R.L. Workman and H.W. Fearing, Phys. Rev. **D37**, 3117(1988).
17. Fred de Jong and Rudi Malfliet, Phys. Rev. C **46** 2567 (1992).
18. J. -F. Zhang, N. C. Mukhopadhyay and M. Benmerrouche, Phys. Rev. **C 52**, 1134 (1995).
19. K. Ohta, Phys. Rev. C **40**, 1335 (1989).
20. H. Haberzettl, Phys. Rev. C **56**, 2041 (1997).
21. H. Haberzettl, C. Bennhold, T. Mart, and T. Feuster, Few-Body Systems Suppl. **99**, 1 (1998).
22. B. S. Han, Ph. D. thesis, Yonsei University, 1999.
23. D. Décamp, B. Dudelzak, P. Eschstruth, and Th. Fourneron, Orsay Linear Accelerator Report LAL-1236, 1970; Th. Fourneron, Thése de Doctrat d'Etat, Orsay Linear Accelerator Report LAL-1258, 1971.
24. M. Bockhorst *et al.*, Z. Phys. **C 63**, 37 (1994).
25. F. James and M. Roos, MINUIT, CERN Report **D506**, 1981.
26. Particle Data Group, Phys. Rev. **D50**, 1994.
27. Particle Data Group, Euro. Phys. Jour. **C3**, 1998.

CASCADE CALCULATION OF EXOTIC HELIUM ATOMS
– s-orbit vs. p-orbit absorption rates –

T. Koike and Y. Akaishi
KEK-IPNS, Tanashi-Branch, Tanashi 188-8501, Japan
E-mail: tkoike@tanashi.kek.jp

We construct a new model for the Stark-mixing process of exotic helium atoms using the impact-parameter method, and compared it with a phenomenological one used so far (sliding transition model). It turns out that the sliding transition model is justified only for low-n states and largely overestimates the Stark-mixing transition rate at high-n states. As a result of the atomic-cascade calculation, the s-(p-)orbit absorption rates in our new model are considerably smaller (larger) than those in the phenomenological one, although both our new model and old one well reproduce the experimental x-ray yields.

1 Introduction

The stopped K^- and Ξ^- reactions are utilized as one of useful tools for the hypernucleus formation in strangeness $S = -1$ and -2 sectors, respectively. When we analyze these reactions, we always encounter the following problem: From which atomic orbit is K^- or Ξ^- absorbed into the nucleus by strong interaction? Atomic cascade calculations have been done in order to answer this question, however, we have not yet come to the definite answer due to ambiguities about the absorption process from high-n atomic orbits. Especially, in the case of exotic helium atoms, α-X^- where X^- denotes a negatively-charged particle such as K^-, $\bar{p}$ and Ξ^-, the Stark-mixing process causes difficulties for the determination of the s- and p-orbit absorption ratio. The phenomenological treatment for the Stark-mixing process, which is called as "sliding transition" [1], succeeded in reproducing the experimental x-ray yields of the α-X^- atoms. However, it is doubtful whether such phenomenological treatment correctly describes the absorption process from high n-atomic orbits since it can not be checked by fitting the x-ray yields which become dominant only at low-n states.

Our purpose is to construct a realistic cascade model for the α-X^- atoms beyond phenomenological ones used so far and to answer the problem of the s- and p-orbit absorption ratio. We newly derived the Stark-mixing transition rate formula for α-X^- atoms using the impact-parameter method [2]. The results of the atomic-cascade calculation are compared between our new model and the phenomenological one use so far, namely sliding transition model [1].

2 Cascade Model

Processes included in our cascade calculation are the followings;

1. Radiative transition: $(X^-\alpha)_i \rightarrow (X^-\alpha)_f + \gamma$

2. External Auger transition [3]: $(X^-\alpha)_i + {}^4He \rightarrow (X^-\alpha)_f + {}^4He^+ + e^-$

3. Formation of molecular ion and subsequent Auger transition [3]:
$(X^-\alpha)_i + {}^4He + {}^4He \rightarrow [{}^4He\text{-}(X^-\alpha)]^+ + {}^4He$,
$[{}^4He\text{-}(X^-\alpha)]^+ \rightarrow (X^-\alpha)_f + {}^4He^+ + e^-$

The formation of the molecular ion $[{}^4He\text{-}(\alpha\text{-}X^-)]^+$ involves the 3-body collisions. Here, the formation probability of $[{}^4He\text{-}(\alpha\text{-}X^-)]^+$, which is denoted as P^{MIF}, is introduced as a free-parameter. The Auger decay rate of $[{}^4He\text{-}(\alpha\text{-}X^-)]^+$ is calculated according to the formulation in Ref.3.

4. Stark mixing process: $(X^-\alpha)_i + {}^4He \rightarrow (X^-\alpha)_f + {}^4He$

(a) Sliding transition [1], which is a phenomenological model used so far:

$$\Gamma^{\text{Sliding}}_{n,l\rightarrow n,l\pm1} = \zeta \, \frac{N}{N_0} \mid < n \; l \pm 1 \mid r \mid n \; l > \mid^2 \mid (l \; 0 \; 1 \; 0 \mid l \pm 1 \; 0) \mid^2. \tag{1}$$

where ζ is a free-parameter, N the target helium atom density and N_0 the liquid helium atom density.

(b) New formula based on the impact parameter method [2]:

$$\Gamma^{\text{Stark}}_{n,l\rightarrow n,l'} = k_{\text{STK}} \; N v \pi \sum_{\nu=|l-l'|}^{l+l'} \frac{\nu+1}{2\,\nu+1}$$
$$\times \int_{a_0}^{\infty} \mid < n \; l' \mid j_\nu (\; 2C(\rho) r \;) \mid n \; l > \mid^2 \rho d\rho \mid (l \; 0 \; \nu \; 0 \mid l' \; 0) \mid^2, \tag{2}$$

where k_{STK} is a free-parameter, v the velocity of α-X^-, and $C(\rho)$ is a function of the impact parameter ρ which has kinematical information (details are omitted here). For simplicity, we assume the straight trajectory approximation with a constant velocity v (the α-X^- kinetic energy = 1 eV). The k_{STK} can be interpreted as a correction factor for such kinematical assumption.

The sliding transition corresponds to the Born approximation, while our new model fully take the higher order contributions into account. Here, Born approximation corresponds to including only the direct transition like Fig.1-(a), and the higher orders denote the multi-step transition like Fig.1-(b). The sliding transition can be derived from our new model, that is, Eq.(2) can be transformed into the form of Eq.(1) in the limit of the weak electric field. However, this is justified only for low-n states, as is shown in Fig.2. At high-n states, the higher order contributions drastically hinder the transition rate by the interference effect.

5. Nuclear absorption

For α-K^- and α-$\bar{p}$ atoms, the nuclear absorption rate is calculated using a relevant optical potential [1,4]. For α-Ξ^- atom, it is taken from the theoretical estimation based on DWBA [5] (see Table 2 in Sec.4).

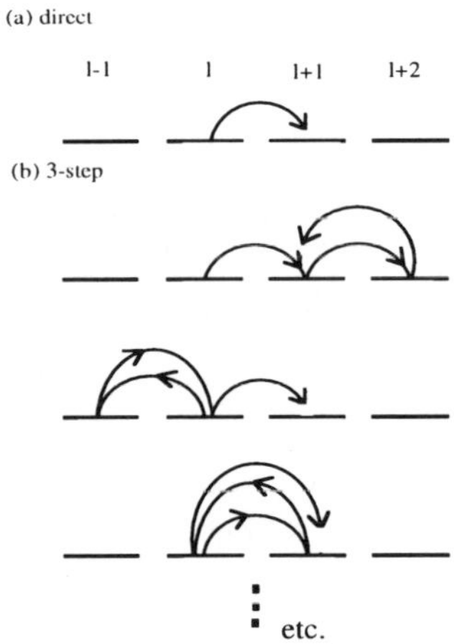

Figure 1: Meaning of higher order contributions in the case of $l \to l+1$ transition.

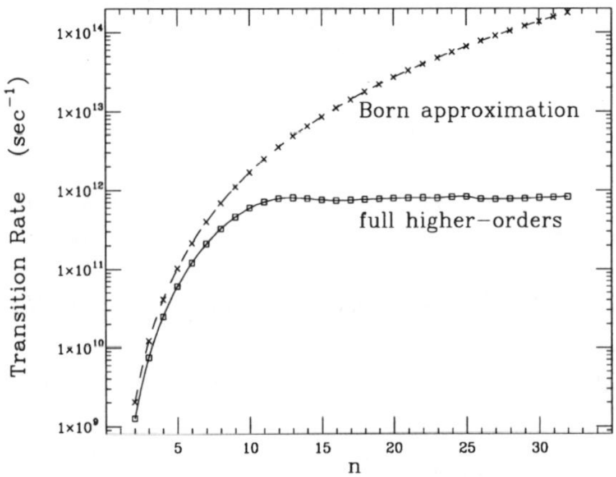

Figure 2: Comparison the exact calculation of the Stark-mixing transition rate with the Born approximation. $\Gamma^{\rm Stark}_{ns\to np}$ in the case of $\bar{p}$-α atom is plotted as a function n. The kinetic energy is taken to be 1 eV and $k_{\rm SKT}$ in Eq.(2) is taken to be 1.

6. Weak decay of X^- on atomic orbit (except for α-$\bar{p}$ atom).

The initial distribution of X^- is taken to be statistical at such states that the ionized α-X^- atom is formed by successive ejection of two electrons ($n \sim 22$ for α-K^-, 32 for α-$\bar{p}$, and 38 for α-Ξ^-). Note that the cascade calculation using any initial angular momentum distribution gives the almost same results as that using the statistical one due to the strong shuffling of the population by the Stark-mixing. The initial kinetic energy of α-X^- atom is taken to be 1 eV, and assumed to be constant during the cascade process. The free-parameters are adjusted in next section to reproduce the experimental x-ray yields.

3 Results for α-K^- and α-$\bar{p}$ atoms

Hereafter, using Eq.(1) for the Stark-mixing process is referred to as "sliding model", and using Eq.(2) as "new model". In Fig 3, the fitting results to the experimental x-ray yields in the liquid helium target are shown. Fitted parameters are the followings; in the case of the sliding model,

$$\zeta = 2\times 10^9, \quad P^{\rm MIF} = 0.8 \quad \text{for } \alpha\text{-}K^- \text{ atom},$$
$$\zeta = 5\times 10^8, \quad P^{\rm MIF} = 0.8 \quad \text{for } \alpha\text{-}\bar{p} \text{ atom},$$

and in the case of the new model,

$$k_{\rm STK} = 4.0, \quad P^{\rm MIF} = 0.3 \quad \text{for } \alpha\text{-}K^- \text{ atom},$$
$$k_{\rm STK} = 5.0, \quad P^{\rm MIF} = 0.3 \quad \text{for } \alpha\text{-}\bar{p} \text{ atom},$$

respectively. As is shown in Fig 3, both the new model and the sliding model well reproduce the experimental x-ray yields of α-K^- and α-$\bar{p}$ atoms.

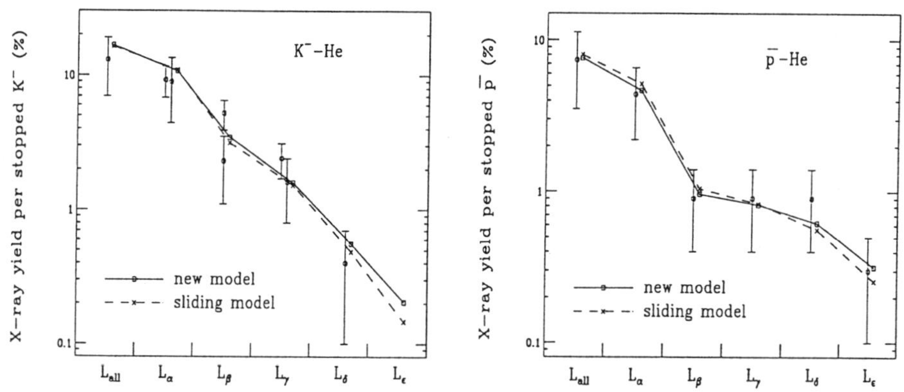

Figure 3: Fitting to the K^--α and $\bar{p}$-α atom x-ray yields in the liquid helium target. L_α (L_β, L_γ, ...) denote the x-ray for 3d→2p (4d→2p, 5d→2p, ...) transition. The experimental data are taken from Ref.4 and references therein.

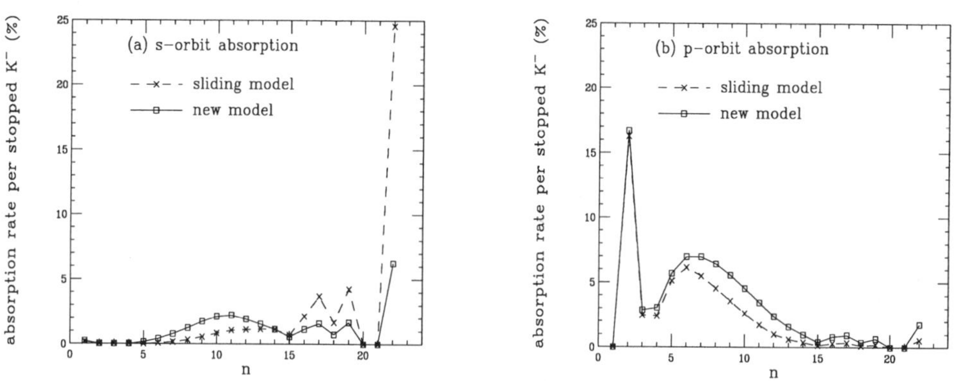

Figure 4: Absorption rate fraction from each atomic state of α-K^- atom. (a) s-orbit absorption and (b) p-orbit absorption. Rates are given per stopped K^-.

Table 1: Absorption rates from atomic s-, p- and d-orbits for α-K^- and α-$\bar{p}$ atoms in liquid helium targets. The rates are given per stopped K^- and $\bar{p}$, respectively.

	α-K^- atom		α-$\bar{p}$ atom	
	sliding model	new model	sliding model	new model
s-orbit absorption	44%	26%	30%	13%
p-orbit absorption	54%	72%	45%	62%
d-orbit absorption	0.4%	0.4%	25%	25%

As far as the x-ray yields are concerned, no essential difference can be seen between two models. Nevertheless, total s-(p-)orbit absorption rates in the new model are considerably smaller (larger) than those in the sliding model, as is shown in Table 1. In Fig 4, the absorption rate fraction from each atomic state is shown in the case of α-K^- atom. In our new model, the high-ns state absorption is suppressed by comparing with that in the sliding model, and the high- and middle-np state absorption is increased alternatively. No x-ray is observed in such region where the difference of nuclear absorption process appears between two models. At the low-n states where the x-ray is observed, the s-/p-orbit absorption rate fraction is almost identical between two models. The situation for α-$\bar{p}$ atom is same as the α-K^- atom case.

In summary, the sliding transition model does not correctly describes the absorption process from high n-atomic orbits, even if it well reproduce the experimental x-ray yields which are observed only at low-n states.

4 Application to α-Ξ^- atoms

The lightest double-Λ hypernucleus ${}^{4}_{\Lambda\Lambda}$H [6] is particularly important, because not only it provides the information about the $\Lambda\Lambda$ and ΞN interaction but also it is closely related to the existence problem of the H-dibaryon. The stopped Ξ^- reaction on ^{4}He is considered to be a suitable reaction for the formation of ${}^{4}_{\Lambda\Lambda}$H by the following reason; The elementary process of the Ξ^-p $\rightarrow \Lambda\Lambda$ conversion releases the energy 28.33 MeV, which is accidentally very close to the binding energy 28.30 MeV of the alpha particle. If an alpha particle is broken up by absorbing a Ξ^- particle, the reaction Q value becomes small and two Λ's cannot get enough energy to escape from the system. Therefore, a high ${}^{4}_{\Lambda\Lambda}$H formation rate can be expected.

Table 2 shows the DWBA results for the formation rates of various processes from α-Ξ^- atom in the cases of s- and p-orbit absorptions, respectively [5]. As seen in Table 2, the ${}^{4}_{\Lambda\Lambda}$H formation rate is largely suppressed for the s-orbit absorption, while it is strongly enhanced for the p-orbit absorption. Thus, the determination of s-/p-orbit absorption ratio for α-Ξ^- atom is crucial for the absolute value of ${}^{4}_{\Lambda\Lambda}$H formation rate.

The free-parameters in the cascade calculation can not be determined in the case of α-Ξ^- atom since no x-ray data exists for α-Ξ^- atom. Here, we tentatively apply the same parameters to α-Ξ^- atom as the α-$\bar{p}$ atom case. Therefore, it should keep in mind that the results can not be confirmed quantitatively. The results of the cascade calculation are summarized in Table 3. Again, the new model give smaller (larger) s-(p-)orbit absorption rate than the sliding one. In addition, a large difference of weak decay rate appears between two models. From Table 2 and 3, total ${}^{4}_{\Lambda\Lambda}$H formation rate is estimated to be about 8% in the sliding model, while it increases to be about 12% in the new one.

Table 2: Formation rates from the α-Ξ^- atom. The parenthesis denotes branching ratios.

Process	s-orbit absorption (10^{18} sec^{-1})	p-orbit absorption (10^{14} sec^{-1})
${}^{4}_{\Lambda\Lambda}H + n$	0.04 (1.5%)	7.28 (20.1%)
$\Lambda + {}^{4}_{\Lambda}H$	0.41 (15.2%)	6.11 (16.9%)
$\Lambda + {}^{4}_{\Lambda}H^{*}$	1.27 (46.7%)	18.88 (52.2%)
$\Lambda + \Lambda + {}^{3}H$	0.94 (34.7%)	3.89 (10.8%)
$\Lambda + {}^{3}_{\Lambda}H + n$	0.05 (1.9%)	0.02 (0.05%)

Table 3: Results of the α-Ξ^- atom cascade calculation. The rates are given per stopped Ξ^-.

	sliding model	new model
s-orbit absorption	61%	25%
p-orbit absorption	35%	58%
weak decay on atomic orbits	3%	16%

5 Problem and Future Prospect

Of course, our new model is more reliable than the sliding model, but there remain some ambiguities which come from the use of free-parameters. In order to get more quantitative conclusion, it is necessary to eliminate the free-parameters in the cascade model. In principle, an artificial factor k_{STK} in Eq.(2) is unnecessary if the kinematics of α-X^- atoms is strictly treated. Further refinement of our model is now in progress. At present stage, we estimate the uncertainty for the s- and p-orbit absorption rate to be still about $\pm 10\%$.

References

1. C. A. Baker *et al.*, *Nucl. Phys.* A **494**, 507 (1989).
2. T. Koike, *Ph.D.thesis*, Hokkaido University, 1998.
3. J. E. Russell, *Proc. Phys. Soc.* **85**, 245 (1965); *Phys. Rev.* A **18**, 521 (1978).
4. S. Baird *et al.*, *Nucl. Phys.* A **392**, 297 (1983).
5. T. Koike, I. Kumagai-Fuse and Y. Akaishi, *Proceeding of the 23rd INS International Symposium on Nuclear and Particle Physics with Meson Beams in the 1 GeV/c Region*, Universal Academy Press, Inc., 469 (1995).
6. S. Nakaich-Maeda and Y. Akaishi, *Prog. Theor. Phys.* **84**, 1025 (1990).

DECAY RATES OF MEDIUM-HEAVY Λ-HYPERNUCLEI WITHIN THE PROPAGATOR METHOD

W. M. ALBERICO, A. DE PACE, G. GARBARINO

Dipartimento di Fisica Teorica, Universitá di Torino and INFN, Sezione di Torino, 10125 Torino, ITALY

A. RAMOS

Departament d'Estructura i Constituents de la Matèria, Universitat de Barcelona, 08028 Barcelona, SPAIN

The Λ decay rates in nuclei has been calculated in ref. [1] using the Propagator Method in Local Density Approximation. We have studied the dependence of the widths (including the one for the two-body induced process $\Lambda NN \to NNN$) on the NN and ΛN short range correlations. Using a reasonable parametrization of these correlations, as well as realistic nuclear densities and Λ wave functions, we reproduce, for the first time, the experimental non-mesonic widths from medium to heavy hypernuclei.

1 Introduction

The study of hypernuclear physics may help in understanding some present problems related, for instance, to aspects of weak interactions in nuclei, or to the origin of the spin-orbit interaction in nuclei. Besides, it is a good instrument to study the role of quark degrees of freedom in the hadron-hadron interactions at short distances and the renormalization properties of pions in the nuclear medium.

The weak decay of hypernuclei occurs via two channels; the so called mesonic channel:

$$\Lambda \to \pi N \quad (\Gamma_M), \tag{1}$$

and the non-mesonic one, in which the pion emitted from the weak hadronic vertex is absorbed by one or more nucleons in the medium, for example

$$\Lambda N \to NN \quad (\Gamma_1), \tag{2}$$

$$\Lambda NN \to NNN \quad (\Gamma_2). \tag{3}$$

Obviously, the non-mesonic processes ($\Gamma_{NM} = \Gamma_1 + \Gamma_2 + ...$) can also be mediated by the exchange of more massive mesons. The non-mesonic decay is only possible in nuclei and, nowadays, the study of the hypernuclear decay is the only practical way to get information on the weak process $\Lambda N \to NN$, especially on its parity conserving part.

The free Λ decay is compatible with the $\Delta I = 1/2$ isospin rule, which is also valid in other non-leptonic strangeness changing processes. This rule is based on the experimental observation that $\Gamma_{\Lambda\to\pi^- p}/\Gamma_{\Lambda\to\pi^0 n} \simeq 2$, but it is not yet understood on theoretical grounds. From theoretical calculations like the one in ref. [2] and from experimental measurements [3] there is some evidence that the $\Delta I = 1/2$ rule is broken in the nuclear mesonic decay. However, this is essentially due to shell effects and might not be directly related to the weak process. A recent estimate of

$\Delta I = 3/2$ contributions to the $\Lambda N \to NN$ reaction [4] found moderate effects on the hypernuclear decay rates. In the present calculation of the decay rates in nuclei we will assume this rule as valid. The momentum of the final nucleon in $\Lambda \to \pi N$ is about 100 MeV for Λ at rest, so this process is suppressed by the Pauli principle, particularly in heavy nuclei. It is strictly forbidden in infinite nuclear matter (where $k_F^0 \simeq 270$ MeV), but in finite nuclei it can occur because of three important effects: 1) in nuclei the Pauli blocking is less effective and the hyperon has a momentum distribution which allows larger momenta for the final nucleon, 2) the final pion feels an attraction by the medium such that for fixed momentum it has an energy smaller than the free one and consequently, the final nucleon again has more chance to come out above the Fermi surface, and 3) on the nuclear surface the local Fermi momentum is smaller than k_F^0 and favours the decay. Nevertheless, the mesonic width decreases fast as the mass number A of the hypernucleus increases [2]. The mesonic rate is very sensitive to the pion self-energy, so that from the study of the mesonic decays it could be possible to extract important information on the pion-nucleus optical potential, which we do not know, nowadays, in a complete form.

The final nucleons in the non-mesonic process $\Lambda N \to NN$ come out with a momentum $\simeq 420$ MeV, so that this decay is not forbidden by the Pauli principle. On the contrary, apart from the s-shell hypernuclei, it dominates over the mesonic decay. The non-mesonic channel is characterized by large momentum transfers, so that the details of the nuclear structure do not have a substantial influence while the NN and ΛN short range correlations turn out to be very important. There appears to be an anticorrelation between mesonic and non-mesonic decay modes: indeed the total lifetime is fairly constant from light to heavy hypernuclei [3,5]: $\tau_{exp} = (0.5 \div 1)\, \tau_{free}$.

Nowadays, the main problem concerning the weak decay rates is to reproduce the experimental values for the ratio Γ_n/Γ_p between the neutron and the proton induced widths $\Lambda n \to nn$ and $\Lambda p \to np$. All theoretical calculations underestimate the experimental data for all the considered hypernuclei [3,4,6,7,8,9]:

$$\left\{\frac{\Gamma_n}{\Gamma_p}\right\}^{Th} \ll \left\{\frac{\Gamma_n}{\Gamma_p}\right\}^{Exp} \qquad\qquad 0.5 \leq \left\{\frac{\Gamma_n}{\Gamma_p}\right\}^{Exp} \leq 2. \tag{4}$$

In the One Pion Exchange (OPE) approximation the values for this ratio are $0.1 \div 0.2$. On the other hand, the OPE model has been able to reproduce the 1-body stimulated non-mesonic rates $\Gamma_1 = \Gamma_n + \Gamma_p$ for light and medium hypernuclei [6,7,9]. In order to solve this puzzle many attempts have been made up to now, but without success. Among these we recall the inclusion in the $\Lambda N \to NN$ transition potential of mesons heavier than the pion [6,8,9], the inclusion of contributions that violate the $\Delta I = 1/2$ rule [4] and the description of the short range baryon-baryon interaction in terms of quark degrees of freedom [7]. This last calculation is the only one which has found a fairly large (but not sufficient) increase of the neutron to proton ratio with respect to the OPE one. However, this calculation is carried out only for s-shell hypernuclei; moreover, the employed quark-lagrangian does not reproduce the experimental ratio between the $\Delta I = 1/2$ and $\Delta I = 3/2$ transition amplitudes for the Λ free decay.

The analysis of the ratio Γ_n/Γ_p is influenced by the 2-nucleon induced process $\Lambda NN \to NNN$. By assuming that the meson produced in the weak vertex is mainly absorbed by a strongly correlated neutron-proton pair, the 3-body process turns out to be $\Lambda np \to nnp$, so that a considerable fraction of the measured neutrons could come from this channel and not only from the $\Lambda n \to nn$ and $\Lambda p \to np$ ones. In this way it might be possible to explain the large experimental Γ_n/Γ_p ratio, which originally has been analyzed without taking into account the 2-body stimulated process. Nevertheless, the situation is far from being clear and simple [3,10]. In fact, Γ_n/Γ_p is sensitive to the energy spectra of the emitted nucleons, whose calculation also requires a careful treatment of the final state interactions of the nucleons. In ref. [11] the energy distributions were calculated using a Monte Carlo simulation to describe the final state interactions. A direct comparison of those spectra with the experimental ones seems to favour large Γ_n/Γ_p ratios, in strong disagreement with the OPE predictions. However, it was also pointed out the convenience of measuring the number of protons per decay event. In fact, this observable gives a more reliable Γ_n/Γ_p ratio and it is less sensitive to details of the Monte Carlo simulation accounting for the final shape of the spectra.

2 Decay widths

We present here the results of a new evaluation [1] of the decay rates for medium to heavy Λ-hypernuclei based on the Propagator Method introduced in ref. [12]. The parameters of the model were adjusted to reproduce the non-mesonic width of ${}^{12}_{\Lambda}$C. Then the decay rates for heavier hypernuclei have been predicted and compared, when possible, with the experiment.

2.1 *The model*

The propagator technique, estensively explained in refs. [1,12], provides a unified picture of the different decay channels and it is equivalent to the standard Wave Function Method [13]. The calculation of the widths is performed in nuclear matter, and then extended to finite nuclei via the Local Density Approximation (LDA).

The starting point for the calculation is the $\Lambda \to \pi N$ effective lagrangian, parametrized in the form:

$$\mathcal{L}_{\Lambda\pi N} = G m_\pi^2 \overline{\psi}_N (A + B\gamma_5)\vec{\tau} \cdot \vec{\phi}_\pi \psi_\Lambda + h.c., \tag{5}$$

where the values of the weak coupling constants G, A, B are fixed on the observables of the free Λ decay. In order to enforce the $\Delta I = 1/2$ rule, in eq. (5) the hyperon is assumed to be an isospin spurion with $I_z = -1/2$. The decay rate of a Λ in nuclear matter is obtained from the imaginary part of the Λ self-energy through the relation:

$$\Gamma_\Lambda = -2\mathrm{Im}\Sigma_\Lambda. \tag{6}$$

The Λ self-energy, Σ_Λ, contains an infinite series of diagrams in which the pion emitted in the Λ vertex couples to *1p-1h*, *2p-2h*, etc excitations, giving rise to the different decay channels when they are cut in all the possible ways. The structure of the decay width in nuclear matter is the following:

$$\Gamma_\Lambda(\vec{k}) \propto \int d\vec{q}...\theta(|\,\vec{k}-\vec{q}\,| - k_F)\theta(k_0 - E_N(\vec{k}-\vec{q}) - V_N)\mathrm{Im}G_\pi(q)\mid_{q_0=k_0-E_N(\vec{k}-\vec{q})-V_N}, \tag{7}$$

where

$$G_\pi(q) = \frac{1}{q_0^2 - \vec{q}^2 - m_\pi^2 - \Sigma_\pi^*(q)}, \tag{8}$$

is the pion propagators in nuclear matter, Σ_π^* being its proper self-energy; k_F is the Fermi momentum, E_N is the nucleon total free energy and V_N is the nucleon binding energy. Moreover, $k_0 = E_\Lambda(\vec{k}) + V_\Lambda$, is the Λ energy, containing a binding term (which we take from the experiment).

The decay widths in finite nuclei are obtained in LDA; a local Fermi sea of nucleons, related to the nuclear density by

$$k_F(\vec{r}) = \left\{\frac{3}{2}\pi^2\rho(\vec{r})\right\}^{1/3}, \tag{9}$$

is introduced. Besides, the nucleon binding potential V_N also becomes r-dependent in LDA. The decay width in finite nuclei is then obtained through the relations:

$$\Gamma_\Lambda(\vec{k}) = \int d\vec{r}\mid \psi_\Lambda(\vec{r})\mid^2 \Gamma_\Lambda[\vec{k},\rho(\vec{r})], \qquad \Gamma_\Lambda = \int d\vec{k}\mid \tilde{\psi}_\Lambda(\vec{k})\mid^2 \Gamma_\Lambda(\vec{k}), \tag{10}$$

where ψ_Λ ($\tilde{\psi}_\Lambda$) is the Λ wave function in the coordinate (momentum) space.

2.2 Results

In ref. [1] we have studied the influence of the short range correlations and the Λ wave function on the decay width of ${}^{12}_{\Lambda}$C, which we have used as a testing ground for the theoretical framework in order to fix the parameters of our model.

The baryon-baryon correlations are governed by the Landau parameters g' and g'_Λ. The choice of these parameters which reproduce the experimental non-mesonic rate of ${}^{12}_{\Lambda}C$ is $g' = 0.8$, $g'_\Lambda = 0.4$. Neglecting the contribution of the 2-body induced decay channel (namely in ring approximation), with the same g'_Λ, we find that $g' = 0.7$ should de used. This value is in agreement with the fenomenology of the (e, e') quasi-elastic scattering [14].

The Λ wave function is obtained from a Wood-Saxon well which exactly reproduces the first two experimental Λ single particle eigenvalues (s and p levels). The numerical results for ${}^{12}_{\Lambda}C$ are shown in table 1, where they are compared with the experimental data from BNL [15] and KEK [16,17,18]. In addition to the Wood-Saxon wave function that reproduces the s and p levels, we also made a test using the harmonic oscillator wave function with an "experimental" frequency ω obtained from the s-p experimental energy shift, the Wood-Saxon wave function of Dover *et al.* [19] and the microscopic wave function calculated from a non-local self-energy using a realistic ΛN interaction in ref. [20]. We have found that all these different Λ wave functions give rise to total decay widths which may differ at most by 15%.

Table 1. Decay rates for ${}^{12}_{\Lambda}$C

	Calculation	BNL [15]	KEK [16]	KEK New [17,18]
Γ_M	0.25	0.11 ± 0.27	0.36 ± 0.15	> 0.11
Γ_1	0.82			
Γ_2	0.16			
Γ_{NM}	0.98	1.14 ± 0.20	0.89 ± 0.18	< 1.03
Γ_T	1.23	1.25 ± 0.18	1.25 ± 0.18	1.14 ± 0.08

Table 2. Decay rates of medium-heavy Λ-hypernuclei

${}^{A+1}_{\Lambda}Z$	Γ_M	Γ_1	Γ_2	Γ_T
${}^{12}_{\Lambda}$C	0.25	0.82	0.16	1.23
${}^{28}_{\Lambda}$Si	0.07	1.02	0.21	1.30
${}^{40}_{\Lambda}$Ca	0.03	1.05	0.21	1.29
${}^{56}_{\Lambda}$Fe	0.01	1.12	0.21	1.35
${}^{89}_{\Lambda}$Y	$6 \cdot 10^{-3}$	1.16	0.22	1.38
${}^{139}_{\Lambda}$La	$6 \cdot 10^{-3}$	1.14	0.18	1.33
${}^{208}_{\Lambda}$Pb	$1 \cdot 10^{-4}$	1.21	0.19	1.40

Then, using Wood-Saxon wave functions that reproduce the s and p Λ-levels and the Landau parameters $g' = 0.8$, $g'_\Lambda = 0.4$, we have extended the calculation to heavier hypernuclei. We note that, in order to reproduce the experimental s and p levels for the hyperon we must use potentials with a nearly constant depth, around $28 \div 32$ MeV, from medium to heavy hypernuclei.

Our results are shown in table 2. The mesonic rate rapidly vanishes by increasing the mass number A. This is well known and it is related to the decreasing phase space allowed for the mesonic channel, and to smaller overlaps between the Λ wave function ψ_Λ and the nuclear surface, as A increases. The 2-body induced decay is rather independent of the hypernuclear dimension and it is about 15% of the total width. The total width is also nearly constant with A, as we already know from the experiment. In fig. 1 we compare the results from table 2 with recent (after 1990) experimental data for non-mesonic decay [15,16,17,21,22,23]. We remind that the data for nuclei heavier than ${}^{12}_{\Lambda}$C refer to the *total* width. The theoretical results are in good agreement with the data over the whole hypernuclear mass range explored. Moreover, we also see that the saturation of the $\Lambda N \to NN$ interaction in nuclei appears to be well reproduced.

The main open problem in the study of weak hypernuclear decays is to understand the large experimental value of the ratio Γ_n/Γ_p. However, we have to remind that the data for Γ_n/Γ_p have a large uncertainty and they have been analyzed without taking into account the 3-body decay mechanism. The study of ref. [10] showed that, even if the three body reaction is only about 15 % of the total decay rate, this mechanism influences the analysis of the data determining the ratio Γ_n/Γ_p. Energy spectra of neutrons and protons from the non-mesonic decay mechanisms were calculated in ref. [11]. The momentum distributions of the primary nucleons were determined from the Propagator Method and a subsequent Monte Carlo simulation was used to account for the final state interactions. It was shown that the shape of the proton spectrum is sensitive to the ratio Γ_n/Γ_p. In fact, the protons from the three-nucleon mechanism appeared mainly at low energies, while those from the two-body process peaked around 75 MeV. Since the experimental spectra show a fair amount of protons in the low energy region they would favour a relatively large three-body decay rate or, conversely, a reduced number of protons from the two-body process. Consequently, the experimental spectra were compatible with values for Γ_n/Γ_p around $2 \div 3$, in strong contradiction with the present theories.

Within the Propagator Method with modified parameters and realistic Λ wave functions we have then generated new nucleon spectra for various hypernuclei using the Monte Carlo simulation of ref. [11]. Although the new non-mesonic widths are sizably reduced (by about 35%) with respect to those of refs. [10,11], the resulting nucleon spectra are practically identical. The reason is that the ratio Γ_2/Γ_1 is essentially the same in both models, and the momentum distributions for the primary emitted nucleons are also very similar. As a consequence, the conclusions drawn in ref. [11] still hold.

Therefore, the origin of the discrepancy between theory and experiment for the ratio Γ_n/Γ_p still needs to be resolved. However, we must notice that on the experimental side, although new spectra are now available [16,17], they have not been corrected for energy losses inside the target and detector, so a direct comparison with the theoretical predictions is not yet possible. Attempts to incorporate these corrections by combining a theoretical model for the nucleon rescattering in the nucleus with a simulation of the energy losses in the experimental set-up are now being pursued [24]. On the other hand, a forward step towards a clean extraction of the ratio Γ_n/Γ_p would be obtained if the nucleons which come out from the different non-mesonic processes, $\Lambda N \to NN$ and $\Lambda NN \to NNN$ were disentangled [25].

Acknowledgments

We would like to thank H. Noumi and H. Outa for discussions and for giving us detailed information about the experiments.

References

1. W. M. Alberico, A De Pace, G. Garbarino and A. Ramos, nucl-th/9902023.
2. E. Oset and J. Nieves, Phys. Rev. **C47** (1993) 1478.
3. E. Oset and A. Ramos, Prog. Part. Nucl. Phys. **41** (Pergamon, 1998), 191.
4. A. Parreño, A. Ramos, C. Bennhold and K. Maltman, Phys. Lett. **B 435** (1998) 1.
5. J. Cohen, Prog. Part. Nucl. Phys. **25** (Pergamon 1990) 1.
6. A. Parreño, A. Ramos and C. Bennhold, Phys. Rev. **C56** (1997) 339.
7. T. Inoue, M. Oka, T. Motoba and K. Itonaga, Nucl. Phys. **A633** (1998) 312.
8. J. F. Dubach, G. B. Feldman, B. R. Holstein and L. de la Torre, Ann. Phys.

(N.Y.) **249** (1996) 146.
9. K. Itonaga, T. Ueda and T. Motoba, Nucl. Phys. **A585** (1995) 331c.
10. A. Ramos, E. Oset and L. L. Salcedo, Phys. Rev. **C50** (1995) 2314.
11. A. Ramos, M. J. Vicente-Vacas and E. Oset, Phys. Rev. **C55** (1997) 735.
12. E. Oset and L. L. Salcedo, Nucl. Phys. **A443** (1985) 704.
13. E. Oset, P. Fernández de Córdoba, J. Nieves, A. Ramos and L. L. Salcedo, Prog. Theor. Phys. Suppl. **117** (1994) 461.
14. E. Oset, H. Toki and W. Weise, Phys. Rep. **83** (1982) 281.
15. J. J. Szymanski *et al.*, Phys. Rev. **C43** (1991) 849.
16. H. Noumi *et al.*, Phys. Rev. **C52** (1995) 2936.
17. H. C. Bhang *et al.*, Nucl. Phys. **A639** (1998) 269c.
18. Y. Sato *et al.*, Nucl. Phys. **A639** (1998) 279c.
19. C. B. Dover, A. Gal and D. J. Millener, Phys. Rev. **C38** (1988) 2700.
20. J. Vidaña, A. Polls, A. Ramos and M. Hjorth-Jensen, Nucl. Phys. **A644** (1998) 201.
21. P. Kulessa *et al.*, Phys. Lett. **B427** (1998) 403.
22. T. A. Armstrong *et al.*, Phys. Rev. **C47** (1993) 1957.
23. H. Ohm *et al.*, Nucl. Phys. **A629** (1998) 416c.
24. H. Outa and O. Hashimoto, private communication.
25. V. J. Zeps, Nucl. Phys. **A639** (1998) 261c.

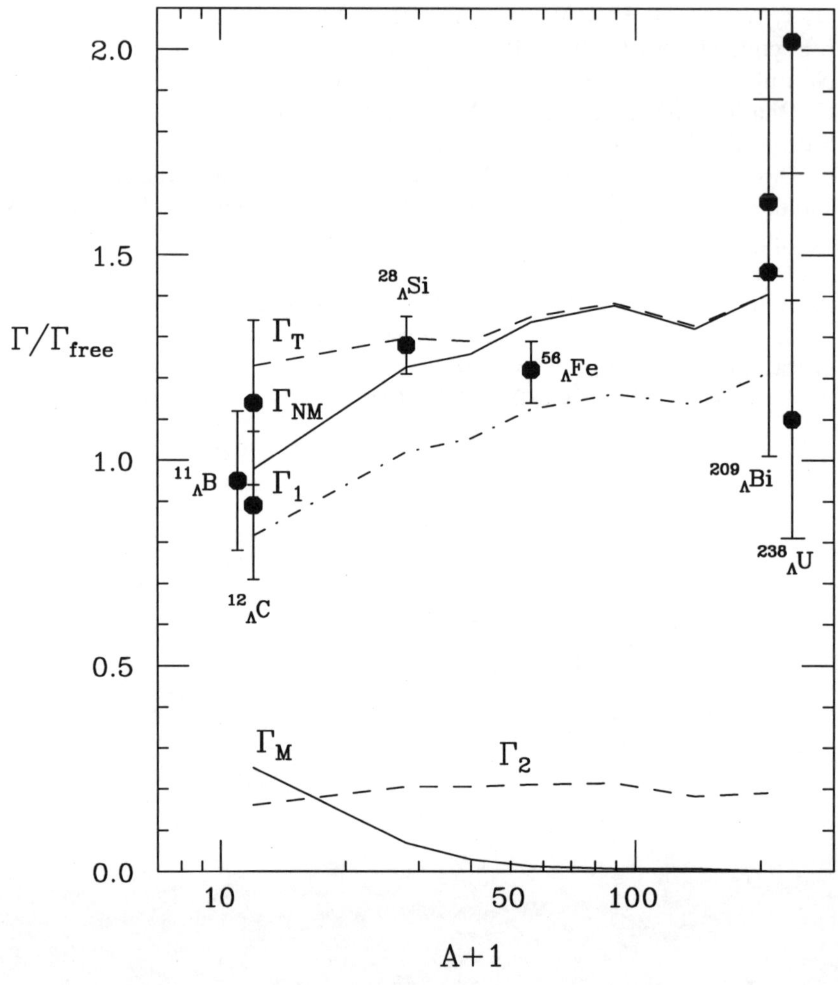

Figure 1. Λ decay widths in finite nuclei as a function of the mass number A.

HYPERNUCLEAR NONMESONIC DECAY AND $\Delta I = 1/2$ AND $\Delta I = 3/2$ INTERACTIONS IN THE MESON EXCHANGE MODEL

K. ITONAGA

Laboratory of Physics, Miyazaki Medical College
Kiyotake, Miyazaki 889-1692, Japan
E-mail: itonaga@physics.miyazaki-med.ac.jp

T. UEDA

Department of Physics, Ehime University
Matsuyama, Ehime 790-8577, Japan

T. MOTOBA

Laboratory of Physics, Osaka Electro-communication University
Neyagawa, Osaka 572-8530, Japan

The hypernuclear weak decay potential is investigated based on the correlated 2 pion exchange diagrams, i.e. 2 pion/rho and 2 pion/sigma exchanges in addition to the one pion exchange mechanism, assuming $\Delta I = 1/2$ rule. These potentials when combined with the one pion exchange have following features : The nonmesonic decay rates of s- and p-shell hypernuclei are reproduced. The n/p ratios of the nonmesonic decay rate can be improved to the numbers as much as 0.3 - 0.4 over the wide mass region. The hypernuclear lifetimes and those A-dependence are evaluated for $A = 4$ to $A = 89$ hypernuclei and are found that the calculated lifetimes tend to be stable for the medium-heavy mass systems. The $\Delta I = 3/2$ weak decay potential is also discussed based on the boson-boson weak coupling model.

1 Introduction

Since the discovery of a hypernucleus in 1953, the weak decay studies have a long history starting with the pioneering works in 1950's and 1960's. Hypernuclear weak decay is still an active field in the nuclear physics. The reasons are, first, that there exist many important problems to be solved in the mesonic and nonmesonic decay mechanism. In the mesonic decay, the $\Delta I = 1/2$ rule is empirically established but its foundation is not clarified yet from the more basic theory. In the nonmesonic decay the interaction is not well understood, though this weak decay process takes predominance over the mesonic one except for the very lightest hypernuclei. The second reason of the activeness of this field lies in the developments in the experimental techniques and facilities. Activities at BNL, KEK, CEBAF, DAΦNE and others are the principal driving force in this field.

Recently hypernuclear lifetimes and nonmesonic and mesonic decay rates have been measured systematically for light to medium-heavy hypernuclei.[1,2,3,4] The striking features are observed that the lifetimes are almost stable for medium-heavy hypernuclei ($A \geq 12$) and the ratios of the neutron-stimulated nonmesonic decay rate to the proton-stimulated one seem generally larger than 1. These data might suggest that the decay interaction could be short-ranged and the weak transitions to the final $I = 1$ states would dominate. In this report we discuss the nonmesonic decay rates and the hypernuclear lifetimes in terms of the meson exchange model.

The $\Delta I = 3/2$ weak decay interaction is also discussed in the weak boson-boson coupling model.

2 Meson exchange model and hypernuclear lifetimes

The one-pion exchange is a good starting point to discuss the nonmesonic decay and its interaction was first applied to the nuclear matter by Adams [5] and later to the finite hypernuclear system by Bandō et al.. [6] Afterward the model was extended to include the full one-boson exchange. The pseudoscalar and vector octet mesons were considered by Parreño et al. [7] and Dubach et al.[8] The present status of the studies are : (1) The one-pion exchange alone is insufficient, especially to explain the large Γ_n/Γ_p ratios observed. (2) As for the full one-boson exchanges, the contributions of ρ and K^*-exchange are dramatic but the total effect does not alter much the results of one-π exchange potential alone.

The two-pion exchange has been considered by Shmatikov [9] and by us. [10] The purposes of introducing the correlated-2π exchange are :

1. One is to see the contribution of the correlated-2π exchange as the next order to the one-boson(π) exchange.

2. Another is to find such a potential that contains the tensor force whose sign is opposite to that of V_π and also to find a potential which has a strong central force at the intermediate range.

We introduce the two types of correlated 2π exchange potentials. [10] One is a 2 pion/rho exchange potential in which 2 pions are correlated and coupled to ρ in the exchange process. The other is a 2 pion/sigma exchange potential. We take a nucleon(N) and/or a sigma(Σ) as the intermediate baryon propagating in between the virtual two pion exchange in the diagrams. Thus the weak vertex corresponds to the $\Lambda N\pi$ and/or $\Sigma N\pi$ for which the $\Delta I = 1/2$ is assumed. We call the type (A) and (B) for the diagram having the N and Σ in the intermediate baryon state, respectively.

The weak coupling constants are determined from the experimental data of mesonic decay rates and asymmetry parameters of Λ and Σ hyperons [11] and with a help of Lee-Sugawara relation. [12,13] The strong coupling constants are basically taken from the work of one-boson exchange NN potentials in Ref. [14] For the $\pi\pi\rho$ coupling constant we adopt the similar value as that determined from the $\rho \to \pi\pi$ decay [15] or those in Ref.[16,17] It is noted that the loop integral for the intermediate momentum in the 2 pion/rho exchange diagram diverges and therefore the monopole-type form factors with cut-off mass are introduced to have the integral to be converged. As for the $\pi\pi\sigma$ coupling constant it is unknown. We determine the coupling strength so that the NN potential obtained from the 2 pion/sigma exchange with a strong $NN\pi$ vertex in place of the weak $\Lambda N\pi$ vertex can simulate the OBE NN potential due to the σ-meson exchange as good as possible, but we modify the strength later.

The $2\pi/\rho$ exchange potential has the following spin-isospin dependent components :

$$\begin{aligned} V_{2\pi/\rho(A,B)} = \{ & V_{\rm S}(r) + V_{\sigma\sigma}(r)\boldsymbol{\sigma}_1 \cdot \boldsymbol{\sigma}_2 + V_{\rm T}(r)S_{12} + V_{\rm LS}(r)\boldsymbol{L}\cdot\boldsymbol{S} \\ & + V_{\rm ALS}(r)(\boldsymbol{\sigma}_1 - \boldsymbol{\sigma}_2)\cdot\boldsymbol{L} + V_{\rm V1}(r)(\boldsymbol{\sigma}_1 \times \boldsymbol{\sigma}_2)\cdot\hat{\boldsymbol{r}} \end{aligned}$$

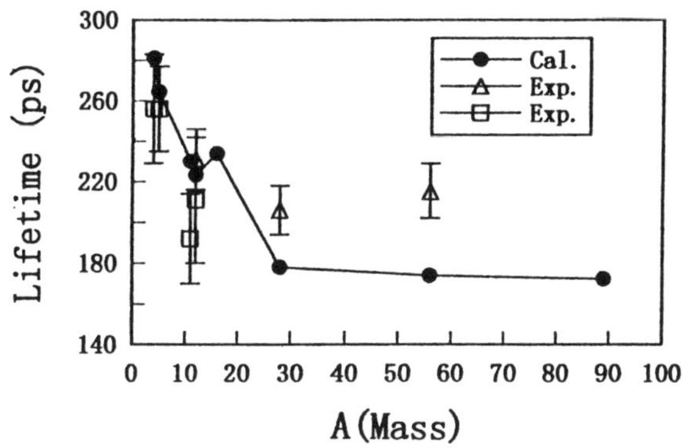

Figure 1: Calculated hypernuclear lifetimes and the experimental data from Refs.1,3,4.

$$+ iV_{V2}(r)\boldsymbol{\sigma}_1 \cdot \hat{\boldsymbol{r}}\} \mathcal{O}_I \ , \tag{1}$$

where $\mathcal{O}_I$ denotes the isospin operator. The type (A) and (B) potentials work additively for the parity-conserving part. It is characteristic that the tensor component is strong enough and its sign is opposite to that of V_π. The $2\pi/\sigma$ exchange potential produces the strong central force at the intermediate range, $r \sim 1$ fm.The potential is expressed as

$$V_{2\pi/\sigma(A,B)} = \{V_S(r) + V_{LS}(r)\boldsymbol{L} \cdot \boldsymbol{S} + iV_V(r)\boldsymbol{\sigma}_1 \cdot \hat{\boldsymbol{r}}\} \mathcal{O}_I \ . \tag{2}$$

Those features of the tensor and central components of the correlated 2π exchange potentials are what we have expected. This is verified in the calculated nonmesonic decay rates of ${}^{12}_{\Lambda}$C, for instance. If one considered the V_π and the correlated 2π exchange potentials, the enhanced Γ_{nm} comparable to the experiment and the improved Γ_n/Γ_p value as much as ~ 0.3 are obtained, though the calculated n/p ratio is still far from the data.

Then we apply these potentials to evaluate the nonmesonic decay rates of light to medium-heavy hypernuclei. The hypernuclei consided here are : ${}^{4,5}_{\Lambda}$He, ${}^{11}_{\Lambda}$B, ${}^{12}_{\Lambda}$C, ${}^{16}_{\Lambda}$O, ${}^{28}_{\Lambda}$Si, ${}^{56}_{\Lambda}$Fe and ${}^{89}_{\Lambda}$Y. Adopted wavefunctions are the following. For the light s- and p-shell hypernuclei (nuclei), the configuration-mixed shell model wavefunctions are employed. The DDHF solutions are, however, used for the Λ wavefunctions to ensure the proper binding effect. For the medium-heavy systems the DDHF and WS solutions are employed for the nucleon and Λ particles. The short range correlations with the tensor coupling both in the initial and the final states are properly considered. Calculated nonmesonic decay rates are increasing with A, but are going to be saturated gradually. The Γ_n/Γ_p ratios are calculated in the range 0.3 - 0.4 over the wide mass system considered.

The hypernuclear lifetimes are evaluated by combining the nonmesonic decay rates and the mesonic decay rates ; $1/\tau = \Gamma_\pi + \Gamma_{nm}$. For the mesonic decay rates we use the results of our previous work. [18] Figure 1 shows the calculated lifetimes together with the experimental data [1,3,4] as a function of the mass A. Evaluated lifetimes tend to be stable but are a bit shorter than the data for the medium-heavy mass systems. The calculations are carried out by employing the DDHF wavefunctions for nucleon and Λ for the medium-mass system.

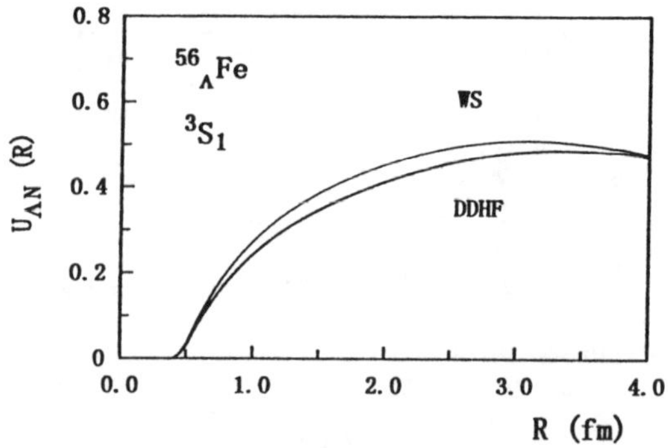

Figure 2: The Λ-N $^1S_0$ relative wavefunctions of DDHF and WS solution in the case of $^{56}_{\Lambda}$Fe

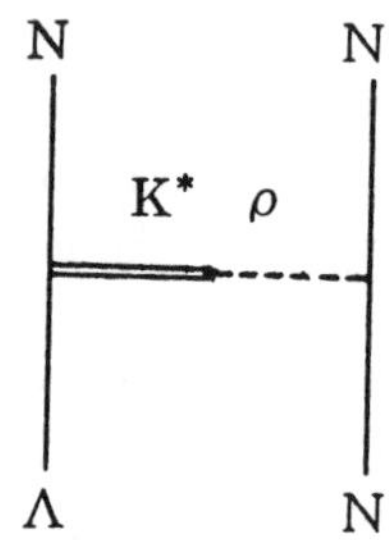

Figure 3: The K^*- ρ exchange diagram in the $\Lambda N \to NN$ process

Calculated nonmesonic decay rates and lifetimes are sensitive to how the Λ and nucleon wavefunctions extend. Figure 2 shows the different behaviour of the Λ-nucleon relative waves for the case of DDHF and WS solutions for $^{56}_{\Lambda}$Fe. Note that the weak decay interaction extends up to $r \sim 1.5$ fm. It turns out that the WS solutions, having large amplitude inside in the present case, produce larger nonmesonic decay rates and thus shorter lifetimes compared with those calculated in the DDHF solutions.

In view of the underestimates of the lifetimes in calculations, a shorter-range decay interaction seems preferable in explaining the experimental data.

3 $\Delta I = 3/2$ weak decay potential

The possibility of the contribution of $\Delta I = 3/2$ interaction to the nonmesonic decay is widely discussed. [19,20,21] We present here the weak $\Lambda N \to NN$ interaction with $\Delta I = 3/2$ based on the boson-boson coupling model.

The Weinberg-Salam or Cabibbo theory tells that K can decay into π and K^* can decay into ρ through the strangeness changing weak processes which involve both $\Delta I = 1/2$ and $3/2$. We mainly discuss the possible $\Delta I = 3/2$ weak potential through such $K\pi$ or $K^*\rho$ processes as depicted in Fig.3.

The $\Delta I = 3/2$ weak hamiltonian is assumed, for the K^*-ρ coupling, in the form,

$$H^{w(\Delta I=3/2)} = g^{w(\Delta I=3/2)}_{K^*\rho} \bar{\phi}_{K^*} \boldsymbol{T} \cdot \boldsymbol{\phi}_\rho \, S \,, \tag{3}$$

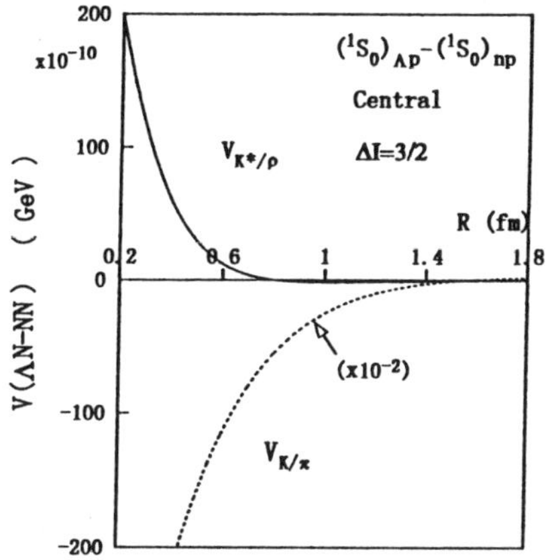

Figure 4: The $\Delta I = 3/2$ weak decay potentials due to the K^*/ρ and K/π exchange in the ${}^1\mathrm{S}_0 \to {}^1\mathrm{S}_0$ channel

where $\mathcal{S}$ is a spurion selecting the third component in the isospin 3/2 spinor and $\boldsymbol{T}$ is a transition operator for $3/2 \to 1/2$ in isospace. The weak coupling constant $g_{K^*\rho}^{w(\Delta I=3/2)}$ is determined as follows. In the boson-boson coupling picture the $\rho^0 \to K^{*0}$ weak transition is written as

$$H_{K^{*0}\rho}^{w} = g_{K^{*0}\rho}^{w} \bar{\phi}_{K^{*0}} \phi_{\rho^0} \ . \tag{4}$$

While in the Weinberg-Salam electroweak theory, the weak $\rho^0 \to K^{*0}$ transition is also expressed in the hamiltonian with the weak coupling constant G_F and the V-A currents having quark-degrees of freedom. This quark hamiltonion is reduced in the non-relativistic form. Imposing the equality condition for the matrix elements of the boson hamiltonian $H_{K^{*0}\rho}^{w}$ and the quark hamiltonian in between the K^{*0} and ρ^0 mesons, we obtain the relation for determining $g_{K^{*0}\rho^0}^{w}$. Then the $g_{K^{*0}\rho^0}^{w}$ can be expressed in terms of G_F and the wavefunction values of ρ^0 and K^{*0} at the origin ($\boldsymbol{r}$ =0) and these are estimated from the experimental data such as the decay width of $\rho^+ \to \mu^+\nu$ and $< r^2 >_{exp}$ of K^+. By carring out the isospin decomposition of the $\rho^0 \to K^{*0}$ transition, the following relations are obtained.

$$g_{K^*\rho}^{w(\Delta I=3/2)} = \sqrt{\frac{2}{3}} g_{K^{*0}\rho^0}^{w} \ , \tag{5}$$

$$g_{K^*\rho}^{w(\Delta I=1/2)} = -\frac{1}{3} g_{K^{*0}\rho^0}^{w} \ . \tag{6}$$

The potential $V_{K^*/\rho}^{(\Delta I=3/2)}$ of the K^*/ρ exchange with $\Delta I = 3/2$ due to Fig. 3 is parity conserving and has central, tensor, LS and ALS components. The potential is shown in Fig. 4 for ${}^1S_0 \to {}^1 S_0$ channel, which is central type.

The features of the potential are: (1) It is rather weak compared with the 1π and the correlated 2π exchange potentials with $\Delta I = 1/2$. (2) It is short ranged. (3)

The relative sign between $V_{K^*/\rho}^{(\Delta I=3/2)}$ and V_π cannot be determined. The $V_{K/\pi}^{(\Delta I=3/2)}$ potential is also evaluated in the similar way. Our work on the weak decay potential with $\Delta I = 3/2$ is very preliminary, but we may say

$$| V_{K/\pi}^{(\Delta I=3/2)} | \ll | V_{K^*/\rho}^{(\Delta I=3/2)} | \tag{7}$$

and

$$| V_{K/\pi}^{(\Delta I=3/2)} | \ll | V_{\pi}^{(\Delta I=1/2)} | \ . \tag{8}$$

The role of K^*/ρ and K/π-exchange potentials with $\Delta I = 3/2$ on the decay rates is under investigation. The parity violating $\Delta I = 3/2$ potentials can be similarly evaluated.

References

1. R. Grace et al., *Phys. Rev. Lett.* **55**, 1055 (1985) ; J.J. Szymanski et al., *Phys. Rev.* C **43**, 849 (1991).
2. H. Noumi et al., *Phys. Rev.* C **52**, 2936 (1995).
3. H. Outa et al., *Nucl. Phys.* A **639**, 251c (1998).
4. H. Bhang et al., *Phys. Rev. Lett.* **81**, 4321 (1998).
5. J.B. Adams, *Phys. Rev.* **156**, 1611 (1967).
6. K. Takeuchi, H. Takaki and H. Bandō, *Prog. Theor. Phys.* **73**, 841 (1985).
7. A. Parreño, A. Ramos and C. Bennhold, *Phys. Rev.* C **56**, 339 (1997).
8. J.F. Dubach, G.B. Feldman, B.R. Holstein, L. de la Torre, *Ann. Phys.* (N.Y.) **249**, 146 (1996).
9. M. Shmatikov, *Nucl. Phys.* A **580**, 538 (1994).
10. K. Itonaga, T. Ueda and T. Motoba, *Nucl. Phys.* A **585**, 331c (1995) ; *Weak and Elecromagnetic Interactions in Nuclei*, ed. H. Ejiri, T. Kishimoto and T. Sato, 546 (World Scientific, Singapore, 1995) ; *Nucl. Phys.* A **639**, 329c (1998).
11. Particle Data Gruop, *Phys. Rev.* D **54**, 1 (1996).
12. B.W. Lee, *Phys. Rev. Lett.* **12**, 83 (1964).
13. H. Sugawara, *Prog. Theor. Phys.* **31**, 213 (1964).
14. T. Ueda, F.E. Riewe and A.E.S. Green, *Phys. Rev.* C **17**, 1763 (1978).
15. T. Ericson and W. Weise in *Pions and Nuclei* (Oxford, 1988).
16. H. Nielsen and G.C. Oaders, *Nucl. Phys.* B **55**, 301 (1973).
17. N. Hedelgard-Jensen, *Nucl. Phys.* B **77**, 173 (1974).
18. T. Motoba and K. Itonaga, *Prog. Theor. Phys.* Suppl. **117**, 477 (1994).
19. K. Moltman and M. Shmatikov, *Phys. Lett.* B **331**, 1 (1976) ; *Phys. Rev.* C **51**, 1576 (1995).
20. M. Oka, T. Inoue and S. Takeuchi, *Properties and Interactions of Hyperons*, ed. B.F. Gibson, P.D. Barnes and K. Nakai, 119 (World Scientific, Singapore, 1994) ; T. Inoue, S. Takeuchi and M. Oka, *Nucl. Phys.* A **597**, 563 (1996).
21. A. Parreño, A. Ramos, C. Bennhold and K. Maltman, *Nucl. Phys.* A **639**, 333c (1998).

The π^+ DECAY OF LIGHT HYPERNUCLEI

B. F. GIBSON
Theoretical Division, Los Alamos National Laboratory
Los Alamos, NM 87545, USA
email: gibson@t5.lanl.gov

The observed π^+ emission from the weak decay of the ${}^4_\Lambda$He hypernucleus has been an intriguing puzzle for more than 30 years, because the Lambda decays in free space only by emission of a π^- or a π^0. We re-examine this puzzling weak decay with our focus upon a decay mechanism involving the $\Sigma^+ N \to \pi^+ nN$ decay of a virtual Σ^+, stemming from ΛN to ΣN conversion (mixing) within the hypernucleus. We emphasize the observed energy distribution of the observed π^+s compared to that of π^-s in standard mesonic decay as well as the isotropic angular distribution of the π^+s. Competing suggestions to explain the positive pion weak decay have been offered. A possible search for π^+ decay from the other Λ hypernuclei is explored as means to test our hypothesis.

1 Introduction

Hypernuclei introduce a third dimension into our investigation of nuclear structure and the strong baryon-baryon force. Strangeness represents a nontrivial additional degree of freedom. Injecting one or more hyperons into the nucleus creates nuclear systems that exhibit novel aspects which do not appear in conventional, nonstrange nuclei: anomalous binding energies, new dynamical symmetries, non mesonic weak decay. Moreover, unraveling the SU(3) nature of the baryon-baryon interaction is a basic challenge to our developing a fundamental understanding of the nuclear force.

Hypernuclear physics provides a test of whether the sophisticated models that we create from our experience in the conventional (non strange) sector, models that describe the structure of many-body systems of neutrons and protons and the reactions which produce them, are capable of extrapolation beyond that $S = 0$ realm in which they were developed. Or are these models merely exquisite interpolation tools? Does the successful picture of few-nucleon systems in terms of the physically observable nucleons and mesons (meson exchange forces) extrapolate to the world of hypernuclei? What are the relevant degrees of freedom, the proper constituents with which to describe the observed experimental phenomena? What are the forces that act between those constituents?

Strangeness physics is more than a simple extension based on conventional $S = 0$ phenomena. The physics is truly different. One contributing factor is $\Lambda \leftrightarrow \Sigma$ conversion, which leads to significant ΛN-ΣN mixing. The few-body hypernuclear data [1,2,3] suggest that ΛN-ΣN coupling (mixing) is important. ${}^3_\Lambda$H would almost certainly be unbound in its absence.[4,5] Charge symmetry breaking (CSB) in the $A =$ 4 isodoublet is clear: ${}^4_\Lambda$He is more bound than ${}^4_\Lambda$H by $\sim$0.35 MeV, some three times the 0.12 MeV estimated for the conventional ^{3}H-^{3}He isodoublet after accounting for the $A = 3$ Coulomb energy difference. The fact that the Λp interaction is stronger than the Λn interaction arises from a CSB in the hyperon-nucleon interaction; the CSB difference can be understood in part as due to the $\Sigma^+ - \Sigma^-$ mass difference. (The Σ mass difference is 8 MeV or some 10% of the 80 MeV $\Lambda - \Sigma$ mass difference.)

The Λp couples to Σ^+n, whereas the Λn couples to Σ^-p. Finally, the anomalously small binding of ${}^5_\Lambda$He seems also related to ΛN-ΣN mixing. Let us compare Λ separation energies for ${}^3_\Lambda$H, ${}^4_\Lambda$He, and ${}^5_\Lambda$He. Simple, central-force potentials fitted to the low energy ΛN scattering parameters,

$$V^{eff}_{\Lambda N} \sim a_{\Lambda N}, r_{\Lambda N},$$

which reproduce the Λ separation energy

$$B_\Lambda({}_\Lambda \mathrm{A}) = B({}_\Lambda \mathrm{A}) - B(\mathrm{A}-1) \tag{1}$$

of $B_\Lambda({}^4_\Lambda \mathrm{H}) = 2.04(4)$ MeV, *predict*:

- $B_\Lambda({}^3_\Lambda \mathrm{H})$ too large[4,6,7] by a factor of 5,
- $B_\Lambda({}^5_\Lambda \mathrm{H})$ too large[8,9,10,11] by a factor of 2.

From the available studies, one can infer that ΛN-ΣN mixing (coupling) is an important aspect of modeling the hyperon-nucleon interaction and that it plays an important role in the binding of light hypernuclei.

Can the observed π^+ decay of Λ hypernuclei be a physical signature of the presence of a virtual Σ in hypernuclei?

2 The π^+ Decay Puzzle

Unexpected observations such as the π^+ decay[12,13,14] of hypernuclei illustrate why hypernuclear physics goes well beyond traditional neutron-proton phenomena. The π^+ emission from the weak decay of ${}^4_\Lambda$He has been a puzzle to theorists for more than 30 years. The ratio of π^+ decay to π^- decay for ${}^4_\Lambda$He is defined as

$$R(\pi^+/\pi^-) = \frac{\Gamma({}^4_\Lambda \mathrm{He} \to \text{ all } \pi^+ \text{ modes})}{\Gamma({}^4_\Lambda \mathrm{He} \to \text{ all } \pi^- \text{ modes})}\,. \tag{2}$$

Results from Mayeur *et al.*[12] and from Bohm *et al.*[13] lie within the range

$$5.4^{+1.5}_{-1.7}\% \le R(\pi^+/\pi^-) \le 6.9^{+1.8}_{-2.1}\%\,. \tag{3}$$

The measurement coming from the bubble chamber study by Keyes *et al.*[14] yielded a value of $R(\pi^+/\pi^-) = 4.3(1.7)\%$. Thus, we see that experimentally the π^+ to π^- ratio for ${}^4_\Lambda$He decay seems to fall in the range of 3-7%. Furthermore, no unambiguous π^+ events have been observed from other hypernuclei. Because the mesonic decay modes of the free Λ

$$\Lambda \to p + \pi^- \quad \text{and} \quad \Lambda \to n + \pi^0,$$

which are first cousins to the well known neutron β decay, produce only π^-s and π^0s, it would appear that more complex mechanisms must be responsible for the π^+s in the decay of the Λ bound in ${}^4_\Lambda$He.

As noted above, the strangeness -1 conversion process $\Lambda \leftrightarrow \Sigma$, which produces strong mixing of the ΛN and ΣN channels, appears to have an important effect on the binding of the light Λ hypernuclei. It seems that (octet-octet) ΛN-ΣN coupling plays a much more significant role in hypernuclei than does the (octet-decuplet) coupling of the nucleon to the $\Delta(1232)$ in conventional nuclear physics. The Σ is stable with respect to the strong interaction, whereas the Δ decays strongly into a nucleon and a pion. Moreover, the mass difference between the Λ and Σ is only 80 MeV, much less than the NΔ mass difference. Finally, the octet-decuplet coupling in the nonstrange sector (NN-ΔN) can be at least partially subsumed in the one-boson-exchange (OBE) model of the NN interaction, because of duality, without the requirement of explicit Δs in the model. Our proposal[15] for a possible solution to the π^+ decay puzzle involves the virtual Σ^+ component of the ${}^4_\Lambda$He wave function.

Dalitz and von Hippel[16,17,18] explored several issues in depth, when the π^+ decay phenomenon was discovered. They considered decay processes of the type:

- $\Lambda \to \pi^0 + n$ decay followed by a $\pi^0 + p \to \pi^+ + n$ charge–exchange reaction,
- $\Sigma^+ \to \pi^+ + n$ decay following a $\Lambda + p \to \Sigma^+ + n$ conversion.

Dalitz and von Hipple concluded that the charge exchange process could account for no more than about a 1/2% π^+ decay rate. Von Hippel then suggested[18] that an s-wave Σ^+ decay might yield a sufficiently high rate to explain the data. However, Dalitz pointed out[19] that the subsequent experimental observation identifying $\Sigma^+ \to \pi^+ + n$ decay as a p-wave process ruled out von Hipple's promising explanation. Recently, Cieplý and Gal[20] re-examined the charge-exchange contribution to π^+ emission in ${}^4_\Lambda$He decay. They concluded that, even though their improved calculation with up-to-date input parameters yields a 1.2% branching ratio, essentially a factor of 2 larger than that originally estimated by Dalitz and von Hippel, the charge-exchange mechanism alone cannot account for the experimental value of some 3-7%. Nonetheless, 1% would be a significant contribution.

3 The Experimental Pion Spectra

Sacton reviewed[21] the experimental situation in detail in 1966. His cogent summary includes the results from Mayeur *et al.*[12] and from Gajewski *et al.*[22]. The data are reproduced in Fig. 1 in which we present the pion kinetic energy spectra for the following four decay processes:

$$\begin{aligned}
{}^4_\Lambda\text{He} &\to \pi^- + p + {}^3\text{He} \quad (a)\\
{}^4_\Lambda\text{He} &\to \pi^+ + n + {}^3\text{H} \quad (b)\\
{}^4_\Lambda\text{H} &\to \pi^- + n + {}^3\text{He} \quad (c)\\
{}^4_\Lambda\text{H} &\to \pi^- + p + {}^3\text{H} \quad (d)
\end{aligned}$$

The π^- spectrum arising from the ${}^4_\Lambda$He decay into p + ^{3}He, process (a), is peaked near a kinetic energy of some 30 MeV, similar to what one might expect for π^-s coming from $\Lambda \to \pi^- +$ p decay in free space. However, the energy distribution tails down to 15 MeV. The π^- spectrum arising from the decay of ${}^4_\Lambda$H into $p+$ ^{3}H, process (d), is peaked in a similar manner. Most of the strength from the ${}^4_\Lambda$H decay lies in the analog mode[13], ${}^4_\Lambda\text{H} \to \pi^- + {}^4\text{He}$, so that the spectrum arising from process (d) contains many fewer events than that arising from (a). (Krecker *et al.*[23] provide a later discussion of the π^- decay data.) Our interpretation of (a) is that the process is one dominated by $\Lambda \to \pi^- + p$ decay embedded within a very

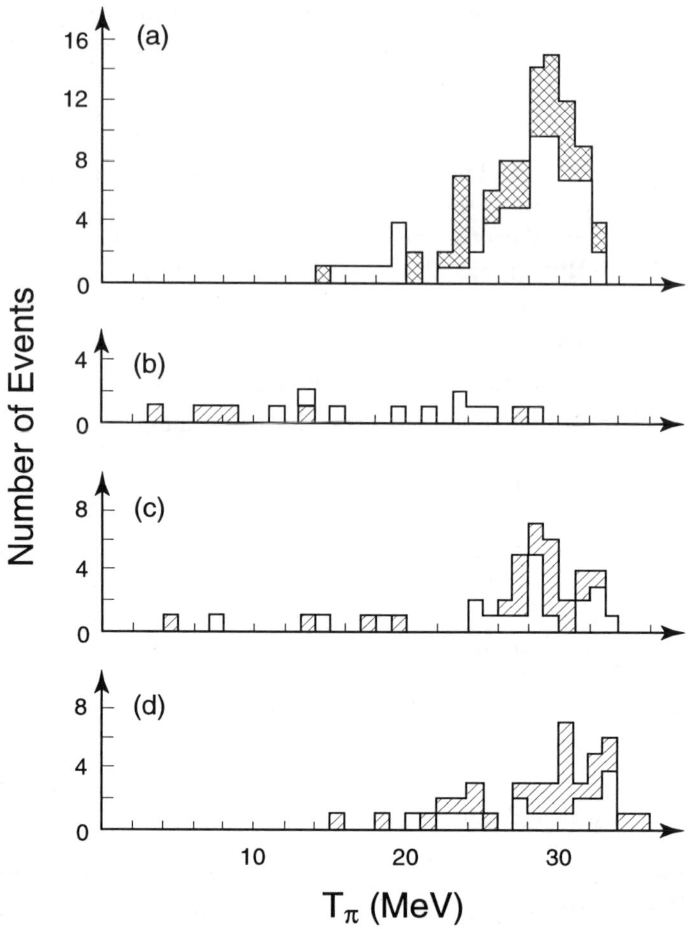

Figure 1: The $\pi^{\pm}$ kinetic energy distribution from the decays: (*a*) ${}^4_\Lambda\mathrm{He} \to \pi^- + \mathrm{p} + {}^3\mathrm{He}$, (*b*) ${}^4_\Lambda\mathrm{He} \to \pi^+ + \mathrm{n} + {}^3\mathrm{H}$, (*c*) ${}^4_\Lambda\mathrm{H} \to \pi^- + \mathrm{n} + {}^3\mathrm{He}$, and (*d*) ${}^4_\Lambda\mathrm{H} \to \pi^- + \mathrm{p} + {}^3\mathrm{H}$, as illustrated in Ref. [21].

light nucleus, so that Fermi smearing of the peak is limited. The low-energy side of the spectrum would appear to come from final-state rescattering. The possible three-body $\Lambda + \mathrm{N} \to \pi^- + \mathrm{p} + \mathrm{N}$ decay mechanism is less likely to result a residual bound trinucleon. Specific decay events involving two protons in the final state have actually been identified [13]. However, were π^-s from three-body decay processes of consequence, one would expect to see π^- events with kinetic energies of less than 15 MeV, which are not readily apparent in Fig. 1*a*.

The π^- kinetic energy spectra for the decay of ${}^4_\Lambda\mathrm{H}$ into $\mathrm{n} + {}^3\mathrm{He}$, process (*c*), and into $\mathrm{p} + {}^3\mathrm{H}$, process (*d*), are statistically equivalent. Moreover, they are quite similar in appearance to that for (*a*). They exhibit a peak in the region corresponding to the $\Lambda \to \pi^- + \mathrm{p}$ free decay peak. In addition, there exist four events at lower energy, below 15 MeV. We interpret the two-body ($\Lambda \to \pi^- + \mathrm{p}$) decay appearance as coming from

$$^4_\Lambda\mathrm{H} \to \pi^- + {}^4\mathrm{He}^* \quad (e)$$

where the $T = 0$ states of ^{4}He* decay equally into n + ^{3}He and p + ^{3}H. It is clear from Sacton's Table I that the two-body decay $^4_\Lambda$H $\to \pi^-$ + ^{4}He generates some ten times the number of events that is produced by the three-body decay of $^4_\Lambda$H $\to \pi^- + p +$ ^{3}H. Therefore, we infer that

- the decay modes (c) and (d) are closely related, as is implied by the number of events for each in Fig. 1,
- such decays come primarily from the decay of the $T = 0$ ^{4}He excited states, following π^- emission.

These conclusions about strong final-state interactions dominating in processes (c) and (d) are supported by the study in Ref. [24], where it is argued that a naive calculation not taking into account resonant final states fails. Nonetheless, the low-energy π^-s most likely do result from multi-nucleon final states. We will return to this point in the discussion below.

The picture drawn here does not contradict the physics of process (a), where the π^- decay of $^4_\Lambda$He is by means of the $T = 1$ excited states of ^{4}Li. The four-nucleon interactions in the $T = 1$ state are much weaker than are the four-nucleon interactions in the $T = 0$ state. Hence, the final-state interactions should be of less consequence in process (a). For that reason the π^- spectrum is observed to be more in agreement with our naive expectation for a decay process dominated by the free $\Lambda \to \pi^- + \mathrm{p}$ process with minimal final-state rescattering.

The puzzle that we wish to explore is the $^4_\Lambda$He $\to \pi^+ + \mathrm{n} +$ ^{3}He decay mode (b). We see from Fig. 1 that, unlike the π^- decay spectra which are peaked according to expectations for a two-body ($\Lambda \to \pi^- + p$) decay, the π^+ spectrum from the decay $^4_\Lambda$He $\to \pi^+$ is uniform in terms of the π^+ energy distribution. Furthermore, it is stated in Ref. [12] that multi–neutron final states are likely for π^+ kinetic energies below 22 MeV. Therefore, the label "^{3}H" in process (b) and in the caption for Fig. 1b should be interpreted as "^{3}H or $n +$ ^{2}H or $n + n +$ ^{1}H." This is clearly stated in Ref. [14], where the final state is identified as "nnnp." Moreover, there is a paucity of events for pion kinetic energies in the allowed region for four-nucleon decay. The low-energy π^-s seen in the π^- spectra in Fig. 1 are likely to result from similar multi-nucleon final states. Examination of the more complete π^+ decay data compilation in Ref. [13] confirms the flat character of the π^+ spectrum. In addition, Keyes *et al.* [14] suggest that the data indicate the π^+ emission process is predominantly s–wave. Hence, we conclude there is no evidence that a two-body decay of the virtual Σ^+, such as $\Sigma^+ \to \pi^+ + \mathrm{n}$, can account for the π^+ spectrum.

4 Theoretical Considerations

Dalitz and von Hippel examined a number of second-order processes which might explain the π^+ weak decay of $^4_\Lambda$He. The possibility that charge exchange $\pi^0 + \mathrm{p} \to \pi^+ + \mathrm{n}$ following $\Lambda \to \pi^0 + \mathrm{n}$ decay is an obvious candidate. However, this was estimated [17] to provide at most a 0.6% decay rate probability. Moreover, Dalitz argued that $\Lambda + \mathrm{p} \to \pi^- + \mathrm{p} + \mathrm{p}$ and $\Lambda + \mathrm{n} \to \pi^- + \mathrm{n} + \mathrm{p}$ decays, which were considered by Dalitz and Rajasekjaran[16], do not appear to significantly contribute to the π^- decay spectra for pion kinetic energies below 15 MeV, the kinetic energy region in which a significant number of the π^+ events lie. Thus, π^+ emission would need to occur from a different three-body mechanism. Otherwise it would be ruled out by the π^- decay spectrum observed in the region of pion kinetic energies where three-body decay processes are expected to dominate. Such a virtual process ($\Lambda + \mathrm{p} \to \Sigma^+ + \mathrm{n} \to \pi^+ + \mathrm{n} + \mathrm{n}$) was investigated by von Hippel. However, his estimate[18] indicated that $\Sigma^+ \to \pi^+ + \mathrm{n}$ decay contributes only some 0.2% to the π^+/π^- ratio, *when* the decay $\Sigma^+ \to \pi^+ + \mathrm{n}$ proceeds via a relative *p*-state in the $\pi^+\mathrm{n}$ system, as was experimentally observed to be the case for free Σ^+ decay. The numerical estimate of Von Hippel was based upon an assumed 10 MeV π^+ kinetic energy in his closure approximation. The mean π^+ kinetic energy is actually about 18 MeV, which whould raise von Hippel's estimate by about a factor of two. That is still too small an effect to account for the experimental observations.

Nonetheless, the large virtual Σ^+ probability is unique to the ${}^4_\Lambda$He hypernucleus. The ${}^4_\Lambda$He wave function can be described schematically as

$$|\,{}^4_\Lambda\mathrm{He}\rangle = \alpha\,|\Lambda\otimes {}^3\mathrm{He}\rangle + \beta\,(-\sqrt{\frac{1}{3}}\,|\Sigma^0\otimes {}^3\mathrm{He}\rangle + \sqrt{\frac{2}{3}}\,|\Sigma^+\otimes {}^3\mathrm{H}\rangle)\ . \qquad (4)$$

Because of charge excess in the ^{3}He nuclear core, the Λ in ${}^4_\Lambda$He can make a virtual transition to a Σ^+ without altering the structure of the ^{3}He core, whereas one would anticipate only analogous Σ^- (and Σ^0) transitions in ${}^4_\Lambda$H as well as other Λ hypernuclei without a proton excess. This suggests that the $\Lambda + \mathrm{p} \to \Sigma^+ + \mathrm{n}$ transition may be the key to understanding the π^+ emission.

In addition, we interpret the flat *s*-wave π^+ spectrum as evidence for a *three-body* decay mechanism of the type $\Sigma^+ + \mathrm{N} \to \pi^+ + \mathrm{n} + \mathrm{N}$. The virtual $\Sigma^+\,\mathrm{N}$ system is "off–shell." There must be a $\Sigma^+ + \mathrm{N} \to \pi^+ + \mathrm{n} + \mathrm{N}$ rescattering process to restore the system to the "on-shell" situation and "free" the observed π^+. One would then expect both nucleons to carry away kinetic energy. That would produce a pion spectrum essentially of uniform distribution over the allowed pion kinetic energy range – from zero to the maximum allowed by an $\mathrm{n} + \mathrm{n} + {}^2\mathrm{H}$ final state. In the process one of the two neutrons in the $\pi^+\mathrm{nn}$ rescattering could be "picked up" by the spectator deuteron to form a triton. Alternatively, the proton in the $\pi^+\mathrm{np}$ rescattering process could be picked up by the spectator di-neutron to form a triton. However, if the ^{3}H final state were to play a significant role in the π^+ decay process, one would expect to see a significant number of π^+ events above the threshold for nn ^{2}H in Fig. 1*b*. That is not the case. Therefore, we conclude[15] that a three-body decay amplitude of the $\Sigma^+ + \mathrm{N} \to \pi^+ + \mathrm{n} + \mathrm{N}$ type, normalized to the theoretically estimated Σ^+ probability in ${}^4_\Lambda$He, could explain both the π^+ decay branching ratio and the isotropic angular distribution as well as the flat energy distribution of the π^+s.

One estimate of the Σ^+ probability in ${}^4_\Lambda$He comes from a model calculation[10] for the ${}^4_\Lambda$He-${}^4_\Lambda$H system. The Σ probability β^2, defined in Eq. (4) was found to be as large as 14%. Such estimates carry a large uncertainty, even in microscopic calculations with sophisticated contemporary potentials, as was shown for the hypertriton where $\mathrm{P}(\Sigma)$ varies by a factor of 10 for different hyperon-nucleon potential models.[5]

Nonetheless, we assume for purposes of illustration the Σ^+ probability in ${}^4_\Lambda$He to be

$$P(\Sigma^+) = \frac{2}{3}\,|\beta|^2 = \frac{2}{3} \times 14(7)\% = 9(5)\% \,. \tag{5}$$

In order to translate this Σ^+ probability into an estimate for the branching ratio $R(\pi^+/\pi^-)$, we should take into account suppression effects due to the Pauli principle for the in-medium decay rates compared to the Λ and Σ^+ lifetimes in vacuum. According to Dalitz and Liu [25] the π^- decay rate of ${}^4_\Lambda\text{He} \to \pi^- + \text{p} + {}^3\text{He}$, which has three protons in the final state, is strongly Pauli suppressed with respect to the $\Lambda \to \pi^- + \text{p}$ decay rate in vacuum. They esimate a suppression to ~40-45% of the free decay rate for the s-channel and to ~30-35% for the p-channel. Also, Dalitz and von Hippel [17] estimated that, in comparison with π^- decay, Pauli suppression is perhaps twice as strong in the π^+ decay of ${}^4_\Lambda$He. The reason for this is that π^+ decay requires both the Λ and a proton to reside inside the nuclear core, whereas in the case of π^- decay, which involves only the Λ, the decay can take place outside the nuclear core. We use this value to estimate the relative importance of Pauli suppression. However, we note that Coulomb repulsion, which was not considered by Dalitz and von Hippel, does favor π^+ decay (three neutrons in the final state) over π^- decay (three protons in the final state).

The $\Sigma^+ + \text{N} \to \pi^+ + \text{n} + \text{N}$ transition is assumed to be an s-wave three-body decay. This is indicated by the experimental π^+ kinetic energy distribution, as discussed above. Additional evidence exists for this hypothesis. The virtual Σ must reside in close proximity to one of the nucleons; this is illustrated for ${}^3_\Lambda$H in Fig. 5 of Ref. [5]. That Σ^+N pair forms a tightly bound system; the binding energy is some 80 MeV. The strong correlation of the Σ^+ with a nucleon strongly suggests that the relevant decay is of the three–body type: $\Sigma^+ + \text{N} \to \pi^+ + \text{n} + \text{N}$. We assume that, except for the reduction in phase space, the Σ^+ decay rate is unmodified in the medium. That is, the Σ^+ in-medium three-body decay rate is presumed to be approximately equal to the two-body free decay rate, apart from the difference in phase space because the Σ^+ is highly virtual.

The relevant in vacuum decay ratio is

$$\Gamma(\Sigma^+ \to \pi^+ + \text{n}) \,/\, \Gamma(\Lambda \to \pi^- + \text{p}) = \frac{1}{2}\Gamma(\Sigma^+) \,/\, \frac{2}{3}\Gamma(\Lambda) \simeq 2.5 \,. \tag{6}$$

The $\Delta I = 1/2$ rule, which is well satisfied experimentally, was assumed. The appropriate phase space provides an additional factor of 70/185, which is the ratio of the average π^+ momentum for the in-medium π^+ decay of ${}^4_\Lambda$He to the value of the average π^+ momentum from the $\Sigma^+ \to \pi^+ + \text{n}$ decay in vacuum. Combining these factors, we obtain an estimate for the three–body virtual Σ^+ decay contribution to the value of the π^+ to π^- branching ratio

$$R(\pi^+/\pi^-) \simeq \frac{1}{2} \times \frac{P(\Sigma^+)}{1 - P(\Sigma^+)} \times 2.5 \times \frac{70}{185} = 5(3)\% \,; \tag{7}$$

the additional 1/2 factor comes from the (relative) Pauli suppression effect outlined above. In addition, we assume that the decay via the virtual Σ^0 component is counted primarily in the π^- decay rate. The principal uncertainty in our estimate arises from the uncertainty in the initial estimate for the Σ^+ probability $P(\Sigma^+)$.

The final result $R(\pi^+/\pi^-) = 5(3)\%$ should be compared with the estimate of 0.2% given by von Hippel[18] (or the 0.4% using the higher average π^+ momentum), which is an order of magnitude smaller. We suggest that a π^+ to π^- branching ratio $R(\pi^+/\pi^-)$ comparable to the experimental limit is a plausible result for a model calculation that includes $\Lambda - \Sigma$ conversion, especially when combined with the charge exchange[20] process. Clearly, a more realistic model calculation, one which includes the charge-exchange channel as well as the $\Sigma^0 + \mathrm{p} \rightarrow \pi^+ + \mathrm{n} + \mathrm{n}$ decay mechanism, is required.

In order for the picture drawn here to be valid, one should explain why there is an absence of low-energy π^-s in the decay of ${}^4_\Lambda$He, shown in Fig. 1*a*. In the data depicted there, two-body decay ($\Lambda \rightarrow \pi^-$+p leading to process *a*) is by far the dominant decay mode. Observation of three-body contributions would require a larger data sample to see the few low-energy π^-s anticipated from a few % decay branch that results from the analogous Σ^0 and other three-body decay mechanisms. The ${}^4_\Lambda$He hypernucleus has no virtual Σ^- component to provide π^-s. In contrast, the ${}^4_\Lambda$H wave function should also have $P(\Sigma^-) \simeq 9(5)\%$. Therefore, for π^- decay of ${}^4_\Lambda$H, low-energy pions are expected from the reaction $\Sigma^- + \mathrm{p} \rightarrow \pi^- + \mathrm{n} + \mathrm{p}$ following the $\Lambda + \mathrm{n} \rightarrow \Sigma^- + \mathrm{p}$ virtual transition, although 3-7% would normally be a small branch to observe. However, the π^- decay of ${}^4_\Lambda$H is dominated by the analog transition to $\pi^- + {}^4$He. Indeed, the analog transition absorbs some 70% of the decay strength[3]; that is, the analog transition acts as a two-body decay filter. The remaining 30% of the π^- spectrum, which leads to the data in Fig. 1*c* and 1*d*, is sufficiently small compared to the ~5% of the π^- decays expected from the virtual Σ^- plus other three-body decay mechanisms, that the three-body decay π^-s should be seen. Indeed, a few low-energy π^- events are observed.

5 Summary

Our model of the π^+ decay of ${}^4_\Lambda$He is based upon the proton number being greater than the neutron number such that $\Lambda \leftrightarrow \Sigma$ conversion can occur without requiring a structure change in the underlying nuclear core. Extending that reasoning to other hypernuclei, we would argue that the main reason π^+ emission is not observed is the absence of a significant Σ^+ admixture, which is related to lack of an excess of positive charge in the system. An experimental test which would destroy our hypothesis would be the observation of

$$R(\pi^+/\pi^-) \gg 1\% \tag{8}$$

for either ${}^4_\Lambda$H or ${}^5_\Lambda$He. Currently there is no evidence for $R(\pi^+/\pi^-) > 0$.

We have presented a plausible explanation of the long-standing enigma involving the π^+ emission in the weak decay of ${}^4_\Lambda$He. We argue that in the spectrum of π^+s one sees the three-body decay ($\Sigma^+\mathrm{N} \rightarrow \pi^+\mathrm{nN}$) of a highly-virtual Σ^+ arising from the in-medium $\Lambda \leftrightarrow \Sigma$ conversion. This resolution of the π^+ decay puzzle is, in fact, consistent with the significant Σ^+ probability that is unique to the ${}^4_\Lambda$He hypernucleus. It also explains in a natural way the experimentally observed π^+

kinetic energy spectrum and isotropic angular distribution. Finally, the π^+ emission from ${}^4_\Lambda$He would appear to provide direct confirmation of the existence of virtual baryons within hypernuclei and the important role played by virtual transitions among members of the baryon octet, which in this case leads to ΛN-ΣN mixing.

Acknowledgments

The author wishes to acknowledge the important collaboration in this research of Rob Timmermans of the Kernfysisch Versneller Instituut, Groningen. He acknowledges the role that C. Dover played in bringing this question to his attention, and he thanks A. Gal for discussions relating to the role of the triton final state in the π^+ decay process. He also thanks the organizers of the APCTP workshop for their invitation to speak and their kind hospitality during the meeting. The author's work was performed under the auspices of the U. S. Department of Energy, contract W-7405-ENG-36.

References

1. M. Juric, G. Bohm, J. Klabuhn, U. Krecker, F. Wysotzki, G. Coremans–Bertand, J. Sacton, G. Wilquet, T. Cantrell, F. Esmael, A. Montwill, D. H. Davis, Nucl. Phys. B **52**, 1 (1973).
2. D. H. Davis and J. Sacton, in *High Energy Physics*, Vol. II (Academic Press, New York, 1967) p. 365.
3. D. H. Davis, in the Proceedings of the *LAMPF Workshop on* (π, K) *Physics*, A.I.P. Conf. Proc. **224**, edited by B. F. Gibson, W. R. Gibbs, and M. B. Johnson (American Institute of Physics, New York, 1991), p. 38.
4. I. R. Afnan and B. F. Gibson, Phys. Rev. C **40**, R7 (1989); *ibid.* **41**, 2787 (1990).
5. K. Miyagawa, H. Kamada, W. Glöckle, and V. Stoks, Phys. Rev. C **51**, 2905 (1995).
6. J. Dabrowski and E. Fedorynska, Nucl. Phys. A **210**, 509 (1973).
7. B. F. Gibson and D. R. Lehman, Phys. Rev. C **11**, 29 (1975).
8. A. R. Bodmer, Phys. Rev. **141**, 1387 (1966).
9. R. C. Herndon, Y. C. Tang, Phys. Rev. **153**, 1091 (1967); *ibid.* **159**, 853 (1967); **165**, 1093 (1969); R. H. Dalitz, R. C. Herndon, and Y. C. Tang, Nucl. Phys. B **47**, 109 (1972).
10. B. F. Gibson, A. Goldberg, and M. S. Weiss, Phys. Rev. C **6**, 741 (1972); B. F. Gibson, A. Goldberg, and M. S. Weiss, in *Few Particle Problems in Nuclear Interactions* (North Holland, Amsterdam, 1972), p. 188.
11. A. Gal, Adv. Nucl. Phys. **8**, 1 (1975).
12. C. Mayeur, J. Sacton, P. Vilain, G. Wilquet, D. O'Sullivan, D. Stanley, P. Allen, D. H. Davis, E. R. Fletcher, D. A. Garbutt, J. E. Allen, V. A. Bull, A. P. Conway, and P. V. March, Nuovo Cim. **44**, 698 (1966).
13. G. Bohm, J. Klabuhn, H. Krecker, F. Wysotzki, G. Coremans, W. Gajewski, C. Mayeur, J. Sacton, P. Vilain, G. Wilquet, D. O'Sullivan, D. Stanley, D. H. Davis, E. R. Fletcher, S. P. Lovell, N. C. Roy, J. H. Wickens, A. Filipkowski,

K. Garbowska–Pniewska, T. Pniewski, E. Skrzypczak, T. Sobczak, J. E. Allen, V. A. Bull, A. P. Conway, A. Fishwick, and P. V. March, Nucl. Phys. B **9**, 1 (1969).
14. G. Keyes, J. Sacton, J. H. Wickens, and M. M. Block, Nuovo Cim. **31A**, 401 (1976).
15. B. F. Gibson and R. G. E. Timmermans, Nucl. Phys. A **628**, 417 (1998).
16. R. H. Dalitz and G. Rajasekharan, Phys. Lett. **1**, 58 (1962).
17. R. H. Dalitz and F. von Hippel, Nuovo Cim. **34**, 799 (1964).
18. F. von Hippel, Phys. Rev. **136**, B455 (1964).
19. R. H. Dalitz, in the Proceedings of the International School of Physics "Enrico Fermi" XXXVIII, *Interaction of High–Energy Particles with Nuclei*, edited by T. E. O. Ericson (Academic Press, New York–London, 1967), p. 89.
20. A. Cieplý and A. Gal, Phys. Rev. C **55**, 2715 (1997).
21. J. Sacton, in Ref.[19], p. 77.
22. W. Gajewski, J. Sacton, P. Vilain, G. Wilquet, D. Stanley, D. H. Davis, E. R. Fletcher, J. E. Allen, V. A. Bull, A. P. Conway, and P. V. March, Phys. Lett. **21**, 673 (1966).
23. U. Krecker, D. Kielczwska, and T. Tymieniecka, Nucl. Phys. A **236**, 491 (1974).
24. O. Adamovic, U. Krecker, G. Coremans–Bertrand, J. Sacton, T. Cantwell, C. Ni Ghogain, D. H. Davis, D. Kielczewska, T. Tymieniecka, and Z. Zakrzekski, Lett. Nuovo Cim. **6**, 9 (1973).
25. R. H. Dalitz and L. Liu, Phys. Rev. **116**, 1312 (1959).

Mass Dependence of Lifetimes of Λ hypernuclei

H. Bhang, H. Park, Y.D. Kim and M. Youn
Department of Physics, Seoul National University, Seoul 151-742, Korea

O. Hashimoto, K. Maeda, Y. Sato and T. Takahashi
Physics Department, Tohoku University, Sendai 980-8578, Japan

K. Aoki, H. Noumi, K. Omata, H. Outa, M. Sekimoto and T. Shibata,
High Energy Accelerator Research Organization (KEK), Tsukuba, Ibaraki 305-0801, Japan

T. Hasegawa, H. Hotchi and Y. Ohta,
Graduate School of Science, University of Tokyo, Tokyo 113-0033, Japan

S. Ajimura and T. Kishimoto
Department of Physics, Osaka University, Toyonaka, Osaka 560-0043, Japan

We have determined the lifetimes of Λ hypernuclei over a broad mass range, A=11∼56, from which its characteristic mass dependence is clearly observed. The lifetime decreases slightly from that of free Λ and quickly saturates at ∼80 percent of τ_Λ in free space as the mass number increases. The measured lifetimes of Λ hypernuclei are ; $\tau({}^{11}_{\Lambda}\mathrm{B})$ = 211±13 ps, $\tau({}^{12}_{\Lambda}\mathrm{C})$ = 231±15 ps, $\tau({}^{27}_{\Lambda}\mathrm{Al})$ = 203±10 ps, $\tau({}^{28}_{\Lambda}\mathrm{Si})$ = 206±12 ps and $\tau({}^{56}_{\Lambda}\mathrm{Fe})$ = 215±14 ps. Short range nature of the non-mesonic weak decay, which is dominent over the pionic weak decay except for very light nuclei, is possibly responsible for the observed mass dependence of the lifetime.

1 Introduction

In this workshop we have seen much of concerns and activities on the Σ- and double Λ-hypernucleus. However, so far the only hypernuceus where the realistic spectroscopic and decay studies are possible seems to be the Λ-hypernucleus. Already in Λ hypernuclei, the good quality results from the high resolution reaction and gamma ray spectroscopic study began to produce long awaited informations on the spin-orbit and spin-spin interactions. Here we will be concerned with the decay aspects of the Λ hypernuclei.

Since the advent of QCD, there have been expectations that where the high momenta or short range correlations are important, the signiture of quarks and gluons would appear. Non-mesonic weak decay(NMWD) of Λ hypernuclei is one of the promising such areas. The large momentum transfer of the ΔS=1 non-mesonic weak decay(NMWD) process of Λ hypernuclei implies that it probes short distances and might expose the role of explicit quark-gluon substructure of the baryons. The NMWD process also provides our primary means of exploring the four fermion, strangeness changing Λ + N → N + N weak interaction. In this study, we focus on the NMWD of Λ hypernuclei via the measurement of its lifetime, namely the total decay width of heavy Λ hypernuclei.

The Λ hyperon bound in a nucleus will eventually decay from the ground state through a weak interaction process, either a π mesonic or a non-mesonic decay process. Through a π mesonic process, a Λ in a hypernucleus decays to a nucleon and a pion as $\Lambda \rightarrow N\pi$, which is the decay mode of free Λ. The π mesonic decay is strongly suppressed in hypernuclei except very light ones due to its small momentum release. Instead, the non-mesonic weak decay (NMWD) process which is opened up for Λ in the nucleus, $\Lambda + N \rightarrow N + N$, dominates in most of hypernuclei due to its large momentum transfer, ~ 400 MeV/c. The branching ratio for NMWD already reaches ~ 78 percent for ${}^{12}_{\Lambda}C$ [1]. For heavier masses, the total widths are essentially due to NMWD processes.

The understanding of NMWD mechanism is still not satisfactory and its branching ratio has been a long standing issue for years. There have been several microscopic model calculations for NMWD processes which were based on either various meson exchange interactions [2,3,4,5] or quark-hadron hybrid interactions [6,7]. But none of them reproduced both total and partial decay rates successfully.

So far the experimental data for the partial decay rates have large uncertainties and fails to give clear guide for the theories. Among the weak decay observables, the lifetime can be measured most accurately free from nuclear final state interactions and the material effects. A high precision measurement for lifetime of heavy Λ hypernuclei where NMWD dominates would be the first step to test the calculations.

Fig.1 shows the status of the lifetime results from the recent measurements. The most recent and reliable measurements for the hypernuclear lifetime, for relatively light hypernuclei from ${}^{4}_{\Lambda}H$ to ${}^{12}_{\Lambda}C$, were carried out at KEK [8] and BNL [9] directly measuring production and decay time of a Λ hypernucleus. There have been a few lifetime measurements for the heavier Λ hypernuclei whose masses are greater than 12. Lifetime of mass number 16 hypernuclei was measured at the LBL Bevalac with 22 events only [10]. In the lifetime measurements of very heavy hypernuclei, delayed fissions after proton or anti-proton beam on targets, like Pb, Bi or U, were observed and attributed to a hypernuclear production. [11,12,13] Their lifetimes are scattered in a wide range of 100-250ps. However, these measurements are subject to criticism since they did not identify the produced hypernuclei explicitly. Due to the large uncertainties of the data and the failures of the explicit identificaiton of the formation of Λ hypernuclei, it has been difficult to see the mass dependence of lifetime for the mass heavier than 12. Essentially the huge region of mass greater than 12 is an open area where unambiguous new data are very much needed.

We have recently extended the accurate measurements of the lifetimes to heavier Λ hypernuclei to fill the huge unexplored area of mass number larger than 12. The previous measurement of the lifetime of ${}^{12}_{\Lambda}C$ was made with (K^-, π^-) reaction. However, it was difficult to extend the (K^-,π^-) reaction to heavier hypernuclei due to inherent large background from in flight kaon decays and its small momentum transfer. The (π^+,K^+) reaction is better suited for measuring the lifetimes of bound states in the heavier hypernuclei, since it gives almost background free spectrum [14].

2 Experiment

The present experiment was carried out with (π^+, K^+) reaction at K6 beamline of KEK 12 GeV PS. π^+ beam of 1.05 GeV/c at which the elementary reaction, $\pi^+ n \rightarrow K^+ \Lambda$, has the largest cross section were incident on ^{12}C, ^{28}Si and ^{56}Fe targets.

The decay particle measurements were carried out in coincidence with hypernuclear production by setting gates on peaks or a region in the hypernuclear mass spectrum. The production of hypernuclear bound states via the (π^+, K^+) reaction was identified using the high resolution, large acceptance scattered particle spectrometer SKS (Superconducting Kaon Spectrometer).[15] Due to the good momentum resolution and the cleanness of (π^+, K^+) spectrum, we were able to separate the ground states of $^{12}_{\Lambda}$C and $^{28}_{\Lambda}$Si reasonably well, even though we used thick targets. It was not possible to seperate the ground state of $^{56}_{\Lambda}$Fe.

Fig. 1 shows the schematics of the coincidence decay spectrometer. In coincidence with K^+ detected by SKS, decay products from the Λ-hypernuclei were measured with two coincidence decay spectrometers placed symmetrically above and below the target. Each spectrometer was composed of a timing counter, a drift chamber, a range counter and a veto counter. Each of them covered about 15 percent of whole solid angle.

The delayed time, τ, can be determined from the four measured quantities as

$$\tau = (T_2 - tof_d) - (T_1 + tof_b). \qquad (1)$$

τ is the interval between a production time (T_1 + tof_b) and a decay time (T_2 - tof_d) of the Λ hyperon. T_1 and T_2 are the timing read from the beam (π^+) timing and decay product (proton) timing counters. The tof_b and tof_d are time of flights of a beam particle and a decay product particle. Here the decay product particle was proton since we selected only the protons for the lifetime measurements.

The major sources of the lifetime uncertainty are those from intrinsic timing resolution of the timing counters (T_1 and T_2)[16] and from the travel time of a decay product, tof_d. To reduce the uncertainty of tof_d, we adopted high resolution drift chambers, placed between the timing counter T_2 and range counters for the energy measurement as shown in Fig. 2. One range counter was composed of 20 layers of scintillator with the total thickness of 10 cm. A typical energy resolution of the range counter for the normal incident protons of energy ~100 MeV is about a couple of MeV. Veto counter was used to reject high energy particles not stopped inside the range counter. The range of the proton was 30-140 MeV and 12-70 MeV for pions.

To determine the velocity of a decay particle, particle identificaion had to be made in addition to the tracking and energy measurement. The particle identification (PID) was made by the dE/dx-Range method. The light output from the 6mm plastic scintillator T_2 timing counter was used for dE/dx. The obtained PID for p and π was good enough and intermixing between p and π group was negligible.

From the kaon momentum measured with SKS, we can deduce the mass of the residual Λ hypernucleus from the equation $M_{HY} = [(E_\pi + M_A - E_K)^2 - (p_\pi^2 + p_K^2 - 2p_\pi p_K cos\theta)]^{1/2}$. Fig. 4(a-c) show the mass spectra of the produced Λ hypernuclei, $^{12}_{\Lambda}$C, $^{28}_{\Lambda}$Siand $^{56}_{\Lambda}$Fe, gated with decay protons and pions. The 1s and 1p Λ single particle states of the hypernuclei were reasonably well resolved for $^{12}_{\Lambda}$Cand poorly resolved for $^{28}_{\Lambda}$Si but distinguishable. 1p Λ states are proton unbound and make strong decay emitting protons to $^{11}_{\Lambda}$B and $^{27}_{\Lambda}$Al which will eventually decay via weak interaction. For the exclusive identification of the mother hypernucleus we set the gate windows for the resolved peaks except for $^{56}_{\Lambda}$Fe. The 1s ground state of

${}^{56}_{\Lambda}$Fe was not seperable anymore. We integrate most of the Λ bound region to obtain the lifetime of ${}^{56}_{\Lambda}$Fe. The pion gated spectra shown as dot lines are dominated with the quasi free pionic decay where pions are produced from the decays of unbound Λs. This clearly can be seen from the rise of π spectrum around B_Λ=0.

To get the timing response function of our system, we used (π,2p) strong reaction events. Fig. 5a shows a typical (π,2p) timing spectrum for ^{12}C target. The timing resolution was about 85 ps in σ which is much smaller than a typical lifetime, $\tau \sim 200$ ps, so that the final error of the measured lifetime is not limited by this resolution but by the statistics.

Fig. 5b-d show the delay time spectra of ${}^{12}_{\Lambda}$C, ${}^{28}_{\Lambda}$Si and $\{{}^{56}_{\Lambda}\text{Fe}\}$ respectively. The gate windows and the number of events are listed in Table 1. To extract the lifetime, this delay time spectrum was fitted with the statistical distribution of

$$D(t) = \int_0^\infty R(t-t')e^{-t'/\tau}dt' \tag{2}$$

which is convoluted distribution of the exponential decay fuction with our response fuction, R(t) of the above mentioned (π,2p) timing spectrum. For fitting procedure, we used maximum likelyhood method adopting Poisson statistics.

3 RESULTS

The lifetime results obtained from the delay time spectra of are listed in the Table 1. Some of these data have been published previously.[17] Various systematic dependences of lifetime, such as the decay particle energy dependence, independent analysis of up and down decay spectrometer, beam momentum dependence, etc. have been studied. There is no significant systematic error found due to them to the final lifetime.

Table 1: The summary of the present lifetime measurement

	Gate window $M_{HY} - M_A(MeV)$	number of events	Lifetime (ps)	Γ (Γ_Λ)
${}^{11}_{\Lambda}B$	188 – 197	338	211±13	1.25±0.08
${}^{12}_{\Lambda}C$	175 – 186	323	231±15	1.14±0.08
${}^{27}_{\Lambda}Al$	171 – 181	661	203±10	1.30±0.07
${}^{28}_{\Lambda}Si$	183 – 188	527	206±12	1.28±0.08
${}^{56}_{\Lambda}Fe$	160 – 185	309	215±14	1.22±0.08

In Fig. 5 present results of lifetimes (dark circles) are compared with the previous data (open circles). Here only the recent data are shown. Note that the Λ bound region of ${}^{56}_{\Lambda}$Fe is open for many particle decay channels. Therefore a Λ bound but excited ${}^{56}_{\Lambda}$Fe nucleus first decays through one or a few particle emissions, then eventually reaches the ground state of a nearby hypernucleus where the Λ makes a weak decay. The quite probable Λ hypernuclei which can be reached via particle decay cover the mass region of $54 \leq A \leq 56$. (These hypernuclear group is represented as $\{^{56}_{\Lambda}\mathrm{Fe}\}$.) The horizontal error bar for $^{56}_{\Lambda}$Fe data represents this range. However, one does not expect much variation of lifetime over this mass region especially since NMWD is dominent and should not be sensitive to the valence nucleon distribution much.

Taking the previous results of $^{9}_{\Lambda}$Be and $^{11}_{\Lambda}$B from BNL into account, we can see that the lifetimes of Λ hypernuclei are almost constant over the mass range of $\sim$10 $\leq$ A $\leq$ 56. As the mass number increases from 5 to $\sim$10, a new lifetime, which is about 20 percent shorter, sets in. And it remains there within statistics over the broad mass range, at least up to A=56. The lifetime of Λ in nuclear medium saturates at $\sim$80 percent of the lifetime of free Λ.

The solid line in Fig. 5 shows the recent result of the calculation based on one pion and correlated two pion exchange model, which is presented in this workshop by Itonaga et al. [18]. The calculation reproduces the lifetimes up to $^{12}_{\Lambda}$C well, but begins to underestimate them from $^{28}_{\Lambda}$Si. The calculation shows a saturation starting at $^{28}_{\Lambda}$Si. The dashed line is the calculation of Galbarino et al. [19], in which the Λ decay rates are calculated in nuclear mattter using the propagator method [20] and extended to finite nuclei in local density approximation. In this calculation, the contribution from two nucleon induced NMWD channel, which remains about 15 percent of the total width without much mass dependence, is also included. After fixing parameters of the two baryon correlation function to $^{12}_{\Lambda}$C data, they reproduce the data resonably well now. This indicates that the mass dependence of the lifetime is reproduced with the local density approximation, which is basically density overlap and a very short range approximation. However, these model calculations are yet to explain the Γ_n/Γ_p ratio.

The saturation we observed in the mass dependence of the lifetime seems to indicate that the short range component of baryon-baryon weak interaction needs to be treated more carefully. One pion exchange model for NMWD, based on the diagram of Fig. 6b, was first introduced by Adams, motivated by the success of One Boson Exchange model (OBE) for the analysis of parity non-conserving interactions in the nucleus (Fig. 6a). Later this one pion exchange model was extended to include heavier mesons like ρ, ω,.. For energies characteristic of bound nucleons, the N-N interaction takes place primarily at distances large compared to the nucleon size, due to the strong repulsion in the N-N interaction at short distances. Thus we expect long-range contributions to dominate the low energy parity non-conserving interaction. However, in NMWD much higher momentum is involved. OBE by itself for the NMWD may not be adequate. We can appreciate it as a starting point, but not as an ultimate model for NMWD. In addition to the above calculations which are based on one boson exchange model, there in this meeting have been introduced new approaches to deal with the short range interactions explicitly such as those of direct quark interaction by Oka et al. [21] and the phenomenological four point interaction by Jun et al. [22]. Though these approaches are yet limited to the application for 1s shell hypernuclei, they showed some promises for the understanding of NMWD processes.

4 SUMMARY

In summary, we have obtained lifetimes of Λ hypernuclei over a broad mass range, A=11$\sim$56. These are the first results in this broad mass region with the explicit identification of the produced Λ hypernuclei.

Saturation of hypernuclear lifetime in nuclear matter at $\sim$80 percent of τ_Λ in free space is our observation. The lifetimes of ${}^{28}_{\Lambda}$Si and $\{{}^{56}_{\Lambda}$Fe$\}$ differ from that of ${}^{12}_{\Lambda}$C only slightly. This indicates that the short range nature of NMWD and the baryon-baryon correlation is responsible for the observed mass dependence of the lifetime. It is still the standing urgent issue to understand the mass dependence of the lifetime and the Γ_n/Γ_p ratio of Λ hypernucleus in a unified picture.

H. B. acknowledges partial supports from the Korea-Japan collaborative research program of KOSEF and the Ministry of Education through the Research Institute of Basic Science of Seoul National University.

References

1. H. Noumi *et al.*, Phys. Rev. C52 (1995) 2936
2. J.B. Adams, Phys. Rev. 156 (1967) 1611.
3. J.F. Dubach, Nucl. Phys. A450 (1986) 71c,
4. A. Ramos *et al.*, Nucl. Phys. A585 (1995) 129c.
5. K. Itonaga *et al.*, Nucl. Phys. A639 (1998) 329c.
6. C.Y. Cheung *et al.*, Phys. Rev. C27 (1983) 335.
7. M. Oka *et al.*, Nucl. Phys. A597 (1996) 563.
8. H. Outa *et al.*, Nucl. Phys. A547 (1992) 109c.
9. R. Grace *et al.*, Phys. Rev. Lett. 55 (1985) 1055. J.J. Szymanski *et al.*, Phys. Rev. C43 (1991) 849.
10. K.J. Nield *et al.*, Phys. Rev. C13 (1976) 1267.
11. J.P. Bocquet *et al.*, Phys. Lett. B182 (1986) 146, Phys. Lett. B192 (1987) 312.
12. T.A. Armstrong *et al.*, Phys. Rev. C47 (1993) 1957.
13. H. Ohm *et al.*, Phys. Rev. C55 (1997) 3062.
14. H. Hasegawa *et al.*, Phys. Rev. C53 (1996) 1210.
15. T. Fukuda *et al.*, Nuc. Inst. Meth. A361 (1995) 485.
16. Y.D. Kim *et al.*, Nuc. Inst. Meth. A372 (1996) 431.
17. H. Bhang *et al.*, Phys. Rev. Lett. 81 (1998) 4321.
18. K. Itonaga *et al.*, Presented in this meeting.
19. G. Garbarino *et al.*, Presented in this meeting.
20. W. Alberico *et al.*, nucl-th/9902023.
21. M. Oka *et al.*, Presented in this meeting.
22. J. Jun *et al.*, Presented in this meeting.

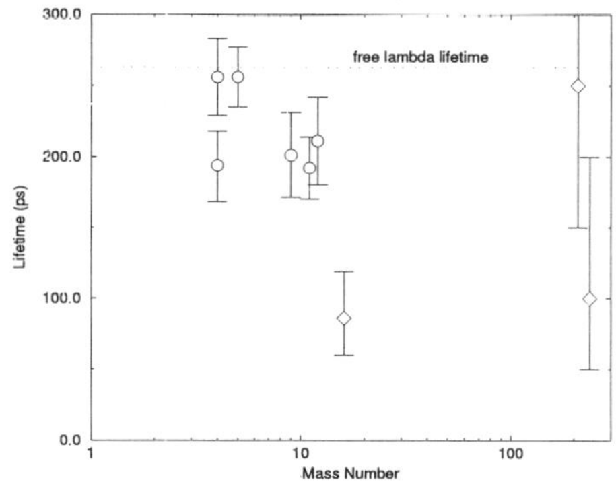

Figure 1: The lifetimes measured in the previous experiments. Open circles are those from ${}^{4}_{\Lambda}$H to ${}^{12}_{\Lambda}$C measured with explicit identification of the produed Λ hypernuclei at BNL and KEK, and open diamonds are those without.

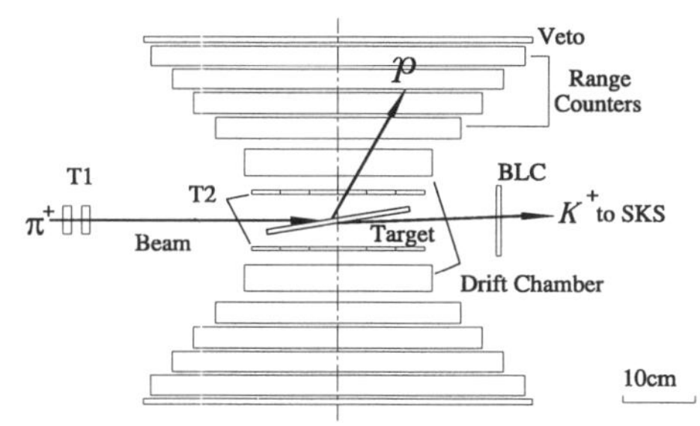

Figure 2: The schematic view of the decay particle spectrometer used in E307 experiment for the lifetime measurement of Λ hypernuclei.

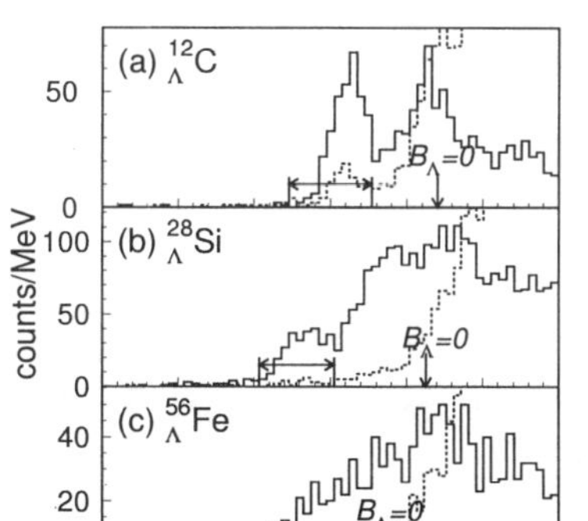

Figure 3: The hypernuclear mass spectra for (a)${}^{12}_{\Lambda}$C, (b)${}^{28}_{\Lambda}$Si and (c)${}^{56}_{\Lambda}$Fe in coincidence with proton (solid line) and pions (dot line). M_{HY} and M_{A} mean the masses of the produced hypernuclei and normal nuclei. The horizontal arrows represent the gate for the lifetime analysis.

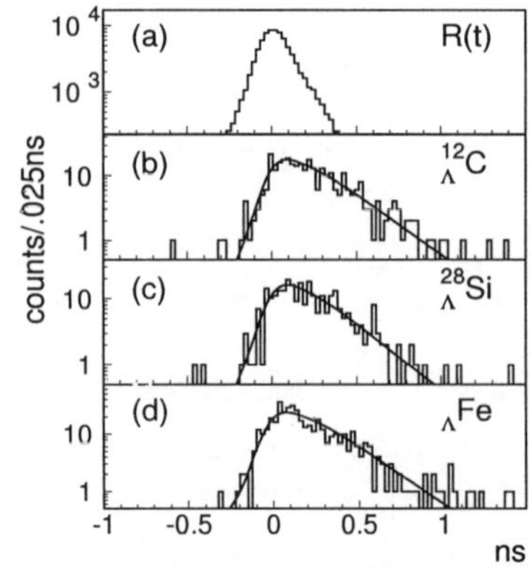

Figure 4: Delay time spectra for $^{12}_{\Lambda}$C(b), $^{28}_{\Lambda}$Si(c) and $^{56}_{\Lambda}$Fe(d) along with the prompt time spectrum from ^{12}C(π,2p)^{11}B reaction (a).

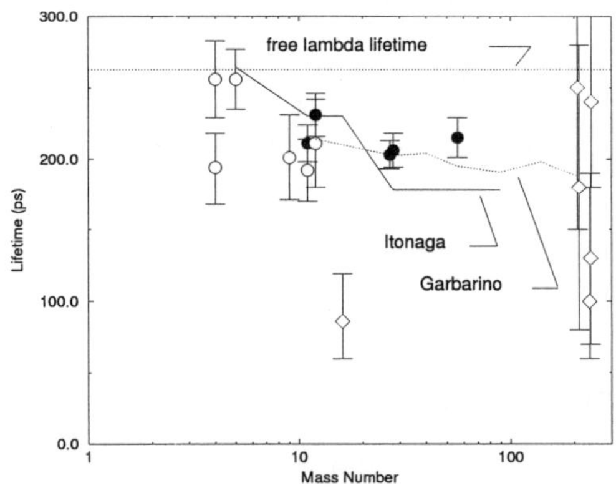

Figure 5: The mass dependence of Λ hypernuclear lifetimes. The dark circles are the results from the present measurements where open circles and the diamonds are from the previous ones. Recent calculations by Itonaga et al.(solid line) and Galbarino et al. (dash line) are also shown.

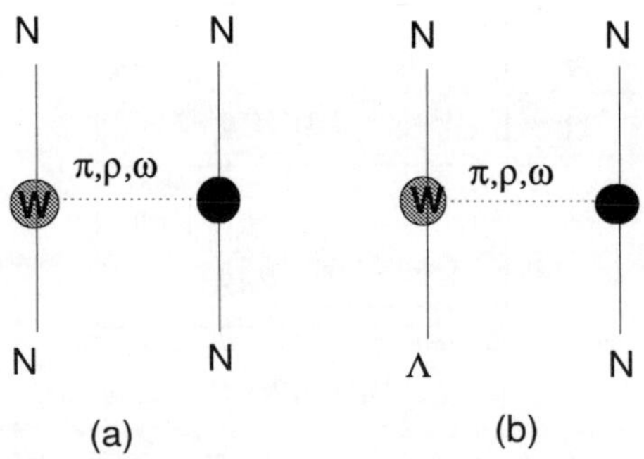

Figure 6: Diagrams for one boson exchange contribution to the PNC N-N interaction (a) and ΔS=1 Λ-N weak interaction (b).

WEAK DECAY OF Λ IN NUCLEI: DIRECT QUARK MECHANISM vs MESON EXCHANGE

Makoto Oka, Kenji Sasaki
Department of Physics, Tokyo Institute of Technology
Meguro, Tokyo 152-8551 Japan
E-mail: oka@th.phys.titech.ac.jp

Takashi Inoue[a]
Department of Physics, Univ. of Tokyo
Hongo, Tokyo 112-0033 Japan

Nonmesonic decays of Λ in nuclear medium and light hypernuclei are studied by using the $\Lambda N \to NN$ weak transition potential derived from the meson exchange mechanism and the direct quark mechanism. The long range part of the transition potential is described by exchanges of the pseudoscalar mesons (π, K, η), while the vector mesons (ρ, ω, K^*) may be considered as the medium- and short-range part in the meson exchange picture. We propose the direct quark transition potential as the short range part, which is derived from the matrix elements of the $\Delta S = 1$ effective weak Hamiltonian in the two baryon states. The results indicate that the direct quark contribution is significantly large and its behavior is qualitatively different from the vector meson exchanges. We also find that the decay rate is sensitive to the choice of form factor and that a soft cutoff must be used for the pion-baryon verteces so that the strong tensor transition is suppressed. We find that the $\pi + K + DQ$ results are compatible with experiment although the n/p ratio is still too large. The π^+ decays of light hypernuclei are related to the $\Delta I = 3/2$ amplitudes of the nonmesonic decay. The role of chiral symmetry for the pionic decays are discussed.

1 Introduction

Mesonic decays of Λ are suppressed in heavy hypernuclei by the Pauli exclusion principle of the outgoing nucleon and instead nucleon induced decays become dominant,

$$\Lambda + p \to n + p \quad : \quad \text{proton induced decay} \tag{1}$$

$$\Lambda + n \to n + n \quad : \quad \text{neutron induced decay} \tag{2}$$

Such a process is regarded as a weak $\Delta S = 1$ reaction of two baryons, which is analogous to the weak NN interaction. While the weak NN interaction is masked by the strong interaction, the $\Lambda N \to NN$ transition provides us with a unique opportunity to study the mechanism of the baryon-baryon weak interaction. Many theoretical and experimental studies have been done for the nonmesonic decays of light and heavy hypernuclei. Yet mechanisms of the nonmesonic weak decay are not clear, especially as the predicted proton-induced decay rates are too large and consequently the n/p ratio, the ratio of the neutron-induced decay rate, Γ_n, to the proton-induced one, Γ_p, is too small compared with the observed value.

A conventional picture of $\Lambda N \to NN$ process is the one-pion exchange, where $\Lambda N \pi$ vertex is induced by the weak interaction.[1,2] In $\Lambda N \to NN$, the relative

[a]Present address: Department of Physics, Tokyo Institute of Technology, Meguro, Tokyo 152-8551 Japan.

momentum of the final state nucleon is about 400 MeV/c, or $(0.5 \text{ fm})^{-1}$. The nucleon-nucleon interaction at this momentum is dominated by the short-range repulsion due to heavy meson exchanges and/or to quark exchanges between the nucleons.[3] It is therefore expected that the short-distance interactions will play significant roles in the two-body weak decay as well. Exchanges of K, ρ, ω, K^* mesons and also correlated two pions have been studied,[4,5,6,7] and it is found that the kaon exchange is significant, while the other mesons contribute less.[6]

Several studies have been made on effects of quark substructure.[8,9,10] In our recent analyses,[10,11,12] we employ a $\Delta S = 1$ effective weak hamiltonian, which consists of various four quark weak verteces derived in the renormalization group approach to include QCD corrections to the $su \to ud$ transition mediated by the W boson.[13] Such an effective weak hamiltonian has been applied to the decays of kaons and hyperons with considerable success.[14] Our aim is to explain the short range part of the weak baryonic interaction by using the same interaction so that we are able to confirm the consistency between the free hyperon decays and decays of hypernuclei.

We proposed to evaluate the effective hamiltonian in the six-quark wave functions of the two baryon systems and derived the "direct quark" (DQ) weak transition potential for $\Lambda N \to NN$.[10,11] Our analysis shows that the direct quark contribution largely improves the discrepancy between the meson-exchange theory and experimental data for the n/p ratios. It is also found that the $\Delta I = 3/2$ component of the effective hamiltonian gives a sizable contribution to $J = 0$ transition amplitudes,[9] although we cannot determine the $\Delta I = 3/2$ amplitudes unambiguously from the present experimental data.

2 Direct quark mechanism

In our recent papers,[10,11,12] we derived the $\Lambda N \to NN$ transition potentials in the direct quark (DQ) mechanism. The transition potential is calculated by evaluating the effective hamiltonian in the nonrelativistic valence quark model.[10] The valence quark wave function of two baryon system is taken from the quark cluster model, which takes into account the antisymmetrization among the quarks. Then the obtained transition potential has been applied to the weak decay of light hypernuclei.

In ref[10], the DQ transition potential is calculated for all the two-body channels with the initial $\Lambda N(\ell = 0)$ and the final $NN(\ell' = 0, 1)$ states. The explicit forms of the transition potential are given in the momentum space representation. In ref.[12], it is expressed in the coordinate space formulation, so that realistic nuclear wave functions with short range correlations can be easily handled. The DQ transition potentials in the coordinate space contain nonlocal terms as a result of the quark antisymmetrization. It also has terms with a derivative operator acting on the initial relative coordinate. Thus the general form of the transition potential is

$$\begin{aligned} V_{DQ}{}^{\ell\ell'}_{ss'J}(r,r') &= \langle NN : \ell' s' J | V(\vec{r}', \vec{r}) | \Lambda N : \ell s J \rangle \\ &= V_{loc}(r)\,\frac{\delta(r-r')}{r^2} + V_{der}(r)\,\frac{\delta(r-r')}{r^2}\partial_r + V_{nonloc}(r,r') \quad (3) \end{aligned}$$

where r (r') denotes the relative radial coordinate of the initial (final) two-baryon state, and ∂_r stands for the derivative of the initial ΛN relative wave function.

We later combine the direct quark (DQ) and meson exchange (ME) potentials, giving the total transition potential as

$$V(r,r') = V^{ME}(r)\frac{\delta(r-r')}{r^2} + V^{DQ}(r,r') \tag{4}$$

The relative phase between ME and DQ is fixed so that the weak quark hamiltonian gives the correct amplitude of $\Lambda \to N\pi$ transition.[11] Note that the relative phases among various meson exchange potentials are determined according to the $SU(6)_w$ symmetry.

3 Meson exchange mechanism

The one pion exchange (OPE) process is expected to give a main part of the $\Lambda N \to NN$ transition potential. The weak $\pi N \Lambda$ vertex is parametrized as

$$\mathcal{H}_w^\pi = iG_F m_\pi^2 \overline{\psi}_N (A_\pi + B_\pi \gamma_5)\vec{\tau}\cdot\vec{\phi}_\pi \psi_\Lambda . \tag{5}$$

where the coupling constants, A_π and B_π, are determined from the $\Lambda \to N\pi$ decay amplitudes. As the momentum transfer in this transition is fairly large, the tensor part plays an important role. As a result, the $\Lambda p(^3S_1)$ to $pn(^3D_1)$ transition becomes dominant. This tensor dominance causes the n/p ratio problem, namely, the observed n/p ratio ($\simeq 1$) can not be explained by the OPE transition.

While the OPE is significant for the long-distance baryon-baryon interaction, the short range reaction mechanism is also important in $\Lambda N \to NN$ due to the large energy transfer involved. Within the meson exchange model, a shorter range contributions may come from the exchange of the heavier mesons[4,5,6] and multi-correlated meson.[7] In ref. [6] the authors consider the exchange of all the octet pseudoscalar and vector mesons, π, K, η, ρ, ω, and K^*. While the $\Lambda N\pi$ weak coupling constant is determined phenomenologically, all the other weak couplings in ref. [6] are only estimated theoretically by assuming the $SU(6)_w$ symmetry.[5] For the strong vertices, the coupling constants are taken from the Nijmegen YN potential model D.

As the baryons and mesons have finite size and structure, we consider a form factor at each meson-baryon vertex. If we assume the same form factor $F(\mathbf{q}^2)$ both for the strong and weak vertices, the potential is given as

$$V(\mathbf{r}) = \int \frac{d^3q}{(2\pi)^3}\frac{e^{i\mathbf{q}\cdot\mathbf{r}}}{\mathbf{q}^2+\mu^2-q_0^2}\mathcal{O}(\mathbf{q})F^2(\mathbf{q}^2) \tag{6}$$

where $\mathcal{O}$ is a product of vertex operators and coupling constants. A standard choice of $F^2(\mathbf{q}^2)$ is the square of the monopole form factor,

$$F_{DP}^2(\mathbf{q}^2) = \left(\frac{\Lambda_{DP}^2-\mu^2}{\Lambda_{DP}^2+\mathbf{q}^2}\right)^2 \tag{7}$$

which we call "double pole" (DP) form factor. On the other hand, a simpler form is often used in literatures,

$$F_{SP}^2(\mathbf{q}^2) = \frac{\Lambda_{SP}^2-\mu^2}{\Lambda_{SP}^2+\mathbf{q}^2} \tag{8}$$

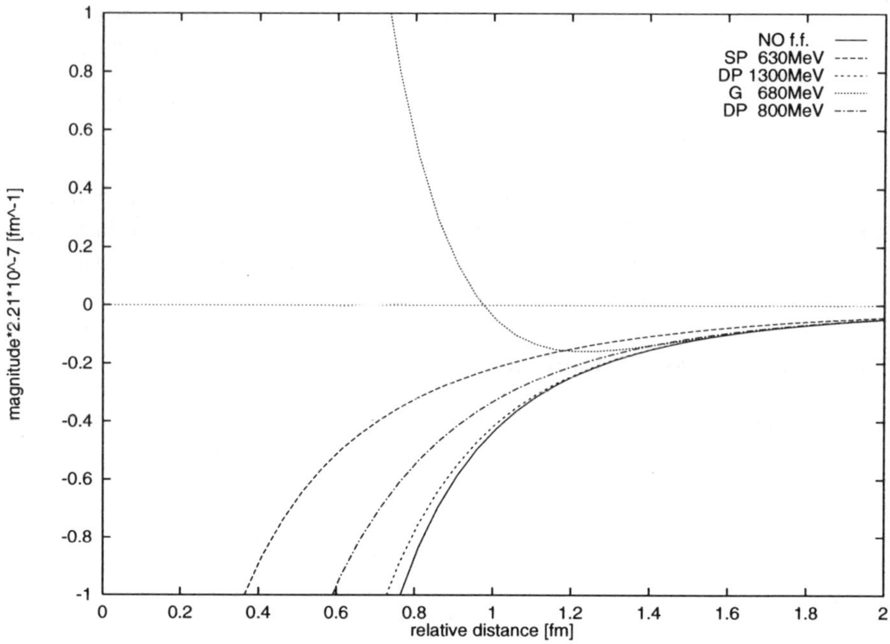

Figure 1: Potentials for the ${}^3S_1 - {}^3D_1$, with relative momentum of $k_r = 1.97\text{fm}^{-1}$

which we call "single pole" (SP) form factor. The cutoff parameter Λ is chosen for individual mesons independently. Another type of form factor considered here is the "Gaussian" (G),

$$F_G^2(\mathbf{q}^2) = \exp\left(-\frac{\mathbf{q}^2}{\Lambda^2}\right), \tag{9}$$

which has the advantage of the consistency with the quark structure for the baryon, if the cutoff parameter is taken according to the size of quark distribution in the baryon. While the SP and DP form factors give similar effects, the G form factor behaves differently at short distances. Fig. 1 shows the behaviors of the OPE potential in the $\Lambda N({}^3S_1)$ - $NN({}^3D_1)$ channel at the relative momentum of $k_r = 1.97\text{fm}^{-1}$. As can be seen clearly, the amplitude has a node for the G form factor, while the others do not behave like that. This oscillation suppresses the tensor transition although the long distance behavior is similar to the "hard" DP form factor or that without the form factor. We test these three form factors and compare the results.

We consider the effect of the finite energy transfer q_0 in eq. (6). Suppose that the initial $\Lambda - N$ pair is at rest, then the energy transfer is given by $q_0 \simeq (m_\Lambda - m_N)/2 \simeq$ 88MeV. Taking this q^0 as an average, we introduce the effective mass of the meson, $\tilde{\mu}_i = \sqrt{\mu_i^2 - (m_\Lambda - m_N)^2/4}$ and make the following replacement,

$$\frac{1}{\mathbf{q}^2 + \mu^2 - q_0^2} \to \frac{1}{\mathbf{q}^2 + \tilde{\mu}^2} \tag{10}$$

This effect leads to measurable changes. For example, the pion mass is reduced by about 25%, and the range of OPE becomes longer.

4 Decay of Λ in nuclear medium

First we study the Λ decay in nuclear matter in order to investigate our approach to $\Lambda N \to NN$ without ambiguities from the nuclear structure. It is regarded as an approximation of decays of heavy hypernuclei. Recent experimental data of the life-times of heavy hypernuclei indicate that the nonmesonic decay rate of Λ in heavy hypernuclei is only 20% larger than the free Λ decay rate.[15] This saturation suggests the short-range nature of the $\Lambda N \to NN$ decay.

We assume that nuclear matter is symmetric ($N_n = N_p$) and the Λ is at rest in it. We employ the plane wave with the short range correlation for the wave function of both the initial and final states. The correlation function, $f(r)$, proposed in ref.[6], is multiplied to the relative wave functions. For the initial ΛN states, we employ

$$f_i(r) = \left(1 - e^{-\frac{r^2}{a^2}}\right)^n + br^2 e^{-\frac{r^2}{c^2}} \tag{11}$$

where $a = 0.5$, $b = 0.25$, $c = 1.28$, $n = 2$, and for the final NN state,

$$f_f(r) = 1 - j_0(q_c r) \tag{12}$$

where $q_c = 3.93\ \text{fm}^{-1}$. The initial state correlation is obtained from a macroscopic finite nucleus G-matrix calculation, and the final state correlation gives a good description of nucleon pairs in ^{4}He.

In the present calculation, we only consider the relative $L = 0$ for the initial $\Lambda - N$ system. It has been pointed out by explicit calculation that contributions of the higher partial waves are of 5% order. The decay rate is given by

$$\Gamma_{NM} = \frac{m_N}{(2\pi)^5 \mu^3} \int_0^{\mu k_F} k^2 k_0 dk \tag{13}$$

$$\times \frac{1}{4} \sum_{p,n} \left(|a|^2 + |b|^2 + 3|c|^2 + 3|d|^2 + 3|e|^2 + 3|f|^2\right) \tag{14}$$

where $k_0^2 \equiv m_N(m_\Lambda - m_N) + k^2/2\mu$ and $\mu \equiv m_\Lambda/(m_N + m_\Lambda)$. The a, b, $\cdots$, f are the matrix elements of the transition potential in each channel designated by its final orbital angular momentum, the initial and final spins.

4.1 Form factors

As the $\Lambda N \to NN$ scattering involves a large momentum transfer, $\sim 400 MeV/c$, the amplitudes are rather sensitive to the form factor. Here we study choices of the form factor for OPE. Table 1 shows the calculated decay rates of Λ in OPE. The SP form has often been used in literatures.[4,2] Ref.[4] employed the SP form factor with $\Lambda_{SP}^2 = 20 m_\pi^2$, or $\Lambda_{SP} \sim 630$MeV. This form factor is based on the dispersion relation analysis with the semi-pole approximation to the πNN form factor. It is a very soft form factor, which cuts off the short-range part rather drastically and the resulting decay rate becomes small. The tensor transition, $\Lambda p(^3S_1) \rightarrow pn(^3D_1)$, seems most affected by such form factor. We find that the tensor transition amplitudes are reduced by a factor two or more by the form factor. Therefore total decay rate is much reduced.

Table 1: Decay rates of Λ in nuclear matter (in units of $\Gamma_\Lambda = (263 \times 10^{-12} sec)^{-1}$) for various choices of the form factors in OPE.

			total	Γ_p	Γ_n	Γ_n/Γ_p	PV/PC
π	no-f.f.		2.819	2.571	0.248	0.097	0.358
	SP	$\Lambda_\pi = 630$MeV	1.103	0.989	0.114	0.116	0.408
		$\Lambda_\pi = 920$MeV	2.332	2.116	0.216	0.102	0.376
	DP	$\Lambda_\pi = 1300$MeV	2.575	2.354	0.221	0.094	0.337
		$\Lambda_\pi = 800$MeV	1.850	1.702	0.148	0.087	0.282
	G	$\Lambda_\pi = 680$MeV	1.514	1.129	0.386	0.342	0.580

The DP form factor is popular in meson exchange potential models of the nuclear force. The Bonn potential, for instance, employs the DP with $\Lambda_{DP}(\pi) = 1300$MeV. This form factor is extremely hard so that the short range part of the meson exchange potential becomes relevant. Lee and Matsuyama carried out an analysis of the $NN \rightarrow NN\pi$ processes and pointed out [16] that a soft form factor, such as $\Lambda < 750$MeV, is preferable. Recent analysis in the QCD sum rule also suggests a soft πNN form factor of cutoff $\Lambda \simeq 800$[MeV] [17]. From the quark substructure point of view, it is also natural that the cutoff Λ is of order $\sim 1/R_h \sim 400-500$MeV. In Table 1, we find that the soft form factor reduces the decay rates by about 40%, while the hard one changes only by $\leq 10\%$.

The G form factor is related to the Gaussian quark wave function of the baryon. Corresponding to the size parameter $b \simeq 0.5$fm, we employ $\Lambda_G(\pi) \simeq 680$MeV because the relation between the size and cutoff parameters is $\Lambda = \sqrt{3}/b$.

The above sensitivity to choice of the form factor is quite annoying because it is not easy to judge which of the form factors is the right one. It should also be noted that the weak vertex form factor can be different from the strong one, although the pole dominance picture of the parity-conserving weak vertex leads to the identical form factor to the strong one. In the meson exchange potential models of the nuclear force, there seems a tendency to choose a hard form factor because the heavy meson exchanges are often important for spin-dependent forces. On the other hand, the quark model approach to the short range nuclear force gives significant spin dependencies comparable to the heavy meson exchanges and therefore the meson exchanges can be cut off rather sharply with a soft form factor. In the present approach to the weak $\Lambda N \rightarrow NN$ interaction, whose short range part is described by the direct quark mechanism, we thus may follow the quark model approach to the nuclear force and take the "soft" form factor for OPE potential as a standard.

Table 2: Nonmesonic decay rates of Λ in nuclear matter (in units of Γ_Λ). The "all" includes π, K, η, ρ, ω, and K^* meson exchanges. The "DP hard (soft)" assumes $\Lambda_\pi = 1300(800)$ MeV.

		$total$	Γ_p	Γ_n	Γ_n/Γ_p	PV/PC
π	DP hard	2.575	2.354	0.221	0.094	0.337
	DP soft	1.850	1.702	0.148	0.087	0.282
	no-f.f.	2.819	2.571	0.248	0.097	0.358
π+K	DP hard	1.099	1.075	0.024	0.022	0.631
	DP soft	0.695	0.674	0.021	0.031	0.632
	no-f.f.	1.143	1.111	0.032	0.028	0.745
all	DP hard	1.672	1.571	0.101	0.064	1.468
	DP soft	1.270	1.152	0.117	0.101	1.537
	no-f.f.	1.744	1.704	0.040	0.024	2.952
DQ		0.418	0.202	0.216	1.071	6.759
DQ+π	DP hard	3.609	2.950	0.658	0.223	0.856
	DP soft	2.726	2.202	0.523	0.238	0.896
DQ+π+K	DP hard	1.766	1.495	0.271	0.181	1.602
	DP soft	1.204	0.998	0.206	0.207	1.954

4.2 Decay rates in nuclear matter

The calculated decay rates of Λ in nuclear matter for several models are listed in Table 2. We compare the results of the soft DP form factor and those of the hard DP form factor. One sees that the even the soft-cutoff OPE predicts a total decay rate much larger than the observed one, about $1.2 \times \Gamma_\Lambda$, for the heavy hypernuclei. This is due to the large tensor transition d_p, $\Lambda p(^3S_1) \to pn(^3D_1)$, which is allowed only in the proton-induced decay. This tensor dominance property leads to the small Γ_n/Γ_p and PV/PC ratio.

The kaon exchange contribution reduces Γ_p by more than factor two. This mainly comes from the cancelation in the channel d_p. At same time, the kaon exchange contribution reduces Γ_n. Therefore, the n/p ratio remains small. The decay is dominated by the $J = 1$ channels.

When we include η, ρ, ω, and K^* ("all") mesons, both the proton and neutron induced decays increase but the n/p ratio is still small ($\simeq 0.1$). It is interesting to see that the PV/PC ratio becomes large, when we include heavier mesons.

One sees in Table 2 that the magnitude of DQ itself is small compared to that of π or $\pi + K$. However, the DQ has a large Γ_n and a large n/p ratio. It is also shown that the DQ mechanism is dominated by the parity violating channels and thus produces a large PV/PC ratio. Such charasteristic behaviors of DQ will distinguish DQ from the other mechanisms.

In *DQ*+meson exchanges, the pion contribution is again senseitive to the choice of the form factor. One sees that the softer form factor is more appropriate to reproduce experimental values of Γ_p. As the DQ has little contribution to the tensor channel, Γ_p is still too large in "$DQ + \pi$". Again, the K exchange reduces the tensor amplitude and thus Γ_p is suppressed by a factor 2. However Γ_n is reduced, at the same time, which results in the n/p ratio$\simeq 0.2$. This value is still smaller than the observed values for the medium and heavy hypernuclei.

Contribution beyond the π and K mesons does not improve the situation. It is also questionable whether the DQ mechanism and the vector meson exchanges are independent and can be superposed. The double counting problem for the vector mesons and DQ contribution in nuclear force is pointed out in ref[18]. We thus take the "$DQ + \pi + K$" with the soft π form factor as our present best model for the nonmesonic Λ decay.

4.3 Decays of light hypernuclei

The same $\Lambda N \to NN$ transition potential has been applied to the study of the nonmesonic weak decays of light s-shell hypernuclei, ${}^5_\Lambda He$, ${}^4_\Lambda He$, and ${}^4_\Lambda H$.[12] The results are summarized in Table 4.3.

Table 3: Nonmesonic decay rates (in units of Γ_Λ) of light hypernuclei. The DP (soft) form factor is used for OPE.

		total	Γ_p	Γ_n	Γ_n/Γ_p
${}^5_\Lambda He$	π	0.740	0.654	0.087	0.133
	$\pi + K$	0.350	0.331	0.028	0.055
	$\pi + K + DQ$	0.521	0.435	0.085	0.195
	exp. [19]	0.41±0.14	0.21±0.07	0.20±0.11	0.93±0.55
	exp. [20]	0.50±0.07	0.17±0.04	0.33±0.04	1.97±0.67
${}^4_\Lambda He$	π	0.542	0.498	0.044	0.089
	$\pi + K$	0.252	0.233	0.019	0.082
	$\pi + K + DQ$	0.309	0.302	0.007	0.024
	exp. [20]	0.19±0.04	0.15±0.02	0.04±0.02	0.27±0.14
${}^4_\Lambda H$	π	0.080	0.022	0.056	2.596
	$\pi + K$	0.020	0.010	0.010	1.099
	$\pi + K + DQ$	0.120	0.060	0.059	0.983
	exp. [20]	0.15±0.13	—	—	—

We find that the "π + K + DQ" picture gives us a fair agreement with experiment for the total decay rate, while the n/p ratios are still too small. As seen in the nuclear matter calculation, the main contribution comes from the OPE mechanism, which produces the large proton-induced rate. Comparing with the experimental data we find that the proton-induced rate is overestimated in all the pictures, while the neutron-induced rate is underestimated.

5 π^+ decays and $\Delta I = 3/2$ amplitudes

The pionic decay is another interesting decay mode of light hypernuclei. While the free Λ decays into $p\pi^-$ or $n\pi^0$, the π^+ decay requires an assistance of a proton, *i.e.*, $\Lambda + p \to n + n + \pi^+$. Some old experimental data suggest that the ratio of π^+ and π^- emission from ${}^4_\Lambda He$ is about 5%.[21] The most natural explanation of this process is $\Lambda \to n\pi^0$ decay followed by $\pi^0 p \to \pi^+ n$ charge exchange reaction. It was evaluated for realistic hypernuclear wave functions and found to explain only 1.2% for the π^+/π^- ratio.[22] Another possibility is to consider $\Sigma^+ \to \pi^+ n$ decay after the conversion $\Lambda p \to \Sigma^+ n$ by the strong interaction such as pion or kaon exchanges. It was found, however, that the free Σ^+ decay which is dominated by P-wave amplitude, gives at most 0.2% for the π^+/π^- ratio. Recently, it was proposed that a two-body process $\Sigma^+ N \to nN\pi^+$ must be important in the ${}^4_\Lambda He$ decay.[23] But its microscopic mechanism is not specified.

In order to solve this problem, we have applied the soft-pion technique to the π^+ decay of light hypernuclei.[24] The soft-pion theorem to the process $\Lambda p \to nn\pi^+(q \to 0)$ reads

$$\lim_{q\to 0}\langle nn\pi^+(q)|H_W|\Lambda p\rangle = -\frac{i}{\sqrt{2}f_\pi}\langle nn|[Q_5^-, H_W]|\Lambda p\rangle \tag{15}$$

Again, because of

$$[Q_5^-, H_W] = -[I_-, H_W] \tag{16}$$

it discriminates the isospin properties of H_W. Similarly to the case of Σ^+ decay, we see that the $\Delta I = 1/2$ part vanishes as

$$[I_-, H_W(\Delta I = 1/2, \Delta I_z = -1/2)] = 0 \tag{17}$$

$$[I_-, H_W(\Delta I = 3/2, \Delta I_z = -1/2)] = \sqrt{3}H_W(\Delta I = 3/2, \Delta I_3 = -3/2) \tag{18}$$

We then obtain

$$\lim_{q\to 0}\langle nn\pi^+(q)|H_W|\Lambda p\rangle = \frac{i\sqrt{3}}{\sqrt{2}f_\pi}\langle nn|H_W(\Delta I = 3/2, \Delta I_3 = -3/2)|\Lambda p\rangle \tag{19}$$

Thus we conclude that the soft π^+ emission in the Λ decay in hypernuclei is caused only by the $\Delta I = 3/2$ component of the strangeness changing weak hamiltonian. In other words, the π^+ emission from hypernuclei probes the $\Delta I = 3/2$ transition of $\Lambda N \to NN$.

6 Future prospects

The Λ decays iin nuclear medium is a unique reaction as a weak interaction which involves two baryons. Its study may reveal roles of quark substructures in weak processes and the mechanism of the $\Delta I = 1/2$ enhancement in nonleptonic weak decays of strange hadrons. There are many unknowns, such as the weak meson-baryon coupling constants, the validity of $SU(6)_w$ symmetry, the magnitude of the parity violation, and so on. Both experimental and theoretical efforts may produce a lot of useful and interesting information in understanding hadronic weak interactions.

Here we considered the new direct quark (DQ) mechanism for the $\Lambda N \to NN$ process, which induces decays of hypernuclei. Combining the meson exchange processes for a long range part, we find that the "$\pi + K + DQ$" mechanism with soft DP form factor for OPE explains the total decay rates of both light and heavy hypernuclei rather well. There, however, remains a difficult problem. That is, the proton-induced decay rate is overestimated and thus we predict a small n/p ratio.

The ratio is imporoved by the DQ contribution. It is yet too small ($\simeq 0.2$) for "$\pi + K + DQ$". The experimental values are not completely fixed, but they suggest $\simeq 1$ for heavy hypernuclei. The situation is similar for ${}^{5}_{\Lambda}$He. It is thus urgent and important to find out what causes this discripancy.

1. M. M. Block, R. H. Dalitz, Phys. Rev. Lett. **11** 96 (1963);
2. K. Takeuchi, H. Takaki, H. Bando, Prog. Theor. Phys. **73** 841 (1985);
3. M. Oka and K. Yazaki, Phys. Lett. **B90** (1980) 41; Prog. Theor. Phys. **66** (1981) 556; *ibid.* **66** (1981) 572; in *Quarks and Nuclei*, ed. by W. Weise (World Scientific, 1985); K. Shimizu, Rep. Prog. Phys. **52** (1989) 1.
4. B. H. J. McKeller, B. F. Gibson, Phys. Rev. C**30**, 322 (1984).
5. J. F. Dubach, G. B. Feldman, B. R. Holstein, L. de la Torre, Ann. Phys. **249**, 146 (1996).
6. A. Parreño, A. Ramos, C. Bennhold, Phys. Rev. C**56**, 339 (1997).
7. M. Shmatikov, Nucl. Phys. **A580** 538 (1994); K. Itonaga, T. Ueda, T. Motoba, Nucl. Phys. **A585** 331c (1995).
8. C.Y. Cheung, D.P. Heddle, L.S. Kisslinger,Phys. Rev. **C27** (1983) 335.
9. K. Maltman and M. Shmatikov, Phys. Lett. **B331** (1994) 1.
10. T. Inoue, S. Takeuchi and M. Oka, Nucl. Phys. **A577** (1994) 281c; Nucl. Phys. **A597** (1996) 563.
11. T. Inoue, M. Oka, T. Motoba and K. Itonaga, Nucl. Phys. **A633** (1998) 312.
12. K. Sasaki, T. Inoue, and, M. Oka, preprint, TIT/HEP-424/NP, nucl-th/9906036.
13. M.K. Gaillard and B.W. Lee, Phys. Rev. Lett. **33** (1974) 108; A.I. Vainshtein, V.I. Zakharov and M.A. Shifman, Sov. Phys. JETP **45** (1977) 670; F.J. Gillman, M.B.Wise, Phys. Rev. **D20** (1979) 2382; E.A. Paschos, T. Schneider and Y.L. Wu, Nucl. Phys. **B332** (1990) 285.
14. J.F. Donoghue, E. Golowich and B. Holstein, Phys. Rev. **131** (1986) 319.
15. H.C. Bhang, in this Proceedings; H.C. Bhang *et al.*, Phys. Rev. Lett. **81** (1998) 4321.
16. T. S. H. Lee, A. Matsuyama, Phys. Rev. C**36** 1459 (1987).
17. T. Meissner, Phys. Rev. C**52** 3386 (1995).
18. K. Yazaki, Prog. Part. Nucl. Phys. **24** 353 (1990).
19. J. J. Szymznski et al., Phys. Rev. C**43** 849 (1991).
20. H. Noumi et al,. in proceedings of the IV International Symposium on Weak and Electromagnetic Interactions in Nuclei, edited by H. Ejiri, T. Kishimoto and T. Sato (World Scientific,1995) p.550
21. C. Mayeur, *et al.*, Nuovo Cim. **44** (1966) 698; G. Keyes, J. Sacton, J.H. Wickens, M.M. Block, Nuovo Cim. **31A** (1976) 401.
22. R.H. Dalitz, F. von Hippel, Nuovo Cim. **34** (1964) 779; F. von Hippel, Phys. Rev. **136** (1964) B455. A. Cieplý, A. Gal, Phys. Rev. **C55** (1997) 2715.
23. B.F. Gibson, R.G.E. Timmermans, Nucl. Phys. **A628** (1998) 417; B.F. Gibson, in this Proceedings.
24. M. Oka, Nucl. Phys. **A647** (1999) 97.

FOUR-BARYON POINT INTERACTIONS FOR NONMESONIC WAEK DECAY OF HYPERNUCLEI

JUNG-HWAN JUN, HEECHANG NA
Department of Physics, Chung-Ang University,
Seoul 156-756, KOREA
E-mail: jhjun@cau.ac.kr

HYOUNG CHAN BHANG
Department of Physics, Seoul National University,
Seoul151-742, KOREA
E-mail: bhang@phya.snu.ac.kr

In an effort to resolve the theoretical difficulties in explaining the total and partial decay rates Γ_{nm}, Γ_p, Γ_n and the ratio Γ_n/Γ_p, the four-baryon $\Lambda N \to NN$ point interactions are introduced as a new model for the short-range part of the non-mesonic decay of Λ-hypernuclei. Based on the traditional one pion exchange model for the long-range part and our model, the nonmesonic decay rates of ${}^4_\Lambda$H, ${}^4_\Lambda$He, ${}^5_\Lambda$He and ${}^{12}_\Lambda$C are calculated, which are in fair agreement with the experimental data.

1 Introduction

Understanding the nonmesonic weak decay (NMWD) is one of the most urgent issues in the Λ-hypernuclear physics today. Especially explaining the total decay rates $\Gamma_{nm} = \Gamma_n + \Gamma_p$ and the ratio Γ_n/Γ_p has become a theoretical challenge. The traditional treatment for the nonmesonic decays of Λ in hypernuclei is obviously the so called one-pion-exchange model (OPE). The OPE interaction along with $\Delta I = 1/2$ rule is based on the lagrangians $\mathcal{L}_\pi$ for the weak $\Lambda N\pi$ vertex and $\mathcal{L}_S$ for the strong $NN\pi$ vertex which are given by

$$\mathcal{L}_\pi = G_\Lambda \bar{N}(1+\alpha\gamma_5)\tau\cdot\vec{\pi}\Lambda \qquad \mathcal{L}_S = g_S\bar{N}i\gamma_5\tau\cdot\vec{\pi}N \tag{1}$$

where $G_\Lambda = 2.4\times 10^{-7}$, $\alpha = -6.9$ and $g_S = 13.26$. Based on this OPE the total and partial decay rates $\Gamma_{nm} = \Gamma_p + \Gamma_n$, Γ_p, Γ_n and the ratio Γ_n/Γ_p were initially calculted by Adams, McKeller, Dubach [1,2,3] and later by Bando and others [4,5,6,7,8,9] since Adams first proposed it [1]. All the calculations based on OPE disagree with the experimental results [10,12,13,11]. First of all, the total decay rates Γ_{nm} are too small and the proton-induced partial decay rates Γ_p are too large in comparison with the neutron-induced decay rates Γ_n. Especially, the ratios Γ_n/Γ_p are too small $0.1\sim 0.2$ even though the experimental results yield about ~ 1.0. Since then intense theoretical as well as experimental investigations have been going on to understand the nonmesonic decays of hypernuclei. In order to resolve the issue various models have been proposed. One is the meson exchange description which incorporates the weak and strong meson-baryon vertices along with the $\Delta I = 1/2$ rule. This meson exchange model has been most extensively studied [6,7,8,9]. Even though some partial success in explaining the total decay rates $\Gamma_{nm} = \Gamma_p + \Gamma_n$ has been reported by adding mesons such as ρ [7,8], there has been no success in explaining the ratio Γ_n/Γ_p. Numerous models such as the quark-hadron hybrid model [14,15], and some phenomenological models [16,17] have been tried. The complete understanding of the nonmesonic decay mode has not been achieved. Other appproaches might still be needed to resolve the issue.

2 Four-Baryon Point Interactions

We still believe the OPE interaction given by Eq.(1) plays a significant role in the long range part of NMWD. For the short range part of NMWD instead of adding another heavy mesons, we would like to suggest the following four-baryon point (4P) interactions [18] in an effort to resolve the present NMWD problems.

$$\begin{aligned}\mathcal{L}_{4P} &= G_s(\bar{N}N)(\bar{N}\Lambda) + G_v(\bar{N}\vec{\tau}N)\cdot(\bar{N}\vec{\tau}\Lambda) \\ &+ G_t[-(\bar{n}p)(\bar{p}\Lambda) + (\bar{n}n)(\bar{n}\Lambda) - (\bar{p}p)(\bar{n}\Lambda)] \qquad (2)\end{aligned}$$

where the $\Delta I = 1/2$ rule holds in the first two terms but the last term is the $\Delta I = 3/2$ component. G_s and G_v are the isoscalar and isovector coupling constants while G_t is the coupling constant for the $\Delta I = 3/2$ component. The $\Delta I = 3/2$ component is necessary, because without it enough enhancement in the ratio Γ_n/Γ_p cannot be obtained [18]. It turns out to be a little more convenient to write Eq.(2) as

$$\mathcal{H}_{4\text{B}} = G_N(\bar{n}p)(\bar{p}\Lambda) + aG_N(\bar{p}p)(\bar{n}\Lambda) + (1+b)G_N(\bar{n}n)(\bar{n}\Lambda) \qquad (3)$$

In Eq.(3) $G_N = 2G_v - G_t$, $aG_N = G_s - G_v - G_t$ and $(1+b)G_N = G_s + G_v + G_t$. The suppressed Dirac matrices in 4P interactions Eq.(2) are given by

$$(\bar{N}_1 N)(\bar{N}_2\Lambda) = [\bar{N}_1\gamma_\mu(1+\eta_{0,1}\gamma_5)N][\bar{N}_2\gamma^\mu(1+\lambda\gamma_5)\Lambda] \qquad (4)$$

In Eq.(4) the coupling constant η_1 is the axial vector coupling constant relative to the vector coupling constant, g_A/g_V or simply g_A for the hadronic weak currents of $n(p) \to n(p)$ transitions. From the nuclear β-decay, we take 1.25 for η_1. λ is the axial vector coupling constant for $\Lambda \to p$. From the decay $\Lambda \to p+e+\nu$, we take 0.75 for λ. For the current $\Lambda \to n$, it is still the same $\Delta S = 1$ and $\Delta I = 1/2$ transtion as $\Lambda \to p$. Hence in the similar vein of the free Λ decay, it is reasonable to take 0.75 for the corresponding axial vector coupling constant. For η_0, only recently, it is suggested that it may be small in comparison with η_1 [19]. They actually estimated it as $-0.1 \sim 0.3$. Thus we leave it as a free parameter which is somewhat smaller than η_1. At this point, one may argue that those axial vector coupling constants may be different in nonleptonic decays. But, it is even more difficult to imagine that the axial vector coupling constant in the hadronic weak current depends on whether the same weak current couples with a hadronic weak current or a leptonic current. For the overall coupling constants G_N, aG_N and $(1+b)G_N$ or equivallently G_s, G_v and G_t we may take them as completely free parameters and then determine them by fitting the decay rates in the experimental data. However, we will estimate them from the simplified effective nonleptonic quark interaction with $\Delta S = 1$.

3 Calculation of Decay Rates

The decay width Γ arising from OPE and 4P for the transition from a hypernuclear state $|{}^A_\Lambda Z; J_i>$ into a nuclear state $|{}^{A-2}Z'; J_f>$ and two outgoing nucleons with momenta and spins $(\vec{q}s_1)$ and $(-\vec{q}s_2)$ becomes

$$\Gamma = \frac{1}{2J_i+1}\left(\frac{p_N^3}{6\pi^3}\right)\sum_{J_f M_f M_i}\frac{qM^2}{M(^A_\Lambda Z)-M(^{A-2}Z')} \times \int \frac{d\Omega_q}{4\pi}| < {}^{A-2}Z'; J_f M_f|\mathcal{H}^\pi + \mathcal{H}^{4P}|^A_\Lambda Z; J_i M_i > |^2 \tag{5}$$

In Eq.(5), p_N is the maximum interacting nucleon momentum relative to Λ at rest, *i.e.*, $p_N = p_F M_\Lambda/(M_\Lambda + M)$, where p_F is the Fermi momentum. $\mathcal{H}^\pi$ is the well-known traditional OPE potential coming from Eq.(1) and $\mathcal{H}^{4P}$ is the nonrelativistic reduction of Eq.(3). $\mathcal{H}^{4P}$ is the nonrelativistic reduction of Eq.(3), which is given by

$$\begin{aligned}\mathcal{H}^{4\mathrm{P}} &= G\left(\frac{E_M}{2M}\right)\{[\chi_2^+(s_2)\chi_N(s_N)\chi_1^+(s_1)\chi_\Lambda(s_\Lambda) - \eta\lambda\sum_{k=1}^{3}\chi_2^+(s_2)\sigma^k\chi_N(s_N) \\ &\times \chi_1^+(s_1)\sigma^k\chi_\Lambda(s_\Lambda) - \eta\chi_2^+(s_2)i\frac{\vec{\sigma}\cdot\nabla_2}{E_M}\chi_N(s_N)\chi_1^+(s_1)\chi_\Lambda(s_\Lambda) \\ &- \lambda\chi_2^+(s_2)\chi_N(s_N)\chi_1^+(s_1)i\frac{\vec{\sigma}\cdot\nabla_1}{E_M}\chi_\Lambda(s_\Lambda) + \lambda\sum_{k=1}^{3}\chi_2^+(s_2)i\frac{\vec{\sigma}\cdot\nabla_2}{E_M}\sigma^k\chi_N(s_N) \\ &\times \chi_1^+(s_1)\sigma^k\chi_\Lambda(s_\Lambda) + \eta\sum_{k=1}^{3}\chi_2^+(s_2)\sigma^k\chi_N(s_N)\chi_1^+(s_1) \\ &\times i\frac{\vec{\sigma}\cdot\nabla_1}{E_M}\sigma^k\chi_\Lambda(s_\Lambda)]\phi_1^*(\vec{r})\phi_2^*(\vec{r})\},\end{aligned} \tag{6}$$

where $\phi_1(\vec{r})$ and $\phi_2(\vec{r})$ are the wave functions of two outgoing nucleons, and $\chi_1(s_1)$, $\chi_2(s_2)$, $\chi_N(s_N)$, and $\chi_\Lambda(s_\Lambda)$ are the nonrelativistic two-component spinors of outgoing nucleons 1 and 2, the interacting nucleon, and the Λ-hyperon respectively. $\nabla_{1\mathrm{or}2}$ acts on $\phi_1^*(\vec{r})$ or $\phi_2^*(\vec{r})$. Also in Eq.(6) the G and η are the shorthand notations of G_N and η_1 for $(np)(p\Lambda)$, aG_N and η_0 for $(pp)(n\Lambda)$, and $(1+b)G_N$ and η_0 for $(nn)(n\Lambda)$ as those in Eqs.(3,4). E_M is the sum of the mass and energy of each outgoing nucleon $E_M = M + E$, which is determined in terms of the masses of the hypernucleus and the residual nucleus, $M(^A_\Lambda Z)$ and $M(^{A-2}Z')$, by the energy conservation, *i.e.*, $M(^A_\Lambda Z) = 2E + M(^{A-2}Z')$. The state of the final nucleus is presumably formed by removing one nucleon and the Λ particle from the initial hypernucleus to be coupled with the outer shell of a nuclear core by cfp's.

Using the shell model wave functions of the harmonic oscillator potential for the Λ-hyperon and the initial interacting nucleon and plane waves for outgoing nucleons, let us apply Eq.(5) to the NMWD of $^4_\Lambda$H, $^4_\Lambda$He, $^5_\Lambda$He, and $^{12}_\Lambda$C in their gound states. In the case of $^4_\Lambda$H and $^4_\Lambda$He, both hypernuclei decay into three channels. $^4_\Lambda$H decays into $(2n)_{\mathrm{SLOW}} + p + n$, $(pn)_{\mathrm{SLOW}} + n + n$ and $d + n + n$, and $^4_\Lambda$He decays into $(2p)_{\mathrm{SLOW}} + n + n$, $(pn)_{\mathrm{SLOW}} + p + n$ and $d + p + n$. The partial decay rates for $^4_\Lambda$H, $^4_\Lambda$He due to OPE, 4P and interference are given by:

$$\Gamma^{\pi,4P.Int}_{n\text{ or }P}(^4_\Lambda\mathrm{H}/^4_\Lambda\mathrm{He})_{\mathrm{SLOW}} = \left(\frac{p_N^3}{24\pi^3}\right)\frac{qE_M^2}{M(^A_\Lambda Z)-2M}R^2 F^{\pi,4P.Int}_{n\text{ or }p}(^4_\Lambda\mathrm{H}/^4_\Lambda\mathrm{He})$$

$$\Gamma_d^{\pi,4P.Int}({}^4_\Lambda\mathrm{H}/{}^4_\Lambda\mathrm{He}) = (\frac{p_N^3}{24\pi^3})\frac{qE_M^2}{M({}^A_\Lambda Z)-2M}R^2F_d^{\pi,4P.Int}({}^4_\Lambda\mathrm{H}/{}^4_\Lambda\mathrm{He}) \quad (7)$$

In Eq.(7) the R is present because the wave functions of nucleons in the initial hypernuclei ${}^4_\Lambda$H and ${}^4_\Lambda$He, which do not take part in the decay process are neither orthogonal to nor the same as the wave functions of slowly moving residual nucleons. For other decays such as ${}^4_\Lambda\mathrm{H} \to d+n+n$, ${}^4_\Lambda\mathrm{He} \to d+p+n$ and the decays of ${}^5_\Lambda$He and ${}^{12}_\Lambda$C, the wave functions of the nucleons in those initial hypernuclei which do not take part in the decay process are assumed to be the same as the wave functions of the final residual nucleons, $1S_{1/2}$, and hence $R = 1.0$. The partial decay rates from the interaction between Λ and the closed s-shell nucleons in ${}^5_\Lambda$He and ${}^{12}_\Lambda$C are

$$\Gamma_{n \text{ or } p}^{\pi,4P.Int} = (\frac{p_N^3}{24\pi^3})\frac{qE_M^2}{M({}^A_\Lambda Z)-M({}^{A-2}Z')}R^2F_{n \text{ or } p}^{\pi,4P.Int}({}^4_\Lambda\mathrm{He}/{}^{12}_\Lambda\mathrm{C}) \quad (8)$$

In Eqs.(7,8), the $F_{n,p,d}^{\pi,4P,Int}$'s are the dynamical factors due to OPE,4P and interference. Here we present the $F_{n,p,d}^{4P}$'s only, the dynamical factors due to 4P. They are given by

$$F_n^{4P}\ ({}^4_\Lambda\mathrm{H}/{}^5_\Lambda\mathrm{He}/{}^{12}_\Lambda\mathrm{C}) = G_N^2I^2(1+b)^2\{2(1+3\eta_0\lambda)^2+16(\frac{q}{E_M})^2(3\eta_0^2+3\lambda^2+2\eta_0\lambda)\}$$
$$F_p^{4P}\ ({}^4_\Lambda\mathrm{H}) = G_N^2I^2\{[(1+a)^2+6(1+a)\lambda(\eta_1+a\eta_0)+9\lambda^2(\eta_1+a\eta_0)^2]$$
$$+\ 16(\frac{q}{E_M})^2[(\eta_1+a\eta_0)^2+(1+a)^2\lambda^2+2(1+a)\lambda(\eta_1+a\eta_0)]\}$$
$$F_d^{4P}\ ({}^4_\Lambda\mathrm{H}) = G_N^2I^2(1+b)^2\{16(\frac{q}{E_M})^2(\eta_0-\lambda)^2\}$$
$$F_d^{4P}\ ({}^4_\Lambda\mathrm{He}) = G_N^2I^2\{[3-3a+(a\eta_0-\eta_1)\lambda)]^2+8(\frac{q}{E_M})^2[(1+a)\lambda-(a\eta_0+\eta_1)]^2\}$$
$$F_n^{4P}\ ({}^4_\Lambda\mathrm{He}) = G_N^2I^2(1+b)^2\{(1+3\eta_0\lambda)^2+32(\frac{q}{E_M})^2(\eta_0+\lambda)^2\}$$
$$F_p^{4P}\ ({}^4_\Lambda\mathrm{He}/{}^5_\Lambda\mathrm{He}/{}^{12}_\Lambda\mathrm{C}) = G_N^2I^2\{4[(1-a+a^2)+3a\lambda(\eta_1+\eta_0)+3\lambda^2(\eta_1^2+a\eta_0\eta_1$$
$$+\ a^2\eta_0^2)]+8(\frac{q}{E_M})^2[3(\eta_1+a\eta_0)^2+3(1+a)^2\lambda^2+2(1+a)\lambda(\eta_1+a\eta_0)]\} \quad (9)$$

In Eq.(9) the I's are the overlap integrals arising from 4P. For the decay rates of ${}^{12}_\Lambda$C due to the interaction between Λ and the p-shell nucleons, it turns out that the decay rates via the 4P interactions are less than a few percent of those due to the interactions between Λ and the s-shell nucleons [18]. Hence, it is sufficient to consider the decay rates via OPE only, which are given by

$$\Gamma_{n \text{ or } p}^{\pi} = (\frac{p_N^3}{24\pi^3})\frac{qE_M^2}{M({}^A_\Lambda Z)-M({}^{A-2}Z')}F_{n \text{ or } p}^{\pi}(\mathrm{p-shell}), \quad (10)$$

where the $F_{n \text{ or } p}^{\pi}(\mathrm{p-shell})$ are given by

$$F_p^\pi = \frac{1}{3}G_\Lambda^2g_S^2[336K_2^2+112\alpha^2K_1^2+\frac{144}{5}\alpha^2P_1^2+\frac{1296}{5}\alpha^2K_3^2]$$
$$F_n^\pi = \frac{1}{6}G_\Lambda^2g_S^2[18K_2^2+\frac{22}{3}\alpha^2K_1^2+16\alpha^2K_1P_1+\frac{456}{5}\alpha^2P_2^1+\frac{1944}{5}\alpha^2K_3^2] \quad (11)$$

In Eq.(11) K_0, K_1, P_1, K_2, and K_3 are various overlap integrals. All the overlap integrals in OPE are modified by the momentum-dependent weak and strong vertex form factor $(\Lambda^2 - m_\pi^2)/(\Lambda^2 + \vec{k}^2)$. We used $\Lambda^2 = 20m_\pi^2$. Also all the overlap integrals in both OPE and 4P are subject to the short range correlations (SRC). There is no problem of enforcing SRC in OPE. However, to enforce SRC in 4P one should relax the 4P interactions *i.e.*, the δ-function interactions by introducing the gaussian type structure function *i.e.*, the form factor for the nucleons [18]. Then the following simple standard SRC [2] for both OPE and 4P is introduced:

$$1 - exp(-|\vec{r}_1 - \vec{r}_2|^2/C^2), \tag{12}$$

where $C = 0.75 \sim 1.0$fm. We take $C = 0.1$fm for both OPE and 4P.

4 Coupling Constants of Four-Baryon Point Interactions

Now we are ready to compute the decay rates, provided the overall coupling constants G_N, aG_N and $(1+b)G_N$ are determined. Since we do not know them we may treat them as completely free parameters and then determine them by fitting the decay rates in the experimental data just as Block and Dalitz did [16]. However, we will try to estimate them using the present effective theory of the nonleptonic quark interaction with $\Delta S = 1$, which is given by

$$\mathcal{H}_{\text{eff}}(\Delta S = 1) = -\sqrt{2}G_F \sin\theta\cos\theta \sum_{i=1}^{6} c_i O_i, \tag{13}$$

where $G_F = 10^{-5}/M$ is the Fermi coupling constant. We take the above effective hamiltonian from Okun's book [20]. Of course, the present stage of QCD is not perfect enough to obtain the coupling constants accurately. For instance, G_Λ the coupling constant of free Λ decay is about 3.6 times greater than the theoretical value which is obtained from Eq.(13) along with the factorization scheme [20]. Nevertheless, it is still desirable to obtain them based on the effective hamiltonian Eq.(13), because the relative ratios and signs of physical constants obtained from Eq.(13) are often in fair agreement with experimentally determined values [20].

Now, with the following simple factorization assumption:

$$\begin{aligned}
< np|\bar{u}_L\gamma^\mu s_L \bar{d}_L\gamma^\mu u_L|\Lambda p> &= < p|\bar{u}_L\gamma^\mu s_L|\Lambda >< n|\bar{d}_L\gamma^\mu u_L|p> \\
< np|\bar{d}_L\gamma^\mu s_L \bar{u}_L\gamma^\mu u_L|\Lambda p> &= < n|\bar{d}_L\gamma^\mu s_L|\Lambda >< p|\bar{u}_L\gamma^\mu u_L|p> \\
< np|\bar{d}_L\gamma^\mu s_L \bar{d}_L\gamma^\mu d_L|\Lambda p> &= < n|\bar{d}_L\gamma^\mu s_L|\Lambda >< p|\bar{d}_L\gamma^\mu d_L|p> \\
< nn|\bar{d}_L\gamma^\mu s_L \bar{u}_L\gamma^\mu u_L|\Lambda n> &= < n|\bar{d}_L\gamma^\mu s_L|\Lambda >< n|\bar{u}_L\gamma^\mu u_L|n> \\
< nn|\bar{d}_L\gamma^\mu s_L \bar{d}_L\gamma^\mu d_L|\Lambda n> &= < n|\bar{d}_L\gamma^\mu s_L|\Lambda >< n|\bar{d}_L\gamma^\mu d_L|n>
\end{aligned} \tag{14}$$

the computations of the overall coupling constants G_N, aG_N and $(1+b)G_N$ of Eq.(3) are straightforward. Following the exactly parallel precedure of Ref.[20], we obtain:

$$G_N = -0.368G_F, \quad aG_N = 0.420G_F, \quad (1+b)G_N = 0.715G_F \tag{15}$$

Certainly, the values for G_N, aG_N and $(1+b)G_N$ are less than experimentally desired values in the sense that the decay rates calculated with the values given by

Eq.(15) are too small to fit in the error range of the experimental data. This is expected and it is really due to the limitations of the factorization scheme. Therefore, while maintaining $G_N : aG_N : (1+b)G_N = -0.368 : 0.420 : 0.715 = -1.0 : 1.14 : 1.94$ for the ratio between coupling constants, we will take $G_N = -G_F$, which is about 2.7 times the G_N in Eq.(15).

When we use Eqs.(1,3) to calculate the nonmesonic decay rates of a hypernucleus, we are faced with the ambiguity regarding the relative phase between OPE and 4P, *i.e*, the relative signs between G_Λ, g_S and G_N. To determine the relative phase between OPE and 4P, we take the estimate G_Λ of Ref.[20] which was computed in terms of f_π using the effective hamiltonian Eq.(13).

$$G_\Lambda = -f_\pi \times (\text{positive} \quad \text{quantity}) < 0 \tag{16}$$

Then from the Goldberg-Treiman relation [21], $f_\pi = Mg_A/g_S$, the relative signs of G_Λ and g_S are determined, which are given by Eq.(1). This way the relative phase is unamguiously determined.

5 Results and Discussions

The experimental data of BNL [10,11] and KEK [12,13] and the calculated nonmesonic decay rates, Γ_n, Γ_p and $\Gamma_{nm} = \Gamma_n + \Gamma_p$ (in unit of Γ_Λ) and the ratio Γ_n/Γ_p are shown in Table 1. The data indicated by EXP1 and EXP2 were taken from BNL and KEK data. The KEK data of ${}^4_\Lambda$H and ${}^4_\Lambda$He are obtained by Outa *et al*, but the KEK data of ${}^{12}_\Lambda$C are obtained by Noumi *et al.* The BNL data of ${}^4_\Lambda$He are obtained by Zeps *et al*, while the BNL data of ${}^5_\Lambda$He and ${}^{12}_\Lambda$C are obtained by Szymanski *et al.* The decay rates of ${}^4_\Lambda$H, ${}^4_\Lambda$He, ${}^5_\Lambda$He and ${}^{12}_\Lambda$C listed in the rows Γ^π and Γ^{4P} are the rates due to OPE and 4P. The values in the rows $\Gamma^{\pi+4P}$ are the final nonmesonic decay rates, which also include the contribution from the interference between OPE and 4P. In Table 1, first we note that the Γ^π_n of ${}^4_\Lambda$He are quite smaller than previously obtained values [8]. This is because they did not take into account the fact that the wave functions of nucleons in the initial hypernuclei ${}^4_\Lambda$He which do not take part in the decay process are neither orthogonal to nor the same as the wave functions of slowly moving residual nucleons. Secondly, for ${}^4_\Lambda$H only the total decay rate is available and our result is slightly outside the error range. However, it is premature to conclude anything on ${}^4_\Lambda$H at this point because of the insufficiency in data. We need the data of Γ_p and Γ_n of ${}^4_\Lambda$H. Finally, the Γ_n of ${}^{12}_\Lambda$C is quite smaller than the KEK data even though the Γ_p is well within the error range, and hence the total decay rate of ${}^{12}_\Lambda$C is outside the error range. This is because our calculation shows that the total decay rate from the *p*-shell contributions via OPE is about 1/5 of that from the *s*-shell contributions. Anyway it is worth comparing our result with other people's calculations. Ramos *et al* [7], Oset *et al* [5] and Dubach *et al* [9] obtained 0.89, 1.45 and 0.5 respectively for the total decay rate of ${}^{12}_\Lambda$C via OPE, while our Γ^π_{nm} is ~ 0.27. However, the partial decay rates Γ_n and Γ_p, and the ratio Γ_n/Γ_p of ${}^4_\Lambda$He and ${}^5_\Lambda$He, and Γ_p and Γ_n/Γ_p of ${}^{12}_\Lambda$C are in good agreement with the experimental data. To our best understanding, no other model results in such a good agreement with the experimental data despite of large error ranges. As for the coupling constant η_0, we take $\eta_0 = 0.3$. From the values of $G_N = -G_F$,

$aG_N = 1.14G_F$, $(1+b)G_N = 1.94G_F$, the coupling constants in terms of G_s, G_v, and G_t are then $G_s = 1.54G_F$, $G_v = -0.2G_F$ and $G_t = 0.6G_F$. Hence in NMWD we can say that the $\Delta I = 3/2$ component is larger than the conventional value. Also if we vary the coupling constants within reasonable ranges of their relative ratio, we obtain similarly encouraging results. At this point one could argue that the current stage of QCD is not perfect enough to yield those coupling constants accurately or one may not even believe in the factorization scheme, it might be useful to treat them as completely free parameters which should be determined by fitting the data [18]. Actually, we have done it too. After trying many values for them and fitting the data, we found that many values for those coupling constants along with suitable values of η_0 between -0.1 and 0.3 can be fitted to the data fairly well.

In this article, using the traditional one pion exchange model for the long-range part and the four-baryon $\Lambda N \to NN$ point interactions for the short-range part the nonmesonic decay rates of ${}^4_\Lambda$H, ${}^4_\Lambda$He, ${}^5_\Lambda$He and ${}^{12}_\Lambda$C are calculated. The results have shown that the four-baryon $\Lambda N \to NN$ point interactions prescribed by Eq.(3) can be a physically reasonable model for the short range part of the nonmesonic weak decay of hypernuclei. Hence in the nonmesonic decay of hypernuclei the four-baryon point interactions are worth being considered seriously along with the interesting view that the NMWD may be a testing ground for the non-leptonic baryon-baryon weak interactions. Of course, further experimental as well as theoretical works are needed. Especially, more improvement in experimental data is strongly desired.

Acknowledgements

Initially this work was presented in the 25th INS International Symposium on Nuclear and Particle Physics at Tokyo in 1996 and later in the International Conference on Hypernuclear and Strange Particle Physics at BNL in 1997.

References

1. J. Adams, *Phys. Rev.* **156**, 1611(1967)
2. B. McKellar and B. Gibson, *Phys. Rev.* **C30**, 322(1984)
3. J. Dubach, *Nucl. Phys.* **A450**, 71c(1986)
4. H. Bando, T. Motoba and J. Ẑofka, *Int. J. Mod. Phys.* **A5**, 4021(1990)
5. E. Oset and L. Salcedo, Nucl. Phys. **A443**, 704(1985)
6. K. Takeuchi, H. Takaki and H. Bando, *Prog. Theor. Phys.* **73**, 841(1985)
7. A. Ramos, E. Meijgaard, C. Bennhold and B. Jennings, *Nucl. Phys.* **A544**, 703(1992) A. Ramos, E. Oset and L. Salcedo, *Phys. Rev.* **C50**, 2314(1994); A. Parreno, A. Ramos and C. Bennhold, *Phys. Rev.* **C52**, 1768(1995)
8. K. Itonaga, T. Ueda and T. Motoba, *Nucl. Phys.* **A585**, 331c(1995)
9. J. Dubach, G. feldman and B. Holstein, *Ann. Phys.* (N.Y.), 249(1996)
10. J. Szymanski et al., *Phys. Rev.* **C43**, 849(1991)
11. V. Zeps *et al.*, *Nucl. Phys.* **A639**, 261c(1998)

Table 1. Experimental data and calculated nonmesonic decay rates, Γ_n, Γ_p and $\Gamma_{nm} = \Gamma_n + \Gamma_p$ (in unit of Γ_Λ) and the ratio Γ_n/Γ_p. The data indicated by EXP1 and EXP2 were taken from BNL [10,11] and KEK [12,13] respectively. The final decay rates are in the rows $\Gamma^{\pi+4P}$, which include the contributions from the interference between OPE and 4P.

		Γ_n	Γ_p	Γ_{nm}	Γ_n/Γ_p
	EXP2	–	–	0.17(0.11)	–
	Γ^π	0.009	0.000	0.009	28.0
${}^4_\Lambda$H	Γ^{4P}	0.023	0.003	0.026	7.3
	$\Gamma^{\pi+4P}$	0.032	0.002	0.034	17.6
	EXP1	0.04(0.02)	0.15(0.02)	0.19(0.04)	0.27(0.14)
	EXP2	0.01(0.05)	0.16(0.02)	0.17(0.05)	0.06(0.4)
	Γ^π	0.001	0.125	0.126	0.006
${}^4_\Lambda$He	Γ^{4P}	0.020	0.035	0.055	0.58
	$\Gamma^{\pi+4P}$	0.024	0.162	0.186	0.15
	EXP1	0.21(0.11)	0.20(0.07)	0.41(0.14)	0.93(0.55)
	Γ^π	0.015	0.154	0.169	0.10
${}^5_\Lambda$He	Γ^{4P}	0.142	0.037	0.179	3.82
	$\Gamma^{\pi+4P}$	0.170	0.184	0.354	0.92
	EXP1	$0.65^{+0.2}_{-0.3}$	$0.49^{+0.3}_{-0.2}$	1.14(0.2)	$1.33^{+1.12}_{-0.51}$
	EXP2	0.58	$0.31^{+0.18}_{-0.11}$	0.89(0.18)	$1.87^{+0.91}_{-1.59}$
	Γ^π	0.028	0.237	0.265	0.12
${}^{12}_\Lambda$C	Γ^{4P}	0.184	0.048	0.232	3.83
	$\Gamma^{\pi+4P}$	0.229	0.276	0.505	0.83

12. H. Noumi *et al.*, Phys. Rev. **C52**, 2936(1995)
13. H. Outa *et al*, *Nucl. Phys.* **A639**, 251c(1998)
14. C. Cheung, D. Heddle and L. Kisslinger, *Phys. Rev.* **C27**, 335(1983)
15. T. Inoue, S. Takeuchi and M. Oka, *Nucl. Phys.* **A597**, 563(1996)
16. M. Block and R. Dalitz, *Phys. Rev. Lett.* **11**, 96(1963); V. N. Fetisov, *Nucl. Phys.* **A585**, 197c(1995)
17. B. Gibson, *Il Nouvo Cimento*, **102A**, 367 (1989)
18. J-H. Jun and H. C. Bhang, *Proc. 25th INS Int. Symp. on Nucl. and Particle Phys.*, World Scientific, Tokyo 457 (1996); *Genshikaku Kenkyu* **41**, 103 (1997); J-H. Jun, D. Kim and H. C. Bhang, *Nucl. Phys.* **A639**, 337c(1998); J-H. Jun and H. C. Bhang, *J. Korean Phys. Soc.* **34**, 181(1999)
19. M. J. Savage and J. Walden, *Phys. Rev.* **D55**, 5376(1997)
20. L. Okun, *Leptons and Quarks*, North-Holland, (N.Y.), (1982); A. I. Vainstein, V. I. Zakharov and M. A. Shifman, *JETP* **45**, 670(1997)
21. M. L. Goldberg and S. B. Treiman, *Phys. Rev.* **110**, 1178(1958)

QUARK DESCRIPTION OF INTERACTION FOR PIONIC DECAY OF Λ PARTICLE

S. Y. YOO AND IL-TONG CHEON

Department of physics, Yonsei University, Seoul, 120-749, Korea

The $\Lambda \to N\pi$ process is investigated with the four-point effective Lagrangian and baryons are described by a bag model. It is shown that a right relative sign between the parity-conserving and parity non-conserving amplitudes can be obtained by introducing an effective mass into the propagating quark.

1 Introduction

The non-leptonic decay of lambda particle is usually described by the general form of Lagrangian consisting of the parity-violating and parity-conserving terms,

$$\mathcal{L}_{\pi N\Lambda} = Gm_\pi^2 \Psi_N (A + B\gamma_5)\tau\Psi_\Lambda \cdot \phi_\pi + h.c., \tag{1}$$

where $Gm_\pi^2 = 0.2211 \times 10^{-6}$, $A = 1.06$ and $B = -7.10$ are determined by fitting the experimental values of the free lambda life-time $\tau = (2.632 \pm 0.020) \times 10^{-10} sec$ and asymmetry in the emitted pion angular distribution, $\lambda = 0.36 \pm 0.06$ [1].

For low energy region, the Lagrangian (1) can be reduced to the non-relativistic Hamiltonian,

$$H = \int \bar{\Psi}_N(\mathbf{r}) \left[s + p\frac{\boldsymbol{\sigma}\cdot\mathbf{q}}{q_0} \right] \exp(i\mathbf{q}\cdot\mathbf{r})\Psi_\Lambda(\mathbf{r}) d^3r + h.c., \tag{2}$$

where the momentum $q_0 = 100.4 MeV/c$ is calculated by the energy relation $M_\Lambda = (m_\pi^2 + q_0^2)^{1/2} + (M^2 - p + q_0^2)^{1/2}$.

Since the relation between the asymmetry of pion angular distribution α and p/s is defined as

$$\alpha = \frac{2Resp^*}{|s|^2 + |p|^2}, \tag{3}$$

we find

$$\alpha = \frac{2\lambda}{1+\lambda^2}, \tag{4}$$

where $\lambda = p/s$. The measured values of α is -0.62 ± 0.07 [1], from which $\lambda = -0.36 \pm 0.06$ is obtained, and then values of A and B are determined as given above. Here, the negative sign for λ should be noticed, consequently p is a negative value. Thus, coupling constants s and p are expressed as

$$s = \frac{Gm_\pi^2 A}{\sqrt{2\pi}} = 0.96 \times 10^{-7}, \tag{5}$$

$$p = \frac{Gm_\pi^2 B}{\sqrt{2\pi}}\frac{q_0}{2M_p} = -0.32 \times 10^{-7}. \tag{6}$$

2 Theory

In this work, the weak interaction is given by the effective Lagrangian derived from Weinberg-Salam-Glashow theory and the baryon states are described by MIT bag.

Now, the Lagrangian describing the non-leptonic weak process, which is evaluated with one gluon exchange and renormalization group scaling [578] , is

$$\mathcal{L} = G \sin\theta_C \cos\theta_C (a_3\ \mathcal{L}_3 + a_6\ \mathcal{L}_6), \tag{7}$$

where θ_C is the Cabibo angle and $\mathcal{L}_{3,6}$ are given as

$$\mathcal{L}_{3,6} = (\bar{u}^i \gamma_L^\mu s_i)(\bar{d}^j \gamma_{\mu L} u_j) \mp (\bar{u}^i \gamma_L^\mu u_i)(\bar{d}^j \gamma_{\mu L} s_j), \tag{8}$$

with $\gamma_L^\mu = \gamma^\mu (1 - \gamma_5)$.

In eq.8, the negative(positive) sign refers to $\mathcal{L}_3(\mathcal{L}_6)$. In the above equation, the subscripts 3 and 6 indicate the dimensions of SU(3) color group of finial state $\mathbf{3} \bigotimes \mathbf{3} = \mathbf{6} \bigoplus \bar{\mathbf{3}}$, and i, j represent the color indices. $a_3 \simeq 3$ and $a_6 \simeq 0.6$ which are evaluated at $\mu \simeq m_\pi$. Group structure tells us that $\mathcal{L}_3$ allows $\Delta I = 1/2$ transition while $\mathcal{L}_6$ includes both $\Delta I = 1/2$ and $3/2$.

For the $\Lambda \to p\pi^-$process, we have to consider quark diagrams shown in fig.1. Among those diagrams, (a)~(c) are called internal graphs and (d) is external graph. The internal graphs contribute only to the $\mathcal{L}_3$ transition channel, because minus signs (originated from color, anti-commuting of fermions and Fierz rearrangement) appear when the initial quark legs s, u in $(\bar{u}s)(\bar{d}u)$ are exchanged. However, in the case of external diagram, the hadronic matrix element is $\langle p|\bar{u}s|\Lambda\rangle\langle\pi^-|\bar{d}u|0\rangle$ and the factor $+1/3$ appears due to the color degree of freedom when the initial quark legs are exchanged. Thereby, the first and second terms of eq. (8) do not cancel each other and accordingly this external diagram contributes both to $\mathcal{L}_3$ and $\mathcal{L}_6$ channels. Thus the $\Lambda \to N\pi$ matrix element can be written as

$$\begin{aligned} \langle p\pi^-|\Lambda\rangle &= \ a_3(A_i + A_e) + 2a_6 A_e \\ \sqrt{2}\langle n\pi^0|\Lambda\rangle &= -a_3(A_i + A_e) + 2a_6 A_e\ , \end{aligned} \tag{9}$$

where A_e and $-A_i$ stand for contributions from external and internal graphs of $(\bar{u}s)(\bar{d}u)$ in $\Lambda \to n\pi^0$process. Needless to say, the small portion of $\Delta I = 3/2$ comes from the A_e in eq. (9).

To evaluate the internal and external contributions to the $\Lambda \to N\pi$ process in a consistent way, we make use of the formalism of reference[3]. In this scheme the $B' \to B\pi$ amplitude is expressed as

$$\begin{aligned} A_{B'B\pi} &= \langle B, \pi_{out}|B'_{in}\rangle \\ &= \int dx^4 dy^4 \phi_\pi^*(x) \langle B_{out}|T\mathcal{L}(y) J_{\pi_i}(x) \delta_S|B'_{out}\rangle\ , \end{aligned} \tag{10}$$

where J_π is a pion source term:

$$J_{\pi_i}^i(x) = (i/2f_\pi)\bar{q}(x)\gamma_5 \tau^i q(x) \tag{11}$$

and $\phi_\pi(x)$ is the pion wave function in spherical cavity mode and δ_s is the surface delta function.

The diagrams to be considered are presented in (a) $\sim$ (d) of fig. 2 which correspond to each graph in the fig.1. Hence the (d) diagram is the newly considered process in this paper which is named as external diagram.

With the usual bag solution in ref. [9] and detailed calculation method given in refs.[3] and [10], we can write the matrix element as

$$A_{\Lambda\to p\pi^-} = 2\pi\delta(E_\pi + E_p - E_\Lambda)\chi_P^\dagger(S - P\boldsymbol{\sigma}\cdot\hat{\boldsymbol{k}})\chi_\Lambda \ , \tag{12}$$

where S and P denote parity violating and conserving part expressed as

$$S(P) = \frac{G}{2f_\pi}\sin\theta_C\cos\theta_C\Big[a_3\{S_i(P_i) + S_e(P_e)\} + 2a_6 S_e(P_e)\Big]. \tag{13}$$

The subscripts in $S(P)_{i,e}$ denote the internal and external contributions. The internal parts calculated in the forms

$$S_i = \frac{R^2\varphi_0(kR)}{4\pi\sqrt{6}}\sum_n\Big\{T_s(\omega_n^{'}) + T_s(-\omega_n)\Big\}$$

$$P_i = \frac{R^2\varphi_1(kR)}{4\pi\sqrt{6}}\sum_n\Big\{T_p(\omega_n) + T_p(-\omega_n^{'})\Big\}, \tag{14}$$

where $\varphi_l(kR)$ is the radial part of the pion wave function with angular momentum l and R is the bag radius. While the first term in summation of eq. (14) is result from the process including quark propagator, the second one is result from the process including anti- quark propagator. ω_n is the nth term of the series of eigenvalues satisfying bag boundary condition, $j_0(\omega_n R) = j_1(\omega_n R)$ and $\omega_n^{'}$ is $j_0(\omega_n^{'}R) = -j_1(\omega_n^{'}R)$.

In eq. (14), $T_s(\omega_n)$ and $T_p(\omega_n)$ are given as

$$\begin{aligned} T_s(\omega_n) &= 4\Big\{(\omega_0,\omega_0)(\omega_n,p_s) - (\omega_0,\omega_n)(\omega_0,p_s)\Big\}/D(\omega_0,-\omega_n) \\ &\quad -6(\omega_0,\omega_0)(\omega_n,p_s)/D(-\omega_0,\omega_n), \\ T_p(\omega_n) &= \Big\{\frac{8}{3}(\omega_0,\omega_0)(\omega_n,p_s) + \frac{4}{3}(\omega_0,\omega_n)(\omega_0,p_s)\Big\}/D(\omega_0,-\omega_n) \\ &\quad -6(\omega_0,\omega_0)(\omega_n,p_s)/D(-\omega_0,\omega_n), \end{aligned} \tag{15}$$

where the term which has $D(\omega_0,-\omega_n)$ in denominator comes from the diagram (c) and the term which has $D(-\omega_0,\omega_n)$ is from (b) in fig. 2. Here the following notations are used,

$$\begin{aligned} &(\alpha,\beta)(\gamma,p_s) = N_\alpha N_\beta N_\gamma N_s \\ &\times\int_0^R drr^2\Big(j_0(\alpha r)j_0(\beta r) + j_1(\alpha r)j_1(\beta r)\Big)\Big(j_0(\gamma r)j_0(p_s r) + \Omega_s j_1(\gamma r)j_1(p_s r)\Big), \end{aligned} \tag{16}$$

and

$$\frac{1}{D(\alpha,\beta)} = N_\alpha N_\beta \frac{2j_0(\alpha R)j_0(\beta R)}{E_\pi + \alpha + \beta}. \tag{17}$$

N_α is a normalizing constant of bag solution with energy ω_α and j_l is the spherical Bessel function. The p_s, m_s and ω_s stand for the momentum, mass and energy of s quark respectively, and Ω_α is $p_\alpha/(\omega_\alpha + m_\alpha)$. The external parts calculated in the same way are given

$$\begin{aligned} S_e &= \frac{6R^2\varphi_0(kR)}{4\pi\sqrt{6}} \sum_{m,n} \Big\{ L_s(\omega_m,\omega_n) - L_s(\omega_m^{'},\omega_n^{'}) - L_s(-\omega_m,-\omega_n) + L_s(-\omega_m^{'},-\omega_n^{'}) \Big\} \\ P_e &= \frac{4R^2\varphi_1(kR)}{4\pi\sqrt{6}} \sum_{n,m} \Big\{ -L_p(\omega_m,\omega_n^{'}) + L_p(-\omega_m,-\omega_n^{'}) \Big\}, \end{aligned} \tag{18}$$

where,

$$\begin{aligned} & L_s(\omega_n,\omega_m) \\ &= \Big\{ (\omega_0,\omega_m)(\omega_n,-p_s) + (\omega_0,\omega_n)(\omega_m,-p_s) - (\omega_m,\omega_n)(\omega_0,-p_s) \Big\} / D(-\omega_m,-\omega_n) \\ & L_p(\omega_m,\omega_n^{'}) \\ &= \Big\{ (\omega_0,\omega_m)(\omega_n^{'},-p_s) + (\omega_0,\omega_n^{'})(\omega_m,-p_s) + (\omega_m,\omega_n^{'})(\omega_0,-p_s) \Big\} / D(-\omega_m,-\omega_n^{'}) . \end{aligned} \tag{19}$$

On the other hand, the diagram (a) in fig. 2, which gives only p wave, can be calculated as

$$\begin{aligned} P_i^{'} &= \frac{N_0^6 R^2 j_0(\omega_0 R)^2}{\sqrt{6}E_\pi} 6(\omega_0,\omega_0)(\omega_0,p_s) \leftarrow (x_0 < y_0) \\ &- \frac{N_0^6 R^2 j_0(\omega_0 R)^2}{\sqrt{6}E_\pi} 6(\omega_0,\omega_0)(\omega_0,p_s) \leftarrow (y_0 < x_0), \end{aligned} \tag{20}$$

and consequently, this term vanishes as was seen in ref.[3].

In this stage, we have to make some comment on the above results. The numerical estimate of S and P does not well reproduce the experimental data. This situation was seen in ref.[3] either. The differences between our result and that of [3] are whether the diagram (d) of fig. 2 is taken into account in the effective Lagrangian. But these effects can not be so large to alter the sign of p part changes because they relate to the $\Delta I = 3/2$ transition which should be small in context of the $\Delta I = 1/2$ rule.

To describe the pole structure consistently, we do not alter the above equation except introducing an effective mass m^* to the quarks in propagator. It means that, when the process (b) in fig. 2 is considered, the series of eigenvalues are shifted by linking up with the boundary condition $j_0(p_n R) = \Omega_n j_1(p_n R)$ and the shifted ground state energy $\omega_0^* = \sqrt{p_0^2 + m^{*2}}$ satisfies,

$$\omega_0^* - \omega_0 \simeq M_\Sigma - M_\Lambda, \tag{21}$$

R	0.6	0.7	0.8	0.9	1.0	1.1	exp. in ref. [11]
$\langle n\pi^0\|\Lambda\rangle^2/\langle p\pi^-\|\Lambda\rangle^2$	0.561	0.563	0.558	0.547	0.531	0.509	0.560

Table 1. The ratio of $\Lambda \to p\pi^-$ and $\Lambda \to n\pi^0$.

which is difference between initial(Λ) and intermediate(Σ) state energies. This alters the second term of $T_p(\omega_n)$ as,

$$-6(\omega_0, \omega_0)(p_n, p_s)/D(-\omega_0, \omega_n^*) \tag{22}$$

with $j_1(p_n r) \to \Omega_n j_1(p_n r)$ and $N_n \to N_{\omega_n^*, p_n, m^*}$.

The first term of $T_p(\omega_n)$ is not affected by this procedure because intermediate(n) and final(p) state have almost same static energy.

Note that $T_p(-\omega_n^{'})$ term in eq. (15) is not affected by the mass shift, because this contains anti-quark propagation and therefore, it does not form the baryon pole. Since $P_i^{'}$ in eq.(21) has no quark propagator, it has nothing to do with the effective quark mass shift and therefore, it vanishes also.

In numerical estimate, it is enough to consider first few terms of the summation in evaluating the internal amplitudes, because it converges very rapidly. However in case of the external amplitudes, the double summation makes very slow convergence and we have to carry out the summation over somewhat large number of terms. Actually, we do not worry about the divergence of external amplitudes.

3 Result and Discussion

Table 1 shows the ratio of the decay rate $\langle p\pi^-|\Lambda\rangle^2/\langle n\pi^0|\Lambda\rangle^2$ versus of R. The experimental value is well reproduced at about $R \simeq 0.7 \sim 0.8fm$ and hence $\Delta I = 3/2$ channel is sufficiently suppressed around that range of R about $\langle\Lambda|\Delta I = 3/2\rangle/\langle\Lambda|\Delta I = 1/2\rangle \simeq 3\%$. Fig. (3) shows the S and P as a function of the bag radius R together with experimental values. In this figure, we can see that experimental values of S and P are well reproduced at about $R = 0.7 \sim 0.8fm$, and they have right relative signs.

From the above analysis, one can see two difference features between results of our scheme and those of ref. [4]. Firstly, the description of pole structure is different. Although the pole approximation in ref. [4] only consider baryon pole in propagator, our effective quark mass description alters not only the propagator but also the form factor of $T_p(\omega)$. This makes the sum of two contribution (b) and (c) in fig. 2 small. Secondly, while the amplitude of (a) in fig. 2 vanishes in our case, it gives an important contribution in ref.[4](In fact, $P_i^{'}$ and P_{i0} are not considered separately in ref. [4]).

By those two reasons mentioned above, contributions of diagrams (a),(b) and (c) in fig. 2 to P_i become smaller than those of ref.[3], [4]. It can be seen In table 2, i.e. P_i is much suppressed compared with those values of refs. [3],[4] and therefore, the external contribution happens to be important in treatment of the p part within our scheme.

R	0.60	0.65	0.70	0.75	0.80	0.85
$P_i \times 10^7$	0.15	0.14	0.12	0.11	0.09	0.08
$P_e \times 10^7$	0.24	0.20	0.18	0.15	0.14	0.12
$a_3 P_i + (a_3 + 2a_6) P_e \times 10^{-7}$	1.45	1.26	1.10	0.98	0.84	0.75
$\Lambda^0_- \times 10^7$	0.77	0.61	0.48	0.39	0.32	0.26

Table 2. P_i is result of the diagrams (a), (b) and (c) of fig. 2 and P_e is obtained by the diagram (d) of fig. 2. The Λ^0_- is p part of ref. [4].

Acknowledgment

This work was supported by Korean Ministry of Education(98-015-D00061 & 1998-001-D00249) and the Korean Science and Engineering Foundation (976-0200-002-2 & 961-0204-018-2)

References

1. J.W. Cronin et al., Phys. Rev. **129**, 1795 (1963).
2. J.F. Donoghu and E. Golowich, Phys. Rev. **D 14** , 1386 (1976); H. Galic, D. Tadic and J. Trampetic, Nucl. Phys. **B 158** 306 (1979); J. F. Donoghu et al. Phys.Rev. **D 21** 186 (1980); D. Tadic and J. Trampetic, Phys. Rev. **D 23** 144 (1981).
3. P. Gonzalez, S. Noguera, J. Bemabeu and V. Vento, Nucl. Phys. **A 423**, 477 (1984).
4. J.A. Oteo and S. Noguera, Phys. Lett. **B 161** 377 (1985).
5. M.K. Gaillard and B.W. Lee, Phys. Rev. Lett. **33**, 108 (1974).
6. M.A. Shifman, A.I. Vainshtein and V.J. Zakharov, Nucl. Phys. **B 120**, 315 (1977).
7. C.G. Callan, R.F. Dashen and D.J. Gross, Phys. Rev. **D 19**, 1826 (1979).
8. L.B. Okun, "Leptons and Quarks" (North-Holland publishing company, NY 1982).
9. G.A. Miller, A.W. Tomas and S. Theberge, Phys. Lett. **B 91** (1980) 1912.
10. O.V. Maxwell and V. Vento, Nucl. Phys. **A 407** (1983) 366.
11. The American Physical Society, Phys. Rev. **D 50** (1994).

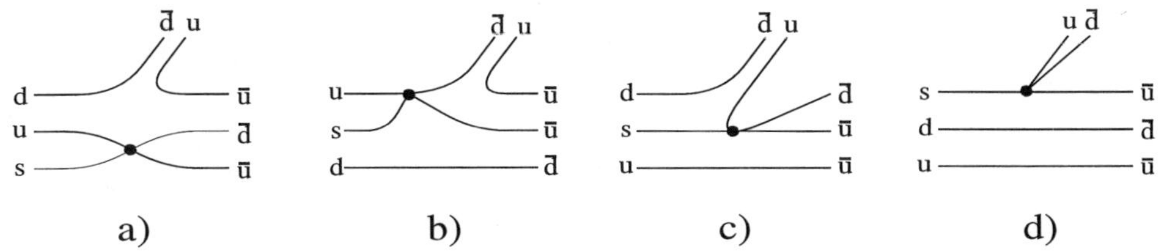

Figure 1. These graphs are possible processes we considered here for the $\Lambda \to p\pi^-$ decay, (a),(b) and (c) represent internal processes and (d) is external process.

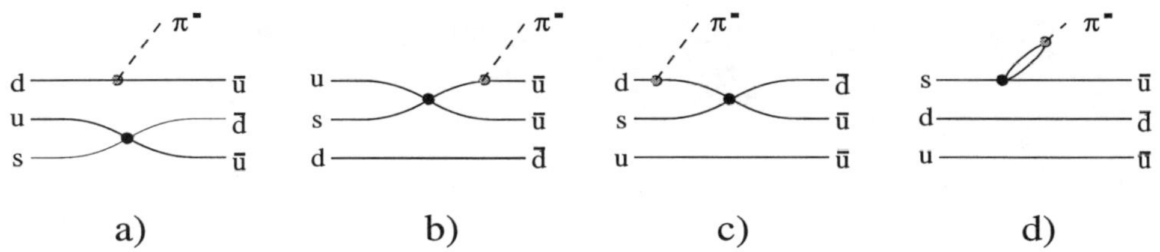

Figure 2. Diagrams generated by the time ordering in the bracket of eq. (4). Each diagram represents two cases $x_0 > y_0$ and $y_0 > x_0$, where x and y are the interaction points of $\bar{d}u\pi^-$ and $(\bar{u}s)(\bar{d}u)$ respectively.

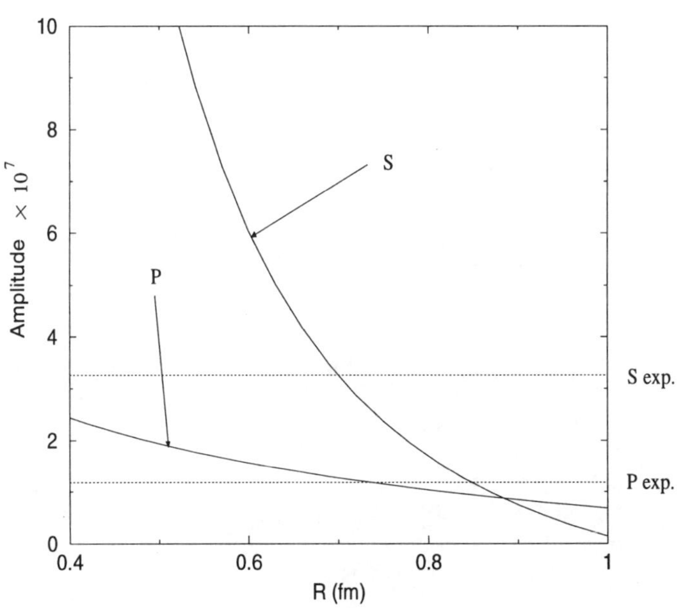

Figure 3. S and P amplitudes versus the bag radius.

Λ-HYPERON SUPERFLUIDITY IN NEUTRON STARS AND IN HYPERONIC NUCLEAR MATTER

T. TAKATSUKA
Faculty of Humanities and Social Sciences, Iwate University,
Morioka 020-8550, JAPAN
E-mail: takatuka@iwate-u.ac.jp

R. TAMAGAKI
Kamitakano Maeda-Cho 26-5, Sakyo-ku,
Kyoto 606-0097, JAPAN
E-mail: tamagaki@yukawa,kyouto-u.ac.jp

Neutron star problems relevant to strange hadronic matter are briefly introduced, and then superfluidity of Λ-hyperons due to the $^1S_0$ pairing is discussed on the basis of the results obtained in a realistic approach. It is found that Λ-superfluid is realized in a restricted density region $\rho \sim (\rho_t - \rho_d)$ of neutron star cores with $\rho_t \simeq 2\rho_0(\rho_0$ being the standard nuclear density) and $\rho_d \simeq (2.6-3.7)\rho_0$ depending on the hyperon core model. This restriction suggests that the neutron stars compatible with hyperon cooling scenario would be not so massive. $^1S_0$ pairing effects in hyperonic nuclear matter associated with Λ hyernuclei is also discussed and found to generate the Λ energy gap of $(0.3-0.4)$ MeV, in contrast with the vanishing of $^1S_0$ energy gap in usual symmetric nuclear matter with ρ_0. Some comments are given to the $^1S_0$ pairing of Σ^-.

1 Introduction

Study of neutron stars from the view of hyperon degrees of freedom is of increasing interest, being stimulated by the recent advances in hypernuclear physics. The data from scattering experiments and hypernuclei are providing us with precious opportunities to have the knowledges of YN and YY, as well as NN, interactions. On the other hand, neutron stars bring about another interesting system to check our knowledges and extract new information about the interactions, since they include a wide density region ranging from subnuclear density to several times nuclear density ρ_0 and thereby are expected to enlarge the properties of interaction latent for finite nuclei.

As the total baryon density ρ of neutron-star matter composed of n, p, e^- and μ^- goes higher toward a center of neutron stars, the Fermi energy of neutrons, a dominant component, increases and eventually exceeds the mass difference between n and Y. Then it becomes energetically possible to replace n by Y as new constituents. The density fraction $Y_Y(\equiv \rho_Y/\rho)$ of admixed Y gets larger according to further increase of ρ. Actually, the threshold density ρ_t for the appearance of Y is made lower by the attractive interaction felt by Y in medium. This problem of hyperon mixing has been investigated by several authors[1~7]. At present, there seems a general agreement that the hyperon-mixed phase would set on at not so high density ($\rho_t \sim 2\rho_0$), although the threshold density $\rho_t(Y)$ and the ρ-dependent population $Y_Y(\rho)$ of respective hyperon species depend notably on the baryon-baryon interactions.

In this paper, we discuss the pairing effects of hyperons. Particularly we direct our attention to the Λ-hyperon pairing, because the uncertainties in YY interaction is relatively small for $\Lambda\Lambda$ case on account of the data from double Λ hypernuclei. First we ask whether the Λ hyperons in the hyperon-mixed phase mentioned above could be superfluid state or not [8]. Next we ask whether a sizable energy gap due to $\Lambda\Lambda$ pairing could exist in hyperonic nuclear matter (hypernuclear matter) closely related to Λ hypernuclei. Finally, we add some comments on the $\Sigma^-\Sigma^-$ pairing.

2 Strange hadronic matter and neutron stars

Before going to our subjects, here we briefly introduce some problems of neutron stars in the context of strange hadronic matter, which include a background of our present study. The interests are conceptually sketched in Fig.1.

2.1 Hyperon mixing and softened EOS (Fig.1(a))

The occurence of hyperon-mixed phase in neutron star cores causes a remarkable softening of the equation of state (EOS) at high densities and thereby makes lower the maximum mass M_{max} sustainable by the EOS. For example, M_{max} based on the EOS with Y-mixing is smaller by about $0.3M_\odot$ than that neglecting the mixing[3]. On the other hand, the mass of the neutron star PSR1913+16 is definitly known from observations as M(PSR1913+16)$\simeq 1.44M_\odot$. This gives a severe constraint to the equation of state (EOS) of neutron stars, i.e., it should not be so soft that it cannot sustain the mass of $1.44M_\odot$. That is, the EOS, which can satisfy the above condition ($M_{max} > 1.44M_\odot$) by neglecting hyperon mixing effects but cannot ($M_{max} < 1.44M_\odot$) by including the effects, should be ruled out. This leads to an interesting test to the baryon-baryon interaction underlying the EOS.

2.2 Late-time collapsing into BH (Fig.1(b))

Another problem is concerned with kaon-condensed phase, a strange hadronic matter with Bose condensation of kaons. Kaon condensation, firstly proposed by Kaplan and Nelson [9], has been discussed to occur at $\rho \gtrsim (3-4)\rho_0$, namely in the core region of neutron stars, and possible consequences on neutron star properties have gathered much attention. Of particular interest are a new rapid cooling mechanishm and a dramatic softening effect on EOS. In relation to the latter property, Brown and Bethe [10] pointed out an interesting possibility that newborn neutron stars might collapse into black holes (BH) several seconds after the birth.

This late-time collapsing scenario is as follows. Similarly to hyperon core case in 2.1, when kaon-condensed phase is realized in neutron star cores, M_{max} would be reduced to about $1.5M_\odot$ due to the significantly softened EOS. This point is consistent with the fact that the observed masses are concentrated around $M \sim 1.4M_\odot$. However, at the birth stage, M_{max}^{hot} for newborn hot neutron stars would be much higher, i.e., $M_{max}^{hot} > M_{max} \sim 1.5M_\odot$, because of the sifffening effects coming from high lepton fraction and high temperature. Then it could happen that newborn neutron stars with M between M_{max} and M_{max}^{hot} is metastable at the birth stage but go into BH's as the cooling proceeds and the lepton fraction decreases.

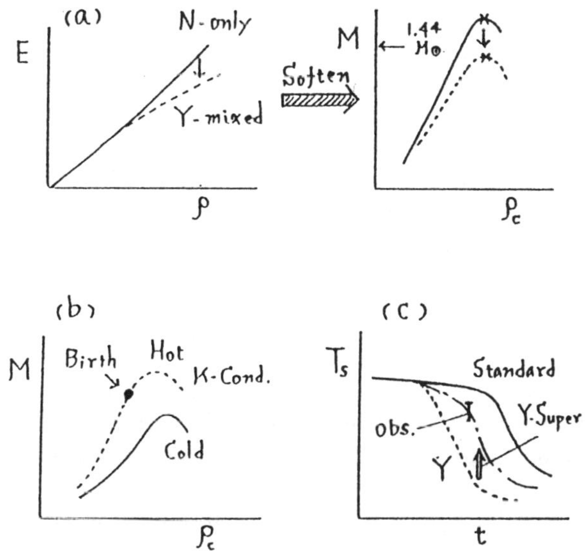

Figure 1: Schematic illustration for the effects of strange hadronic matter on neutron stars. (a) Reduction of the maximum mass due to hyperon mixing (total energy per baryon E versusu density ρ, mass M versus central density ρ_c); (b) metastable state and late time collapsing into black holes based on the EOS with kaon condensation; (c) rapid hyperon cooling controled by hyperon superfluidity (surface temperature T_s versus age t).

They attribute no finding of a neutron star in SN 1987A to this late-time collapsing scenario. So far very few studies have been done on hot neutron star models [11], although many interesting problems are there. The above scenario is one of them but was discussed on the basis of cold neutron star models. Of course, more realistic studies extended into hot cases are necessary [12].

2.3 Hyperon cooling (Fig. 1(c))

Finally we introdue a problem of neutron star cooling which is closely related to the motivation of our present study. Surface-temperature observations, combined with theoretical studies, suggest that some neutron stars are cooled much faster than expected by a standard cooling mechanism, namely, modifid URCA process (e.g., $n + n \to n + p + e^- + \bar{\nu}_e$ and the inverse processes), and demand more rapid cooling mechanism [13~15]. Theoretically several candidates for fast cooling mechanism have been proposed in relation to the exotic phases such as pion-condensed, kaon-condensed, hyperon-mixed phases and so on. For instance, the hyperon-mixed phase, if it occurs, makes possible the neutrino emission of usual β-decay type (e.g., $\Lambda \to p + e^- + \bar{\nu}_e, \Sigma^- \to \Lambda + e^- + \bar{\nu}_e$ and so on) without bystander nucleons needed in the standard process and extremely enhances the neutrino emissivity, providing a rapid cooling mechanism called "hyperon cooling" [16].

However, direct action of such a fast cooling leads to a surface temperature much lower than that observed. Therefore some suppression mechanism, most naturally, the coexistence of baryon superfluidity with these exotic phases would be necessary. Then, observations can be explained by an operation of fast cooling scenario dependent on M. That is, neutron stars with larger M (hence higher central density ρ_c exceeding the onset density of an exotic phase) possess a core with rapid cooling mechanism but those with smaller M do not (i.e., only by a standard scenario). The coexistence problem is of great interest and has been investigated for various exotic phases [17]. In the next section, we discuss the hyperon superfluidity associated with hyperon cooling scenario.

3 Λ superfluidity in neutron stars

Λ particles appear in a dense core of neutron stars ($\rho \gtrsim \rho_t \simeq 2\rho_0$) but the density $\rho_\Lambda (= Y_\Lambda \rho)$ is relatively low because of not so large Y_Λ((10 − 25)% at most) and so the $\Lambda\Lambda$ pairing responsible for the Λ superfluidity would primarily be in the $^1S_0$ pair state most attractive at low scattering energies. Quite similarly to the well-known $^1S_0$ pairing of nucleons [18], we solve the gap equation of $^1S_0$-type and obtain the energy gap Δ.

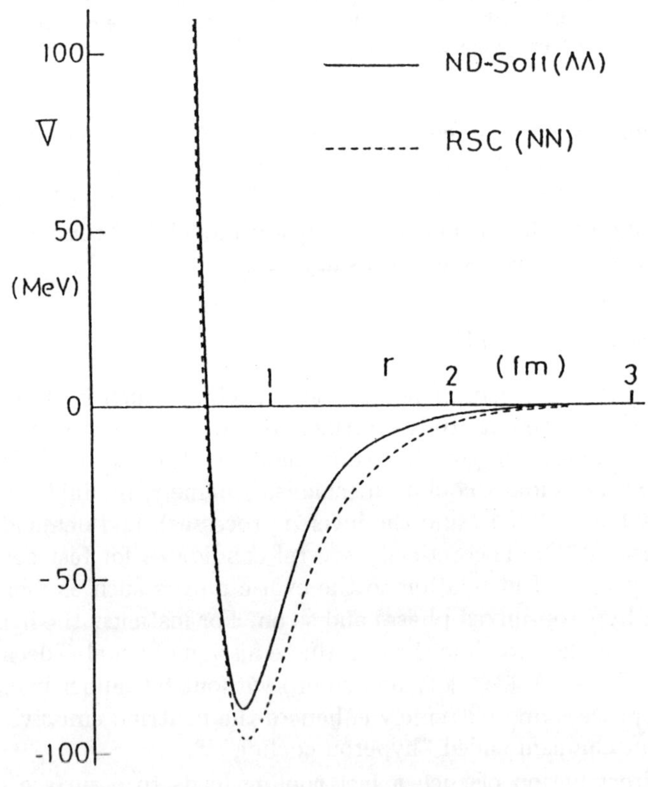

Figure 2: Comparison of $^1S_0$ $\Lambda\Lambda$ potential for ND-Soft with $^1S_0$ NN potential for RSC.

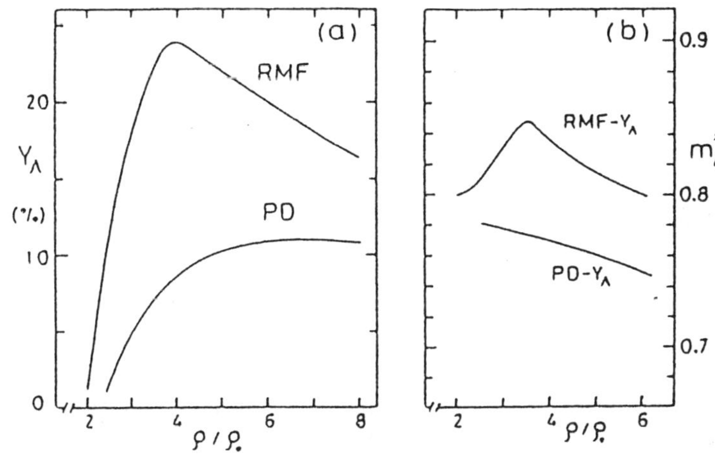

Figure 3: (a) Λ fraction Y_Λ versus total baryon density ρ for PD and RMF hyperon core models. (b) Corresponding effective mass parameter $m^*_\Lambda(\rho)$. ρ_0 denotes the nuclear density.

Very recently, Balberg and Barnea [19] calculated the energy gap and discussed the $^1S_0$ Λ-superfluidity. They used an effective $\Lambda\Lambda$ interaction $\tilde{V}_{\Lambda\Lambda}$ instead of a "bare" $\Lambda\Lambda$ one $V_{\Lambda\Lambda}$. However, their approach to use an effective interaction for the pairing interaction is not justified and results in giving too large energy gaps, as clearly shown in Ref.8). Here we make a realistic approach to pay attention to the following points [A]~[C] [8].

[A]: For a $\Lambda\Lambda$ pairing interaction $V_{\Lambda\Lambda}(^1S_0)$ in the $^1S_0$ state, we use the OBEP-type potential named here as ND-soft. This potential is a gaussian soft core version of Nijmegen-D hard core potential [20], constructed by Yamamoto [21] so as to fit the t-matrix from the original Nijmegen-D. For convenience, it is expressed by the gaussian superposition as

$$V_{\Lambda\Lambda}(r;{}^1S_0) = \sum_{\gamma=c,ss}\sum_{i=1}^{3} v_i^{(\gamma)}(\Lambda\Lambda)\exp(-(r/\beta_i)^2) \tag{1}$$

with $v_i^{(c)}(\Lambda\Lambda) = (-21.337, -187.01, 10853.3)$, $v_i^{(ss)}(\Lambda\Lambda) = -3\times(0.19321, 32.166, 2035.1)$ and $\beta_i = (1.342, 0.777, 0.35)$ corresponding to $i = (1,2,3)$. In the treatment of the gap equation, the soft-core potential is more suitable than the hard-core one. Of course, this potential reproduces well the data of double Λ hypernuclei. In Fig.2, the $V_{\Lambda\Lambda}(^1S_0)$ is compared with usual $^1S_0$ NN potential from Reid-Soft-Core (RSC) [22] case. The $\Lambda\Lambda$ interaction in the $^1S_0$ state are less attractive than NN one, suggesting that Λ superfluidity is less likely compared with the nucleon case, as far as the pairing interaction is concerned.

[B]: The ρ-dependent population of Y depends largely on the baryon-baryon interaction and the calculational method employed. Therefore we consider two representative cases for the hyperon core model (HCM). The one (PD) is from the conventional potential description approach by Pandharipande[2] and the other (RMF) from the relativistic mean-field approach by Schaffner and Mishustin[5]. They are illustrated in Fig.3(a). We see the differnce between the two as a gradual (rapid) rise of Y_Λ with ρ for PD (RMF) and the largest Y_Λ of about 10%(25%) for PD (RMF). It is of interest to see how the difference affects the Λ superfluidity.

[C]: In solving the gap equation, we adopt the effective mass (M_Λ^*) approximation for the single particle energy of Λ and use the effective mass parameter $m_\Lambda^*(\equiv M_\Lambda^*/M_\Lambda)$ derived from the G-matrix calcualtions with Nijmegen-D potential for $\{n+\Lambda\}$ matter specified by Y_Λ. For higher densities ($\rho \gtrsim 3.5\rho$), we use m_Λ^* extrapolated from those by G-matrix calculations. The resulting m_Λ^* depends both on ρ and Y_Λ. It decrease with ρ for a given Y_Λ and increases with Y_Λ for a given ρ. For instance, $m_\Lambda^* \simeq (0.80 \to 0.75)$ for $Y_\Lambda = 0.05$ and $m_\Lambda^* \simeq (0.84 \to 0.77)$ for $Y_\Lambda = 0.15$, according to the increase of ρ as $\rho = (2 \to 6)\rho_0$. The ρ-dependent m_Λ^* taking into account the $Y_\Lambda - \rho$ relation from two HCM's are shown in Fig.3(b). We remark that Λ hyperons admixed have the effective mass by far larger than protons admixed in usual neutron star matter; e.g. m_Λ^*(PD)$\simeq$ (0.8 $\to$ 0.75) for $\rho = (2 \to 6)\rho_0$ in comparison with $m_p^* \simeq (0.6 \to 0.5)$ for $\rho = (1 \to 3)\rho_0$ and RSC potential. The large m_Λ^* is expected to act for the occurence of Λ superfluidity.

3.1 Numerical results and discussion

For convenience, we present numerical results in terms of the critical temperature T_c for Λ superfluidity which is related to Δ (in MeV) as $T_c \simeq 0.66\Delta \times 10^9$ K. When T_c exceeds the internal temperature $T_{in} \simeq 10^8$ K of neutron stars, the Λ superfluid

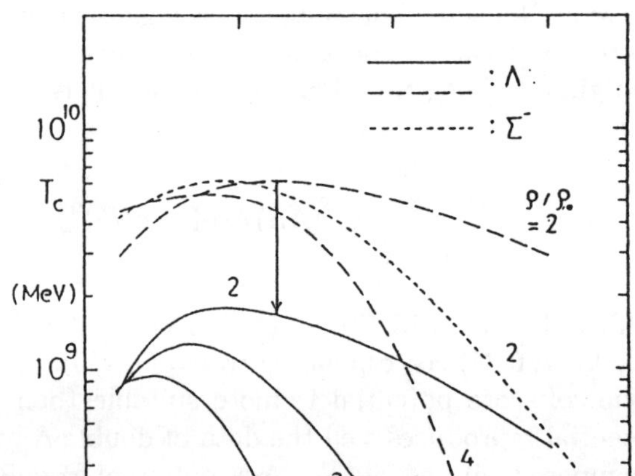

Figure 4: Critical temperature T_c of Λ superfluidity versus Y_Λ illustrated for several ρ (solid lines). Dashed lines show the case where $m_\Lambda^* = 1$ is taken ρ- and Y_Λ- independently. That of Σ^- superfluidity for $\rho = 2\rho_0$ case (dotted lines) is also shown for comparison.

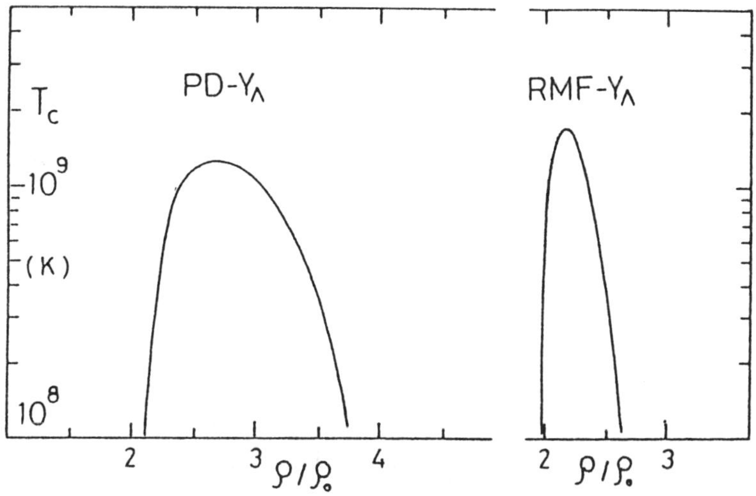

Figure 5: T_c versus ρ taking account of $Y_\Lambda - \rho$ relation from two hyperon core models (PD-Y_Λ and RMF-Y_Λ).

comes into existence. Calculated T_c is shown in Fig.4 as a function of Y_Λ, for $\rho = 2\rho_0, 3\rho_0$ and $4\rho_0$. We note the following points:
(i) Comparison between solid lines and dashed ones shows that T_c is much affected by the ρ- and Y_Λ- dependences of m^*_Λ.
(ii) Except the cases with smaller $Y_\Lambda(\lesssim 5\%)$, T_c decreases as ρ and/or Y_Λ increases. This is because ρ_Λ gets larger both with ρ and Y_Λ and a net attraction felt by the $\Lambda\Lambda$ pair is diminished due to the growth of short-range repulsion. For $Y_\Lambda < 5\%$, the increase of T_c with Y_Λ is due to the fact that Δ has a peak value when the Fermi energy $\epsilon_F \sim 15$ MeV.
(iii) For the parameter-range of interes, illustrated in Fig.3(a), we can expect that Λ superfluidity with $T_c \gtrsim T_{in}$ can surely exist in a hyperon core of neutron stars.

The results taking account of the Y_Λ-ρ relation (i.e., HCM) are shown in Fig.5, which indicates the point:
(iv) The Λ superfluidity disappears at high densities and so the existence density region becomes wider or narrower, depending strongly on the HCM adopted; $\rho \simeq (2-3.7)\rho_0$ for PD whereas $\rho \simeq (2-2.6)\rho_0$ for RMF.

4 $^1S_0$ pairing of Λ in hyperonic nuclear matter

Here we consider the hyperonic nuclear matter with $\rho = \rho_0$ composed of equal number neutrons and protons and in addition Λ hyperons with the mixing ratio Y_Λ, namely the so-called Λ hypernuclear matter ($\rho_n+\rho_p+\rho_\Lambda = \rho_0, \rho_n = \rho_p, \rho_\Lambda = Y_\Lambda\rho_0$). Our main concern is to ask whether the $^1S_0$ pairing correlation of Λ could generate the sizable energy gap. We study the problem by an approach quite similar to that in the preceding section. However, an important difference is that we have

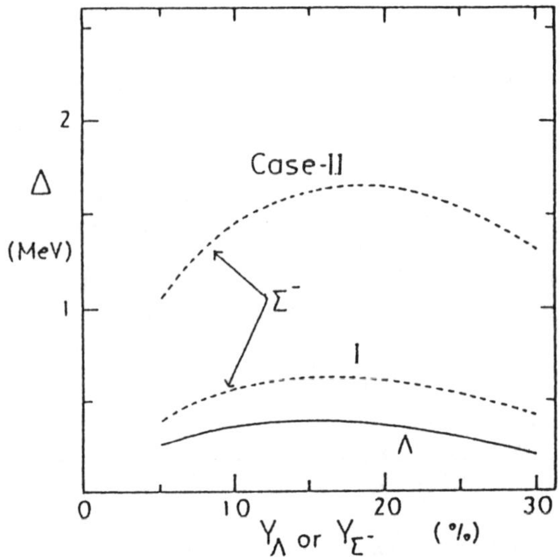

Figure 6: $^1S_0$ energy gap Δ of Λ versus Y_Λ in Λ hypernuclear matter. That of Σ^- in Σ^- hypernuclear matter is also shown for comparison.

to use m^*_Λ derived from the G-matrix calculation performed for Λ hypernuclear matter mentioned above. It is notable that m^*_Λ thus obtained are much larger than $m^*_n = m^*_p \sim 0.6$ in usual symmetric nuclear matter with $\rho = \rho_0$; e.g., $m^*_\Lambda \sim 0.84, 0.89$ and 0.92 corresponding to $Y_\Lambda = 10, 20$ and 30%.

The results for $^1S_0$ gap Δ are shown in Fig.6 as a function of Y_Λ. It is remarked that the energy gap of $(0.3 - 0.4)$ MeV is realized for the Y_Λ-range $(10\% - 30\%)$ of interest. This feature is in contrast with the well-known fact, i.e., the non-exsitence of $^1S_0$ gap for nucleons in usual symmetric nuclear matter [18] and is a consequence that in comparison with the latter case, the effect coming from remarkably larger m^*_Λ in the former overwhelms the disadvantage of weaker attraction of $V_{\Lambda\Lambda}(^1S_0)$ (Fig.2).

5 Comments on $\Sigma^-\Sigma^-$ pairing

At present, $\Sigma^-\Sigma^-$ interaction is more uncertain than $\Lambda\Lambda$ one, due to the lack of experimental information. However it is of interest and even instructive to address what can be said about the effects of $\Sigma^-\Sigma^-$ pairing. Here we treat the pairing in the $^1S_0$ state as in the preceding sections and consider two cases for $V_{\Sigma^-\Sigma^-}(^1S_0)$. In the first case (Case-I) we simply assume that $V_{\Sigma^-\Sigma^-} = V_{\Lambda\Lambda}$. In the second case (Case-II), we use $V_{\Sigma^-\Sigma^-}$ based on the OBEP. In practice we try to use the one constructed by Yamamoto [21]. It corresponds to ND-Soft (Eq.(1)) with potential parameters multiplied by factors; $v_i^{(\gamma)}(\Sigma^-\Sigma^-) = C_a v_i^{(\gamma)}(\Lambda\Lambda)$ for $i = 1, 2$ and $v_i^{(\gamma)}(\Sigma^-\Sigma^-) = C_r v_i^{(\gamma)}(\Lambda\Lambda)$ for $i = 3$, with $C_a = 1.37$ and $C_r = 1.85$. Similarly to $V_{\Lambda\Lambda}$, this $V_{\Sigma^-\Sigma^-}$ is constructed so as to reproduce the t-matrix from Nijmegen-D potential.

In Fig.6, the preliminary results for Σ^- hypernuclear matter are compared with Δ in Λ hypernuclear matter by taking simply $m^*_{\Sigma^-} = m^*_\Lambda$. We see that both of Case-I and Case-II give larger Δ to Σ^- hypernuclear matter. It is remarked that the Σ^- energy gap for Case-II amounts to $\sim$ 1.5 MeV. The result of $\Delta(\Sigma^-) > \Delta(\Lambda)$ even for the Case-I is due to the fact that $M_{\Sigma^-} > M_\Lambda$ and thereby the Fermi energy is relatively low under the condition $\rho_{\Sigma^-} = \rho_\Lambda$. Of course, we can expect that the Σ^- superfluidity in neutron stars is more likely to occur than the Λ superfluidity, as indicated by a dotted line in Fig.4. To have a final conclusion it is of course nesessary to use a realistic $m^*_{\Sigma^-}$ obtained from G-matrix calculations for the Σ^--mixed phase.

6 Concluding remarks

By a realistic approach to use $\Lambda\Lambda$ interactions compatible with experiments and to use the ρ- and Y_Λ-dependent m^*_Λ, we have calculated the $^1S_0$ energy gap for $\Lambda\Lambda$ pairing. It is found that the $^1S_0$ Λ-superfluidity with $T_c \sim 10^{8-9}$ K does indeed take place in hyperon cores of neutron stars. The existence density region of the Λ superfluid, however, is limited as $\rho \simeq (2\rho_0 - \rho_d)$ with $\rho_d \simeq (2.6 - 3.7)\rho_0$ depending on the hyperon core model employed. The appearance of upper limit ρ_d means that neutron stares consistent with hyperon cooling scenario are not so massive, e.g., $M \lesssim 1.5 M_\odot$ for the BJ-1H EOS [23], a standard one. This is because massive stars with $\rho_c > \rho_d$ have normal fluid in the central region and there loose a suppression mechanism indispensable for the too rapid hyperon cooling.

In Λ hypernuclear matter, the effect of $\Lambda\Lambda$ pairing is found to generate the $^1S_0$ energy gap of (0.3 – 0.4) MeV, in contrast with no $^1S_0$ gap in usual symmetric nuclear matter. This is due to that although the $^1S_0$ pairing interaction is less attractive, the effective mass in the former is remarkably larger than that in the latter.

We have also calculated the $^1S_0$ energy gap for Σ^-. Preliminary results shows the possibility that the $^1S_0$ $\Sigma^-\Sigma^-$ pairing produces larger energy gap than the $\Lambda\Lambda$ one; the superfluidity of Σ^- in neutron star cores is more likely to exist than that of Λ and the $^1S_0$ energy gap of Σ^- in Σ^- hypernuclear matter amounts to $\sim$ 1.5 MeV whereas that of Λ in Λ hypernuclear matter is (0.3 – 0.4) MeV. This comes from the stronger attraction in $^1S_0$ $\Sigma^-\Sigma^-$ interaction enhanced by the contributions from isovector mesons (ρ, π) absent in $\Lambda\Lambda$ channel. More detailed investigations on the $\Sigma^-\Sigma^-$ pairing and also on the $\Xi^-\Xi^-$ pairing are in progress and reported in near future.

Acknowledgments

The authors acknowledge Y. Yamamoto for providing the ND-Soft $\Lambda\Lambda$ and $\Sigma^-\Sigma^-$ $^1S_0$ interactions and useful discussions. They also thanks S. Nishizaki and Y. Yamamoto for the use of m^*_Λ from G-matrix calculations. This work is indebted to the Grant-in-Aid for Scientific Research of Education, Science and Culture (No. 1164025).

References

1. W.D. Langer and L. Rosen, *Astrophys. and Space Sci.* **6**, 217 (1970).
2. V.R. Pandharipande, *Nucl. Phys.* **A178**, 123 (1971).
3. N.K. Glendenning, *Phys. Lett.* **B114**, 392 (1982); *Nucl. Phys.* **A493**, 521 (1989).
4. F. Weber and M.K. Weigel, *Nucl. Phys.* **A505**, 779 (1989).
5. F. Schaffner and I.N. Mishustin, *Phys. Rev.* **C53**, 1416 (1996).
6. S. Balberg and A. Gal, *Nucl. Phys.* **A265**, 435 (1997).
7. M. Baldo, G.F. Burgio and H.J. Schultz, *Phys. Rev.* **C58**, 3688 (1998).
8. T. Takatsuka and R. Tamagaki, Proc. Int. Sym. on *Physics of Hadrons and Nuclei*, Tokyo, Dec.14-17, 1998, ed. Y. Akaishi, O. Morimatsu and M. Oka (to appear in Nucl. Phys. A); submitted to *Prog. Theor. Phys.*(1999).
9. D.B. Kaplan and A.E. Nelson, *Phys. Lett.* **B175**, 57 (1986).
10. G.E. Brown and H.A.Bethe,*Astrophys. J.* **423**, 659 (1994).
11. T. Takatsuka, *Prog. Theor. Phys.* **82**, 475 (1989);**85**, 427(1991); **95**, 901 (1996).
12. T. Tatsumi and M. Yasuhira, *Phys. Lett.* **B441**, 9 (1998); APCTP Workshop on *Strangeness Nuclear Physics*, Soeul, Feb. 19-22, (World Scientific 1999) to be published.
13. S. Tsuruta, *Comm. Astrophys.* **56**, 237 (1986).
    H. Umeda, K. Nomoto, S. Tsuruta, T. Muto and T. Tatsumi, *Astrophys. J.* **431**, 309 (1994).
    H. Umeda, S. Tsuruta and K. Nomoto, *Astrophys. J.* **433**, 256 (1994).
    S. Tsuruta, *Phys. Reports* **292**, 1 (1998).
14. D. Page and E. Baron, *Astrophys. J.* **354**, L17 (1990).
    D. Page and J.H. Applegate, *Astrophys. J.* **394**, L17 (1992).
    D. Page, *Astrophys. J.* **428**, 250 (1994).
15. C. Schaab, F. Weber, M.K. Weigel and N.K. Glendenning, *Nucl. Phys.* **A605**, 531 (1996).
16. Madappa Prakash, Manju Prakash, J.M. Lattimer and C.J. Pethick, *Astrophys. J.* **390**, L77 (1992).
17. T. Takatsuka and R. Tamagaki, *Prog. Theor. Phys.* **94**, 457 (1995); **97**, 263(1997); **97**, 345(1997); **98**, 393(1997); **101**, 1043(1999).
18. T. Takatsuka and R. Tamagaki, *Prog. Theor. Phys. Suppl.* **No.112**, 27 (1993).
19. S. Balberg and N. Barnea, *Phys. Rev.* **C57**, 409 (1998).
20. M.M. Nagels, T.A. Rijken and J.J. de Swart, *Phys. Rev.* **D15**, 2547 (1977).
21. Y. Yamamoto, T. Motoba, H. Himeno, K. Ikeda and S. Nagata, *Prog. Theor. Phys. Suppl.* **No.117**, 361 (1994); private communication with Y. Yamamoto.
22. R.V. Reid, *Ann. Phys.* **50**, 411 (1968).
23. R.C. Malone, M.B. Johnson and H.A. Pethe, *Astrophys. J.* **199**, 741 (1975).

STRUCTURE OF MULTI-STRANGE NUCLEAR MATTER

D. E. LANSKOY
Institute of Nuclear Physics, Moscow State University, 119899 Moscow, Russia
E-mail: lanskoy@npi.msu.su

Infinite hypernuclear matter consisting of nucleons, Λ, and Ξ hyperons is considered at various strangenesses s and electric charges q per baryon. The bound state region in the $s-q$ plane is infered. Ξ hyperons appear in the neutron-rich matter at rather low s. Strongly asymmetric matter is bound at $s \approx 1$ and, otherwise, only nearly isosymmetric systems can be probably bound at larger s. In the neutral matter, a crystal-like state with protons and Ξ^- hyperons in sites of a cubic lattice is possibly favorable.

1 Introduction

Generally, three types of multi-strange baryonic systems are of current interest. First, it is the neutron star matter, which is stable with respect to the strong as well as to the weak interaction. Secondly, it is highly unstable matter produced in relativistic ion collisions. We concern here in the third kind: multi-strange hypernuclear systems, which are stable with respect to the strong, but not to the weak, interaction.

For a long time, hypernuclei with $|S| \gg 1$ had been believed to consist of nucleons and Λ hyperons only. [1,2,3,4,5] More recently, it was shown [6,7,8,9,10] that ground states of multi-strange hypernuclei contain generally also Ξ hyperons. At some critical strangeness, conversion $\Xi N \Rightarrow \Lambda\Lambda$ becomes Pauli blocked. Hence, allowed baryonic compositions are dictated by chemical equilibrium with respect to the $\Xi N \Leftrightarrow \Lambda\Lambda$ conversion. It was shown [10] that a wide variety of multi-strange hypernuclei is possible including those with rather exotic A, S, and Z. Particularly, neutral hypernuclear systems (containing equal numbers of p and Ξ^-) can be bound. Such nuclei cannot be destroyed by the Coulomb interaction at any A.

So far, specific examples of finite multi-strange hypernuclei have been considered mainly. [10] In order to study average conditions of stability of such systems, we deal with the infinite matter at various strangenesses $s = |S|/B$ and electric charges $q = Z/B$ per baryon, where $S = N_\Lambda + 2N_{\Xi^0} + 2N_{\Xi^-}$, $Z = N_p - N_{\Xi^-}$, and $B = N_p + N_n + N_\Lambda + N_{\Xi^0} + N_{\Xi^-}$.

2 Model

Energy per baryon ε of the matter is a function of total density ρ and partial baryonic fractions $y_i = N_i/B$:

$$\varepsilon \equiv \frac{E}{B} = \varepsilon(\rho, y_p, y_n, y_\Lambda, y_{\Xi^0}, y_{\Xi^-}). \tag{1}$$

Only five of these variables are independent due to the normalization condition:

$$\sum_i y_i = 1. \tag{2}$$

Considering equilibrium states, we should impose the saturation condition:

$$\frac{\partial \varepsilon}{\partial \rho} = 0 \tag{3}$$

and the chemical equilibrium conditions:

$$C_1 \equiv \mu_p + \mu_{\Xi^-} - 2\mu_\Lambda + M_p + M_{\Xi^-} - 2M_\Lambda = 0, \tag{4}$$

$$C_2 \equiv \mu_n + \mu_{\Xi^0} - 2\mu_\Lambda + M_n + M_{\Xi^0} - 2M_\Lambda = 0, \tag{5}$$

where $\mu_i = \frac{\partial E}{\partial N_i}|_V$ is the chemical potential of baryon i. The last two conditions forbid the $\Xi N \Leftrightarrow \Lambda\Lambda$ conversion. If $y_{\Xi^-} = 0$ or $y_p = 0$, Eq. (4) is modified as $C_1 \geq 0$, and similarly for Eq. (5). Therefore, two variables (s and q for our choice) are actually independent. For arbitrary s and q, one can find ρ, y_i, and ε from the set (1)-(5).

Another condition ensures stability with respect to single-baryon emission:

$$\mu_i < 0. \tag{6}$$

As a dynamical model, we use the nonrelativistic effective interaction approach with Skyrme-like baryon-baryon potentials. Certainly, the current knowledge of hyperon-nucleon and especially hyperon-hyperon interactions is rather deficient. We try to utilize available hypernuclear data as fully as possible. However, as a general strategy, we should vary poorly known potentials and deduce qualitative features, which remain stable. Here, we present results obtained with parameter sets Sk3,[11] 1,[12] S$\Lambda\Lambda$2,[13] and SΞN2[14] for NN, ΛN, $\Lambda\Lambda$, and ΞN potentials, respectively (three-body forces are implied in these potentials). The first two parameter sets have been used widely and can be considered as reliable ones. The second two sets were fitted to data on $S = -2$ hypernuclei. The available data are too scarce for a quantitative extraction of the corresponding potentials, but these sets are hopefully at least reasonable.

As we have no information on $\Xi\Xi$ and $\Lambda\Xi$ interactions, we arbitrarily choose

$$V_{\Xi\Xi} = \alpha V_{NN}, \qquad V_{\Lambda\Xi} = \alpha V_{\Lambda N} \tag{7}$$

and show here results obtained with $\alpha = 0.5$.

3 Uniform matter and bound state region

The whole domain of s and q is $0 \leq s \leq 2$, $-s/2 \leq q \leq 1 - s/2$. However, ignoring the mass differences between p and n as well as between Ξ^0 and Ξ^-, we may restrict the consideration to the range $-s/2 \leq q \leq (1-s)/2$ due to isospin symmetry.

In Fig. 1, the bound state region (BSR) of the hypernuclear matter is presented on the $s - q$ plane. Only a half of the BSR is depicted. The full BSR is symmetric with respect to the $q \leftrightarrow 1 - s - q$, $p \leftrightarrow n$, and $\Xi^- \leftrightarrow \Xi^0$ transformation. This half of the BSR is enclosed by the n and Ξ^- drip lines, corresponding to the $\mu_n = 0$ and $\mu_{\Xi^-} = 0$ conditions, respectively. Domains of different baryonic compositions are also shown.

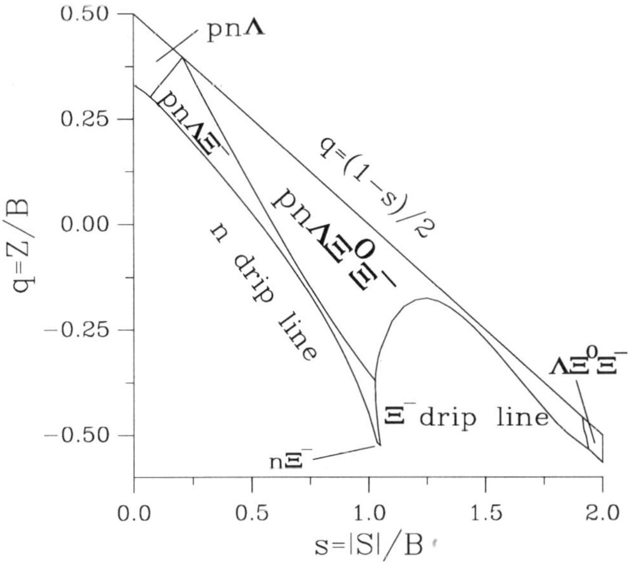

Figure 1: Bound state region.

It is seen that $pn\Lambda$ systems occupy a rather small part of the BSR. The $\Lambda\Lambda \Rightarrow \Xi N$ conversion begins at $s = 0.21$ in the isospin symmetry line and at s as small as 0.07 in the neutron drip line. In neutron-rich systems, μ_p is normally strongly negative, so the $\Lambda\Lambda \Rightarrow \Xi^- p$ conversion is promoted.

Another remarkable feature of the BSR is a pronounced "cape" at $s \approx 1$. For this particular choice of the potentials, the cape reaches the boundary $s = -q/2$, corresponding to purely $n\Xi^-$ systems. Otherwise, the BSR becomes extremely narrow at greater s.

The domain at s close to 2 corresponds to purely hyperonic systems. This area is affected by the potential ambiguities most. Purely Ξ bound systems ($s = 2$) occur up to $\alpha = 0.4$.

In Fig. 2, energy per baryon, density, and Ξ chemical potential are shown as functions of s for the symmetric matter. The binding energy per baryon reaches approximately 20 MeV at $s \approx 1$ compatible with the results of model 2 of Ref. [10] obtained in a rather different framework. The maximal total density is only little more than twice normal nuclear density, so the partial baryonic densities are not higher than those in ordinary nuclear matter. Ξ chemical potential approaches zero at $s \approx 1.4$. A relatively slight variation of the hyperonic potentials is sufficient to ensure positive μ_Ξ in this region. Then the BSR becomes discontinuous, i.e., the matter is bound at $s \leq 1$ and $s = 2$, but it is unbound in an intermediate region (cf. Fig. 17 in Ref. [10] and also Ref. [5]).

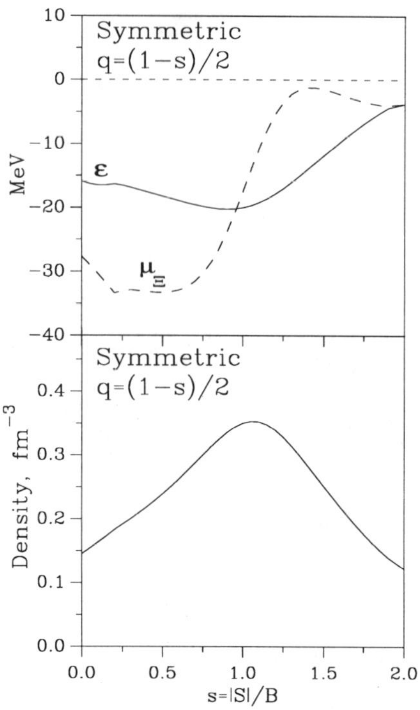

Figure 2: Energy per baryon (ε), Ξ chemical potential (μ_Ξ), and saturation density of the isosymmetric hypernuclear matter as functions of s.

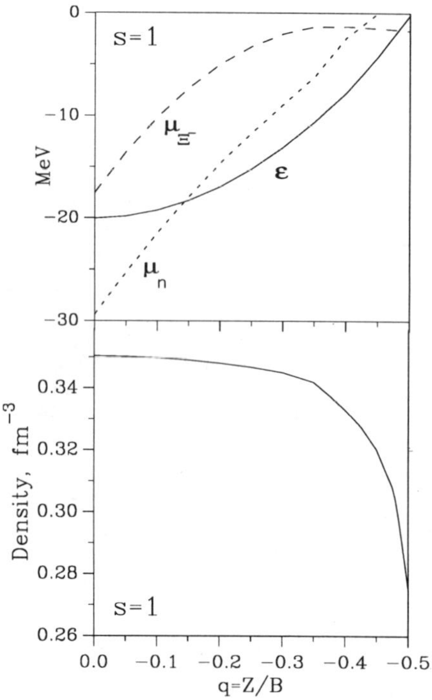

Figure 3: Energy per baryon (ε), n and Ξ^- chemical potentials (μ_n and μ_{Ξ^-}), and saturation density of the hypernuclear matter at $s = 1$ as functions of q.

In Fig. 3, energy per baryon, density, and chemical potentials are shown as functions of q at $s = 1$. It is seen that ε and μ_n grow monotonically with isospin asymmetry, which is natural for baryonic systems. However, μ_{Ξ^-} has a maximum and diminishes slightly in the strongly asymmetric matter. This behavior is originated from the drop of the density in the asymmetric matter. The corresponding decrease of the baryonic Fermi momenta leads to a decrease of the kinetic energies. Since hyperonic interactions are weaker than nucleonic ones, the kinetic energy contributes more substantially to μ_{Ξ^-} than to μ_n. Thus μ_{Ξ^-}, but not μ_n, becomes a nonmonotonic function of q. The complicated behavior of μ_{Ξ^-} is responsible for the shape of the Ξ^- drip line and, therefore, for the cape appearance (see also [5,15]).

Summing up this section, the $\Lambda\Lambda \Rightarrow \Xi^- p$ conversion is promoted in neutron-rich systems, which are Ξ^- abundant at $s > s_0$, where $s_0 \approx 0.06 - 0.08$. Strongly asymmetric systems can exist at $s \approx 1$. Otherwise, at $s \approx 1.2 - 1.6$, only nearly symmetric systems are possible, since Ξ hyperons are bound very loosely, if at all.

4 Crystal-like structure in neutral hypernuclear matter

Here, we concern in the neutral matter ($y_p = y_{\Xi^-}$, $q = 0$). It is seen from Fig. 1 that the neutral matter is bound in a wide range of s. Usually nuclear matter models are based on ignoring of the Coulomb interaction. However, it is not needed at $q = 0$: In the neutral uniform matter, the Coulomb energy vanishes exactly.

But the uniform state is not necessarily the true ground state. In a rather simple model, we consider a simple cubic lattice (typical for ionic crystals like, e.g., KCl) with protons and Ξ^- hyperons in the lattice sites (Fig. 4). The same energy density functional, as for the uniform matter, is used with the additional contribution of the Coulomb interaction. Kinetic energies are treated in spirit of the Thomas-Fermi approximation. For the charged baryons, we choose the harmonic oscillator densities:

$$\rho_p = \frac{1}{\pi^{3/2} r_0^3} \sum_{\xi\eta\zeta} \exp\left[-\frac{(\mathbf{r} - \mathbf{r}_m)^2}{r_0^2}\right], \tag{8}$$

$$\rho_{\Xi^-} = \frac{1}{\pi^{3/2} r_0^3} \sum_{\xi\eta\zeta} \exp\left[-\frac{(\mathbf{r} - \mathbf{R}_m)^2}{r_0^2}\right] \tag{9}$$

with the same parameter r_0. Here $\mathbf{r}_m = \xi\mathbf{i} + \eta\mathbf{j} + \zeta\mathbf{k}$, ξ, η, and ζ are integer numbers varying independently from $-\infty$ to ∞, $\mathbf{R}_m = \mathbf{r}_m + a(\mathbf{i} + \mathbf{j} + \mathbf{k})/2$, and a is the cell size (see Fig. 4). In terms of the averaged densities, $a^{-3} = \overline{\rho}_p = \overline{\rho}_{\Xi^-}$. We take uniform spatial distributions for the neutral baryons, ignoring possible polarization by the localized charged baryons.

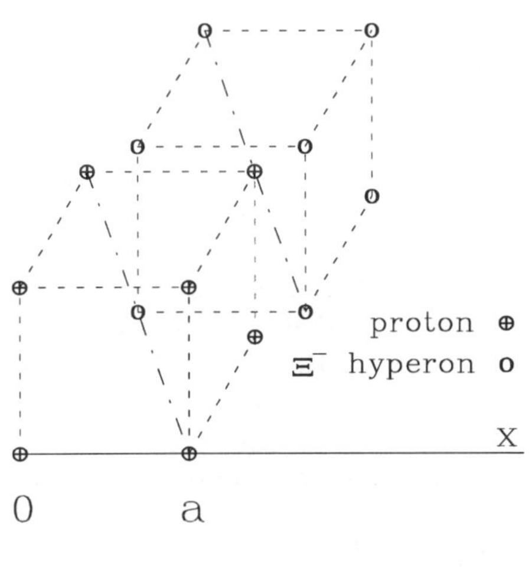

Figure 4: Elementary cells of the simple cubic lattice.

Generally, the set (1)-(5) should be solved again. However, we calculate for the first estimation the energy per baryon with the same average total density and the same baryonic fractions y_i as obtained before for the uniform matter, treating r_0 as the single variational parameter.

In Fig. 5, the total energy per baryon ($\varepsilon_{\rm crystal}$) as a function of r_0 is shown at $s = 1$. Contributions of the strong ($\varepsilon_{\rm strong}$) and Coulomb ($\varepsilon_{\rm Coulomb}$) interactions are depicted also separately. Kinetic energies are included in $\varepsilon_{\rm strong}$ as well. For convenience, $\varepsilon_{\rm total}$ and $\varepsilon_{\rm strong}$ are shifted by $\varepsilon_{\rm uniform} = -20.03$ MeV, which is the energy per baryon of the uniform matter.

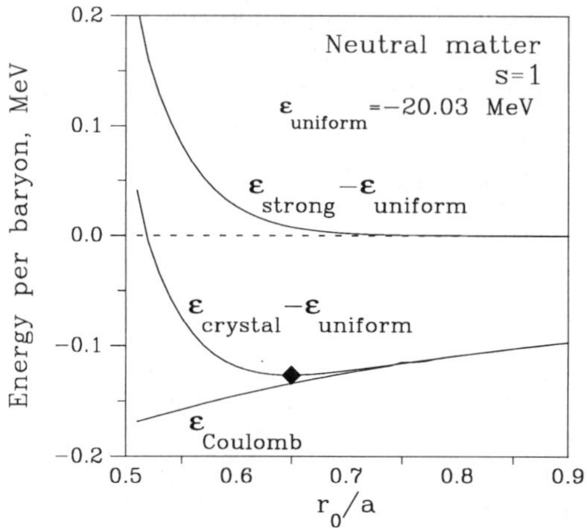

Figure 5: Total energy per baryon ($\varepsilon_{\rm crystal}$) and the contributions of the strong ($\varepsilon_{\rm strong}$) and Coulomb ($\varepsilon_{\rm Coulomb}$) interactions for the crystal-like state of the hypernuclear matter as functions of the oscillator parameter r_0 at $s = 1$. Kinetic energies are included in $\varepsilon_{\rm strong}$. The diamond indicates the minimum position.

It is seen that the uniform state is favorable with respect to the strong interaction, and $\varepsilon_{\rm strong}$ tends to $\varepsilon_{\rm uniform}$ at $r_0 \to \infty$ from above. However, $\varepsilon_{\rm Coulomb}$ tends to zero much more gradually. Hence, a broad minimum appears at $r_0 = 1.60$ fm $= 0.65a$. This means that the crystal-like state can be the genuine ground state of the matter.

Fig. 6 presents the energy gain due to the crystal ordering and the minimum position at various s belonging to the BSR. Both the values are rather stable. Existence of the minimum, its position, and depth are also fairly insensitive to the choice of the potentials since it is dictated mainly by the Coulomb interaction.

The energy gain is small relatively to the total energy per baryon. On the other hand, in a system of, for instance, several hundreds of baryons, the energy gain as large as tens of MeV may be expected between the crystal-ordered ground state and the lowest usual shell state.

However, we cannot claim conclusively that the crystal-like structure in the hypernuclear matter really occurs. The main hindrance is very high ratios r_0/a (Fig. 6). Note that a similar ratio in quantum crystals known in solid state physics is not greater than ~ 0.3 for solid ^{3}He (e.g., [16]). It is seen from Fig. 7 that actual modulations of p and Ξ^- densities are rather slight. In particular, the harmonic

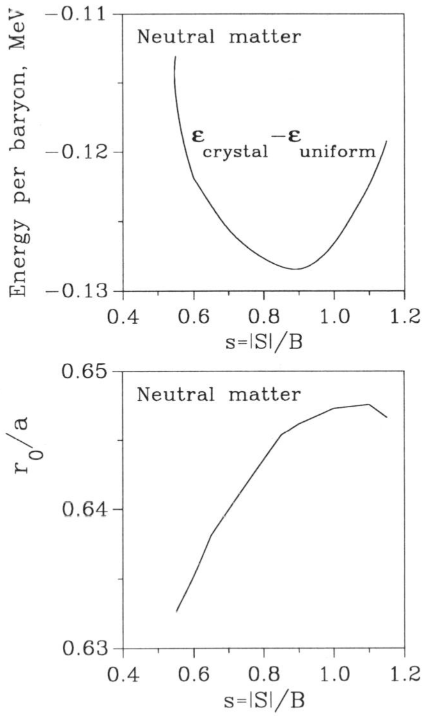

Figure 6: Energy gain due to the Coulomb interaction ($\varepsilon_{\text{crystal}} - \varepsilon_{\text{uniform}}$) and the minimum position (r_0/a) at various s belonging to the bound state region.

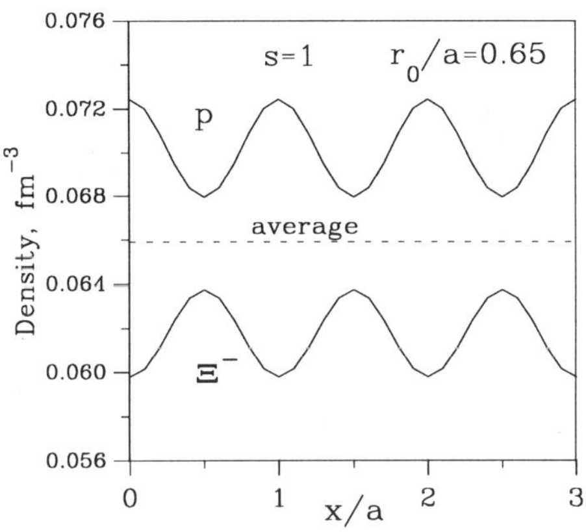

Figure 7: Proton and Ξ^- densities along a protonic axis (as shown in Fig. 4) at $s = 1$.

approximation (8)-(9) can be inadequate, and anharmonicity effects (known as important in quantum crystals) can be substantial.

On the other hand, high values of r_0/a justify some other approximations used here. In this case, the polarization of the neutral baryon distribution is likely small. Differences between saturation densities and baryonic fractions of the uniform and the lattice states are probably small too. Note that both the effects, which are neglected here, generally enhance the energy gain. Also the usage of the same energy density functional for the lattice and the uniform states is reasonable since the baryon density differences are small.

Note that the polarization of the neutral baryon distribution means in other words that the neutral baryons move in periodic fields. So the single-baryon spectra become generally band-structured, which modifies the chemical equilibrium conditions (4)-(5). It is very interesting to study these band spectra. However, the effect is possibly not considerable.

Summarizing, the lattice spatial structure in neutral (and, maybe, nearly neutral) multi-strange hypernuclear systems is possible. For definite conclusions, more comprehensive further study is needed. Nevertheless, it may be suggested that multi-strange hypernuclei are probably not balls, but rather cubes in some cases.

References

1. S. Nagata and H. Bandō, *Prog. Theor. Phys.* **72**, 113 (1984).
2. M. Rufa, J. Schaffner, J. Maruhn, H. Stöcker, W. Greiner, and P.-G. Reinhard, *J. Phys.* G **13**, 143 (1987); *Phys. Rev.* C **42**, 2469 (1990).
3. J. Mareš and J. Žofka, *Z. Phys.* A **333**, 209 (1989); A **345**, 47 (1993).
4. M. Barranco, R.J. Lombard, S. Marcos, and S.A. Moszkowski, *Phys. Rev.* C **44**, 178 (1991).
5. D.E. Lanskoy and T.Yu. Tretyakova, *Z. Phys.* A **343**, 355 (1992).
6. J. Schaffner, C. Greiner, and H. Stöcker, *Phys. Rev.* C **46**, 322 (1992).
7. C.B. Dover and A. Gal, *Nucl. Phys.* A **560**, 559 (1993).
8. J. Schaffner, C.B. Dover, A. Gal, C. Greiner, and H. Stöcker, *Phys. Rev. Lett.* **71**, 1328 (1993).
9. S. Balberg, A. Gal, and J. Schaffner, *Prog. Theor. Phys. Suppl.* **117**, 325 (1994).
10. J. Schaffner, C.B. Dover, A. Gal, C. Greiner, D.J. Millener, and H. Stöcker, *Ann. Phys.* **235**, 35 (1994).
11. M. Beiner, H. Flocard, N.V. Giai, and P. Quentin, *Nucl. Phys.* A **238**, 29 (1975).
12. Y. Yamamoto, H. Bandō, and J. Žofka, *Prog. Theor. Phys.* **80**, 757 (1988).
13. D.E. Lanskoy, *Phys. Rev.* C **58**, 3351 (1998).
14. D.E. Lanskoy, *Few-Body Systems Suppl.* **9**, 277 (1995).
15. D.E. Lanskoy, in *Proc. of the 23rd INS Intern. Symp., Tokyo*, eds. S. Sugimoto and O. Hashimoto (Universal Academy Press, Tokyo, 1995), p. 475.
16. R. Guyer, *Sol. Stat. Phys.* **23**, 413 (1969).

THE ELECTROPRODUCTION OF LIGHT HYPERNUCLEI †

E. V. HUNGERFORD
Department of Physics, University of Houston, Houston, TX 77204, USA
E-mail: hunger@uh.edu

It is anticipated that the first (e,e'K$^+$) experiments will introduce sub-MeV hypernuclear spectroscopy. By carefully crafting the experimental geometry, the signal to noise ratio in these experiments may be maximized as well as the resolution, and this low-luminosity layout may also allow future coincidence experiments to be implemented.

1 Introduction

The production of hypernuclei by secondary hadronic beams, particularly beams of kaons and pions, is well established[1]. It is less appreciated that $\approx$ 2 GeV electron beams are also capable of competitively producing hypernuclei in certain situations[2]. Indeed the new CW electron machines, of which the Continuious Electron Beam Acceleration Facility, CEBAF, at the Jefferson Laboratory, Jlab, is the best example, can produce light hypernuclei at reasonable rates, and with resolutions a factor of 3 to 5 times better that is presently possible with hadronic beams[3]. On the other hand, electroproduction experiments are somewhat more complicated, and it is not obvious that these experiments can be extended to observe particles emitted in coincidence with those involved in the production process.

In this brief paper I intend to develop a few thoughts on the use of electron machines to investigate hypernuclear production with high resolution. In fact the first high resolution electroproduction experiments[3] will be undertaken at Jlab at the beginning of the year, 2000. These experiments are intended to demonstrate the feasibility of an (e,e'K$^+$) experimental program at CEBAF. I will also present some preliminary ideas about experiments which could simultaneously detect particle emission from the de-exitation or decay of hypernuclear systems after they are formed. If these experiments are practical at Jlab, then the precision of electron beams can be exploited to develop a number of interesting measurements, including the study of hypernuclear weak decay[4], and the use of peturbed angular correlations to measure hypernuclear electromagnetic moments[5,6].

2 The High Resolution Hypernuclear Spectrometer System at Jlab

I will use the layout of CEBAF experiment E89-009 to discuss several issues in the design of a high resolution system to invesitgate the (e,e'K$^+$) reaction on nuclei. The kinematic parameters which define this reaction are illustrated in Figure 1. The precision of the electron beam, having a spatial resolution of about 40μm and an energy resolution, $\Delta E_e/E_e$, of $\leq 10^{-4}$, can translate into a missing mass resolution in the final hypernuclear system some 3-5 times better that experiments using hadronic beams. The reaction may be described in terms of the energy and

†Supported in part by the US DOE

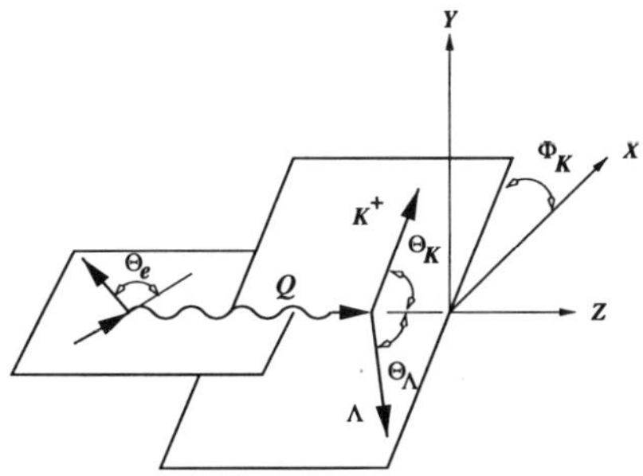

Figure 1: The Scattering Geometry for Electroproduction.

momentum transfers, $\omega = E_e - E'_e$, and $\mathbf{k} = \mathbf{p}_e - \mathbf{p'}_e$ respectively. From these definitions the 4-momentum transfer is written as, $Q^2 = \mathbf{k}^2 - \omega^2$

Now the cross section for electroproduction can be written in a very intuitive form by separating out a factor, which under certain conditions, may be identified as the virtual photon flux produced by (e,e') scattering[2]. Using this notation the electroproduction cross section may be written;

$$\frac{\partial^3\sigma}{\partial E'_e \partial\Omega'_e \partial\Omega_k} = \Gamma\left[\frac{\partial\sigma_T}{\partial\Omega_k} + \epsilon\frac{\partial\sigma_L}{\partial\Omega_k} + \epsilon\cos(2\phi) + \cos(\phi_k)\sqrt{2\epsilon(1+\epsilon)}\frac{\partial\sigma_I}{\partial\Omega_k}\right].$$

The factor, Γ, is the virtual flux factor evaluated with electron kinematics in the lab frame. It has the form;

$$\Gamma = \frac{\alpha}{2\pi^2 Q^2}\frac{E_\gamma}{1-\epsilon}\frac{E'_e}{E_e}.$$

In the above equation, ϵ is the polarization factor;

$$\epsilon = \left[1 + \frac{2|\mathbf{k}|^2}{Q^2}\tan^2(\Theta_e/2)\right]^{-1}.$$

For those virtual photons almost on the mass shell, ($Q^2 = \mathbf{p}_\gamma^2 - E_\gamma^2 \rightarrow 0$), the differential cross sections in the above expression can be approximated by the real photo-production cross sections. The label on each of cross section expressions (T, L, P, and I) represent transverse, longitudinal, polarization, and interference terms. For real photons of course, $Q^2 \rightarrow 0$, so only the transverse cross section is non-vanishing.

The virtual flux factor rises very steeply to a maximum at an electron scattering angle of $m_e\omega/4EE'$, Figure 2. For the energies involved here, this is very small, ≈ 2 milliradians, and falls to less than one-tenth its maximum value in 10 mr. Therefore if the electron spectrometer is placed at zero degrees, it need only have a small acceptance to assure that essentially all the scattered electrons are transported to the focal plane. It also means that aberations in the optics of this magnetic transport system will be small, and easily corrected.

On the other hand, zero-degree-electron tagging uses essentially transverse photons which precludes the investigation of all but transverse reactions. In addition it results in very high rates in the electron spectrometer due to bremsstrahlung in the target. In fact it is the bremsstrahlung rate which limits the luminosity. This

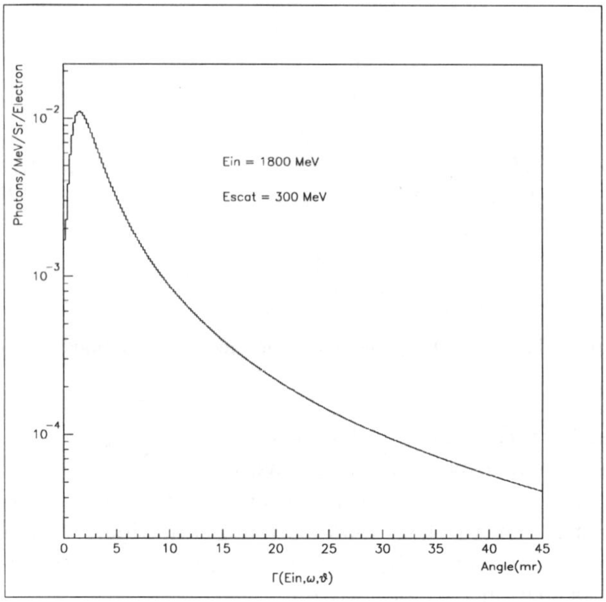

Figure 2: The Virtual Flux Factor

geometry can, however, maximize the resolution, signal-to-noise, and benifit/cost ratio of of these experiments.

The layout of the E89-009 experiment is shown in Figure 3. A 1-2 μA electron beam of primary energy $\approx$ 1.8 GeV strikes a thin target placed just before a small zero-degree dipole magnet. This magnet splits the scattered particles; the e^- of about 300 MeV/c into a split-pole spectrometer and the K^+ of about 1.2 GeV/c into the Short Orbit Spectrometer, SOS, fixed to the Hall C pivot at CEBAF. The magnetic optics of the splitting magnet/SOS system must be designed so that there is a virtual focus along the optic axis extension of the SOS magnet. In E89-09 the image occurs somewhat further away than for a system without the splitting magnet, reducing slightly the acceptance, but preserving the optics. The splitting magnet is necessary in order to capture the scattered electrons and reaction kaons at very forward angles ($\leq 3°$).

Because the target/beam intersection is approximately a point source, and almost all the electrons are contained in a small cone about zero degrees, tracking in the electron spectrometer is not required, and high rates, a few x 10^8 can be accepted. The maximum hypernuclear photoproduction rate occurs at photon energies of about 1.5 GeV, so the scattered electron energy will be approximately 300 MeV.

Hadronic rates in the SOS are small because the target luminosity is limited to approximately $10^{33}\text{cm}^{-2}\text{s}^{-1}$, and so the system can be triggered when a kaon is observed in the SOS. Note that this trigger does not require a coincidence with an electron, which requires careful application of time-of-flight in the high-rate, electron focal-plane detector. Such coincidences will still be made in the off-line analysis. The expected rates of positrons/pions/protons/kaons are $4x10^4/2x10^2/20/0.4$ s^{-1}, and this requires a system of velocity-selecting aerogel Cherenkov detectors, time-

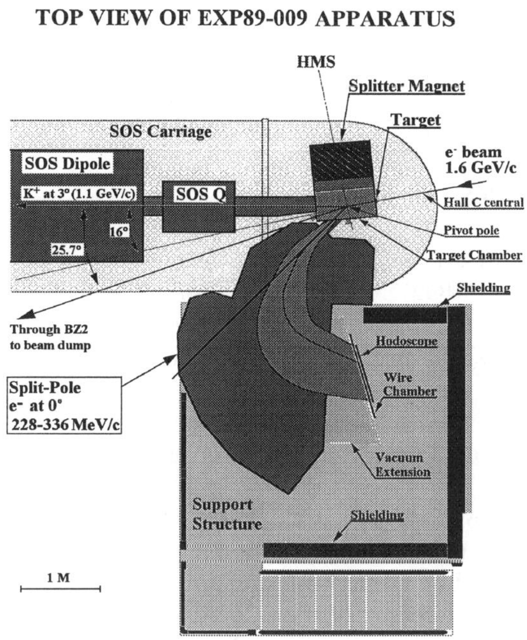

Figure 3: The Experimental Geometry for CEBAF E89-009.

of-flight, and shower counters to identify kaons within the background of other particles in the SOS. This particle identification system, PID, has been adequately demonstrated in previous (e,e'K) experiments at CEBAF[7].

It is also important to keep the incident electron beam energy below $\approx$ 1.8 GeV. A low beam momentum increases the experimental resolution, given a fixed Δp/p in the spectrometer magnets, and minimizes the sizes of these magnets and their field strengths. In addition, it limits kaon production only to elementary processes involving either Λs or Σs, as opposed to unwanted hyperon and meson resonances. Table I shows the primary energy thresholds for 1.2 GeV/c momentum kaon production at 10°.

The expected resolution for this experimental arrangement is estimated to approach 600 keV. The various contributions to this resolution are shown in Table 2, and are combined as if they were statistically independent. The system resolution is dominated by the SOS magnet, which is not designed for high resolution spectroscopy, as it has low dispersion and large momentum bite. However it is a short-orbit magnet (important as the decay length for a 1.2 GeV/c kaon is about 9m), is already mounted and tested at the Hall C pivot, and has the sophisticated PID package which is required to identify kaons within the expected large background of pions and positrons. Additional, but preliminary magnet designs, have demonstrated that a dedicated, high resolution replacement for the SOS could im-

Table 1. Representative Thresholds for K Photoproduction from a Nucleon with the Kaon Emitted at 1.2 GeV/c and 10°.

Reaction	Gamma Threshold(GeV)
$\gamma p \to K\Lambda$	1.55
$\gamma p \to K\Sigma$	1.67
$\gamma p \to K^*(892)\Lambda$	1.87
$\gamma p \to K\Lambda^*(1405)$	2.00
$\gamma p \to K^*(892)\Sigma$	2.02
$\gamma p \to K\Lambda^*(1520)$	2.20
$\gamma p \to \phi(1020)p$	2.30

prove the resolution in Table 2 by perhaps a factor of 2, with a substantial increase in rate[8].

Since the above system uses target thicknesses of $\approx$ 10mg/cm^2, almost any separated-isotope target of light atomic mass number can be used. As the target Z increases, electrons produced by bremsstrahlung processes will increase as Z^2/A, but hyperon production should increase as only as Z/A. Since bremsstrahlung limits the luminosity, the hypernuclear production rate should fall as 1/Z. Of course nuclear effects must be applied, but it would appear from these arguments that the geometry employed for E89-09 is suitable only for lighter targets.

Overall hypernuclear production rates for the E89-09 geometry will be approximately 3/hr to the 2^- level of the ground state doublet which has a photoproduction cross section of about 0.25 μb/Sr. The expected signal to noise rate will be approximately 6 for a resolution of 600 keV. The rate is comparable to that from production by meson beams, but here the expected resolution is better by at least a factor of 3, and could be substantially improved if a dedicated high-resolution spectrometer were added to the apparatus.

3 Angular Correlations

Given the low luminosity of the E89-09 experimental geometry as described above, it may be possible to detect particles emitted from a hypernucleus in coincidence with those that define the production kinematics. Still count rates will be so low that in

Table 2. Anticiated Missing Mass Resolution in E89-009.

Component	Resolution Contribution (keV)
Splitter and Split-Pole Magnets	120
Beam	160
Splitter and SOS Magnets	550-800
Target	20
Scattering Angle Kinematics	200
Total	600-1000

order to maintain an acceptable coincidence rate, a large-acceptance, high-efficiency detector must be employed for this purpose. Given the inherent low efficiency of Ge gamma detectors, which in any event may "see" substantial numbers of low-energy photons from bremsstrahlung, high resolution gamma spectroscopy will be difficult if not impossible. However, until particle backgrounds in the vicinity of the E89-09 target can be measured, this will not be known.

Although coincident detection of hypernuclear electromagnetic transitions may be difficult, the observation of coincident charged particles emitted from hypernuclei is certainly possible. Such experiments must couple the E89-09 spectrometer system with a large 4π detector such as the CLAS at Jlab, and by using this hybrid detector system, mesonic and non-mesonic decays of hypernuclei could be studied.

However to measure hypernuclear electromagnetic moments, one would need to impart polarization to the nuclear system. This could be done if the electrons are scattered at angles $\geq 5°$ (obviously precluding the zero-degree scattering geometry used in E89-09), or if the initial nuclear target is polarized. Conceivably it is also possible to statistically infer the spin axis of a nuclear system by observing the directions between spin-dependent decays, i.e. angular correlations. In the appendix, I present angular correlation coefficients for the gamma-gamma and gamma-weak mesonic decays of light hypernuclei.

4 Summary

It is expected that new information about the structure of light hypernuclei will be soon forthcoming from Jlab. The proposal discussed here, E89-09, has several unique features including high resolution and low luminosity. Given the small target thickness which is required, separated isotope targets may be used to study the structure of whatever nucleus is of most interest. In addition the program may be extended to include experiments that detect particles from the decay of a hypernucleus after it is formed. It is expected that a rich experimental program is possible.

Acknowledgments

This paper was developed, in part, through the work of the E89-09 collaboration at the Jefferson National Acceleration Laboratory, and this author acknowledges their significant contributions to the material presented here.

Appendix

The experimental and theoretical aspects of angular correlations between particles emitted from excited nuclei has been well developed over the past 40 years[9]. In particular gamma ray angular correlations have been used to obtain the spin and parity of various nuclear levels, and beta-gamma correlations have been employed to test the elementary properties of the weak interaction in nuclei[10]. It has also been recognized that angular correlations can provide information about the local environment of the intermediate nuclear state before its decay. This process, called perturbed angular correlations, has been used to measure the electromagnetic moments of excited nuclei [11,12], and conversely can be used to extract the local electromagnetic fields acting on a nucleus[13].

I first include for completeness the expression for gamma-gamma perturbed correlations[9]. The directional correlation coefficient is;

$$\mathrm{W}(k_1,k_2,t) = \sum_{k_1 k_2 N_1 N_2} A_{k_1}(1)A_{k_2}(2)\mathrm{G}_{k_1 k_2}^{N_1 N_2}(t)[(2k_1+1)(2k_2+1)]^{-1/2}\mathrm{Y}_{k_1}^{*N_1}\mathrm{Y}_{k_2}^{N_2}.$$

Here the perturbation factor; $\mathrm{G}_{k_1 k_2}^{N_1 N_2}(t)$, reduces to $\delta_{k_1 k_2}\delta_{N_1 N_2}$ for vanishing perturbations. For a free atom in a uniform magnetic field, atomic precession leads to a perturbing factor of the form;

$$\mathrm{G}_{k_1 k_2}^{N_1 N_2}(t) = \sum_{FF'}[(2k_1+1)(2k_2+1)]^{1/2}(2F+1)(2F'+1)\begin{Bmatrix} F & F' & k_1 \\ I & I & J_e \end{Bmatrix}\begin{Bmatrix} F & F' & k_2 \\ I & I & J_e \end{Bmatrix}$$
$$\sum_{ff'}\begin{pmatrix} F & F' & k_1 \\ -f & f' & N_1 \end{pmatrix}\begin{pmatrix} F & F' & k_2 \\ -f & f' & N_2 \end{pmatrix} e^{i/\hbar(E_f-E_{f'})}.$$

In the above expression, I is the nuclear spin of the precessing state, J_e is the atomic angular momentum, and $\mathbf{F} = \mathbf{I} + \mathbf{J_e}$ the total system angular momentum. The energy of the state in the magnetic field is then;

$$E_f = a/2[F(F+1) - I(I+1) - J(J+1)];$$

where $a = \dfrac{\mu B(0)}{IJ_e}$. In this expression $B(0)$ is the magnetic field produced by the electrons at the nucleus. The electromagnetic transition density coefficients, A_k^N have the form;

$$A_k^N = \sum_{LL'}(-1)^{2J_F-J_I-L'}\begin{Bmatrix} J_a & J_a & k \\ L & L' & J_I \end{Bmatrix} C_k(L,L'')\langle J_a\|L\|J_I\rangle\langle J_a\|L'\|J_I\rangle^*.$$

In this expression, C_k, are the radiation parameters of multipole order L and L' between the states, J_a and J_I.

Now it is also possible to work out the Racah algebra for the transition density coefficient for a mesonic weak decay of the ground state of a hypernucleus. We assume, although these conditions may be relaxed, that the recoil proton from the decay of a bound, s-shell Λ, is itself bound to a shell of angular momentum, l, and attached to a nuclear core with angular momentum, J_c. This produces a final nuclear state of angular momentum, J_f. The calculated coefficient, B_K^N, then replaces A_k^N in the above correlation equation for W, or can be used to determine the response from an initially polarized hypernucleus.

$$B_K^N = \sum_{k,k',\epsilon,J,J'}(-1)^{2J_I+J+J'+k+k'+J_f+N}T_\epsilon T_\epsilon^* 2(2J+1)(2J'+1)(2J_I+1)$$
$$(2J_f+1)(2J_p+1)(2l+1)[(2\epsilon+1)(2\epsilon'+1)(2k+1)(2k'+1)]^{1/2}$$
$$\begin{Bmatrix} J_I & J_I & K \\ k & k' & J_f \end{Bmatrix}\begin{Bmatrix} J_I & J_f & k \\ \epsilon & l & J \end{Bmatrix}\begin{Bmatrix} J_I & J_f & k' \\ \epsilon' & l & J' \end{Bmatrix}\begin{Bmatrix} J_p & J_c & J_f \\ 1/2 & 1/2 & \epsilon \\ l & l & J \end{Bmatrix}\begin{Bmatrix} J_p & J_c & J_f \\ 1/2 & 1/2 & \epsilon' \\ l & l & J' \end{Bmatrix}.$$

In the above expression, the value of ϵ is 0(1) for the parity conserving (parity non-conserving) component of the transition, and the coefficients, T_ϵ, represent the parity conserving(partiy non-conserving) amplitudes for the weak, mesonic decay in that particular hypernucleus. Thus in lieu of recoil polarization in the hypernucleus, imparted by either a polarizing reaction or by residual polarization left from an initial polarized system, a spin dependent correlation may be induced by first observing a gamma transition. Of course, rates for such an experiment are be expected to be small.

References

1. B. F. Gibson and E. V. Hungerford, Phys. Rep.**257**(1995)350.

2. E. V. Hungerford, Prog. Theor. Phys. Sup. **117**(1994)135.

3. Jlab Experiment E89-009, R. E. Chrien, E. V. Hungerford, and L. G. Tang, spokespersons

4. AGS Experiment E931, E. V. Hungerford, V. Zeps, and D. Dehnhard.

5. H. Tamura, Nucl. Phys. **A639**(1998)83c.

6. E. V. Hungerford, PILAC Collaboration Meeting, Los Alamos National Laboratory, 1990, LA-UR-90-2230.

7. G. Niculescu, *et. al.*, Phys. Rev. Lett. **80**(1998)1805.

8. E. V. Hungerford and L.G. Tang, private communication; O. Hashimoto, private communication.

9. R. M. Steffen and H. Frauenfelder, "Peturbed Angular Correlations", E. Karlsson, E. Matthias, and K. Siegbahn, eds., pgs 1-89, North Holland, Amsterdam, 1964.

10. E. V. Hungerford, *et. al.*, Phys. Rev. **152**(1966)1068

11. T. K. Alexander, *et. al.*, Phys. Rev. **9C**(1974)1748.

12. J. Christiansen and W. Witthuhn, "Hyperfine Interactions of Radioactive Nuclei", J. Christiansen, ed, Springer-Verlag, pgs1-14, 1983.
13. "Hyperfine Structure and Nuclear Radiations", E. Matthias and D. A. Shirley, eds., North Holland, Amsterdam, 1968.

SUBTHRESHOLD AND NEAR THRESHOLD K^+ MESON PRODUCTION ON LIGHT NUCLEI BY PROTONS

E.Ya. Paryev
Institute for Nuclear Research, Russian Academy of Sciences, Moscow 117312, Russia

Kaon production in proton–nucleus reactions at bombarding energies less than threshold energies in a collision of free nucleons has been extensively studied both experimentally and theoretically in recent years [1–12]. The inclusive and exclusive subthreshold K^+ production in pA interactions is under investigation presently at the accelerators COSY–Jülich [13] and CELSIUS [14]. Since kaons have a long mean free path inside the nucleus, it is expected to extract from these studies valuable information about the nuclear structure at short nucleon–nucleon separations as well as about the dynamics of the reaction, properties of the produced particles in the nuclear environment. However, in spite of great efforts, subthreshold kaon production in proton–nucleus reactions is still far from being fully understood. In order to fully clarify the role played by nucleon–nucleon correlations induced by realistic interactions in the subthreshold kaon production in pA collisions as well as to get a deeper insight into the relative role of the primary and secondary reaction channels, it is obviously necessary to properly analyze within the same approach the existing experimental data both on total [1] and differential [7, 11] kaon production cross sections.

In this study we have performed such analysis of the total and differential K^+ production cross sections from pBe and pC reactions in the near threshold and subthreshold energy regimes using the spectral function approach [9] that has been modified [10] to take into account the new data points for $pp \to K^+\Lambda p$ reaction close to threshold [15] as well as to treat the mean–field potential effects on the primary and secondary creation processes on an equal footing.

Apart from participation in the elastic scattering an incident proton can produce a K^+ directly in the first inelastic pN collision due to nucleon Fermi motion. Since we are interested in a few GeV region (up to 3 GeV), we have taken into account the following elementary processes which have the lowest free production thresholds:

$$p + N \to K^+ + \Lambda + N, \tag{1}$$

$$p + N \to K^+ + \Sigma + N. \tag{2}$$

Following the predictions of the effective chiral Lagrangian approach by Kaplan and Nelson [16], we assume that the mass of the produced kaon is not changed in the nuclear medium due to an approximate cancellation of attractive scalar and repulsive vector mean fields, i.e., the total energy E_{K^+} of the K^+ meson with momentum $\mathbf{p}_{K^+}$ is given by:

$$E_{K^+} = \sqrt{\mathbf{p}_{K^+}^2 + m_K^2}, \tag{3}$$

where m_K is the rest mass of a kaon in free space. The effective masses m_h^* of other final hadrons (nucleon and hyperon) participating in the K^+ production processes

(1), (2) are defined by the following dispersion relation [17]:

$$E_h^2 = \mathbf{p}_h^2 + m_h^{*2} = (\sqrt{\mathbf{p}_h^2 + (m_h + U_s^h)^2} + U_v^h)^2, \tag{4}$$

where m_h is the free space mass of hadron h, U_s^h and U_v^h are the scalar and vector (time–like component) self–energies of hadron h, $\mathbf{p}_h$ denotes its canonical three–momentum. Use the relativistic dispersion relation for the quasiparticle given by the l.h.s. of Eq. (4) enables us to keep on dealing with relativistic kinematics as in the on–mass shell case. The effective masses m_h^* include the effective scalar mean–field potentials U_{eff}^h, needed for our calculations:

$$m_h^* = m_h + U_{eff}^h. \tag{5}$$

Equations (4) and (5) allow to extract these potentials provided that the scalar and vector fields U_s^h and U_v^h are known. In the general case these fields are density– and momentum–dependent [17]. However, for the purposes of the present study as well as for reasons of simplicity it is sufficient to neglect the explicit momentum dependence of U_s^h and U_v^h and to evaluate the effective fields U_{eff}^h knowing the quantities U_s^h and U_v^h only in a very limited density and momentum range relevant for the observed [1, 7, 11, 12] subthreshold kaon production in pA reactions. Since the K^+ creation due to first–chance pN collisions (1), (2) occurs mainly inside the target nucleus and populates as showed our calculations the outgoing nucleon and hyperon in a limited kinematical range with the average kinetic energies $\leq 0.1\ GeV$ for the most part of kinematical conditions of the experiments on subthreshold kaon production which will be analyzed below, the potentials U_s^h and U_v^h should be estimated at the normal nuclear matter density for the above characteristic energy. Employing the recent parametrization for the nucleon scalar and vector potentials from [17], obtained within the self–consistent Dirac–Brueckner approach, it can readily be found that the nucleon effective potential U_{eff}^N at kinetic energy of 100 MeV for normal matter density becomes $U_{eff}^N = -34\ MeV$. In view of the substantial uncertainties of the model hyperon self–energies [18] as well as since at subthreshold energies hyperons from primary pN interactions (1) and (2) are produced, as was noted above, in a limited kinematical range, it is naturally to use for the effective potentials U_{eff}^Λ and U_{eff}^Σ seen by the final low–energy Λ and Σ hyperons the values of the corresponding optical potentials at normal nuclear matter density, extracted from the properties of hypernuclei, namely: $U_{eff}^\Lambda = -30\ MeV$ [18] and $U_{eff}^\Sigma = -26\ MeV$ [19]. The set of parameters:

$$U_{eff}^N = -34\ MeV,\ U_{eff}^\Lambda = -30\ MeV,\ U_{eff}^\Sigma = -26\ MeV \tag{6}$$

we will use throughout our calculations. To see the sensitivity of kaon production cross section from the one–step processes (1), (2) to the effective nucleon and hyperon potentials, we will also ignore these potentials in our calculations. Using (3)–(5), one can easily get the following expression for the in–medium reaction thresholds:

$$\sqrt{s_{th}^*} = m_K + m_Y^* + m_N^* = \sqrt{s_{th}} + U_{eff}^Y + U_{eff}^N, \tag{7}$$

where $\sqrt{s_{th}} = m_K + m_Y + m_N$ are the threshold energies in free space and the effective potentials are given by (6). Hence, the reduction of the K^+ threshold in the medium will be 64 MeV in the case of $pN \to K^+\Lambda N$ reaction and 60 MeV for the $pN \to K^+\Sigma N$ process. This will strongly enhance the K^+ production via first chance pN collisions (see, below).

Kinematical considerations show that in the bombarding energy range of our interest ($\leq$ 3.0 GeV) the following two–step production processes may not only contribute to the K^+ production in pA interactions but even dominate [1, 3–5, 7–9] at subthreshold energies. An incident proton can produce in the first inelastic collision with an intranuclear nucleon also a pion through the elementary reactions:

$$p + N_1 \to N + N + \pi, \tag{8}$$

$$p + N_1 \to N + N + 2\pi. \tag{9}$$

We recall that the free threshold energies for these reactions respectively are 0.29 and 0.60 GeV. Then the intermediate pion, which is assumed to be on–shell, produces the kaon on a nucleon of the target nucleus via the elementary subprocesses with the lowest free production thresholds (respectively, 0.76 and 0.89 GeV):

$$\pi + N_2 \to K^+ + \Lambda, \tag{10}$$

$$\pi + N_2 \to K^+ + \Sigma, \tag{11}$$

provided that these subprocesses are energetically possible. Since the main contribution to the K^+ production at subthreshold incident energies comes from a very fast pions moving in the beam direction, the relevant kinetic energy ϵ_N of each nucleon produced in the reactions (8), (9) together with a high–energy pion can be approximately estimated as $\epsilon_N \approx 0.1$ GeV [9] at beam energies of our interest. As has been mentioned above (see, (6)), such low–energy outgoing nucleons feel inside the nucleus the attractive effective potential $U^N_{eff} = -34$ MeV that reduces their free masses in line with the formula (5). Thus, it is necessary to incorporate properly the effective nucleon mass m^*_N in calculations of the K^+ production cross section from the two–step processes (8)–(11). Moreover, it should be taken into account in these calculations also the same in–medium modification of the masses of hyperons from secondary πN collisions (10) and (11) as that for hyperons from primary pN collisions due to the corresponding effective potentials U^Λ_{eff} and U^Σ_{eff} (see, (5), (6)). We have taken into account in the calculation of the K^+ production cross sections from the one–step (1), (2) and two–step (8)–(11) processes also the influence of the nuclear optical potential V_0 on the incoming proton (see, [10]).

In our method the K^+ production cross sections for pA reactions from primary (1), (2) and secondary (10), (11) reaction channels can be expressed [9, 10] as the respective integrals of the "in–medium" inclusive elementary K^+ production cross sections which include the medium effects under consideration and nucleon spectral function $P(\mathbf{p}_t, E)$ over the struck target nucleon momentum $\mathbf{p}_t$ and removal energy E. The nucleon spectral function $P(\mathbf{p}_t, E)$ represents the probability to find in the nucleus a nucleon with momentum $\mathbf{p}_t$ and removal (binding) energy E. The expressions for the high momentum and high removal energy part (correlated part) of the nucleon spectral function as well as for the single–particle (uncorrelated) part of the one proposed in [20, 21] were used in our calculations of K^+ production in pBe and pC collisions.

The comparison of the calculated invariant cross sections for the production of K^+ mesons with momentum of 1.28 GeV/c at the laboratory angle of 10.5^0 from primary $pN \to K^+YN$ and secondary $\pi N \to K^+Y$ channels with the experimental data [11] for $p + Be^9 \to K^+ + X$ reaction at the various bombarding energies is given in Fig. 1. One can see that:

1) Our model for primary and secondary kaon production processes, based on nucleon spectral function, fails completely to reproduce the experimental data at subthreshold beam energies (at energies ≤ 2.1 GeV for the kinematical conditions of the experiment [11]) without allowance for the influence of the corresponding effective potentials on the one–step (1), (2) and two–step (8)–(11) production processes.

2) Our calculations for the one–step reaction channels (1), (2) with the set of parameters $V_0 = 40$ MeV, $U^N_{eff} = -34$ MeV, $U^\Lambda_{eff} = -30$ MeV, $U^\Sigma_{eff} = -26$ MeV (see, (6)) reproduce quite well the experimental data [11] in the energy region far below the lowest threshold, but overestimate the data by a factor of 2 at higher bombarding energies, what indicates that there is no need for employing the medium effects considered by us at these energies.

3) The results of our calculations of the kaon yield from the two–step reaction channels (8)–(11) with the same set of parameters for the effective potentials V_0, U^N_{eff}, U^Λ_{eff}, U^Σ_{eff} as that used above in the calculating the K^+ yield from the primary reaction channels (1), (2) underestimate essentially the data [11] both at subthreshold and above threshold incident energies, what means the dominance of the one–step K^+ production mechanism for the considered "hard" kaon production at all beam energies of interest.

4) The primary proton–nucleon production process (1) misses the experimental data in the far subthreshold region when the effective potential U^Λ_{eff} that is seen inside the nucleus by a slow Λ particle increases from the value of $U^\Lambda_{eff} = -30$ MeV, estimated from the study of the binding and decay of hypernuclei, to $U^\Lambda_{eff} = -20$ MeV, what implies the strong sensitivity of "hard" kaon yield at "low" beam energies to the lambda potential in the nuclear matter.

5) The kaon yield from the one–step K^+ production mechanism is entirely governed by the correlated part of the nucleon spectral function only in the vicinity of the absolute reaction threshold (at bombarding energies of $\epsilon_0 \approx$ 1.65–1.70 GeV), what makes difficult to extract the information on the high momentum and high removal energy components within the target nucleus even from the "hard" kaon production experiment [11].

It should be noticed that the overall calculated kaon spectrum (denoted by the dashed line with two dots in Fig. 1) is not very sensitive to the effective potential

U^{Σ}_{eff} in the subthreshold energy region, since the one–step kaon production process (2) plays a minor role here (see, Fig. 1). Thus, changing this potential from the value of $U^{\Sigma}_{eff} = -26\ MeV$ used above to the value of $U^{\Sigma}_{eff} = 20\ MeV$, as showed our calculations, does not affect practically the overall kaon spectrum at beam energies $\epsilon_0 \leq 2.2\ GeV$ and reduces it approximately by 10% at higher incident energies. Therefore, inclusive subthreshold and near threshold kaon production in pA collisions does not permit to differentiate the scenario with an attractive Σ potential from that with repulsive Σ potential.

The results of our calculations for the double differential cross sections for the production of "soft" K^+ mesons from primary $pN \to K^+YN$ and secondary $\pi N \to K^+\Lambda$ channels at an angle of 40^0 in the interaction of protons with energy of 1.5 GeV with C^{12} nuclei and the experimental data [7] are considered in Fig. 2. It is seen that:

1) Our model calculations for proton and pion induced reaction channels underpredict significantly the data without taking into account the medium effects on the hadrons produced in these channels, what is consistent with our previous findings of Figure 1.

2) The inclusion of the in–medium effects under consideration leads to the substantial enhancement of the K^+ cross sections both from primary and secondary kaon production processes as well as to the similar magnitude for these K^+ creation processes at indicated bombarding energy.

3) The "soft" kaon yield from the secondary reaction channel (10) with Λ particle in the final state is not very sensitive to the effective potential U^{Λ}_{eff} that is seen inside the nucleus by a lambda when this potential increases from $U^{\Lambda}_{eff} = -30\ MeV$ to $U^{\Lambda}_{eff} = 0$.

4) Our overall calculations (the sum of results obtained both for the one–step (1), (2) and two–step (8)–(11) reaction channels, thin solid line in Fig. 2) reproduce reasonably the experimental data only if we include the effect of the respective effective potentials on the one–step kaon production processes (1), (2) and on pion production reactions (8), (9).

Finally, in Fig. 3 we compare the results of our calculations for the total cross sections for K^+ production in pC^{12} collisions from primary $pN \to K^+YN$ and secondary $\pi N \to K^+Y$ channels with the experimental data [1, 7, 12]. It is nicely seen that our overall model calculations (the sum of results obtained both for primary and secondary kaon production processes, thin solid line in Fig. 3) with including the same mean–field effects as those, allowed us to describe above the data on differential [7, 11] kaon production cross sections, reproduce fairly good also the measured [1, 7] total K^+ production cross sections at proton energies of 1.5, 2.5 GeV, below 1 GeV and overestimate the data point [12] at 1.7 GeV by a factor of about 3. Whereas the similar calculations with no mean–field effects (thick solid line in Fig. 3) describe reasonably both this data point and that at 2.5 GeV, but fail completely to reproduce the data at lower incident energies, what counts in favour of the former model calculations. It is also seen that at beam energies below about 1.5 GeV the contributions from the two–step and one–step kaon production processes are comparable in both calculations.

Taking into account the considered above, one may conclude that the relative strength of the proton and pion induced reaction channels in light target nuclei in the subthreshold energy regime is mainly governed by the kinematics of experiment on inclusive kaon production in pA interactions. Our results demonstrate also that further measurements of the total (and differential) cross sections for K^+ production on light nuclei in the proton–energy range 1.0–1.7 GeV are extremely needed nowadays to reliably test the spectral function approach presented in this study as well as to deeply elucidate the underlying mechanism of subthreshold kaon production and the role played by nucleon–nucleon correlations in this phenomenon.

References

1. V.P.Koptev et al., *ZhETF.* **94** (1988) 1.
2. A.Shor et al., *Nucl. Phys.* **A514** (1990) 717.
3. W.Cassing et al., *Phys. Lett.* **238B** (1990) 25.
4. A.Sibirtsev and M.Büscher, *Z. Phys.* **A347** (1994) 191.
5. W.Cassing et al., *Z. Phys.* **A349** (1994) 77.
6. H.Müller and K.Sistemich, *Z. Phys.* **A344** (1992) 197.
7. M.Debowski et al., *Z. Phys.* **A356** (1996) 313.
8. A.Sibirtsev, W.Cassing and U.Mosel, *Z. Phys.* **A358** (1997) 357.
9. S.V.Efremov and E.Ya.Paryev, *Eur. Phys. J.* **A1** (1998) 99.
10. E.Ya.Paryev, *Eur. Phys. J.A*, in the press.
11. A.V.Akindinov et al., *APH N.S., Heavy Ion Physics* **4** (1996) 325.
12. Yu.T.Kiselev and V.A.Sheinkman, private communication.
13. D.Grzonka and K.Kilian, *Nucl. Phys.* **A639** (1998) 569c.
14. A.Badala et al., *Phys. Rev. Lett.* **80** (1998) 4863.
15. W.Oelert, *Nucl. Phys.* **A639** (1998) 13c.
16. D.B.Kaplan and A.E.Nelson, *Phys. Lett.* **175B** (1986) 57.
17. C.H.Lee et al., *Phys. Lett.* **412B** (1997) 235.
18. G.Q.Li and C.M.Ko, *Phys. Rev.* **C54** (1996) 1897.
19. C.B.Dover and G.E.Walker, *Phys. Rep.* **89** (1982) 1.
20. C.Ciofi degli Atti et al., *Phys. Rev.* **C41** (1990) R2474.
21. C.Ciofi degli Atti and S.Simula, *Phys. Rev.* **C53** (1996) 1689.

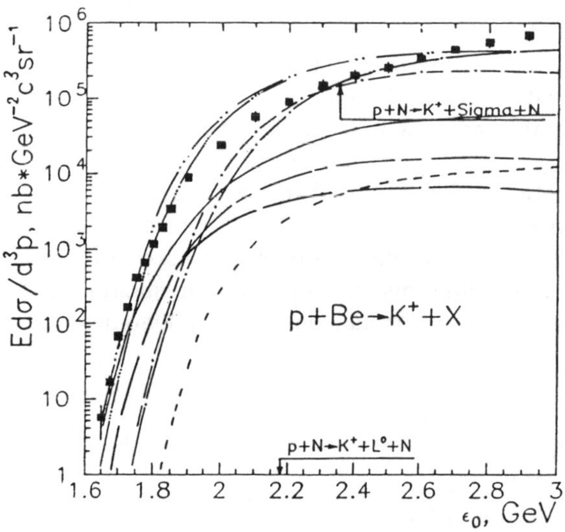

Fig.1

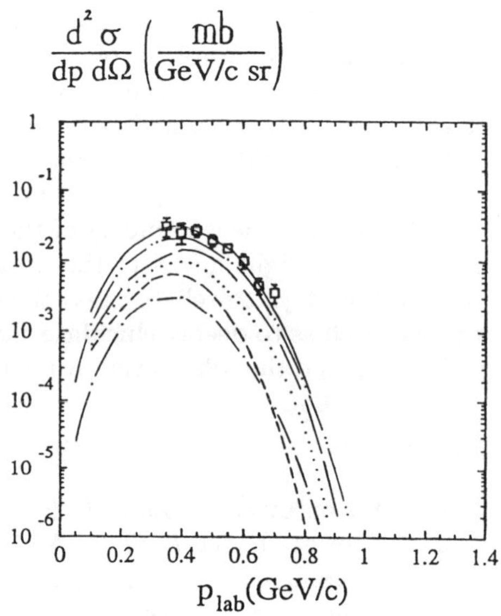

Fig.2

Fig.1. Lorentz invariant cross sections for the production of K^+ mesons with momentum of 1.28 GeV/c at the lab angle of 10.5^0 in $p + Be^9$ reactions as functions of the laboratory energy of the proton. The experimental data (full squares) are from [11]. The curves are our calculation. The dashed lines with one, two and three dots are calculations for primary production processes (1), (2) with the total nucleon spectral function at $V_0 = 0$, $U^N_{eff} = 0$, $U^\Lambda_{eff} = 0$, $U^\Sigma_{eff} = 0$; $V_0 = 40\ MeV$, $U^N_{eff} = -34\ MeV$, $U^\Lambda_{eff} = -30\ MeV$, $U^\Sigma_{eff} = -26\ MeV$ and $V_0 = 40\ MeV$, $U^N_{eff} = -34\ MeV$, $U^\Lambda_{eff} = -20\ MeV$, $U^\Sigma_{eff} = -26\ MeV$, respectively. The two–dash–dotted line represents our calculations for primary production process (2) with the total nucleon spectral function at $V_0 = 40\ MeV$, $U^N_{eff} = -34\ MeV$, $U^\Sigma_{eff} = -26\ MeV$. The solid line denotes the same as the dashed line with two dots, but it is supposed that the total nucleon spectral function is replaced by its correlated part. The short– and long–dashed lines are calculations for the secondary production process (10) at $V_0 = 0$, $U^N_{eff} = 0$, $U^\Lambda_{eff} = 0$ and $V_0 = 40\ MeV$, $U^N_{eff} = -34\ MeV$, $U^\Lambda_{eff} = -30\ MeV$, respectively. The line with alternating short and long dashes represents calculations for the secondary production processes (10), (11) at $V_0 = 40\ MeV$, $U^N_{eff} = -34\ MeV$, $U^\Lambda_{eff} = -30\ MeV$, $U^\Sigma_{eff} = -26\ MeV$. The arrows indicate the thresholds for the reactions $pN \to K^+\Lambda N$ and $pN \to K^+\Sigma N$ occuring on a free nucleon.

Fig.2. Double differential cross sections for the production of K^+ mesons at an angle of 40^0 in the interaction of 1.5 GeV protons with the C^{12} nuclei as functions of kaon momentum. The experimental data (open squares) are from [7]. The curves are our calculation. The dotted line is calculation for the secondary production process (10) at $V_0 = 40\ MeV$, $U^N_{eff} = -34\ MeV$, $U^\Lambda_{eff} = 0$. The thin solid line is the sum of the dashed line with two dots and dotted line. The rest of the notation is identical to that in Fig. 1.

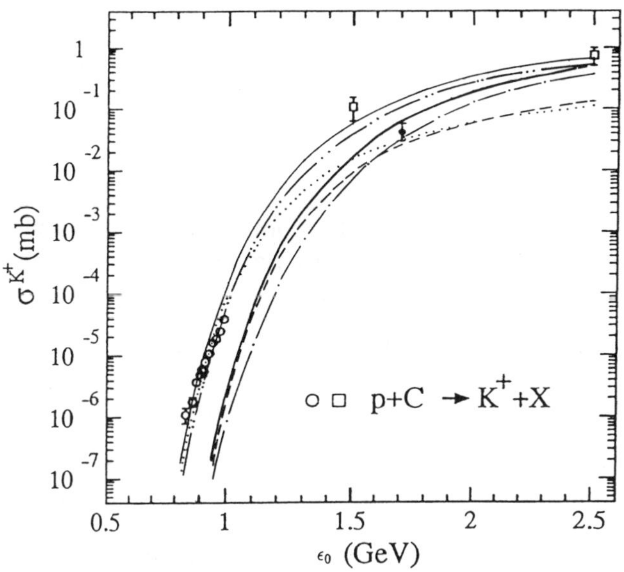

Fig.3. The total cross sections for K^+ production in $p+C^{12}$ reactions as functions of the laboratory energy of the proton. The experimental data are from [1] (open circles), [7] (open squares) and [12] (full dot). The curves are calculation. The dashed lines with one and two dots are calculations for primary production processes (1), (2) with the total nucleon spectral function at $V_0 = 0$, $U^N_{eff} = 0$, $U^\Lambda_{eff} = 0$, $U^\Sigma_{eff} = 0$ and $V_0 = 40\ MeV$, $U^N_{eff} = -34\ MeV$, $U^\Lambda_{eff} = -30\ MeV$, $U^\Sigma_{eff} = -26\ MeV$, respectively. The short–dashed and dotted lines are calculations for the two–step production processes (8)–(11) at $V_0 = 0$, $U^N_{eff} = 0, U^\Lambda_{eff} = 0, U^\Sigma_{eff} = 0$ and $V_0 = 40\ MeV, U^N_{eff} = -34\ MeV, U^\Lambda_{eff} = 0$, $U^\Sigma_{eff} = 0$, respectively. The thin solid line is the sum of the dashed line with two dots and dotted line. The thick solid line is the sum of the dot–dashed and short–dashed lines.

SPIN OBSERVABLES IN HYPERON PRODUCTION

R. BERTINI
DISTO COLLABORATION.*
Dipartimento di Fisica "A. Avogadro" and INFN - Torino, Italy

An overlook of the present experimental and theoretical knowledge on the polarization of the S= -1 hyperons is outlined. The available data from the DISTO set-up are presented. The first data on the spin transfer parameter D_{NN} in the *exclusive* Λ production channel $pp \to pK^+\Lambda$ are shown and future perspectives in the data analysis discussed.

1 Introduction

The discovery that the $\Lambda's$, inclusively produced in the fragmentation region with high transverse momenta p_T, carry [1] a large polarization perpendicular to the Λ production plane, was surprising for the following reasons:

1) The contributions of many different inelastic channels X, each with a polarization of random sign, are added up in an inclusive reaction $a + b \to c + X$. One would then expect the sum of the polarizations to average to zero.

2) In the constituent quark model, assuming the ud system of the Λ to be in a singlet state, the polarisation of the Λ is that of the strange quark. The proton fragmentation into a Λ, for large p_T, corresponds to the replacement of a valence quark u of the proton by a strange quark s from the sea which must be accelerated to get a nonzero p_T . Helicity conservation implies that the amplitude for the quark q, that enters the interaction region with helicity λ_q to emerge with helicity $\lambda_q^{'}$, is proportional to $\delta_{\lambda_q,\lambda_q'} + O(m_q/E_q)\ \delta_{\lambda_q,-\lambda_q'}$. The second term vanishes at high energies and small quark masses, in perturbative QCD. The quark polarisation is then, at most, $P_q \sim m_q/E_q\alpha_s$. Extensive calculations of P_q give values of few percent in strong contradiction with the measured values.

In the last years a large amount of data has been collected [2,3] on inclusive hyperon production (I will limit my discussion hereafter to S=-1 hyperons (Y)), that all show the same general trends in the beam fragmentation region:

- The invariant cross section $Ed^3\sigma/dp^3$ depends, roughly, only on x_F and p_T and *not* on the C.M. energy.

- The transverse polarisation **P** is *negative (positive)* with respect to the direction

*F. BALESTRA, Y. BEDFER, R. BERTINI, L.C. BLAND, A. BRENSCHEDE, F. BROCHARD, M.P. BUSSA, V. CHALYSHEV, SEONHO CHOI, M. DEBOWSKI, M. DZEMIDZIC, J.-CL. FAIVRE, I.V. FALOMKIN, L. FERRERO, J. FORYCIARZ, V. FROLOV, R. GARFAGNINI, D. GILL, A. GRASSO, S. HEINZ, V.V. IVANOV, W.W. JACOBS, W. KÜHN, A. MAGGIORA, M. MAGGIORA, A. MANARA, D. PANZIERI, H.-W. PFAFF, G. PIRAGINO, G.B. PONTECORVO, A. POPOV, J. RITMAN, P. SALABURA, F. TOSELLO, S.E. VIGDOR, G. ZOSI.

$\mathbf{n} = \mathbf{p_b} \times \mathbf{p_Y}$ for $\Lambda(\Sigma)$ production, respectively. Here the sign convention is: positive polarisation in the same direction of **n**.

- **P** is almost energy independent for beam momenta from 12 up to 2000 GeV/c.
- **P** has a linear dependence on p_T, for $p_T \leq 1$ GeV/c, with a slope increasing with x_F.
- For $p_T > 1$ GeV/c **P** does not depend on p_T and is about linear with x_F.

On the theoretical side many attempts have been made to explain these regularities:

- In the Lund model [4] the ud scalar diquark from the incoming proton continues forward as a unit and a color dipole field is streched between the diquark and the central collision region. One then obtains $q\overline{q}$ pairs; the $s\overline{s}$ pair has orbital angular momentum which must be balanced by the s quark spin, which then acquires polarisation. This makes polarised the hyperon.
- In Ref. [5] the s quark of the target proton carries a very small fraction of its momentum. However, at the end of the recombination process, it is a valence quark of the Λ and must carry a large fraction of the Λ momentum. Then the recombination strongly accelerates the s quark, that then feels a Thomas precession. The associated energy is minimized when its spin direction is parallel to **n**. The produced Λ is then polarised along this direction.
- A valence quark core, with the naïve quark model spin content, is coupled to a spin-triplet $s\overline{s}$ pair in which the $\overline{s}$ antiquark is supposed negatively polarised. This model successfully reproduces[6] the negative longitudinal polarisation for Λ found in the $\overline{\nu}$ N deep inelastic scattering.
- A relationship is established[7] between the polarisation of hyperons in inclusive production and the left-right asymmetries observed in single spin inclusive hadron processes. To calculate the contributions of the different channels to the spin observables (P_Λ, A_N and D_{NN}) in the inclusive production, the authors used the direct fusion model[8].
- A completely different approach is assumed in Ref. [9], where it is pointed out that data on hyperon β-decay and deep inelastic scattering from polarised nucleons show that the description of the Λ given in the point 2) above is not true. The contribution to the Λ spin would be for about 60 % on the s (and $\overline{s}$) quarks, while -40 % is on u (and $\overline{u}$) and d (and $\overline{d}$) quarks[10].
- The interference between the various channels X produces asymmetries[11]. For the inclusive Λ production, the $\Sigma^0 \rightarrow \Lambda\,\gamma$ and $\Sigma^*(1385) \rightarrow \Lambda\,\pi$ channels, that have cross sections comparable to that for the direct process, are considered. The Triple-Regge parametrisation and the SU(6) coupling constant are used to fit the unpolarised cross-sections and to predict asymmetries.
- The Reggeized one-pion-exchange model[12], produces cross-sections and Λ polarisation P_Λ, making use of the existing $\pi p \rightarrow K\Lambda$ data.

- The relativistic one meson exchange model is the basis of the calculations[13,14] of cross-sections and spin observables for the exclusive channels $pp \to pK^+\Lambda$, $pp \to nK^+\Sigma^+$ and $pp \to pK^+\Sigma^0$. These calculations include both π and K exchanges and final state interactions (FSI).
- At C.M. energies about 100 MeV above threshold for the reaction $pp \to pK^+\Lambda$, it is expected[15] that the π, η and ρ-meson exchange toghether with the excitation of the $N^*(1650)$, $N^*(1710)$ and $N^*(1720)$ resonances can play an important role in the production mechanism.

2 The DISTO Program

All the quoted models do not give a fully satisfactory explanation, both qualitative and quantitative, for the regularities observed in inclusive hyperon production. It was then suggested[16] that a complete reconstruction of the kinematics with the energy and momenta of all the detected prongs and hence the possibility to select the contribution of each channel for the $\Lambda(\Sigma)$ hyperon production, could be a way to find a satisfactory description of the data. The suggested procedure can be described with the reaction $pp \to pK^+\Lambda$, where the Λ is detected via its decay $\Lambda \to p\pi^-$. The measurement of the momenta of all the four charged prongs (p p $K^+\pi^-$) and a good particle identification gives the possibility to reconstruct:

1) The invariant mass of the Λ hyperon with the (p π^-) tracks.
2) The missing mass of the system using the (p K^+) tracks. (This allows to know if the Λ were directly produced or coming from the decay of other hyperon states like through the $\Sigma^0 \to \Lambda\gamma$ or $\Sigma^*(1385) \to \Lambda\pi$ decays.)
3) The contribution of the $N^*(1650)$, $N^*(1710)$ and $N^*(1720)$ resonances to the hyperon production, from the coupling of the (ΛK^+) tracks.
4) The (p Λ) system to provide information about the FSI.

The DISTO apparatus, described in detail elsewhere [17], allows to measure, at the same time, all the reaction channels with four or more charged prongs. It provides, in addition, a good particle identification, a precise reconstruction of the vertex position of the tracks (to distinguish between the reaction vertex and the vertices of the weak decaying hyperons or strange mesons) and finally the momenta of each track with a moderate resolution, but good enough to separate the various hyperon states and N^* resonances. Its acceptance, both angular and in the momentum space, covers the largest kinematical region. It has been used to provide with data cross-sections and the spin parameters P_Y, A_N and D_{NN} for various reactions producing S = -1 hyperons at the Saturne II accelerator.

3 Hyperon Production Channels Measured by DISTO

Data have been collected with the DISTO apparatus and with the polarised proton beam at three energies: 2.145, 2.50 and 2.85 GeV.

The hyperon production channels, that can be detected at 2.85 GeV, with a trigger selecting four or more charged prongs, are given in Tab. 1. For each raw in the Table are subsequently indicated: the reaction number, the reaction channel, the beam kinetic energy (in GeV) at threshold and the charged prongs detected by the apparatus. Inside parentheses are given the tracks corresponding to a secondary vertex.

In Tab. 2 are given the decay modes, branching ratios and the asymmetry parameter α in the weak decay, when available, for the first S= -1 hyperon states[18].

To describe these tables let make use of Fig. 1. Fig. 1 a) shows the invariant mass spectrum obtained from the $(p\pi^-)$ tracks that stem from a secondary vertex distant more than 1 cm. from the reaction vertex, defined by the (pK^+) tracks. The reactions 1 to 4, 15, 16, 24, 25, 26, 27, and 30 can, in principle, feed that spectrum. Reactions 5 to 10, 13, 14, 17, 19 to 23, 28 and 29 do not contribute because there is not a (pK^+) primary vertex. Reactions 11 to 14 and 18 to 23 do not have a secondary $(p\pi^-)$ vertex. An additional cut that rejects π^+s to suppress the background from the $\vec{p}p \rightarrow pp\pi^+\pi^-$ and $\vec{p}p \rightarrow pp\pi^+\pi^-\pi^0$ reactions rejects also events from the reactions 3, 4, 16 and 26. Of reactions 24 and 25 only events from the $\Sigma^0\pi^0$ decay can be seen. Of reaction 27 only events from the $\Lambda\pi^0$ or the $\Sigma^0\pi^0$ decay can be seen.

Now have a look to the Fig. 1 b), where the missing mass spectrum is presentend, assuming four prongs in the final state and selecting the events falling in the Λ invariant mass range 1100-1130 MeV/c^2 (solid line). These conditions imply that, when additional neutral or charged prongs are present, the corresponding events are scattered over a region more or less wide of the missing mass spectrum.

The first peak, that corresponds to the Λ mass, is due to reaction 1. When events filling this peak and falling into the gate 1040-1150 MeV/c^2 are selected and the Λ invariant mass reconstructed, one gets the spectrum shown in Fig. 1 a) (dashed curve). The ratio peak/background is much larger, when compared to the similar spectrum of Fig. 1 a) (solid curve), and the background is flat showing no Λ momentum dependence. Events from reactions 2 and 30 are scattered over a large region of the missing mass spectrum and can be rejected, requiring the total energy of the $(\vec{p}p)$ system to be equal to that of the detected prongs.

Reaction 30 is of particular relevance for the understanding of the inclusive Λ polarisation production because the K^{*+} resonance has $J^\pi = 1^-$ and the associated Λ is expected to have[7] $\mathbf{P_\Lambda} > 0$, instead of $\mathbf{P_\Lambda} < 0$, as usually seen in inclusive Λ production. This resonance (width 49.8 $\pm$ 0.8 MeV/c^2) decays $K^{*+} \rightarrow K^+\pi^0$ with a branching ratio of 99.899 $\pm$ 0.009 %. However the available C.M. energy $(\sqrt{s} - \sqrt{s_0})$ for this reaction, as well as for reaction 25, is equal only to 31 MeV/c^2. Events of both reactions fall then at the limit of our acceptance and have not yet been investigated at the present stage of our analysis.

The second peak of Fig. 1 b), labeled Σ^0 is feeded by events of reaction 15 through the decay $\Sigma^0 \rightarrow \Lambda\gamma$. Helicity conservation implies the transfer of only -1/3 of the Σ^0 polarisation to the Λ. The Σ^0 polarisation can then been inferred from the measurement of the Λ polarisation from the events contributing to this peak.

The third bump in Fig. 1 b), labeled Σ^*, includes events from reaction 27 $(\sqrt{s} - \sqrt{s_0} = 165\ MeV/c^2)$ and 24 $(\sqrt{s} - \sqrt{s_0} = 145\ MeV/c^2)$. The Σ^{*0} resonance

Table 1. Hyperon production channels detected with DISTO at 2.85 GeV.

#	Reaction	Threshold	Detected Prongs
1	$\vec{p}p \to pK^+\vec{\Lambda}$	1.58	$pK^+(p\pi^-)$
2	$\vec{p}p \to p\pi^0K^+\vec{\Lambda}$	1.96	$pK^+(p\pi^-)$
3	$\vec{p}p \to p\pi^+\pi^-K^+\vec{\Lambda}$	2.38	$p\pi^+\pi^-K^+(p\pi^-)$
4	$\vec{p}p \to p\pi^+\pi^-\pi^0K^+\vec{\Lambda}$	2.80	$p\pi^+\pi^-K^+(p\pi^-)$
5	$\vec{p}p \to n\pi^+K^+\vec{\Lambda}$	1.98	$\pi^+K^+(p\pi^-)$
6	$\vec{p}p \to n2\pi^+\pi^0K^+\vec{\Lambda}$	2.82	$2\pi^+K^+(p\pi^-)$
7	$\vec{p}p \to p\pi^+K^0\vec{\Lambda}$	1.98	$p\pi^+(\pi^+\pi^-)(p\pi^-)$
8	$\vec{p}p \to p\pi^+\pi^0K^0\vec{\Lambda}$	2.38	$p\pi^+(\pi^+\pi^-)(p\pi^-)$
9	$\vec{p}p \to p2\pi^+\pi^-K^0\vec{\Lambda}$	2.83	$p2\pi^+\pi^-(\pi^+\pi^-)(p\pi^-)$
10	$\vec{p}p \to n2\pi^+K^0\vec{\Lambda}$	2.40	$2\pi^+(\pi^+\pi^-)(p\pi^-)$
11	$\vec{p}p \to p\pi^+K^+\vec{\Sigma}^-$	2.21	$p\pi^+K^+(\pi^-)$
12	$\vec{p}p \to p\pi^+\pi^0K^+\vec{\Sigma}^-$	2.62	$p\pi^+K^+(\pi^-)$
13	$\vec{p}p \to n2\pi^+K^+\vec{\Sigma}^-$	2.64	$2\pi^+K^+(\pi^-)$
14	$\vec{p}p \to p2\pi^+K^0\vec{\Sigma}^-$	2.64	$p2\pi^+(\pi^+\pi^-)(\pi^-)$
15	$\vec{p}p \to pK^+\vec{\Sigma}^0$	1.79	$pK^+(p\pi^-)$
16	$\vec{p}p \to p\pi^+\pi^-K^+\vec{\Sigma}^0$	2.62	$p\pi^+\pi^-K^+(p\pi^-)$
17	$\vec{p}p \to p\pi^+K^0\vec{\Sigma}^0$	2.20	$p\pi^+(\pi^+\pi^-)(p\pi^-)$
18	$\vec{p}p \to p\pi^-K^+\vec{\Sigma}^+$	2.19	$p\pi^-K^+(p)$ *or* (π^+)
19	$\vec{p}p \to n\pi^+\pi^-K^+\vec{\Sigma}^+$	2.61	$\pi^+\pi^-K^+(p)$ *or* (π^+)
20	$\vec{p}p \to pK^0\vec{\Sigma}^+$	1.79	$p(\pi^+\pi^-)(p)$ *or* (π^+)
21	$\vec{p}p \to p\pi^0K^0\vec{\Sigma}^+$	2.18	$p(\pi^+\pi^-)(p)$ *or* (π^+)
22	$\vec{p}p \to n\pi^+K^0\vec{\Sigma}^+$	2.20	$\pi^+(\pi^+\pi^-)(p)$ *or* (π^+)
23	$\vec{p}p \to p\pi^+\pi^-K^0\vec{\Sigma}^+$	2.62	$p\pi^+\pi^-(\pi^+\pi^-)(p)$ *or* (π^+)
24	$\vec{p}p \to pK^+\Lambda^*(1405)$	2.40	$pK^+\pi^+(\pi^-)$ from $\Sigma^-\pi^+$ $pK^+(p\pi^-)$ from $\Sigma^0\pi^0$ $pK^+\pi^-(p) or (\pi^+)$ from $\Sigma^+\pi^-$
25	$\vec{p}p \to pK^+\Lambda^*(1520)$	2.75	pK^+pK^- $pK^+(\pi^+\pi^-)$ from $n\overline{K}^0$ $pK^+\pi^+(\pi^-)$ from $\Sigma^-\pi^+$ $pK^+(p\pi^-)$ from $\Sigma^0\pi^0$ $pK^+\pi^-(p) or (\pi^+)$ from $\Sigma^+\pi^-$ $pK^+\pi^+\pi^-(p\pi^-)$ from $\Lambda\pi^+\pi^-$
26	$\vec{p}p \to p\pi^+K^+\Sigma^{*-}(1385)$	2.78	$p\pi^+K^+\pi^-(p\pi^-)$ from $\Lambda\pi^-$ $p\pi^+K^+(\pi^-)$ from $\Sigma^-\pi^0$
27	$\vec{p}p \to pK^+\Sigma^{*0}(1385)$	2.34	$pK^+(p\pi^-)$ from $\Lambda\pi^0 or \Sigma^0\pi^0$ $pK^+\pi^+(\pi^-)$ from $\Sigma^-\pi^+$ $pK^+\pi^-(p) or (\pi^+)$ from $\Sigma^+\pi^-$
28	$\vec{p}p \to nK^+\Sigma^{*+}(1385)$	2.35	$K^+\pi^+(p\pi^-)$ from $\Lambda\pi^+ or \Sigma^0\pi^+$
29	$\vec{p}p \to pK^0\Sigma^{*+}(1385)$	2.35	$p(\pi^+\pi^-)\pi^+(p\pi^-)$ from $\Lambda\pi^+ or \Sigma^0\pi^+$ $p(\pi^+\pi^-)(p) or (\pi^+)$ from $\Sigma^+\pi^0$
30	$\vec{p}p \to pK^{*+}(892)\Lambda$	2.75	$pK^+(p\pi^-)$ from $K^+\pi^0$ $p(\pi^+\pi^-)\pi^+(p\pi^-)$ from $K^0\pi^+$

($J^\pi = 3/2^+$) has a mass of 1383.7 ± 1.0 MeV/c^2 and a width of 36 ±5 MeV/c^2. Only the two decay modes (see Tab. 2) $\Lambda\pi^0$ and $\Sigma^0\pi^0$ contribute to the bump. The $\Lambda^*(1405)$ ($J^\pi = 1/2^-$) has a mass of 1406.5 ± 4.0 MeV/c^2 and a width of

Table 2. Main decay modes of the first S=-1 hyperon states

Y	Decay mode	B. R.	α
Λ	$\Lambda \to p\pi^-$	0.639 ± 0.005	$+0.642 \pm 0.013$
	$\Lambda \to n\pi^0$	0.358 ± 0.005	$+0.648 \pm 0.043$
Σ^-	$\Sigma^- \to n\pi^-$	$(99.848 \pm 0.005)\%$	-0.068 ± 0.008
Σ^0	$\Sigma^0 \to \Lambda\gamma$	100%	
Σ^+	$\Sigma^+ \to p\pi^0$	0.516 ± 0.003	-0.980 ± 0.017
	$\Sigma^+ \to n\pi^+$	0.483 ± 0.003	$+0.066 \pm 0.016$
$\Sigma^*(1385)$	$\Sigma^*(1385) \to \Lambda\pi$	0.88 ± 0.02	
	$\Sigma^*(1385) \to \Sigma\pi$	0.12 ± 0.02	
$\Lambda^*(1405)$	$\Lambda^*(1405) \to \Sigma\pi$	100%	
$\Lambda^*(1520)$	$\Lambda^*(1520) \to N\overline{K}$	0.45 ± 0.01	
	$\Lambda^*(1520) \to \Sigma\pi$	0.42 ± 0.01	
	$\Lambda^*(1520) \to \Lambda\pi\pi$	0.10 ± 0.01	

50 ± 2 MeV/c^2. Only a fraction of the decay mode (the $\Sigma^0\pi^0 \to \Lambda\gamma\pi^0$ branch) can contribute to the bump. The nature of this resonance is still a puzzle. Two different pictures are suggested[19]: a) it is an L =1 SU(3) *uds* state or b) it is an unstable $N\overline{K}$ state. Both pictures are unsatisfactory. If suggestion a) is taken as good, the problem is that current models reproduce all the baryon masses but that of the $\Lambda^*(1405)$, that is too low shifted from the other L = 1 *uds* state (the $\Lambda^*(1520)$). If b) were valid, another $J^\pi = 1/2^-$ is needed with a mass closer to that of the $\Lambda^*(1520)$, that has not be found by experiments. Therefore new data on the $\Lambda^*(1405)$ are badly needed. However, at the present stage of our analysis no clear evidence for this state is present in our data.

The asymmetry in the hyperon weak decay gives a powerfull tool to analyse hyperon polarisation. Tab. 2 shows that α is large but in the $\Sigma^- \to n\pi^-$ and $\Sigma^+ \to n\pi^+$ decays. It is therefore unlikely that we will able to extract, from our data (reactions 11 to 14), Σ^- or Σ^+ (reactions 18 to 23) polarisations with good statistical significance.

Reactions 7 to 10, 14, 17, 20 to 23 and 29, including K^0 production, are detected, in the DISTO apparatus, with a lower efficiency than those involving K^+ production. This because only the $K^0{}_S \to \pi^+\pi^-$ branch is detected giving $\sim$ 34 % of the reaction rate.

Coming back to Fig. 1 b) the background coming from misidentified particles is given in the dashed curve. It is produced selecting events that fall, in the invariant mass spectrum of Fig. 1 a), within the gate 1150 to 1180 MeV/c^2. Once this background is subtracted, the events falling inside the Λ peak have been used to produce our first results on the spin transfer parameter D_{NN} in *exclusive* Λ production, as described in detail in Ref. [20].

4 The D_{NN} spin transfer parameter in *exclusive* Λ production

The polarization of the Λ's can be extracted from the angular distribution of the daughter proton in the Λ rest frame, measured with respect to a spin quantization

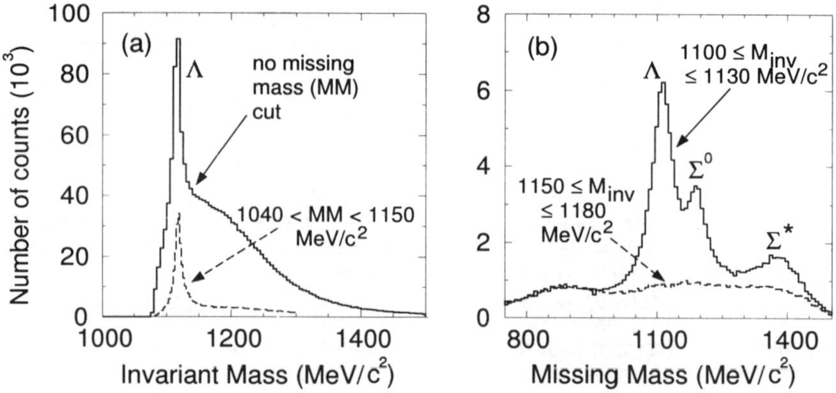

Figure 1. (a) $p\pi^-$ invariant mass ($\sigma \approx 5$ MeV/c^2) with (dashed) and without (solid) a cut on the Λ missing mass. (b) Missing mass spectrum ($\sigma \approx 25$ MeV/c^2) from primary pK^+ pairs for $p\pi^-$ invariant mass in the range 1100–1130 MeV/c^2 (solid line), and 1150–1180 MeV/c^2 (dashed line).

axis parallel to $\mathbf{n}$. The Λ polarisation $\mathbf{P}_\Lambda$, the depolarisation parameter D and the differential cross-section σ are related as follows:

$$\sigma \mathbf{f} \cdot \mathbf{n} = \sigma_0(\mathbf{P}_\Lambda \cdot \mathbf{n} + D\, \mathbf{e} \cdot \mathbf{n})$$

Here $\mathbf{e}$ is the beam polarisation vector, $\mathbf{f}$ is the Λ polarisation vector. The other vectors have been defined in Sec. 1.

We present, as first result on Λ spin observables, the D_{NN} data at 2.85 GeV, because they are sizable and because, to get them, we do not need to apply all the asymmetry and acceptance corrections needed for P_Λ, A_N and $d\sigma/d\omega$.

Fig. 2 shows the D_{NN} results for the exclusive Λ production as a function of both (a) the transverse momentum transfer p_T (from $\vec{p}$ to $\vec{\Lambda}$) and (b) the Λ's longitudinal momentum, expressed as a fraction (x_F) of its maximum kinematically allowed value. Exclusive results are extracted only for $x_\mathrm{F} > 0.1$, since at smaller x_F the rising p and K^+ momenta cause the missing mass resolution to deteriorate. D_{NN} is large and negative ($\simeq -0.4$) over most of the kinematic region. The negative sign signifies that the component of the Λ polarization that is correlated with the beam spin is oriented *opposite* to the beam spin.

The error bars in Fig. 2 include statistical, but not systematic, uncertainties from background subtraction and Λ-Σ^0 peak fitting. We estimate the associated systematic errors to be $\lesssim \pm 0.02$ and ± 0.04, respectively. Overall scale uncertainties in D_{NN} arise from the present uncertainties in the beam polarimeter calibration ($\pm 7\%$) and in the Λ decay asymmetry ($\pm 2\%$ [18]). The systematic errors do not affect the qualitative behavior observed for D_{NN} or the conclusions drawn below.

The D_{NN} parameter is especially sensitive to the hyperon production mechanism [21]. For example, in a meson-exchange framework, D_{NN} distinguishes clearly between π- and K- exchange contributions. This is illustrated in Fig. 2 by theoretical calculations employing the model of Ref. [13,14]. At large positive x_F, D_{NN} is maximally different for the two dominant contributions indicated in the figure.

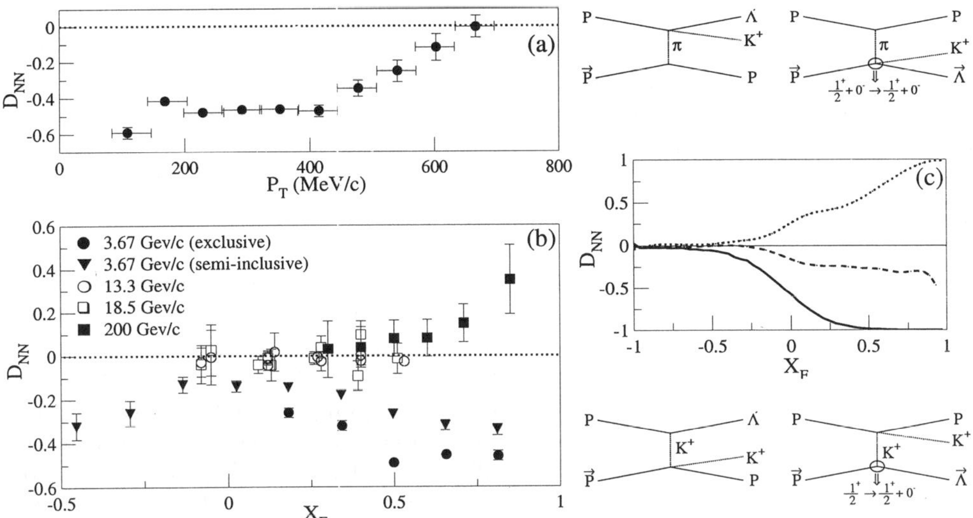

Figure 2. (a) D_{NN} values vs. p_T for the *exclusive* $\vec{p}p \rightarrow pK^+\vec{\Lambda}$ reaction. The horizontal error bars reflect the width of the p_T bins analysed, vertical bars statistical uncertainties only. (b) D_{NN} vs. x_F for our *exclusive* Λ data and for *inclusive* production from Refs. [22,23]. Also shown our *semi*-inclusive results. (c) Theoretical calculation [13] of D_{NN} for various *exclusive* Λ production mechanisms: kaon-exchange (solid curve), pion-exchange (dashed), or both combined with a Λ-p final state interaction (dot-dashed). The Feyman diagrams indicate the dominant exchange contributions for positive vs. negative x_F and spin structures at vertices.

To conserve angular momentum and parity, kaon emission at the polarized proton vertex (bottom right diagram, Fig. 2) causes a spin-flip, yielding $D_{NN} = -1$.

In contrast, the second mechanism (depicted in the top-right diagram of Fig. 2), where a virtual pion undergoes a strangeness-changing rescattering at the polarized proton vertex, has a spin structure $(\frac{1}{2}^+ + 0^- \rightarrow \frac{1}{2}^+ + 0^-)$ for which angular momentum and parity conservation require $D_{NN} = +1$ [a]. The predicted D_{NN} tends toward zero for both mechanisms at the more negative x_F, since the hyperon is then connected preferentially to the *un*polarized (target) proton.

The large negative values observed for D_{NN} at $x_F > 0$ in Fig. 2(b) can thus be interpreted in the framework of kaon-exchange dominance. The full theoretical calculation (dot-dashed curve in Fig. 2) is qualitatively similar to the data.

To compare the present results to polarization transfers measured previously[22,23] for *inclusive* Λ production at higher energies, we include in Fig. 2(b) D_{NN} results for the *semi*-inclusive sample obtained by summing the present data over all missing masses above 1000 MeV/c^2.[b] Inclusion of the Σ^0 and $\Sigma^*(1385)$ yields makes D_{NN} only slightly less negativc than our exclusive Λ results at $x_F > 0$. The present data clearly differ in sign from inclusive D_{NN} results obtained at 200 GeV/c[23] for large

[a] The latter holds rigorously for the on-shell $\pi\vec{p} \rightarrow K\vec{\Lambda}$ process, but only approximately in this case, where the rescattering can have an out-of-plane component of momentum transfer.

[b] The D_{NN} data, obtained in inclusive production at 6 GeV/c in the pioneer work of Ref.[24] have not been included in the figure because of their large error bars. Their general behaviour shows negative D_{NN} values at large x_F.

x_F, successfully interpreted in a simple constituent quark model in Ref. [7].

The present results are not naturally explained within this quark model. They suggest a production mechanism dominated by kaon-exchange, and that differs markedly from high-energy observations for inclusive hyperon production.

5 Future Perspectives

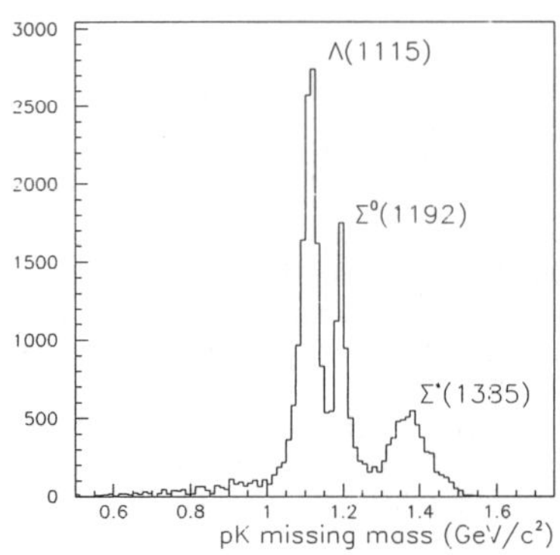

Figure 3. Missing mass spectrum from pK^+ pairs for $p\pi^-$ invariant mass in the range 1100-1130MeV/c^2 after refit.

To progress in the analysis of our data and in particular to make possible to extract from our data spin observables on Σ^0 production we have made an extensive development of the retracking code. The improvements include a refit procedure involving constraints on all the analysed tracks, the search for secondary vertices and the introduction of corrections for multiple scattering in the detectors. The preliminary result of this analysis is shown in Fig. 3, where the missing mass spectrum is presented, assuming four prongs in the final state and selecting the events falling in the Λ invariant mass range 1100-1130 MeV/c^2. The comparison of this figure with Fig. 1 b), shows that the separation between the Λ and Σ^0 peaks is more easily seen. This is not only due to an improvement in the mass resolution, particularly important in the negative x_F regions where rising pK^+ momenta produce worst mass resolution, but also to a strong reduction of the "background" of events that were not properly reconstructed in the previous analysis due to a failure of a correct association of hits and tracks. This improvement will be the basis for the next steps in our analysis:

- Provide with data P_Y, A_N and D_{NN} for *exclusive* Λ *and* Σ^0 production at the three measured energies 2.145, 2.50 and 2.85 GeV.
- Search for reactions with two secondary vertices like reaction 7 or 30.
- Extract from our data all the available information, outlined in Sec. 3.

In addition, the comparison of the data at 2.145 GeV, where the N^* resonance discussed in Sec. 1 can play a role in the Λ production process, with our data at higher energies can clarify if the approach of Ref. [15] is correct.

Recently new data on Λ polarisation at 27.5 GeV/c, with an unpolarised beam have been published[25]. The *exclusive* Λ production has been studied in the reaction channels: $pp \rightarrow p\pi^+\pi^-K^+\Lambda$, $pp \rightarrow p\pi^+\pi^-\pi^+\pi^-K^+\Lambda$, $pp \rightarrow$

$p\pi^+\pi^-\pi^+\pi^-\pi^+\pi^- K^+\Lambda$ and $pp \to p\pi^+\pi^-\pi^+\pi^-\pi^+\pi^-\pi^+\pi^- K^+\Lambda$. An empirical parametrisation is given for the Λ polarisation as a function of x_{F} and p_T: $\mathbf{P}_\Lambda = (-0.443 \pm 0.037)x_F p_T$. This parametrization is independent of the final state and provides a good description of the data. The first of these channels is also present in our data (reaction 3 of Tab. 1). We are going to check if such a parametrisation holds also at 2.85 GeV.

Finally, it could be pointed out that the amount of data, collected with the DISTO apparatus, is so rich to remind the bubble chamber experiments of the 70's, that so much contributed to the understanding of many physical problems. We had here a kind of "electronic bubble chamber", that provided data with a much higher statistics, and, we hope, will be also so successfull in clarifying the puzzle of the Λ polarisation, at least in the DISTO energy domain.

References

1. G. Bunce *et al.*, *Phys. Rev. Lett.* **36**, 1113 (1976)
2. K. Heller, in *Proc. Intl. Symposium on Spin Physics, Amsterdam, 1996* (World Scientific, Singapore, 1997), p. 23, and references therein.
3. J. Lach, in *Flavour and Spin in Hadronic and Electromagnetic Interactions*, eds. F. Balestra *et al.*, (Editrice Compositori, Bologna, 1993), p. 3.
4. B. Andersson, G. Gustafson, and G. Ingelman, *Phys. Lett.* B **85**, 417 (1979).
5. T. Degrand, J. Markkanen, and H. Miettinen, *Phys. Rev.* D **32**, 2445 (1985), and references therein.
6. J. Ellis, D. Kharzeev and A. Kotzinian, Z. Physik C**69**, 467 (1996).
7. Liang Zuo-tang and C. Boros, *Phys. Rev. Lett.* **79**, 3608 (1997), and quot. ref.
8. C. Boros and Liang Zuo-tang, *Phys. Rev.* D **53**, R2279 (1996), and ref. therein.
9. R.L. Jaffe, *Phys. Rev.* D **54**, 6581 (96).
10. M. Burkardt and R.L. Jaffe, *Phys. Rev. Lett.* **70**, 2537 (1993).
11. R. Barni, G. Preparata and P.G. Ratcliffe, *Phys. Lett.* B **296**, 251 (1992).
12. J. Soffer and N. Törnqvist, *Phys. Rev. Lett.* **68**, 907 (1992).
13. J.M. Laget, *Phys. Lett.* B **259**, 24 (1991).
14. J.M. Laget, in *Flavour and Spin in Hadronic and Electromagnetic Interactions*, eds. F. Balestra *et al.*, (Editrice Compositori, Bologna, 1993), p. 81.
15. A. Sibirtsev, K. Tsushima and A.W. Thomas, *Phys. Lett.* B **421**, 59 (1998).
16. R. Bertini, *Nucl. Phys.* A **497**, 307c (1989).
17. F. Balestra *et al.*, Nucl. Instr. Meth. A**426**, 385 (1999).
18. Particle Data Group, Eur. Phys. J. **3**, 1 (1998).
19. R.H. Dalitz, Eur. Phys. J. **3**, 676 (1998).
20. F. Balestra *et al. Phys. Rev. Lett.* **83**, 1534 (1999).
21. S. E. Vigdor, in *Flavour and Spin in Hadronic and Electromagnetic Interactions*, eds. F. Balestra *et al.* (Editrice Compositori, Bologna, 1993), p. 317.
22. B. E. Bonner *et al.*, *Phys. Rev. Lett.* **58**, 447 (1987); B. E. Bonner *et al.*, *Phys. Rev.* D **38**, 729 (1988).
23. A. Bravar *et al.*, *Phys. Rev. Lett.* **78**, 4003 (1997).
24. A. Lesnik *et al.*, *Phys. Rev. Lett.* **35**, 770 (1975).
25. J. Felix *et al.*, *Phys. Rev. Lett.* **82**, 5213 (1999).

THE FINUDA EXPERIMENT AT DAΦNE

T. BRESSANI, D. BERGO, E. BOTTA, L. BUSSO, D. CALVO, P. CERELLO, A. FELICIELLO, A. FILIPPI, S. MARCELLO, O. MORRA, A. PANZARASA, A. RAIMONDO

Dipartimenti di Fisica Generale e Sperimentale, Università di Torino, Via Giuria 1, I-10125 Torino, Italy and INFN, Sezione di Torino, Via Giuria 1, I-10125 Torino, Italy

M. AGNELLO, F. IAZZI, B. MINETTI

Dipartimento di Fisica del Politecnico di Torino, C.so Duca degli Abruzzi, 24, I-10124 Torino, Italy and INFN, Sezione di Torino, Via Giuria 1, I-10125 Torino, Italy

L. CELANO, G. D'ERASMO, E.M. FIORE, L. FIORE, A. PANTALEO, V. PATICCHIO, G. TAGLIENTE

Dipartimento di Fisica, Università di Bari, Via Amendola 173, I-70126 Bari, Italy and INFN, Sezione di Bari, Via Amendola 173, I-70126 Bari, Italy

A. BIANCONI, G. BONOMI, L. VENTURELLI, A. ZENONI

Dip. di Chimica e Fisica per l'Ingegneria e per i Materiali, Università di Brescia, Via Valotti 9, I-25133 Brescia, Italy and INFN, Sezione di Pavia, Via Bassi 6, I-27100 Pavia, Italy

L. BENUSSI, M. BERTANI, S. BIANCO, F.L. FABBRI, P. GIANOTTI, M. GIARDONI, V. LUCHERINI, E. PACE, M. PALLOTTA, S. SARWAR, V. SERDYUK, L. TRASATTI

INFN, Laboratori Nazionali di Frascati, Via Fermi 40, I-00044 Frascati, Italy

V. FILIPPINI, M. MARCHESOTTI, G. MARCIANO, A. ROTONDI

INFN, Sezione di Pavia, Via Bassi 6, I-27100 Pavia, Italy and Dipartimento di Fisica Nucleare e Teorica, Università di Pavia, Via Bassi 6, I-27100 Pavia, Italy

N. MIRFAKHRAI

Department of Physics, Shahid Beheshty University, 19834 Teheran, Iran

P. BOTTAN, P. CAMERINI, N. GRION, R. RUI, R. WHEADON

INFN, Sezione di Trieste, Via Valerio 2, I-34127 Trieste, Italy

P. AMAUDRUZ, G. BEER, A. OLIN

TRIUMF, 4004 Wesbrook Mall, Vancouver, BC, V6T 2A3, Canada

The FINUDA experiment at DAΦNE, the recently commissioned Frascati ϕ–Factory, is now ready to start the data taking. Main expected performances of the detector are reviewed. The initial physics program, aiming to study the spectroscopy and the decay of Hypernuclei formed following K^- capture at rest, is described. Further developement of the detector capabilities are outlined.

1 Introduction

The DAΦNE (e^+e^-) collider[1], built at the Laboratori Nazionali di Frascati, has successfully accomplished the commissioning phase at the end of past year. The design of the collider is optimized for operation at a center of mass energy of 1020 MeV with a top luminosity of $\mathcal{L} = 5 \times 10^{32} cm^{-2} s^{-1}$. The selected center of mass energy corresponds to the rest mass of the ϕ–meson, decaying mainly into kaons. The most frequent decay, with a B.R. of 49.1% is into (K^+, K^-) pairs, emitted back–to–back with a momentum of 127 MeV/c. The (K^+, K^-) pairs are emitted isotropically in the azimuthal plane and with a $sin^2\theta$ dependence with respect to the colliding beams axis. The K^- may be stopped in very thin solid nuclear targets and provide the best experimental situation for the production of Hypernuclei following K^- capture at rest:

$$K^-_{stop} + {}^A X \rightarrow {}^A_\Lambda X + \pi^- \quad (1)$$

This circumstance allows also to study in a clean way the decay of the Hypernuclei, if a suitable vertex detector is utilized. These remarks[2] led to propose the construction of FINUDA, a powerful detector optimized for the study of Hypernuclear Physics at DAΦNE.

2 FINUDA: a new experiment on Hypernuclear Physics

FINUDA is the acronym for FIsica NUcleare a DAΦNE, a new experiment which was proposed in 1993[3] and soon approved for installation at one of the two interaction regions of DAΦNE. The term "new" can be adopted for at least three reasons:

i) it is a new experiment at a new machine;

ii) it is the first example of a Nuclear Physics experiment, which is inherently fixed target Physics, carried out at a collider. Quite amusing: fixed target Physics was shut off the door by Collider Physics and returns now at a collider!

iii) it is the first experiment of Hypernuclear Physics in which both aspects of detection and decay of Hypernuclei are studied with the same apparatus and with the same precision, allowing a complete reconstruction of the events. This is perhaps the most relevant novelty for Hypernuclear Physics.

We will very briefly describe the main characteristics of FINUDA; a detailed description is given in[4]. Fig. 1 shows a sketch of the detector, immersed in a superconducting solenoid which provides a homogeneous magnetic field of 1.1 T inside a cylindrical volume of 146 cm radius and 211 cm length.

The (e^+, e^-) collide at the center of the magnet and the current distribution in the coil as well as the shape of the iron yoke and of the end-caps were designed in such a way as to have a 1% homogeneity inside a tracking volume shaped as a clepsydra.

Fig. 2 shows a sketch of the interaction/target region. The rôle of the detectors arranged around the Be beam pipe (radius = 5.5 cm, thickness = 400 μm) is to select the $(K^+,\ K^-)$ pairs from ϕ-decay and to provide the first level trigger. There is a first barrel of scintillators (Tofino), 2 mm thick, sliced into 12 slabs viewed at both ends by Hybrid PhotoDiodes (HPD) through suitably shaped Lucite light guides. $(K^+,\ K^-)$ pairs crossing the barrel are very efficiently selected at trigger level by $\Delta E/\Delta x$ measurements and back-to-back correlation. There is also an array of Si μ-strip detectors (ISIM), arranged along the sides of an octagonal prism, with bi-dimensional read-out.

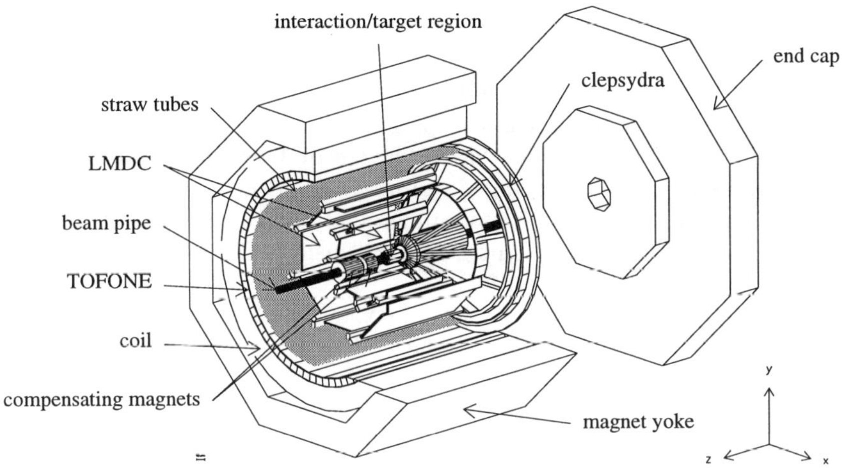

Figure 1. Schematic view of the FINUDA apparatus; the "right" end-cap has been shifted outward, while the "left" end-cap and the "left" part of the mechanical structure (clepsydra) have not been drawn.

It localizes, with a precision better than 50 μm, the K^- (K^+) before stopping in the nuclear target facing ISIM at a distance of less than 2 mm.

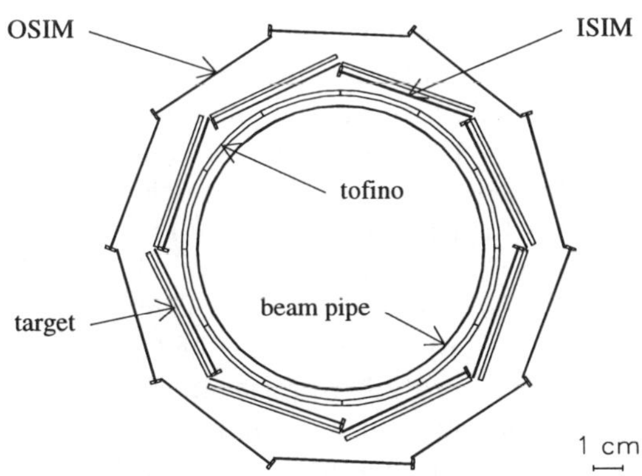

Figure 2. Schematic front view of the interaction/target region.

In addition, ISIM allows a further $\Delta E/\Delta x$ measurement of the particle, which is very useful for pattern recognition and the determination of the interaction vertex. The thickness of the nuclear targets (solid materials, ranging from Li to Pb) is such as to stop the K^- (K^+), with thickness varying between 200 and 300 mg/cm^2. The total length (17 cm) of the detectors surrounding the beam pipe is such that $\sim$ 90% of the K^- from ϕ–decay are detected.

The tracking device must allow the determination of the π^- momenta from the reaction (1) with a $\Delta p/p$ of 0.3% fwhm. It consists of three different types of sub-detectors, all of them immersed in a Helium atmosphere, in order to minimize the multiple Coulomb scattering which strongly affects the momentum resolution in a non-focusing spectrometer like FINUDA. The π^- emitted isotropically from the nuclear targets encounter a first array of detectors, consisting of Si μ-strip (OSIM), arranged following the sides of a decagonal prism.

Beyond these, there are (see Fig. 1) two layers of Low-Mass, high precision planar Drift Chambers (LMDC), arranged again following the sides of an octagonal prism. The two prisms are rotated by an angle of 11° and firmly held in position by a stiff Al structure (clepsydra). This geometry ensures the best acceptance for π^- of 250 $\div$ 280 MeV/c from reaction (1).

The last array of tracking devices consists of six layers of aluminized mylar straw tubes. There are three super-layers (one axial and two $\pm$ 12° stereo super-layers) each with two layers of staggered straws.

A scintillator barrel (TOFONE) composed of 72 elements 10 cm thick, encloses the tracker. It is used for trigger purposes and is very useful in reducing the machine–induced background. Simulations based on calculated beam losses in the target region (mostly Touschek scattering) showed that this background would be lower than 1 Hz at $\mathcal{L} = 10^{32}$ cm^{-2}s^{-1}. Furthermore, TOFONE will detect neutrons emitted following the non–mesonic (NM) decay of the Hypernuclei. The energy of neutrons (a broad spectrum peaked at $\sim$ 80 MeV, but extending up to $\sim$ 150 MeV) will be determined by time-of-flight, with a $\sim$ 500 ps FWHM time resolution and a $\sim$ 12% efficiency. These features will allow a great improvement in the experimental study of these decays.

The most crucial feature of the apparatus is the momentum resolution because it is directly related to the resolution of the hypernuclear energy levels. From extensive and accurate simulations carried out with a Monte Carlo program based on GEANE we estimated that FINUDA would reach a momentum resolution less than 0.3% FWHM, which corresponds to $\sim$ 700 keV in energy. Concerning the production rate of Hypernuclei, it was evaluated that for an Hypernuclear final state produced at a rate of 10^{-3}/(stopped K^-), which is typical for p-shell Hypernuclei[5], 80 events/hour are expected at a luminosity of $\mathcal{L} = 10^{32}$cm^{-2}s^{-1}. The detection rate for non-mesonic decays is, under the above conditions, 5 events/hour for the (n,p) decay and 1.0 events/hour for the (n,n) decay[6]. The lifetime of the Hypernuclei will also be directly measured with very good precision. These rates are higher by an order of magnitude than those planned at existing machines. The reason is that the spectrometer has a very wide acceptance (2π) and, furthermore, that the same detector with the same huge solid angle is used for both the detection of the fast π^- produced in reaction (1) and for the detection of the products of the Hypernuclear decay.

3 The initial physics program of FINUDA

FINUDA will allow an unprecedented step forward in Hypernuclear Physics concerning the spectroscopy, the decay and the production of new species of Hypernuclei. Let us briefly summarize the more relevant items.

a) The Spectroscopy

The expected resolution (700 keV FWHM) is better by more than a factor 2 than the best one achieved with the SKS spectrometer at KEK[7]. The expected counting rates are so large that even Hypernuclear states produced at a rate of 10^{-5}/(stopped K^-) could be isolated. The (K^-_{stop}, π^-) reaction has kinematic features similar to those of the (π^+, K^+) one, even though, expecially for high–A targets, the specific properties of the kaonic atom from which the Hypernucleus formation occurs may alter the population of final hypernuclear states. The most exciting results in Hypernuclear Physics in the last few year were just obtained by experiments based on the (π^+, K^+) reaction, at BNL and afterwards at KEK. Among others, it is enough to recall the excitation spectra of $^{89}_{\Lambda}Y$, $^{139}_{\Lambda}La$ and $^{208}_{\Lambda}Pb$, showing clear peaks due to production of the Hypernuclei in states for which the Λ occupies all the orbitals expected from a simple shell model picture. These spectra were often referred as the textbook example of the validity of the shell model orbital, not just in the valence region, but deep inside the nuclear interior.

With FINUDA it will be possible to continue these investigations with higher statistics and better resolution. The splitting of the peaks due to the spin interactions could also be observed for some selected targets and we may expect that the bulk of these data may contribute to obtain more precise and reliable parameters for the potential well.

For low–A Hypernuclei, an improved resolution will permit the identification of excited states not belonging to those before mentioned (Λ in a shell orbital) but to different excitations of the nuclear core or more complicated configurations[7]. In (π^+, K^+) reactions these states are produced with a strenght considerably lower than the previous ones and we may expect a similar pattern in (K^-_{stop}, π^-) reactions.

However the situation for (K^-_{stop}, π^-) reaction for detecting states produced at a low capture rate is somehow complicated by the presence of a physical background that, in the region where the hypernuclear peaks are expected, is essentially due to the two–step process occurring inside a nucleus, in which the first reaction involves a "quasi–deuteron" of the target nucleus:

$$K^-_{stop} + (np) \rightarrow \Sigma^- + p \tag{2}$$

$$\Sigma^- \rightarrow \pi^- + n \tag{3}$$

Since the Σ^- decays in flight, from the kinematics it is expected that the π^- from (3) and the proton from (2) are emitted roughly back–to–back. The simultaneous detection of both particles, as possible in FINUDA, could allow a substantial reduction of the background and then an improvement of the sensitivity to states produced at a low rate. We recall that this technique was first proposed by Tamura et al.[8] and a detailed simulation of what we may expect in FINUDA was already presented[9].

Fig. 3 summarizes the results of such a simulation. The final π^- spectrum obtained with an integrated luminosity of 2.5 (pb^{-1}) (corresponding to 7 hours at $\mathcal{L} = 10^{32} cm^{-2} s^{-1}$) is shown. π^- peaks at 273 MeV/c (corresponding to ${}^{12}_{\Lambda}C$ formation in ground state, with a capture rate of $10^{-3}/K^-_{stop}$) and 261 MeV/c (corresponding to ${}^{12}_{\Lambda}C$ formation in an excited state at $\sim$ 10 MeV with a capture rate of $2 \times 10^{-3}/K^-_{stop}$) are clearly visible.

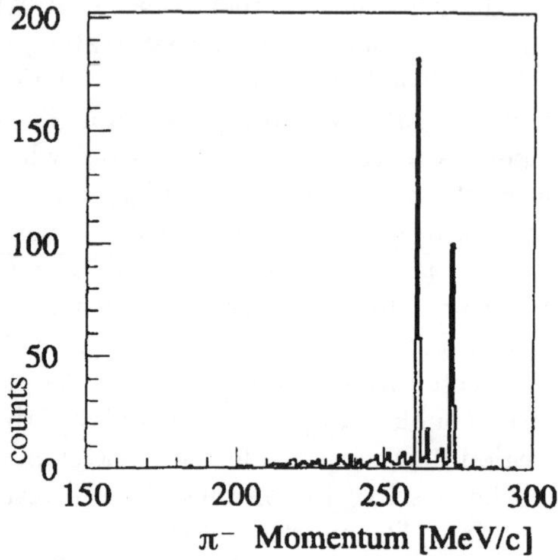

Figure 3. Simulated reconstructed π^- momentum spectrum after the rejection of the correlated ($\pi^- p$) background (from Ref. 9).

But also smaller peaks, corresponding to excited states produced at a lower rate are visible.

b) The decay

Even though the importance of the Hypernucleus decay in weak interaction physics was recognized from a long time, only recently some relevant new experimental data were published, giving input for a renewed interest by theoreticians. The main observables that may be measured in Hypernuclear Physics are:

1) τ_Λ, the life–time of the Λ embedded in a particular nucleus ($\tau_\Lambda = \hbar/\Gamma_\Lambda$)

2) The partial decay widths Γ_π and Γ_{NM} ($\Gamma = \Gamma_\pi + \Gamma_{NM}$), referring to the weak processes $\Lambda \to N\pi$ (mesonic decay) and $\Lambda N \to NN$ (NM decay) in nuclei. It is well known that in all but lightest Hypernuclei Γ_{NM} exceeds Γ_π by a factor more than 100 due to the Pauli blocking. Γ_π and Γ_{NM} may be furthermore splitted into $\Gamma_\pi = \Gamma_{\pi^-} + \Gamma_{\pi^0}$ and $\Gamma_{NM} = \Gamma_p + \Gamma_n + \Gamma_{2p2h}$ referring to charge states $\Lambda \to p+\pi^-$, $\Lambda \to n+\pi^0$, $\Lambda p \to np$, $\Lambda n \to nn$, $\Lambda\mathcal{NN} \to \mathcal{NNN}$. The last term is included in order to account for possible weak interactions involving two correlated nucleons. The measurement of the above partial decay widths

may be performed with an energy resolution able to isolate the final nuclear states or not. In the first case we will deal with "exclusive measurements", in the second with "inclusive measurements". FINUDA will allow to perform a real breakthrough in this subject, thanks to the high production rate of Hypernuclei in their ground states and to the detector's design.

Concerning the measurement of τ_Λ, very recently new data on ${}^{12}_{\Lambda}C$, ${}^{28}_{\Lambda}Si$ ${}_{\Lambda}Fe$ were published[10]. They are very important since they show that τ_Λ in nuclei is definitely lower (more than 20%) than for free Λ. None of the theoretical models suggested for NM decay is able to predict such a behaviour, as shown in Fig .4. There are data also for the high–A region (Bi, U), not reported in the figure.

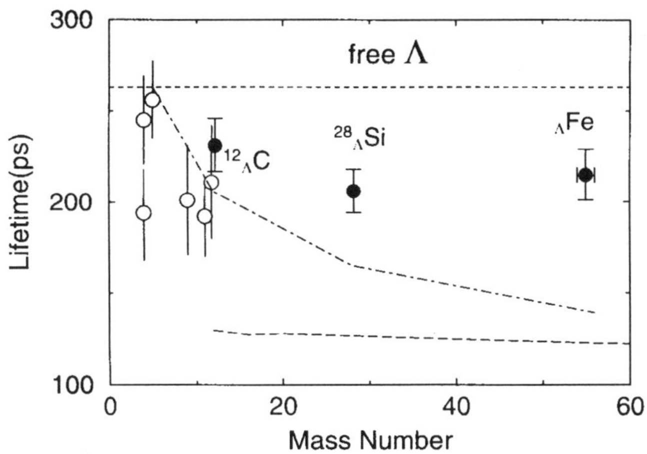

Figure 4. Mass dipendence of hypernuclear lifetimes, from H. Bhang *et al.*[9]. Recent results (closed circles) are compared with previous ones (open circles) and with theoretical calculations (dot–dashed and dashed lines). The dotted line indicates free Λ lifetime.

They were obtained with the delayed–fission technique, using as projectiles p and $\bar{p}$, and are affected by large statistical and systematic errors. It is quite clear the importance of expanding this kind of measurements to a large value of A–number, trying also to improve the precision of the data.

FINUDA is very well suited to accomplish this job. The instrumental time resolution on the delay between the time of detection of the product of the Hypernuclear decay (usually the proton from the NM decay $\Lambda p \rightarrow np$) with respect to the formation time will be worse by a factor at ~ 3 with respect to that at present obtained by SKS[10]. This is due to the circumstance that in FINUDA ~ 3 m long scintillators (TOFONE) will be used instead than thin and small scintillators, like in the SKS. However, the expected statistics exceeds by at least one order of magnitude those at SKS and then total errors on τ_Λ lower than $(2-3\%)$ are expected. Furthermore, the better energy resolution and the use of thin targets would allow the unambiguous mass identification of the decaying Hypernucleus. As a matter of fact, in[10] the measurement on Iron is reported as ${}_{\Lambda}Fe$, since, due to the reduced statistics and energy resolution, it was not possible to isolate unambiguously the final Hypernucleus, but it was necessary to integrate over an interval of (2–3) units on the mass number.

The measurement of Γ_p, Γ_n, perhaps Γ_{2p2h} will be the battle horse of FINUDA. Notwithstanding the big theoretical effort on the subject, recently grown following the suggestion of a possible violation of the $\Delta I = 1/2$ rule occurring in NM decay, no substantial improvement from the experimental side was reported. The aim of FINUDA is that of measuring Γ_p with a total precision (statistics plus systematics) better than 5%, and Γ_n better than 8%. The spectra of proton–neutron and neutron–neutron pairs emitted in coincidence will be measured with an energy resolution better than 1% for protons, 10% for neutrons. The precision on the measurement of the angular correlation will be $\sim$ 4%. This information is essential in order to understand the effect of final state interaction on the weak $\Lambda\mathcal{N} \to \mathcal{N}\mathcal{N}$ process. At present it is difficult to assess wheather Γ_{2p2h} could be really measured, even with coarse precision, or only an upper limit could be given. The real background in running conditions has to be measured before making any prediction.

The measurement of Γ_p and Γ_n with the above precision on some selected targets spanning over the periodic table will provide a unique bank of data for the theoretical interpretation, including a possible violation of the $\Delta I = 1/2$ rule in NM decays. The measurements will be of "inclusive type", apart perhaps the case of ^6Li target, for which an "exclusive" measurement is within reach of the performances of the detector[11]

In the proposal we did not mention the possibility of measuring Γ_{π^-}. The reason is that π^- from mesonic decay, with momenta in the range (100–150) MeV/c, spiralize into the spectrometer, never reaching the outer detectors and cannot be tracked. However, the vertex detector is potentially able to measure, with a coarse precision (a few %) the momenta of π^- from mesonic decay, emitted backward from the target, crossing Tofino, the beam pipe and crossing again ISIM and OSIM. The acceptance of the vertex detector for such events exceeds 2π sr. Taking into account that the total efficiency for measuring (n, n) pairs is $\sim$ 2%, we may not expect that a measurement of Γ_{π^-} at 10% precision is possible, if Γ_{π^-} is $10^{-2}\Gamma_{NM}$.

Other rare NM decay, like ${}^4_\Lambda He \to d + d$ and ${}^4_\Lambda He \to p + {}^3H$, easily recognizable from the monochromatic d and p peaks, well measured by the spectrometer, could be studied with FINUDA. ^4He cannot be used as a target but we may expect copious production of ${}^4_\Lambda He$ from heavier targets, as the case of ${}^4_\Lambda H$ observed at KEK[5]. We have not yet explored this possibility.

c) Production of new species of Hypernuclei

As pointed out by Majling[12], neutron–rich Hypernuclei may be produced following K^- capture at rest by the reaction:

$$K^- + {}^A X_Z \to {}^A_\Lambda X_{Z-2} + \pi^+ \tag{4}$$

corresponding to a two step process (K^-_{stop}, π^0) and $(\pi^0, \pi+)$ on two (correlated) protons in the same nucleus. The importance of observing neutron–rich Hypernuclei is related to the possible existence of neutron halos, and to the different rôle that a Λ could play in stabilizing these configurations. The possibility of studying or not such Hypernuclei with FINUDA is related to their production rate, that may fall in the $(10^{-5} - 10^{-6})/K^-_{stop}$ range. A production rate of $10^{-6}/K^-_{stop}$ could be marginal for the detection capabilities of FINUDA, even though the background in the π^+ spectra is expected to be considerably lower than for the π^- spectra.

An important point that has to be stressed is that all the measurements so far mentioned will be performed on the same target at the same time, since there is no need of a specific hardware optimization. The optimization of a particular production or decay channel will be obtained by software cuts on the raw data.

In conclusion, we may expect that FINUDA will run for the first year with targets of ^6Li, ^{12}C, ^{28}Si, ^{89}Y and perhaps a heavy one (Bi, Pb?), aiming to obtain the following measurements for each target:

1) Excitation energy spectra with up to 10^5 events;

2) τ_Λ with a precision of 2%;

3) Γ_p with a precision of 5%;

4) Γ_n with a precision of 8%;

Additional information could be obtained on Γ_{2p2h}, Γ_{π^-}, rare NM decays and neutron–rich Hypernuclei.

The assumption "one year of operation" is 100 days of running time with a luminosity in the range $(0.5 - 1) \times 10^{32} cm^{-2}s^{-1}$.

4 Possible upgrades of FINUDA

Even though yet not running, we are thinking of possible upgrades of the detector in order to cover an even larger range of observables of Hypernuclear Physics. The present detector does not allow any measurement on photons involved in the formation/decay of Hypernuclei. γ–rays up to a few MeV are expected to be emitted by Hypernuclei formed by the (K^-_{stop}, π^-) reaction in such low–lying excited states. A few of these transitions were already observed by previous experiments at BNL and at KEK, using $(K^-, \pi^-\gamma)$ and $(\pi^+, K^+\gamma)$ reactions.

The detection of a low–energy γ ray in coincidence with the formation product allows to push down the energy resolution on Hypernuclear levels from the value typical of the spectrometer to that of the Photon Detector (a few keV for a Ge detector). All the effects expected from spin–spin terms (LS, spin–spin, tensor) in the $\Lambda - \mathcal{N}$ interaction could be isolated and studied in properly selected Hypernuclei. Notwithstanding this appealing feature, experiments so far performed, even if very successfull and interesting[13], were rather limited. The obvious limiting reason was the counting rate due to total efficiency (detection and geometrical) of the γ–detector array.

When DAΦNE will reach the design luminosity of $5 \times 10^{32} cm^{-2}s^{-1}$ we expect $\sim$ 400 events/hour for a Hypernuclear state produced at a rate of $10^{-3}/K^-_{stop}$, typical of the p–shell Hypernuclei. Then, by installing an array of Ge detectors near the target/interaction region (but condamning inevitably $\sim 1/3$ of the acceptance of the spectrometer for π^-, see Fig. 1) a count rate of up to 10 Hypernuclear γ–rays per hour in coincidence is expected, assuming a total efficiency of (3 – 4%) for the Ge array. This number is very comfortable and the experimental information needed before starting such a project is the knowledge of the background spectrum near the interaction/target region. We have not to forget that we will work at an (e^+, e^-) collider and unexpected electromagnetic backgrounds may be present.

If we consider photons of energy in the range of hundreds of MeV, they are expected as decay products from the π^0 from the formation reaction:

$$K^-_{stop} + ^A X \rightarrow ^A_\Lambda (X+1) + \pi^0 \tag{5}$$

If we were able to measure the energy of both these photons with good resolution ($\sim 1\%$) and spatial accuracy, the π^0 spectroscopy could be performed at a precision even of 1 MeV[14]! From the Physics point of view, the reaction (5) would give access to a completely new class of Hypernuclei, mirrors of those produced in (K^-, π^-) and (π^+, K^+) reactions. However, the modifications of the present FINUDA design in order to be able to study even reaction (5) would be very large, perhaps a complete newconcept of the spectrometer, and we have not precise ideas at present.

5 Conclusions

After a couple of years of delays (magnet, machine) FINUDA is now ready to start the data taking. The magnet has been installed into the experimental pit and the detector was completed and tested with cosmic rays in a dedicated area. DAΦNE is working well and has started the first physics runs for the experiments KLOE and DEAR. We expect that, following this first period, FINUDA will be installed and will start the data taking approximately at the end of this year. We hope then that we will be able to present first physics data at the next Conference/Workshop on Hypernuclear Physics.

References

1. Proposal for a ϕ–Factory, The ϕ–Factory Study Group, Rep. LNF–90/031 (IR)
2. T. Bressani, in *proc. of the Workshop on Physics and Detectors for DAΦNE*, Frascati, April 9-12, 1991, edited by G. Pancheri, p. 475.
3. M. Agnello *et al.*, Rep. LNF–93/021 (IR)
4. M. Agnello *et al.*, Rep. LNF–95/024 (IR)
5. H. Tamura *et al.*, *Nuovo Cimento*, **102A** (1989), 575.
6. M. Agnello *et al.*, *Nucl. Phys.*, **A585** (1995), 271 c.
7. T. Nagae, these Proceedings
8. H. Tamura, *Phys. Lett* **B160** (1985), 32.
9. V. Filippini *et al.*, *Nucl. Phys.* **A639** (1998), 537c.
10. H. Bhang *et al.*, *Phys. Rew. Lett.* **81** (1998), 4321.
11. T. Bressani, *Nuovo Cimento*, **108A** (1995), 917.
12. L. Majiling , *Nucl. Phys.*, **A585** (1995), 211 c
13. For a recent review see e. g. H. Tamura, *Nucl. Phys* **A639** (1998), 83c.
14. A. Rusek, *Nucl. Phys.* **A639** (1998), 111c.

DIRECT MEASUREMENT OF THE LIFETIME OF HEAVY HYPERNUCLEI AT CEBAF USING LOW PRESSURE MWPC TECHNIQUE

L. TANG

Department of Physics, Hampton University, Hampton, VA 23668, USA
and
Thomas Jefferson National Accelerator Facility, Newport News, VA 23606, USA

Representing the Jefferson Lab E99-003 (former E95-002) collaboration[1]

Systematic and precise measurement of the lifetime of heavy Λ-hypernuclei, especially in the very heavy mass region, can provide importance information about the YN interactions (both strong and weak) and its characters in the nuclear medium. However, such measurement has great technical challenge thus the available data are very limited and poor in quality, in particularly in the very heavy region. Here we discuss a new experiment and technique applying Low Pressure MWPC and fission process. This technique combined with the unique characters of the continuous electron beam at Jlab will allow us to measure such lifetime in the very heavy region with a precision of 7 ps or better. This proceeding presents this technique and briefly outlines our experimental approaches.

1 Introduction

The lifetime of heavy Λ-hypernuclei is a specially interesting subject from the viewpoint of the weak decay mechanism of the Λ in a nucleus. High precision measurements of hypernuclear lifetimes and the A dependence will provide valuable constraints in attempts to understand many fundamental issues in hypernuclear physics. Such open issues include the effective YN interaction mechanisms and range; $\Lambda N \to \Sigma N$ conversion strength; and the ΛNN three body interactions. At the quark level, Pauli blocking may lead to an observable deviation from the expected A dependence of the hypernuclear lifetime as calculated by conventional hadronic models, especially in the very high mass region.

It is commonly believed that only the lifetimes of very light Λ hypernuclei are dominantly determined by the mesonic decay, $\Lambda \to p + N(+40\text{MeV})$. As the mass of the hypernucleus increases, the rate of the nonmesonic decay, $\Lambda + N \to N + N(+176\text{MeV})$ increases while that of mesonic decay decreases rapidly because of Pauli suppression of the low momentum final state nucleons. Thus, it is generally believed that the lifetime of Λ hypernuclei is a decreasing function of mass number A. However, the effective YN interaction range may change this description, i.e. additional nucleons may appear to have no effect upon the lifetime of hypernuclei as A reaches a certain level. The very limited data seems to have such suggestion (see later discussion). Theoretically, it is still unclear why should not the lifetime of hypernuclei differ significantly from that of a free Λ.

The $\Delta I = \frac{1}{2}$ rule, until now, is still a key constrain to build up the model for hadronic weak decays. Such rule was empirically obtained from the pionic decay of kaons and then observed also in the mesonic decay of free Y. The origin of the rule is unclear nor suggested by the standard model. Many studies have hinted that

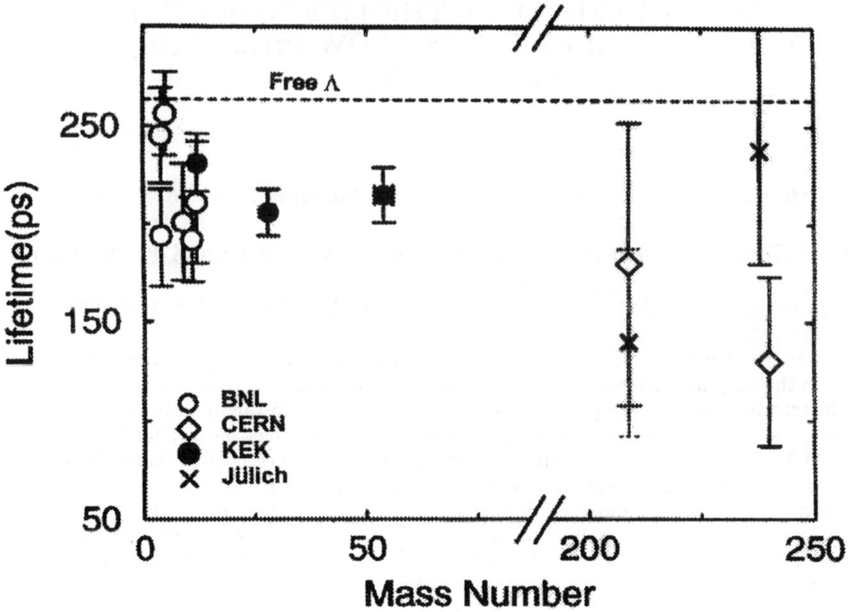

Figure 1. Up-to-date available data on the lifetime of Λ-hypernuclei as a function of A.

ΛN → NN decay will be a powerful tool to study the weak decay involving purely the baryons and observe a violation of such rule. Many references can be found from Ref.2. The lifetime of heavy hypernuclei which is dominated by such decay may give important information for further theoretical investigation.

In very heavy hypernuclei, partial deconfinement may require that the u and d quarks in the Λ be antisymmetrized with the nonstrange quarks in the neighboring nucleons so that observable effects, such as isospin mixing in hypernuclei or modification of the mesonic decay rate in the nuclear medium, may occur. The partial widths of different decay modes and the A dependence are subjects of many theoretical investigations [3-19].

A high precision hypernuclear lifetime measurement for various masses, especially in the heavy region, is necessary and important. However, the available data are scarce and poor in quality (about 100% relative error in most of cases in the very heavy region). Recently, two experiments used proton beam on both Bi and U targets without kaon tagging and positive pion beam on C, Si, and Fe targets using (p^+, K^+) reaction were carried out at Julich [20] and KEK [21], respectively. Fig. 1 shows the combined available data of lifetime as a function of A.

The program at Jlab Hall C is aiming to measure directly the lifetime of heavy Λ hypernuclei, such as Bi, at a precision of about 7 ps with negligibly small systematic error. This can be accomplished by utilizing the high intensity, high duty factor, and fine micro-structure features of the continuous electron beam at Jlab. An unique fission fragment detection technique, Low Pressure MWPC (LPMWPC),

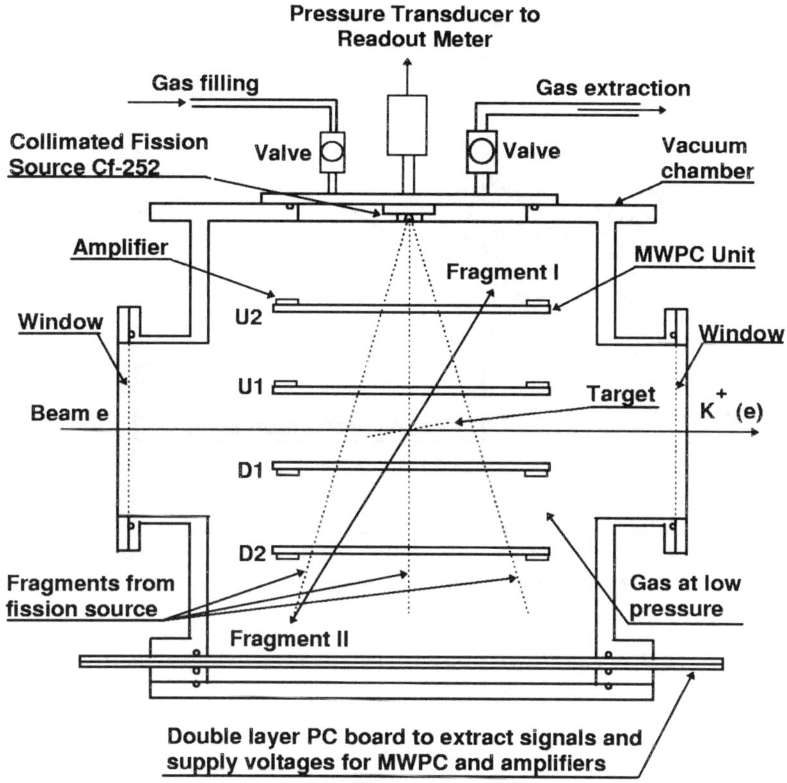

Figure 2. Schematic illustration of LPMWPC system and the experimental technique.

which utilizes the high fission probability from the heavy hypernuclear decay via the nonmesonic decay channel, will be applied. Such detector is completely transparent to all particles except the fragments with large Z. The excellent timing and position resolution from this detector plus the narrow production beam bunch width (1.67 ps) will enable us to measure the lifetime with high precision.

2 Experimental Technique

The principle of this experiment is illustrated in Fig. 2 and is designed to carry out a direct measurement of the lifetime of heavy Λ hypernuclei by fully utilizing the unique features of the CEBAF beam (2 ns beam bunch spacing and 1.67 ps bunch width) and a fission fragment detection technique. The heavy hypernuclei will be produced by (e, K^+) reactions from a very thin target foil $(2mg/cm^2)$ which was tilted to make an effective thickness of about $15mg/cm^2$ for production. Λ production will be ensured once a kaon is detected near production threshold by the Short-Orbit-Spectrometer (SOS). Since the nonmesonic decay of the Λ hypernuclei, in which a large amount of energy (176 MeV) is released, dominantly determines the lifetime, the probability of a time- delayed fission ($10^{-10}sec$) due to Λ hypernuclear

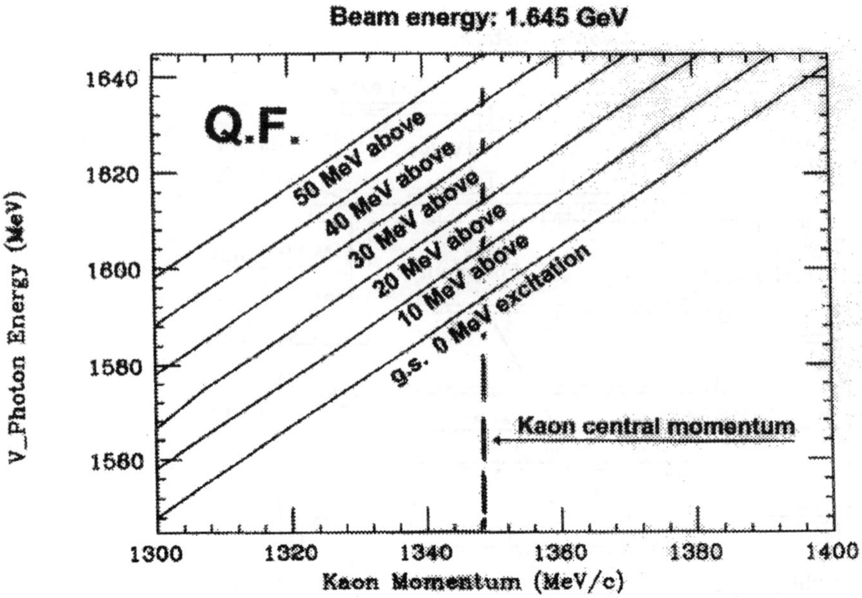

Figure 3. Correlation between virtual photon energy and kaon momentum for various of hypernuclear masses.

decay is sufficiently high to use fission as a gate for the hypernuclear decay. Thus, a coincidence detection of a kaon and fission fragments will select the production and measure the decay of a Λ hypernucleus unambiguously [22].

The Hall SOS spectrometer has been successfully used as a kaon spectrometer in two previous experiments, E93-018 and E91-016, using $(e, e'K^+)$ reaction. The excellent kaon identification system can cleanly separate kaons from all other backgrounds. More importantly, the excellent timing of this system has been capable to reconstruct the the actual production beam bunch. Thus, in this experiment the time zero of the Λ production will be obtained by reconstructing the production beam bunch using the detected kaon. Since the bunch width is only 1.67ps with a precise 2ns separation, the production time-zero can be determined with extremely small error. The detailed discussion of this system can be found in Ref.1 and 23.

Even though the scattered electron, e', is not detected, a chosen kaon momentum acceptance with a fix beam energy will allow us to selectively gate the Λ-hypernuclear production at various of missing mass regions. Fig. 3 shows the correlation for various of specific missing masses between the undetermined virtual photon energy and the chosen kaon momentum acceptance with the beam energy fixed at 1.645 GeV. Thus, by selecting range of kaon momentum, we can effectively gate the lifetime measurement to different missing mass ranges.

The decay time will be reconstructed by detection of the fission fragments released by the heavy hypernuclei undergone nonmesonic weak decay. Such delayed

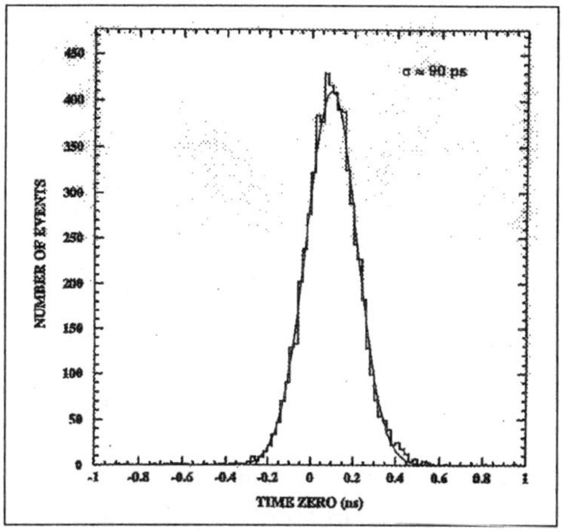

Figure 4. Timing resolution obtained by tracking the fission fragments traveled through the LPMWPC planes from a ^{252}Cf source mounted on top of the system but inside the low pressure chamber.

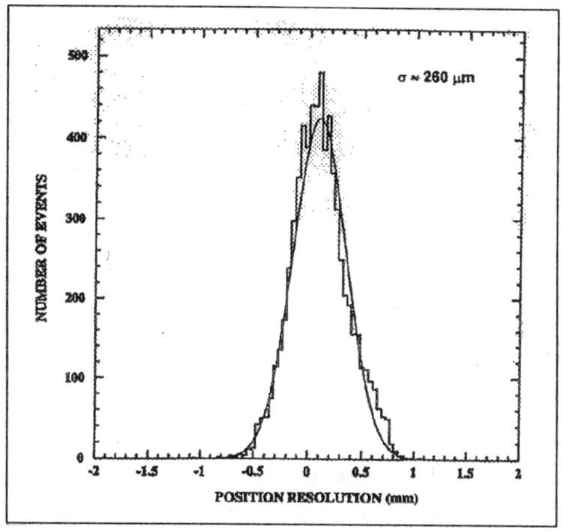

Figure 5. Position resolution obtained by tracking reconstruction of the fission fragments gone through the LPMWPC planes from the fission source.

fission due to nonmesonic hypernuclear decay features by releasing a pair of heavy fission fragments in a binary mode. The fragments will be detected by a Low Pressure MWPC chamber system (LPMWPC) mounted around the target. Under very low pressure ($P < 2Torr$) the specially designed and constructed LPMWPC has excellent position and timing resolution, sensitive only to the desired Z range of fragments, and has about 100% detection efficiency. Fig. 4 shows a system timing reconstruction accuracy tested using a collimated ^{252}Cf fission source mounted at the top center of the chamber system. The accuracy was obtained by analyzing and comparing the difference between the measured time of flight (TOF) from the top two chambers and the reconstructed TOF for those two chambers done by measurement for the top and bottom chamber. Excellent timing accuracy was obtained. The time-zero offset is due to the ionization energy loss of the fragments penetrated through a length of the common gas chamber. One can notice that such ionization loss does not effect on the Gaussian shape of the distribution but only causing a shift of centroid. Since the fragments from prompt fissions will undergo identical processes, so that a measurement of prompt fissions will allow a precise determination of time zero with virtually no systematic errors. Fig. 5 shows the position resolution done by comparing the measured position on a single chamber to that reconstructed by other three planes. The position determination is needed in reconstructing the actual decay time. The obtained accuracy of position is about three times better than we needed. The detailed discussion on construction and electronics readout of this LPMWPC system can be found in Ref. 24. Under low pressure, the LPMWPC is no longer working like the commonly used MWPC for particle trackings. Low Z particles do not have enough ionization to generate a detectable signals thus it becomes completely transparent to those particles. Only those fragments with high Z can produce sufficiently large signals with extremely small time spread. Thus, such device can be used under high radiation background.

3 Discussion about the experiment

With the high counting rate capability and radiation resistance, the position and timing information provided by a double unit system will allow reconstruction of the actual decay time with respect to the production beam bunch with high precision. In addition, the prompt fission time spectra from various channels are measured simultaneously, providing an accurate determination of the prompt fission background shape and an accurate absolute time-zero measurement to cancel almost all possible systematic errors.

Detailed rate estimation can be found in the Ref. 1. With a beam current of about $10\mu A$, an effective Bi target thickness of about $15 mg/cm^2$, and a detection of kaons at small forward angles, we expect to collect more than 1000 events in the decay time spectrum from the bound region only within about 7 days of beam time. The experiment was approved for 7 days of beam.

With a statistics of about or more than a total of 1000 events, the lifetime extraction accuracy, predominantly from statistical error, can be found to be about or better than 7 ps ($< 5\%$) with the timing reconstruction accuracy better than 200 ps in rms. As mentioned above, the identical processes in prompted and delayed fission will cancel almost all systematic errors involved in such a direct lifetime measurement. Thus, we will be able to provide data with the highest ever reached precision in such measurement.

Finally, the measurement of the complete coordinates (positions and timings) at the four separated units allow us to fully reconstruct the information such as the mean velocity (V_{rms}) and kinematics folding angle of the fission fragments. The measurement of V_{rms} of the fragments from the fission event provides us a determination of the final state hypernuclear mass which fissioned by [25]

$$M = [\frac{0.498Z}{V_{rms}}]^{\frac{3}{2}}, where\ V_{rms} = \sqrt{V_1 V_2} \tag{1}$$

with an accuracy of $\delta A \leq 1$. The measured distributions of velocities and folding angles can allow us to investigate the possibility of extracting the branching ratio of Gn/Gp. It was expected that the V_{rms} of the two fission fragments may have significant difference due to the p or n stimulated nonmesonic decays.

4 Status of the experiment

The collaboration has completed the prototype study. A real sized system was then designed and built. A source (^{252}Cf) test and calibration was carried out successfully. The system has been mounted inside the experimental hall (Hall C) near beam dump. It is currently under study of high beam background and radiation effects for various of radiation conditions produced by the running experiments.

We plan to complete the experiment in two phases. In the first phase, we will use the dumped beam with a beam energy of 1.7-2.2 GeV and low beam current, running in a parasitic mode with another experiment. In this case, no kaons will be detected. The parasitic targets will be built with thickness of only $100\mu g/cm^2$, thin enough to allow hypernuclei fly out of target without significant absorption. A relatively thick backing is to support the target, conduct heat, and absorb one of the fragments from the prompt fission. A series of such targets will be mount in parallel with only about 1 mm spacing. Only long lived hypernuclei can fly out of target and then decay in the space between targets, thus producing a pair detection of both fragments at the top and bottom arms. The prompt fission takes place inside the target only, thus only one fragment can reach either top or bottom arm while the other one is absorbed by the backing. Therefore, we will be able to separate the time spectrum for decayed fission from the prompt one which provides us the time zero calibration. The time spectrum will be reconstructed against the accelerator RF timing, so that the lifetime is still directly measured. This technique is rather similar to the experiments carried out at COSY. Since the lifetime is directly measured, the systematic error can be minimized to be negligiblly small. The statistics is high with an expection of obtaining about or more than 10,000 events in the delayed time spectrum in the parasitic run with the hypernuclear spectroscopy experiment which will run about 600 hours of beam time, thus the lifetime of long lived systems produced from Bi will be measured precisely. Also, the parasitic run will allow us to systematically evaluate the performance of the device and the experimental technique. We expect to carry out this investigation in the spring of 2000.

In the second phase, we will carry out this experiment with kaon detection included. By obtaining more than 1000 events in the time spectrum, we will be able to compare the results from the parasitic run. This is aimed to answer the question of whether the long lived system is corresponding to hypernuclei. If fission can be identified directly associated with hypernuclear decay, the fission technique will provide a powerful tool in study the heavy hypernuclei in many other areas and provide a more effective gating to solve problems associated with low statistics.

5 Summary

The most recent KEK results seem to suggest that the lifetime of Λ-hypernuclei has saturated around ^{28}Si. However this behavior can not be explained with the available theoretical model calculations for non-mesonic decay. It is particularly needed for high precision data in the heavy region. The only available data on Bi and U systems are still lacking precision and not clarified to be associated directly with hypernuclear formation. Therefore, good quality data in the very heavy mass region without ambiguity is still needed in helping to understand the non-mesonic decay process.

The Jlab experiment can carry out such task with minimum effort and excellent accuracy. Our equipment has been built and is under systematic test. The physics result can be obtained and examined even at parasitic run level. If fission can be determined clearly associated with heavy hypernuclear decay, then a systematic study in heavy mass region can be carried out easily with high precision using the fission technique at Jlab.

References

1. Jefferson Lab Proposal E99-003/E95-002, L. Tang and A. Margaryan Spokespersons, (1999/1995). The full collaboration list can be found in E99-003 Proposal.
2. B.F. Gibson and Ed V. Hungerford, *Phys. Rep.* **257**, 349-388 (1995).
3. C.Y. Cheung, D.P. Heddle and L.S. Kisslinger,*Phys. Rev.* C **27**, 335 (1983).
4. D.P. Heddle and L.S. Kisslinger, *Phys. Rev.* C **33**, 608 (1986).
5. Y.V. Dubach, *Weak and Electromagnetic interactions in Nuclei.* (Springer, 1986), pp.576.
6. H. Bando and H. Takaki, *Phys. Lett.* B **150**, 409 (1985).
7. H. Bando and T. Motoba, *Prog. Ther. Phys.* **76**, 1321 (1986).
8. H. Bando, T. Motoba and J. Zofka, *Int. J. of Mod. Phys.* A **5**, 1 (1990).
9. F.J. Gilman and Wise, *Phys. Rev.* D **20**, 2392 (1979).
10. M.A. Shifman, et. al., *Nucl. Phys.* B **120**, 316 (1977).
11. C.B. Dover and G.E. Walker,*Phys. Rep.* **89**, 148 (1982).
12. J. Cohen, *Prog. Nucl. Part. Phys.* **25**, 139 (1990).
13. J.B. Adams, *Phys. Rev.* **156**, 1611 (1967).
14. B.H.J. McKellar and B.F. Gibson, *Phys. Rev.* C **30**, 322 (1984).
15. K. Takeuchi, et. al., *Prog. Ther. Phys.* **73**, 841 (1985).
16. J. Dubach, *Nucl. Instrum. Methods* A **450**, 71 (1986).
17. E. Oset and L.L. Salcedo, *Nucl. Phys.* A **333**, 704 (1985).
18. G. Narduli, *Phys. Rev.* C **38**, 832 (1988).
19. E. Oset, et. al., *Phys. Rep.* **188**, 79 (1990).
20. P. Kulessa, et. al, *Nucl. Phys.* A **639**, 283 (1998); and H. Ohm, et. al, *Phys. Rev.* C **55**, 3062 (1997).
21. H.C. Bhang, et. al, *Nucl. Phys.* A **639**, 269 (1998), Y. Sato, et. al, *Nucl. Phys.* A **639**, 279 (1998).
22. A.T. Margarian, *Nucl. Instrum. Methods* A **357**, 495 (1995).
23. G. Niculescu, et. al, *Phys. Rev. Lett.* **81**, 1805 (1998). G. Niculescu, et. al, *Nucl. Phys.* A **639**, 189 (1998). J. Reinhold, et. al, *Nucl. Phys.* A **639**, 197 (1998). W. L. Hinton, et. al, *Nucl. Phys.* A **639**, 117 (1998).
24. K. Assamagan, et. al, , *Nucl. Instr. and Meth.*, accepted for publication.
25. J.C.D. Milton J.S. Fraser,*Can. Jour. Phys.* **40**, 1626 (1962).

STRANGENESS EXPERIMENTS AT COSY

D. GRZONKA
Institut für Kernphysik, Forschungszentrum Jülich
D - 52425 Jülich, Germany
E-mail: d.grzonka@fz-juelich.de

At the cooler synchrotron COSY several experimental facilities are in operation at internal and external target stations. The combination of a high quality low emittance proton beam with high acceptance detector systems allows precise and complete experiments. The momentum range of the COSY beam from 300 MeV/c up to about 3.5 GeV/c opens a large field of experimental studies. One main topic of the present experiments is strangeness production in various reaction channels.

1 Introduction

The Cooler synchrotron and storage ring COSY in Jülich provides unpolarized and polarized beams with momenta up to $\sim$ 3.5 GeV/c for high precision proton-proton and proton-nucleus interaction studies. More than 1 GeV free energy is reachable in proton-proton reactions which allows the crossing of several thresholds for the production of baryon excitations, of long living mesons like π, η, ω and Φ and of associated strangeness production into $K\bar{K}$ and KY.

At COSY much effort is put into the preparation of high quality beams by phase-space cooling. Extremely thin internal targets are in operation which minimize the disturbance of the reaction products and the background due to secondary interactions. They allow nevertheless for high luminosity due to the high revolution frequency of stored particles. Super-slow resonance extraction with stochastic feeding produces high density external beams with an excellent duty factor. In combination with liquid hydrogen targets of only a few mm thickness this gives an excellent definition of the interaction region. On the detector side special efforts are made to allow for kinematically complete multi-particle measurements with full solid angle coverage and with "massless" detectors which do not absorb or disturb the reaction products. Multi-track events occur especially in associated strangeness production and after delayed decays of strange particles. High acceptance becomes therefore very important.

2 Accelerator

The COSY facility [1] with injector cyclotron and experimental installations is shown in fig. 1. For internal experiments about $4 \cdot 10^{10}$ protons were accelerated up to $\sim$ 3.5 GeV/c. Extraction efficiencies of up to 70 % are achieved with a maximum extracted intensity of $1 \cdot 10^{9}$ protons/s. Extraction is working over the full energy range and stochastic feeding of the resonance extraction (super-slow extraction) is routinely used e.g. with 1 minute spill-time. Electron cooling at low energy is used for machine physics studies. Stochastic cooling is applied for internal target experiments. In fig. 2 the effect of stochastic cooling is seen in the event rate at the internal experiment COSY-11. Without cooling the event rate during the

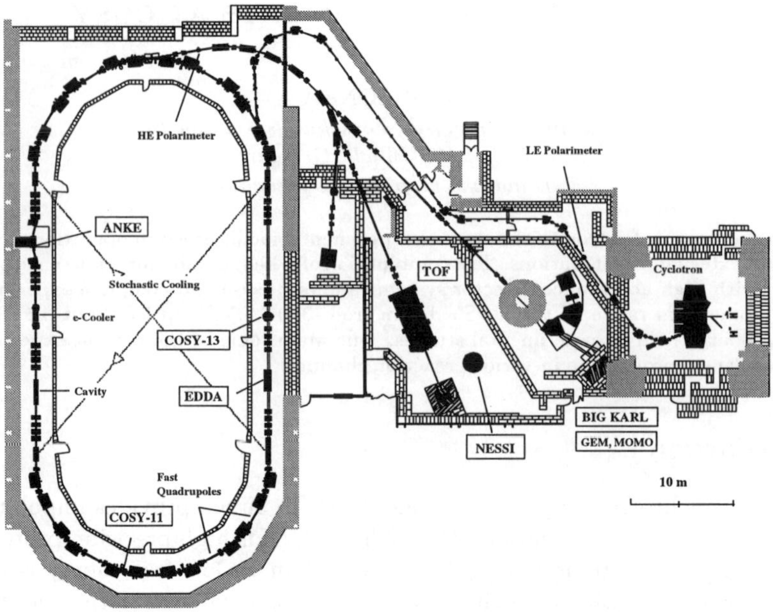

Figure 1: The cooler synchrotron COSY with injection and extraction beam lines, the injector cyclotron and the experimental areas.

experiment decreases due to beam losses. The radial shift of the stored beam by energy loss and the beam blow up by straggling are compensated by the cooling. As a result there is lower background and the energy loss in the target is compensated by the cooling and results in a constant event rate i.e. a higher mean luminosity.

3 Physics at COSY

The main topic of the COSY physics program is the structure of mesons and baryons in the confinement regime of QCD. This field includes the analysis of decay properties and interactions of meson-meson, meson-baryon and baryon-baryon systems.

The reactions under investigation or proposed at the different experimental installations are summarized in table 1.

As experimental installations at COSY there are magnetic spectrometers which are characterized by a high momentum resolution but limited solid angle and non magnetic systems which cover much larger laboratory solid angle but have lower resolution.

The internal experiments are the non magnetic EDDA [2] and COSY-13 [3] and the magnetic detectors COSY-11 [4] and ANKE [5]. For external experiments two stations are available, the area with the non magnetic TOF [6] spectrometer and the non-magnetic Neutron spectrometer NESSI [7] and the area of the focusing magnetic spectrometer BIG KARL where the GEM [8] and MOMO [9] experiments are being performed.

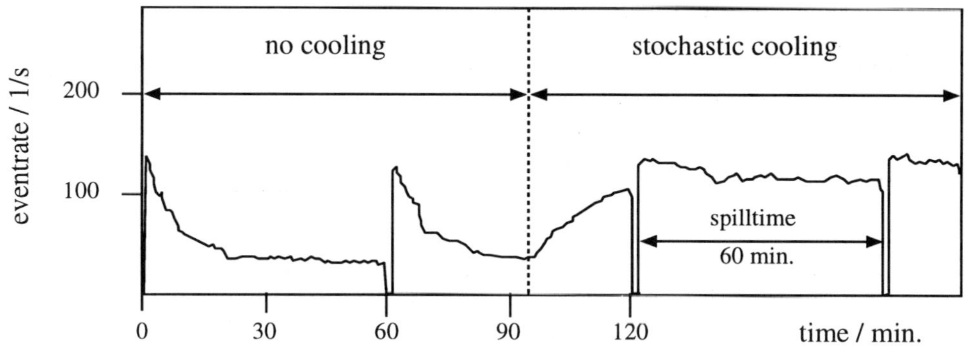

Figure 2: Stochastic cooling at the internal experiment COSY-11. The cooling system is switched on in the middle of one spill. The event-rate is increasing due to the reduction of the beam dimension which increases the beam-target overlap.

Table 1: Present program at the different experimental installations.

internal		external	
ANKE	$p^{12}\mathrm{C} \to K^+X,\ K^+dX$	GEM	$pp \to d\pi^+,\ pp\ \pi^0,\ \eta,\ \eta'$
	$p^{12}\mathrm{C} \to K^+K^-X$		$pd \to {}^3\mathrm{He}\,\eta$
	$pp \to da_0,\ \vec{p}\vec{p} \to pp\Phi$		$pd \to {}^3\mathrm{He}\,\pi^0,\ {}^3\mathrm{H}\,\pi^+$
	$pp \to d\pi^+,\ pn\pi^+,\ pp\pi^0$		$p^{40}\mathrm{Ca} \to pp({}^{39}\mathrm{Ca}\pi^-)_{1s}$
	$pd \to pd\,\omega,\ \Phi$		$pA \to (B\eta)^3\mathrm{He}$
COSY-11	$pp \to pp\ \eta, \eta', \omega, \Phi$	TOF	$pp \to pp,\ pp\ \gamma,\ \pi^0,\ \eta,\ \eta'$
	$pp \to pp\pi^+\pi^-,\ ppK^+K^-$		$pp \to pn\pi^+,\ d\pi^+$
	$pp \to pK^+\Lambda,\ \Sigma^0$		$pd \to {}^3\mathrm{He}\,\eta,\ {}^3\mathrm{He}\,\eta'$
	$pp \to d'\pi^+ \to pp\pi^-\pi^+$		$pp \to d'\pi^+ \to pp\pi^-\pi^+$
	$\vec{p}p \to pp\eta$		$pp\,(\vec{p}\vec{p}) \to pK^+\Lambda,\ \Sigma^0, pK^0\Sigma^+$
	$pd \to {}^3\mathrm{He}\,\eta,\ \eta';\ pnK^+\Lambda$		$p\alpha \to N^*\alpha \to \Lambda K^+\alpha$
	$pd \to ppK^+\Sigma^-,\ pnK^+\Sigma^0$		$pd \to {}^3_\Lambda\mathrm{H}K^+,\ pd \to \Lambda pp$
COSY-13	$p\,(A > 100) \to K^+\,{}_\Lambda(A > 100)$	MOMO	$pd \to {}^3\mathrm{He}\ \pi^+\pi^-$
	delayed fission		$pd \to {}^3\mathrm{He}\ K^+K^-$
EDDA	$\vec{p}\vec{p} \to pp, d\pi^+$	NESSI	$pA \to$ spallation

4 Strangeness Production

The strangeness production is a central topic in the experimental studies at COSY. In table 2 the reaction channels are listed which are presently under investigation. In the following a few results are discussed in more detail.

4.1 Elementary interactions

Studies on the hyperon production are performed at the COSY-11 and the TOF experiment.

The TOF experiments use an external large acceptance time-of-flight spectrometer [6]. It is a flexible modular system of several rotationally symmetric scintillator hodoscopes in vacuum. Finally up to 3 barrel elements can be combined with two planar front hodoscopes of 1.16 and 3 m outer diameter, providing very large solid

Table 2: Reaction channels involving strangeness for which data are available.

Reaction	Elementary Interactions	In Medium Interactions
Meson Production	$pp \to pp\ \eta,\ \eta'$ [10], $\Phi,\ K^+K^-$ $pd \to {}^3\mathrm{He}\ \eta,\ \Phi,\ K^+K^-$	$pC \to K^+X$
Hyperon Production	$pp \to pK^+\Lambda$ [22,6], Σ^+ [12]	$pU, Bi \to K^+\ {}_\Lambda U, {}_\Lambda Bi$ [3,13]

angle coverage without holes. In fig. 3 a sketch of the longest version of the TOF detector is shown with flight path in vacuum of more than 8 m. The present setup consists of central- ring- and one barrel-hodoscope. The central part of the end-cap is the so called "quirl" detector, a three layer scintillator hodoscope with wedges and spirals in two opposite directions [14]. In fig. 4a the pixel structure of such

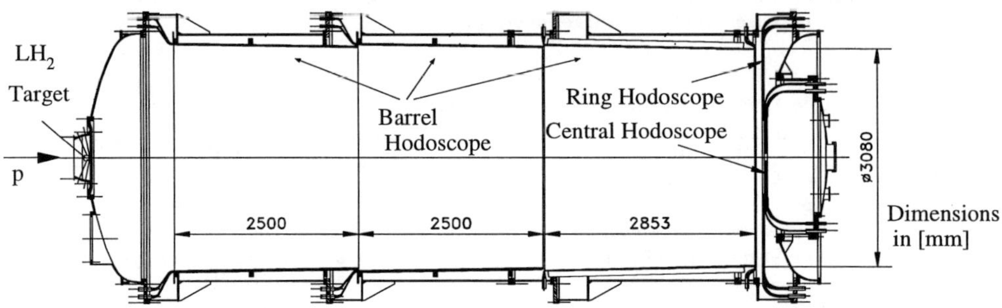

Figure 3: TOF spectrometer in the longest version with 3 barrel elements and the planar front hodoscopes.

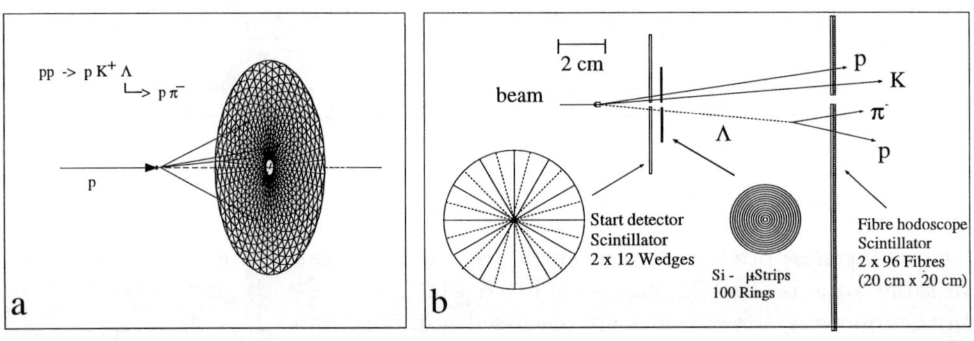

Figure 4: a) Structure of the "Quirl" detector with 3 layers of wedges, left and right wound spirals. A typical $pK^+\Lambda$ event is shown. b)Decay spectrometer for the measurement of the Λ decay in the associated strangeness production $pp \to pK^+\Lambda \to pK^+p\pi^-$.

a detector is shown with the tracks of a typical $pK^+\Lambda$ event. The hit position is defined by the overlap of the fired detector elements. By measuring the time of flight and the direction, the momentum vector can be determined for a certain mass hypothesis. As a start detector a very thin segmented scintillation detector is used. The ring detector is structured like the quirl with 3 layers. The first barrel hodoscope is a single layer device of scintillator bars read out at both ends with ~ 3 m active length and ~ 3 m diameter. A miniaturized liquid hydrogen target[15] with very small amount of inactive material and beam windows of only 0.9 μm Mylar is used. The first data from TOF with the "quirl" have demonstrated that two and three particle reactions like $pp \to d\pi^+$, $pp \to pp\pi^0$ and $pn\pi^+$ can be identified and separated simultaneously.

Close to the target a hyperon decay spectrometer was installed which is shown in fig. 4b. The delayed decay of the Λ hyperon results in a multiplicity increase between start detector and a fibre hodoscope. This increase is used as a very selective trigger for such events. A clean, background free separation of the four fold over-constrained $pK^+\Lambda$ events is achieved. Due to the self-analysing properties of the Λ, from the measurement of the decay particles also the Λ polarization can be determined. The values for the cross section $pp \to pK^+\Lambda$ are (2.7 ± 0.3) μb at 2.50 GeV/c and (12.0 ± 0.4) μb at 2.75 GeV/c[6].

The COSY-11 installation[4], see fig. 5, is designed for meson production studies close to threshold. It is an internal target experiment using a COSY accelerator dipole as a magnetic spectrometer. The track of positively charged ejectiles are measured with drift chambers and scintillator hodoscopes are used for the time of flight information. Furthermore a Si-pad detector is installed in the dipole gap for the detection of negatively charged particles which is important for channels like $pp \to ppK^+K^-$. In case of the hyperon channels the proton and kaon four momenta

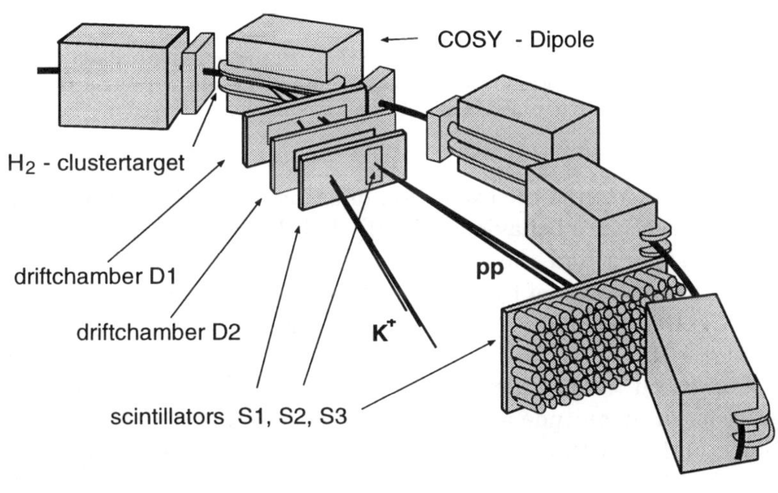

Figure 5: The internal installation COSY-11 in a bending section of the COSY ring with a hydrogen cluster target in front of a machine dipole.

are measured and the hyperon four momentum is given by the missing mass. A very clear separation of the Λ and Σ^0 hyperon events is achieved by this method. In fig. 6 the missing mass distribution for the Σ^0 data at an excess energy of about 8 MeV is shown.

The total cross sections for the hyperon production close to threshold are pre-

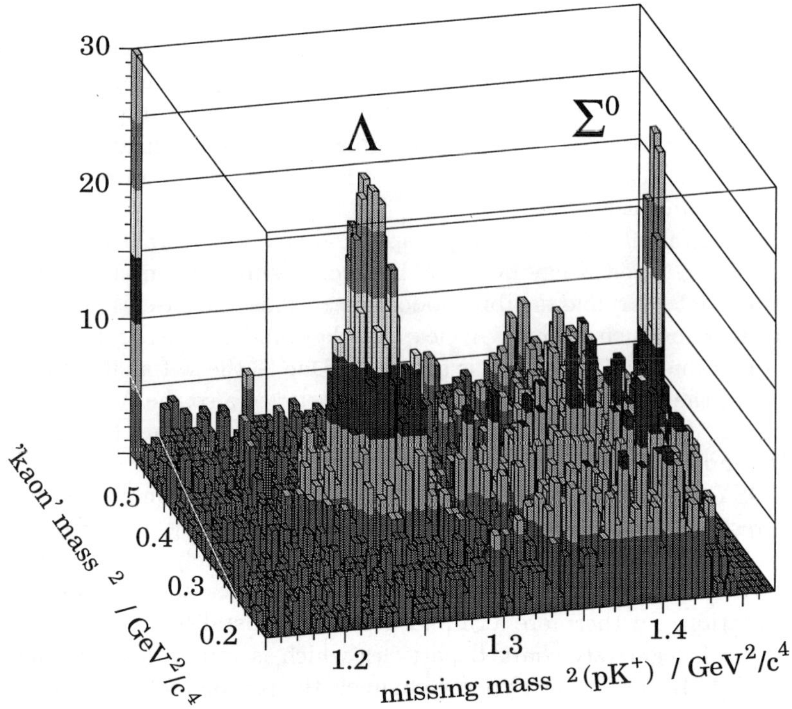

Figure 6: Events with an identified proton track for the Σ^0 production measurement at an excess energy of about 8 MeV. The invariant mass for the second track is plotted versus the missing mass for a pK^+ system. Λ and Σ hyperons are clearly identified.

sented in fig. 7. The form of the excitation function in the threshold region is given by the 3-body phase space behaviour ($d\rho(3)$) slightly modified by the $p\Lambda$ final state interaction ($f_{FSI}(q_{p\Lambda})$) and the pK^+ Coulomb interaction ($f_C(q_{Kp})$). After the publication of the new data of hyperon production from COSY several calculations within meson exchange modes have been published [16], [17], [18], [19] where partly the Λ production via the $S_{11}(1650)$ nucleon resonance is considered as dominant contribution. Compared to the older parametrizations [20], [21] which miss the threshold data by an order of magnitude a drastic improvement was achieved. They describe the experimental data within a factor of 2-5 which is within the uncertainties of the models.

A remarkable result of these hyperon production data is the cross section ratio between Λ and Σ^0 production $\sigma_\Lambda/\sigma_\Sigma$ at the same ϵ in the order of 28 in contradiction to the ratio between 2-3 at higher excess energies. A possible explanation for the drastic enhancement of the Λ production could be a strong conversion from $\Sigma^0 p$ to Λp [12]. To draw definite conclusions further data in this field are needed.

In a three body final state within a certain approximation each two body subsystem can be analyzed separately. This allows the extraction of low energy scattering parameters which are hardly accessible by conventional elastic scattering

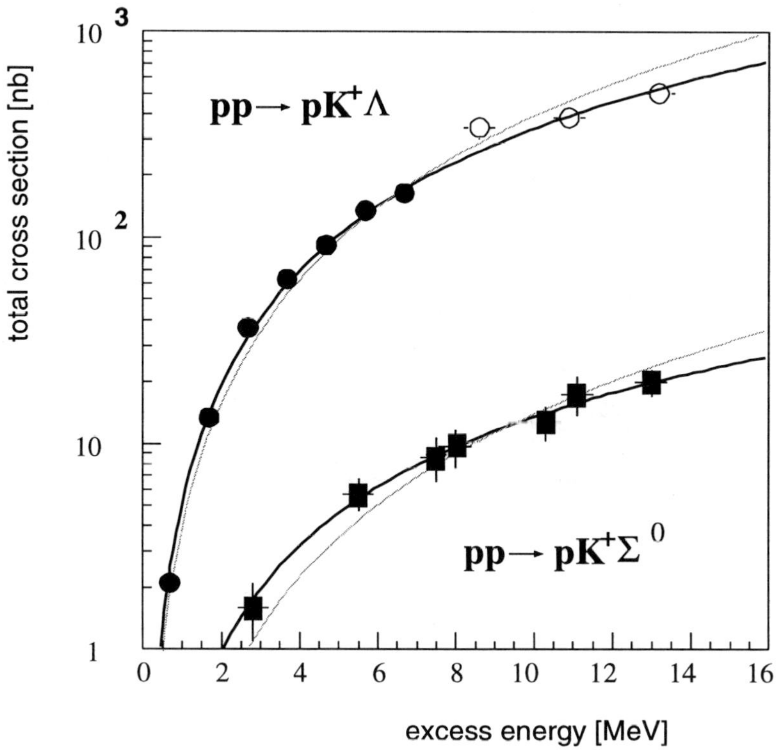

Figure 7: Excitation function for Λ and Σ^0 production in the threshold region. The curves correspond to phase space distribution with (thick lines) and without (thin lines) $p\Lambda$ final state interaction included.

analysis. For the Λ production data a Dalitz plot analysis has been performed [22]. By a comparison between Monte Carlo and acceptance corrected experimental Dalitz plots the scattering parameters can be determined. In the $pK^+\Lambda$ system a factorization of the cross section according to $d\sigma \sim d\rho(3)\, f_C(q_{Kp})\, f_{FSI}(q_{p\Lambda})$ is justified due to the dominant $p\Lambda$ FSI. The final state interaction between proton and Λ can be described in terms of scattering length a and effective range r_0, $f_{FSI}(q, \bar{a}, \bar{r_0}) = 1 \,/\, \bar{a}^2 q^2 + (-1 + \frac{1}{2}\bar{r}\bar{a}q^2)^2$ with a mean a and r_0 because these data include the singlet and triplet sub-states.

Figure 8 shows the χ^2-distribution of a Dalitz plot fit, where a valley is observed, which is consistent with existent a and r_0 parameters extracted from experimental $p\Lambda$ elastic scattering and theoretical calculations represented by the different symbols.

To extract the correct $\bar{a}$ and $\bar{r_0}$ combination further information is needed like elastic Λp scattering data. A fit to low energy Λp-data which can be described by $\sigma = 4\pi/q^2 + (-1/\bar{a} + 1/2\bar{r_0}q^2)^2$, with the restriction that only ($\bar{a}$, $\bar{r_0}$) combinations from the valley in fig. 8 are allowed results in $\bar{a} = 2.0fm$ and $\bar{r_0} = 1.0fm$, represented by the cross symbol in fig. 8.

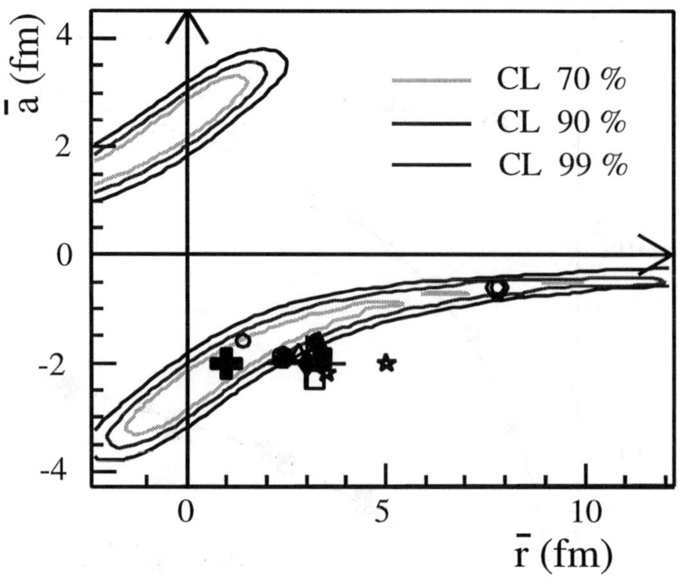

Figure 8: Result of combined Dalitz plot fits for spin averaged scattering length a_0 and effective range r_0 for $pp \to pK^+\Lambda$ at 6 energies up to 10 MeV above threshold. The confidence levels in the χ^2-distribution are given. The cross symbol represent the COSY-11 result.

The next steps in these studies should be the use of polarized beam and target which allows a separation of singlet and triplet states in the entrance and spin transfer studies to the Λ.

The threshold production of K^+K^- in the pp channel is studied also at the COSY-11 experiment. The analysis of K^+K^- data at COSY-11 has shown that the hyperon resonance production like $\Lambda(1405)$ or $\Sigma(1385)$ with subsequent decay to channels including a proton is a strong background source at this low cross section level. The missing mass technique in the ppK^+K^- channel, i.e. measuring of the 4 momenta from the positively charged ejectiles p, p and K^+ and selecting the events from the calculated rest mass, is not sufficient for an unambiguous identification of the K^+K^- events. Additionally the detection of the K^- via the Si-detector in the dipole gap is necessary by which the total detection efficiency is reduced to about 50 % of the efficiency for the ppK^+ detection. Preliminary results indicate a very low cross section in the order of 900 pb ($\pm$ 500 pb) at an excess energy of 17 MeV. The final analysis has to be awaited from which uncertainties below 20 % are expected.

4.2 In medium strangeness production

First results concerning strangeness production in nuclei have been performed at the COSY-13[3] experiment. It is a small internal apparatus for the study of heavy hyper-nuclei by measuring the hyperon induced delayed fission with a low pressure MWPC. To reduce the background from prompt proton induced fission the detector is located in the shadow of the target holder. Only fragments from delayed fission can hit the detector. Measurements with an Uranium and a Bismuth target have been performed from which lifetimes as well as cross sections were extracted. The results are $\tau(U_\Lambda) = 240 \pm 60$ ps, $\tau(Bi_\Lambda) = 161 \pm 21$ ps. The recently published precise data in the medium mass range [23] with lifetimes of 231 $\pm$ 15 for ${}^{12}_{\Lambda}C$, 206 $\pm$ 12 for ${}^{28}_{\Lambda}Si$ and 215 $\pm$ 14 for ${}_{\Lambda}Fe$ indicate a saturation value already reached at A=12. The COSY data especially the Bi data point with a lower error bar give clearly lower values for the lifetimes. Also for U an improvement to an error level of 20 ps is expected from the analysis of new data. In order to get more insight into the in medium decay mechanisms of the Λ more data with higher precision are needed.

At the internal 0° spectrometer ANKE the subthreshold K^+ production in reactions of protons on nuclei is studied. ANKE is a very flexible facility in a straight section of COSY. With a system of 3 dipole magnets an orbit bump of the COSY beam is generated where the central magnet acting as spectrometer is equipped with large exit windows for positively and negatively charged particles. The measurement of K^+ is performed with carefully tuned telescopes of scintillator and Ĉerenkov counters in which the kaons are stopped. This allows a very effective discrimination against the high pion background. By using the delayed signal from muons resulting from the kaon decay a background free separation is possible. Presently data from $pC \to K^+X$ at several momenta down to 1.7 GeV/c are under analysis.

5 Summary and Outlook

The COSY accelerator started tuning up in 1993. Internal and external beams up to the maximum momentum of 3.5 GeV/c are now available for experiments. Presently 1/4 of the beam time is used for machine development and 3/4 is given to the experiments at the internal and external installations. Different experiments at COSY have extracted the first physics results. The COSY data for pp induced reactions are summarized in fig. 9 which demonstrate the high sensitivity of the COSY experiments. Further data are available for *pd* and *pA* induced reactions. Especially in the strangeness sector new precise data are available. These studies will be continued and extended as can be seen from table 1. A central topic in the future will be the use of polarized beams and polarized targets which allows more detailed analyses.

1. R. Maier, *Nucl. Instrum. Methods* A **390**, 1 (1997).
2. D. Albers et al., *Phys. Rev. Lett.* **78**, 1652 (1997) .
3. H. Ohm et al., *Phys. Rev.* C **55**, 3062 (1997).
4. S. Brauksiepe et al., *Nucl. Instrum. Methods* A **376**, 397 (1996).
5. O. Schult et al., *Nucl. Phys.* A **583**, 629 (1995).
6. R. Bilger et al., *Phys. Lett.* B **420**, 217 (1998).
7. D. Hilscher et al., *Nucl. Instrum. Methods* A **414**, 100 (1998).
8. M. Drochner et al., *Phys. Rev. Lett.* **77**, 454 (1996).
9. F. Bellemann et al., nucl-ex/9812011.
10. P. Moskal et al., *Phys. Rev. Lett.* **80**, 3202 (1998).
11. J. Balewski et al., *Phys. Lett.* B **420**, 211 (1998).

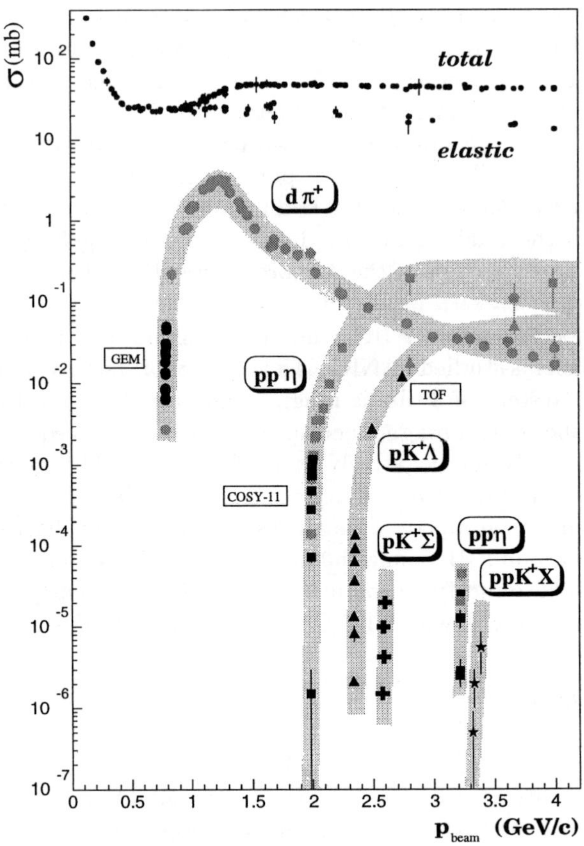

Figure 9: Cross sections for some pp-induced reactions. The light symbols represent data from literature and the black symbols, except for the total and elastic data, are data from COSY which are partially ($pp\eta$, ppK^+X) preliminary.

12. S. Sewerin et al., *Phys. Rev. Lett.* acc. for publ. (1999).
13. P. Kulessa et al., *Phys. Lett.* B **427**, 403 (1998).
14. M. Dahmen et. al., *Nucl. Instrum. Methods* A **348**, 97 (1994).
15. V. Jaeckle et al., *Nucl. Instrum. Methods* A **349**, 15 (1994).
16. G. Fäldt and C. Wilkin, *Z. Phys.* A **357**, 241 (1997).
17. K. Tsushima et al., nucl-th/9801063.
18. A. Sibirtsev and W. Cassing, nucl-th/9802019.
19. G.Q. Li and C.M. Ko, *Phys. Rev.* C **57**, 434 (1998).
20. J. Randrup and C.M. Ko, *Nucl. Phys.* A **343**, 519 (1980).
21. B. Schürmann and W. Zwermann, *Mod. Phys. Lett.* A **3**, 251 (1988).
22. J. Balewski et al., *Eur.Phys.Jou.* A **2**, 99 (1998).
23. H. Bhang et al., *Phys. Rev. Lett.* **81**, 4321 (1998).

FUTURE PLANS OF HIGH-RESOLUTION HYPERNUCLEAR GAMMA-SPECTROSCOPY

H. TAMURA
Department of Physics, Tohoku University, Sendai 980-8578, Japan
E-mail: tamura@lambda.phys.tohoku.ac.jp

Recently we have started a series of experiments of hypernuclear gamma-spectroscopy using a germanium detector system (Hyperball). Experimental plans and a possible upgrade in the future are described. We will continue γ-spectroscopy studies of light Λ hypernuclei for the purpose of determining all the ΛN spin-dependent interactions. We are also planning $B(M1)$ measurements, spectroscopy of neutron rich hypernuclei, study of heavy hypernuclei and double Λ hypernuclei, etc. Some of them are the main subjects in the JHF 50 GeV PS.

1 Introduction

High-resolution γ-ray spectroscopy using germanium (Ge) detectors was almost impossible for hypernuclei before because of technical problems and small hypernuclear γ-ray yields. Recently we solved these problems by constructing a large-acceptance Ge detector system equipped with fast readout electronics (Hyperball), and started a series of experiments.[1]

We have successfully performed the first experiment at KEK (E419) for ${}^{7}_{\Lambda}$Li. In this experiment, well-identified hypernuclear γ-transitions were observed for the first time with Ge detectors. $B(E2)$ of a hypernucleus was derived also for the first time. See ref.[2] for details. The second experiment is under way at BNL-AGS.[2]

High-resolution hypernuclear spectroscopy allows us to investigate detailed structures of various hypernuclei, from which we can study hyperon-nucleon and hyperon-hyperon interactions, various nuclear-structure changes induced by "impurity", and possible changes of baryon properties in nuclear medium. In this article, we will discuss how to approach these physics subjects and describe experimental plans and possibilities in the future.

2 Hyperball and its possible upgrade

Figure 1 shows our Ge detector system (Hyperball). It consists of 14 n-type coaxial Ge detectors of 60% relative efficiency (69mm$\phi\times$69mm in size) arranged at 15 cm from the beam line. The Ge detectors cover a solid angle of about 15%$\times 4\pi$sr, and the photo-peak efficiency including the solid angle is about 3% for 1 MeV γ-rays. The signal of a Ge detector is processed with a transistor-reset preamplifier and a fast shaping amplifier with a gated integrator. Each Ge detector is surrounded by 6 BGO counters which give veto signals to suppress background from Compton scattering, high-energy γ-rays from π^0, and high-energy charged particles. Detailed descriptions are found in ref.[1]

In order to investigate various hypernuclei in a limited beam time available at KEK and BNL, the photo-peak efficiency should be as large as possible. We are planing to upgrade the detector system by adding several large-volume Ge detectors.

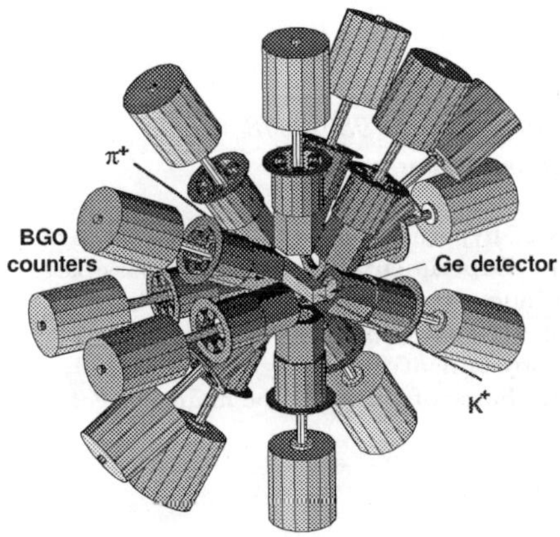

Figure 1: Large-acceptance Ge detector system dedicated to hypernuclear γ-spectroscopy (Hyperball). It consists of 14 Ge detectors of 60% relative efficiency and BGO counters surrounding each Ge detector.

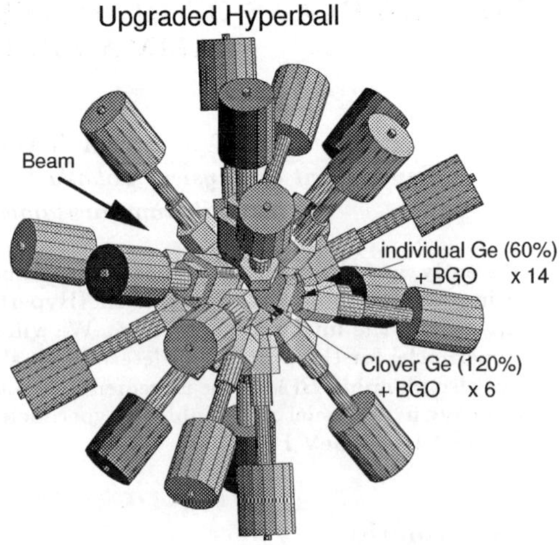

Figure 2: An example of upgraded Hyperball. Six clover-type Ge detectors (120% relative efficiency) are added to the present Hyperball.

The best candidate is the "clover" type detector which consists of four Ge crystals closely mounted together in a common cryostat. It has a relative efficiency of 120% or more when the 4 signals are summed up. It is also effectively used in measurement of linear polarization of γ-rays using compton scattering events in two of the four crystals. The electronodes of new detectors had better be segmented in order to reduce a geometrical error in the Doppler shift correction and to improve tolerance for a high counting-rate condition. Segmented clover-type detector has recently become commercially available. In our case, however, the multiplicity of γ-rays are much less than the case of high-spin state physics, and the requirement for the granularity is not as crucial as in the other Ge ball detectors.

Figure 2 shows an example of the upgraded Hyperball, which will hopefully be realized in near future. Six clover Ge detectors are added and the total photo-peak efficiency will be almost doubled (about 6 % at 1 MeV), being not far from the efficiency of Gammasphere (10% at 1 MeV) which is one of the largest Ge detector system in the world.

3 Experiments at KEK-PS and BNL-AGS

Various experiments are planned for hypernuclear γ-ray spectroscopy employing the present or upgraded Hyperball. In this section, we describe planned experiments which are feasible in the present beam intensities at BNL-AGS or KEK-PS.

3.1 Spectroscopy of light hypernuclei and spin-dependent ΛN interactions

We are now carrying out the second experiment using Hyperball at BNL (E930). The purpose of this experiment is to establish all the ΛN spin-dependent (spin-spin, spin-orbit, tensor) interactions quantitatively from detailed level structures of ${}^{9}_{\Lambda}$Be, ${}^{16}_{\Lambda}$O, ${}^{12}_{\Lambda}$C and ${}^{7}_{\Lambda}$Li studied by using (K^-,π^-) reaction. Then we will extend our studies at BNL or KEK to more light hypernuclei such as ${}^{10}_{\Lambda}$B, ${}^{13}_{\Lambda}$C, ^{20}Ne, etc.

The ΛN LS force is known to be smaller than that of nucleons, but it is not quantitatively established. Some hypernuclear data suggest a rather large LS force of Λ (about 1/3 of that of nucleon), and some others suggest much smaller values. Theoretically, meson exchange models predict a LS force of about 1/3-1/4 of that of NN, while quark models predict much smaller values. Therefore, crucial data for the ΛN LS force is necessary to establish the strength of Λ LS force and to clarify the origin of the nuclear LS force. The ${}^{9}_{\Lambda}$Be data taken in E930 will provide unambiguous information on the ΛN LS force from the energy difference of the $\frac{5}{2}^+$ and $\frac{3}{2}^+$ levels.[1] However, the sign of the LS force may not be determined, because the two $E2$ γ-rays ($\frac{5}{2}^+ \rightarrow \frac{1}{2}^+$ and $\frac{3}{2}^+ \rightarrow \frac{1}{2}^+$) are expected to have almost the same yields. In order to determine the sign, we need a ${}^{7}_{\Lambda}$Li spectroscopy again to observe the $M1$ transition between the upper doublet states ($\frac{7}{2}^+$,$\frac{5}{2}^+$). Since this energy spacing is determined by the spin-spin term plus the LS term,[9] and we know the spin-spin term from the ground-state doublet spacing of ${}^{7}_{\Lambda}$Li, the sign of the LS term can be determined and the consistency to the ${}^{9}_{\Lambda}$Be data can be checked concerning the size of the LS term.

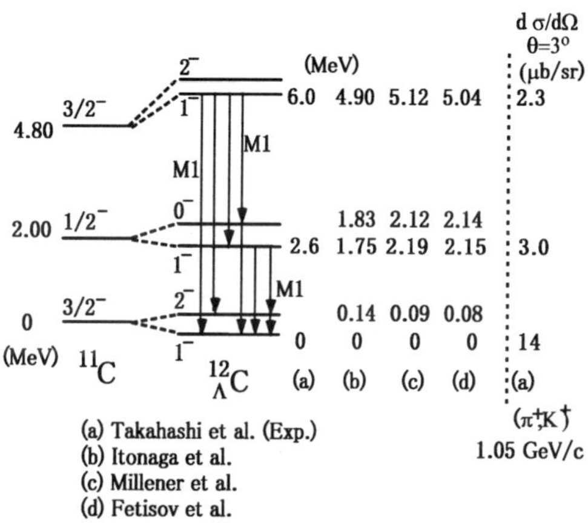

Figure 3: Expected low-lying levels and γ-transitions of ${}^{12}_{\Lambda}$C. (a) shows measured level energies and cross sections,[3] while (b), (c), and (d) are theoretically estimated level energies by Itonaga *et al.*,[4] by Millener *et al.*,[5] and by Fetisov *et al.*,[6] respectively.

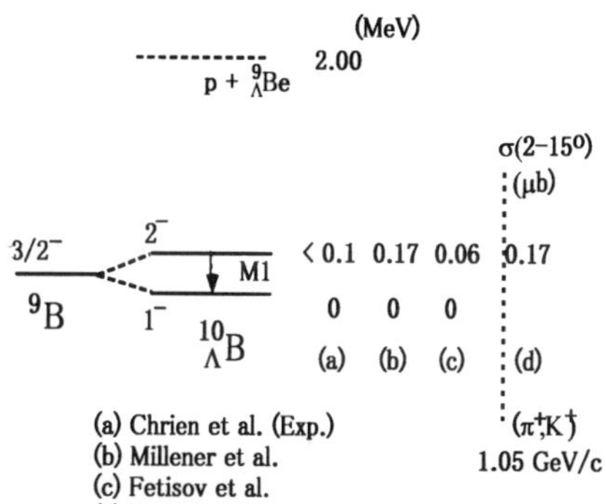

Figure 4: Expected low-lying levels and γ-transitions of ${}^{10}_{\Lambda}$B. (a) shows experimental limit of the level energy by Chrien *et al.*,[7] while (b) and (c) show theoretically estimated level energies by Millener *et al.*[5] and by Fetisov *et al.*,[6] respectively. (d) shows measured cross section by Hasegawa *et al.*.[8]

Figures 3 and 4 show expected low-lying levels of ${}^{12}_{\Lambda}$C and ${}^{10}_{\Lambda}$B, in which the ground-state doublet spacings are also determined by both of the spin-spin and LS interactions. Therefore, their measurements will be a good check of the strengths of

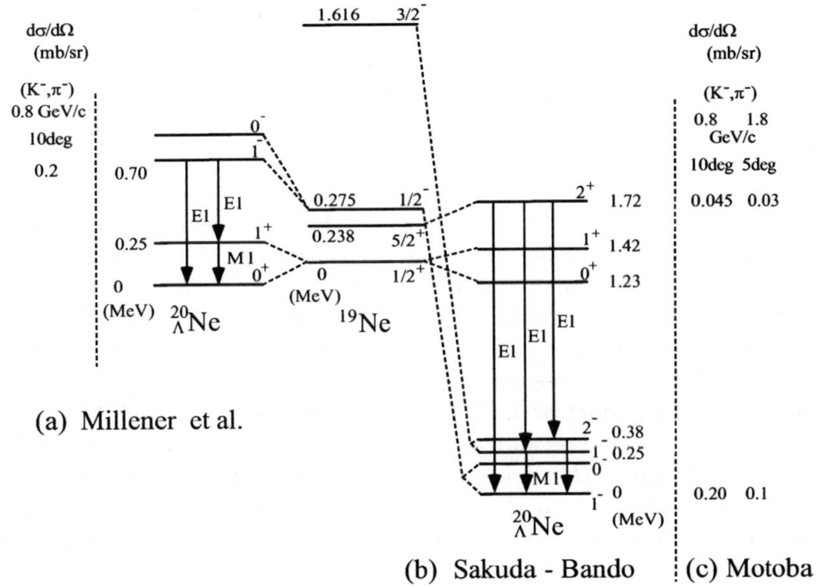

Figure 5: (a) and (b) show expected level scheme of $^{20}_{\Lambda}$Ne predicted by a shell model [5] and by a cluster model. [10] (c) shows cross sections estimated by Motoba. [11]

these interactions derived from $^{7}_{\Lambda}$Li and $^{9}_{\Lambda}$Be. Because of ambiguities in theoretical treatments of nuclear structure, measurements of various hypernuclear levels are necessary to unambiguously determine each term of spin-dependent interactions. Moreover, various hypernuclear data allow us to separately determine the symmetric and anti-symmetric LS forces.

The ground state double spacing of $p_{1/2}$ shell hypernuclei such as $^{16}_{\Lambda}$O and $^{15}_{\Lambda}$N is sensitive to the ΛN tensor interaction.[5] The expected level scheme and experimental details are described in ref. [1] One pion exchange, which is the main source of the NN tensor force, is forbidden between Λ and N. However, the three-body force via intermediate Σ may give a sizable tensor force between Λ and N.

In the case of $^{20}_{\Lambda}$Ne, a negative-parity ground state is predicted by a cluster model calculation due to a drastic change of the cluster structure by a Λ.[10] Figure 5 shows the expected level scheme of $^{20}_{\Lambda}$Ne by a shell model [5] and by a cluster model. In the cluster model, the negative parity states of ^{19}Ne are described by ^{15}O + ^{4}He, and the positive parity states of ^{19}Ne are described by ^{16}O + ^{3}He. When a Λ is added to these systems, it is expected that the ^{15}O + ^{4}He system shrinks and increase the binding energy more than the ^{16}O + ^{3}He system, because the former is more clusterized but the latter has a more shell-like structure. It is an interesting example of a drastic change of nuclear structure induced by presence of a Λ, which can be investigated by γ spectroscopy.

These experiments are designed to use the (K^-,π^-) reaction with intense K^- beam available at BNL, although the future schedule of AGS proton run is unclear. These hypernuclei can also be studied by the (π^+,K^+) reaction at KEK. In this case, however, we have two difficulties: severer Doppler shift coming from much larger momentum transfer, and much longer beam time necessary due to small cross sections and limited beam intensity allowed by Ge detectors. The improvements of the detector system for much higher efficiency and a small segmentation are essential in this case.

3.2 B(M1) measurement

The magnetic moment of Λ in nucleus is an interesting subject related to possible modification of baryon properties in nuclear matter. Measurement of hypernuclear magnetic moments is, however, extremely difficult. On the other hand, we can derive a g-factor of Λ in a nucleus from a probability of the spin-flip $M1$ transition ($B(M1_{sf})$) between hypernuclear spin-doublet states, because it is naively expressed as [12]

$$B(M1_{sf}) \propto |<\phi_{lo}|\mu|\phi_{up}>|^2 = |<\phi_{lo}|g_N\mathbf{J}_N + g_\Lambda\mathbf{J}_\Lambda|\phi_{up}>|^2 \propto (g_N - g_\Lambda)^2$$

where g_N and g_Λ denote effective g-factors of the core nucleus and the Λ, and $\mathbf{J}_N$ and $\mathbf{J}_\Lambda$ denote their spins, respectively. Here the space components of the wavefunctions of the lower and upper states of the doublet (ϕ_{lo}, ϕ_{up}) are assumed to be identical.

Taking advantage of a high resolution of Ge detectors, $B(M1)$ is derived from a lifetime of the excited state which can be measured from partly Doppler-broadened peak shape (Doppler attenuation method), if the lifetime and the stopping time of the recoil excited hypernucleus are close to each other within a factor of 2 or 3. In order to satisfy this condition, the recoil momentum of the produced hypernucleus and the density of the target material have to be carefully selected. Table 1 summarizes the conditions for $B(M1_{sf})$ measurements for ${}^{10}_{\Lambda}$B and ${}^{7}_{\Lambda}$Li. The table shows expected lifetimes of the excited states calculated for some cases of the transition energy using predicted transition rates [12], and the stopping time of the recoil hypernucleus calculated with the SRIM code for the most appropriate target material.

By using the (π^+,K^+) reaction, $B(M1_{sf})$ of the ground state doublets in ${}^{10}_{\Lambda}$B (see Fig.4) can be measured if the level spacing is between 0.15 MeV and 0.3 MeV. The spin-flip $M1$ transition of the ground state doublet of ${}^{10}_{\Lambda}$B was previously searched for at BNL but not observed, which was attributed to a small ($<$ 0.1 MeV) doublet spacing. [7] On the other hand, the ΛN spin-spin force determined from our ${}^{7}_{\Lambda}$Li data as well as the $A = 4$ data (the radial integral of the spin-spin term in the ΛN effective interaction, Δ =0.5 MeV) predicts a spacing of ${}^{10}_{\Lambda}$B about 0.2 MeV. The transition energy has to be measured at first using Hyperball, by which the transition down to 60 keV is detectable. We expect an accuracy of the $B(M1_{sf})$ measurement to be about 10% within 800 hours beam time at KEK, if the level spacing matches the target density.

By using the (K^-,π^-) reaction, $B(M1_{sf})$ of the ${}^{7}_{\Lambda}$Li($\frac{3}{2}^+\rightarrow\frac{1}{2}^+$) which we observed at 0.69 MeV can be measured with a Li_2O target, although a small recoil momentum makes the accuracy of the $B(M1_{sf})$ measurement worse than the case of ${}^{10}_{\Lambda}$B with the (π^+,K^+) reaction. $B(M1_{sf})$ of the ${}^{7}_{\Lambda}$Li($\frac{7}{2}^+\rightarrow\frac{5}{2}^+$) can also be measured with the (K^-,π^-) or (π^+,K^+) reactions if the spacing is between 0.2 and 0.8

Table 1: Conditions for Doppler shift attenuation method in $B(M1_{sf})$ measurement. Expected lifetimes of the upper state (τ) for typical γ-ray energies estimated from transition rates calculated by Dalitz *et al.* or Hiyama *et al.* are shown together with stopping time (t_{stop}) calculated for the (π^+,K^+) or (K^-,π^-) reaction in various target materials. Doppler shift attenuation method works well when τ and t_{stop} are close to each other.

<table>
<tr><td rowspan="2">$^{10}_{\Lambda}$B($2^-\rightarrow1^-$)</td><td>E_γ</td><td>(0.3 MeV)</td><td>(0.2 MeV)</td><td>(0.15 MeV)</td></tr>
<tr><td>τ</td><td>2.0 ps</td><td>6.9 ps</td><td>16 ps</td></tr>
<tr><td rowspan="2">(π^+,K^+)</td><td>mater.(ρ)</td><td>^{10}B metal(2.2)</td><td>B_2H_6(0.45)</td><td></td></tr>
<tr><td>t_{stop}</td><td>1.3 ps</td><td>4.1 ps</td><td></td></tr>
<tr><td rowspan="2">$^{7}_{\Lambda}$Li($\frac{3}{2}^+\rightarrow\frac{1}{2}^+$)</td><td>$E_\gamma$</td><td>0.69 MeV</td><td></td><td></td></tr>
<tr><td>τ_{exp}</td><td>0.54 ps</td><td></td><td></td></tr>
<tr><td rowspan="2">$^{7}_{\Lambda}$Li($\frac{7}{2}^+\rightarrow\frac{5}{2}^+$)</td><td>$E_\gamma$</td><td>(0.80 MeV)</td><td>(0.50 MeV)</td><td>(0.30 MeV)</td></tr>
<tr><td>τ</td><td>0.37 ps</td><td>1.5 ps</td><td>7.0 ps</td></tr>
<tr><td>(π^+,K^+) or</td><td>mater.(ρ)</td><td colspan="2">Li_2O(2.0)</td><td>^{7}Li metal(0.54)</td></tr>
<tr><td>(K^-,π^-)</td><td>t_{stop}</td><td>2.2 ps *</td><td>4.2 ps</td><td>14 ps</td></tr>
</table>

*: (K^-,π^-) reaction at 0.9 GeV/c. Others: (π^+,K^+) reaction at 1.05 GeV/c.

MeV. Because of a small production cross section of the $\frac{7}{2}^+$ state in the (π^+,K^+) reaction, we had better use the (K^-,π^-) reaction at 1.1 GeV/c having a large cross section for such spin-flip states in spite of its smaller momentum transfer.

3.3 Neutron-rich hypernuclei

Neutron-rich and proton-rich hypernuclei are extremely exotic; they may have a Λ-halo as well as n-halos or p-halos, and their nuclear size and structure may be drastically changed by the presence of Λ. Using γ-spectroscopy together with the (K^-,π^0) or (π^-, K^0) reaction, we can study n-rich hypernuclei such as $^{7}_{\Lambda}$He and $^{11}_{\Lambda}$Be. For example, measurements of the energy levels and $B(E2)$ of $^{7}_{\Lambda}$He will provide information on the n-halo structure in ^{6}He. Here, a Λ makes the nucleus more bound than ^{6}He and allows γ-ray emissions.[13]

The (K^-,π^0) or (π^-, K^0) reaction is also employed to study charge symmetry breaking (CSB) in hypernuclei. CSB is clearly seen only in the mass difference between $^{4}_{\Lambda}$H and $^{4}_{\Lambda}$He, but is not very evident for other hypernuclei. Precise measurements of the level energies of several pairs of mirror hypernuclei such as $^{4}_{\Lambda}$H/$^{4}_{\Lambda}$He, $^{10}_{\Lambda}$Be/$^{10}_{\Lambda}$B, and $^{12}_{\Lambda}$B/$^{12}_{\Lambda}$C will give a clue to understand the origin of this interesting phenomenon.

3.4 Other possibilities

Taking advantage of the intense K^- beam at 1.8 GeV/c (2×10^6/pulse) available at the BNL AGS D-line, we can study various hypernuclear levels with rather small cross sections using the (K^-,π^-) reaction. One interesting experiment is being proposed to measure the 4 MeV $E1(p_\Lambda \rightarrow s_\Lambda)$ transition in $^{208}_{\Lambda}$Pb.[14] At the D-line, we can also use (K^-, K^+) reaction to produce Ξ^- and study Ξ-atoms. Shifts and widths of atomic X-rays of $\Xi^- + {}^{208}$Pb, for example, can be measured with Hyperball, and information on the ΞN interactions will be derived.[14]

On the other hand, KEK does not have K^- beam lines with a high intensity and a good resolution necessary for the hypernuclear production by the (K^-,π^-) reaction at present. There is a possibility of using stopped K^- absorption, which has a large yield of producing various hyperfragments, if appropriate tagging methods to select hyperfragment weak decay events are developed with a scintillating fiber target.[15]

4 Experiments at JHF

4.1 *Experimental possibilities in JHF*

The strangeness nuclear physics is one of the major motivations of the JHF 50 GeV proton synchrotron. At JHF, about 10–100 times more intense beam is available than at BNL. In order to avoid too high counting rate, Hyperball has to be upgraded so that Ge detectors have much finer segmentations and keep a much larger distance from the target. Developments of new readout methods are also necessary.

Here, γ-γ coincidence will be a powerful tool to reconstruct a level scheme. We can also employ other traditional techniques in γ-spectroscopy, such as measurements of angular correlations, polarization, and transition probabilities, as well as coincidence with weak-decay products. The angular correlation between γ and π^- is more sensitive to the spins of the hypernuclear states than the correlation between two γ's emitted in a cascade decay.[12]

We will be able to investigate various hypernuclei up to $^{208}_{\Lambda}$Pb and various interesting phenomena will be revealed. As we have observed the shrinking effect of $^{7}_{\Lambda}$Li, impurity of Λ in nuclei might affect nuclear structure, such as shell and cluster structures, collective motions and deformations. The $B(M1_{sf})$ of heavy hypernuclei will be also feasible. If the property of Λ is changed in nuclear matter, the g-factor of Λ may be significantly changed particularly in a heavy nucleus.

4.2 *Study of S=-2 systems*

The intense kaon beam at JHF enables us to study S=-2 systems by γ-spectroscopy. Structures of double Λ hypernuclei will clarify the Λ-Λ interactions in detail including the $\Lambda\Lambda$-ΞN coupling and a possible H dibaryon-like correlation.

The $E1(p_\Lambda s_\Lambda \rightarrow s_\Lambda s_\Lambda)$ transition in $^{13}_{\Lambda\Lambda}$B, for example, is essential to obtain information on the Λ-Λ interaction. Particularly, measurements of $p_{3/2\Lambda} s_\Lambda \rightarrow s_\Lambda s_\Lambda$ and $p_{1/2\Lambda} s_\Lambda \rightarrow s_\Lambda s_\Lambda$ will provide information on the LS force between Λ and Λ.

Another interest is a possible large shrinking effect of nucleus induced by two Λ's. Loosely-bound clusterized nuclei such as ^{6}Li and ^{8}Be are contracted by adding one Λ particle, but shell-like nuclei such as ^{16}O are not expected to be contracted. Whether or not the shell-like nuclei can be compressed by two Λ's is an interesting problem which can be answered at JHF.

Even if a nucleus cannot be compressed by two Λ's, a double hypernucleus should have a slightly higher baryon density than a normal nucleus. Since it is a "stable" higher-density matter unlike the case of relativistic heavy ion collisions, we can investigate its properties in detail by γ-spectroscopy. The investigation of such nuclear matter together with hyperon-nucleon and hyperon-hyperon interactions will provide us with essential information on possible existence of baryonic strange matter as well as precursors of chiral symmetry restoration.

5 SUMMARY

We have started experiments of high-resolution hypernuclear γ-spectroscopy using Hyperball, a large-acceptance Ge detector system for hypernuclear studies. Presently we are investigating various p-shell hypernuclei at KEK-PS and BNL-AGS in order to determine ΛN spin-dependent interactions. Then further studies, such as $B(M1)$ measurements, spectroscopy of n-rich hypernuclei and heavier hypernuclei, etc. will be carried out. Much more intense beam at JHF 50 GeV PS enables us to extend our research into a wide region of hypernuclear systems including double Λ hypernuclei. An upgrade of Hyperball is necessary to improve the efficiency and the segmentation.

ACKNOWLEDGEMENT

The author thanks to all the members of KEK-E419 and BNL-E930 experiments, particularly to K. Imai, M. May and K. Tanida. He is also grateful to T. Motoba, E. Hiyama and D.J. Millener for valuable discussions. This work is supported by Grant-In-Aid for Scientific Research on Priority Areas (Strangeness Nuclear Physics) from the Ministry of Education, Science, Sports and Culture.

References

1. H. Tamura, Nucl. Phys. **A639** (1998) 83c.
2. K. Tanida *et al.*, in this proceedings.
3. T. Takahashi *et al.*, Proc. KEK-Tanashi Int. Symp. "Physics of Hadrons and Nuclei", December 1998, Tokyo, (Nucl. Phys. A, in press).
4. K. Itonaga *et al.*, Prog. Theor. Phys. Suppl. **117** (1994) 17.
5. D.J. Millener, A. Gal, C.B. Dover, R.H. Dalitz, Phys. Rev. **C31** (1985) 499.
6. V.N. Fetisov, L. Majling, J. Žofka and R.A. Eramzhyan, Z. Phys. **A339** (1991) 399.
7. R.E. Chrien *et al.*, Phys. Rev. **C41** (1990) 1062.
8. T. Hasegawa *et al.*, Phys. Rev. Lett. **74** (1995) 224.
9. E. Hiyama *et al.*, Nucl. Phys. **A639** 173c.
10. T. Sakuda and H. Bandō, Prog. Theor. Phys. **78** (1987) 1317.
11. T. Motoba, private communication (1996).
12. R.H. Dalitz and A. Gal, Ann. Phys. **116** (1978) 167.
13. E. Hiyama, M. Kamimura, T. Motoba, T. Yamada and Y. Yamamoto, Phys. Rev. **C53** (1996) 2075.
14. M. May, private communication (1998).
15. K. Imai, private communication (1998).

PROSPECTS OF JAPAN HADRON FACILITY

Ken'ichi Imai
Department of Physics, Kyoto University, Kyoto 606, Japan

Japan Hadron Facility (JHF) is the accelerator complex which provides high intensity secondary beams such as pion, kaon, anti-proton, neutrino, muon, neutron and unstable nuclei. The JHF now becomes a joint project of KEK and JAERI (Japan Atomic Energy Research Institute), and the facility is planned to be built at Tokai Laboratory of JAERI. In this talk, JHF and the current status of the project is briefly described. We will mainly discuss the physics prospects with the 50 GeV synchrotron on the emphasis on the strangeness nuclear physics.

1 Overview of the JHF

Japan Hadron Facility (JHF) is an accelerator complex which consists of a 200 MeV LINAC, 3 GeV synchrotron and 50 GeV synchrotron. The 3 GeV synchrotron provides an average beam current of 200μA with 25Hz high repetition rate. The total beam power is 0.6 MW. The 50GeV PS provides an average beam current of 10μA at the repetition rate of 0.3Hz with beam spill length of 0.7sec. The total beam power is 0.5 MW. The circumference of the ring is about 1.4 km. The beam powers of both synchrotron are largest among the world in GeV energy region.

High intensity pulsed neutrons and muons are abundantly produced with the 3 GeV synchrotron. The neutrons are used for wide range of the material science such as magnetic materials, high-Tc superconductors, fractals, liquid, polymers etc. Biological science is expected to be another big user of neutrons. High intensuty muons are used for μSR studies on material science, muonium physics and muon catalyzed fusion. Unstable nuclei are also produced in the spallation reactions with high intensity protons. Physics with ISOL-type unstable nuclear beam, such as search for superheavy nuclei and study of nucleosynthesis, is anticipated.

At 50 GeV PS, various high intensity secondary beams, such as K, π, p-bar, neutrino etc., will be used for nuclear and particle physics. Major topics to be studied in nuclear and hadronic physics are: 1) Spectroscopy of hypernuclei (both single and double hypernuclei), 2) Hyperon-nucleon scattering, 3) Hadron spectroscopy, 4) Hadrons in nuclear matter.

The strangeness nuclear physics is discussed in the next section. The gluonic degree of freedom should play important roles in the hadron states. The exotic hadrons such as glueballs (0^{++} and 2^{++}) and hybrids ($s\bar{s}g, c\bar{c}g$ etc.) can be studied as well as charmonium and charmed baryons. For these studies, a separated kaon beam up to 10 GeV/c and a pion beam up to 25 GeV is anticipated, where anti-protons are also available. In future, we plan to build an anti-proton storage ring for the precision spectroscopy of charmonium and charmed hadrons, as well as various physics with cooled anti-proton beams. The anti-proton acumulator ring which had been used at CERN has already been sent to KEK.

Major topics in particle physics are: 1) Kaon rare decays to study CP violation, 2) Flavor mixing and topics beyond the standard model, 3) Neutrino oscillation experiment using Super-Kamiokande. The evidence of the oscillation of the atomospheric neutrino observed at Super-Kamiokande pushed the long baseline neutrino oscillation experiments using the accelerators. The K2K experiment which will detect the oscillation between KEK and Super-Kamiokande is now ready to take data. It will detect ν_μ disappearance and ν_e appearance around the region of $\Delta m^2 = 10^{-2} eV^2$. The JHF provides high intensity neutrino beam of which peak energy is about 6 GeV, whereas it is 1.5 GeV at the present KEK-PS. Since the threshold of the τ appearance is 3.5 GeV, at the JHF it is possible to measure the $\nu_\mu \to \nu_\tau$ oscillation by the appearance method at the sensitivity to $\Delta m^2 = 10^{-2} eV^2$.

Primary beams such as polarized proton beam and heavy ion beam are also planned in future for nuclear physics. The energy is optimum to produce high density (baryonic) matter.

One of the difficulties of the high intensity synchrotron is that one needs high field gradient RF cavity for the quick accerelation. Recently, Y.Mori et al., succeeded to test the high field gradient of 50kV/m with use of new magnetic alloy called FINEMET [1]. It provides five times higher gradient than currently used ferrite cavities. It was quite important progress of accelerator technology to achieve rapid cycling synchrotron and high intensity synchrotron, namely JHF.

2 Strangeness Nuclear Physics at JHF

At JHF the intensity of kaon beams become 10 times than that of AGS and 1000 times than KEK-PS. It greatly enhances the capability of the strangeness nuclear physics by order of magnitudes. Quantitative improvement by the order of magnitudes can give a breakthrough in physics. The expected intensity of 0.8 GeV K^- beam can be a few times 10^6 and that of 1.8 GeV K^- beam is a few times 10^7.

2.1 Spectroscopy of Λhypernuclei

High resolution spectrometer including the beam line is designed for the spectroscopy of Λ hypernuclei by Noumi et al.,[2]. The energy resolution of 300 keV is expected. It is almost one order better than the current value of the SKS (2 MeV). The high precision γ spectroscopy is anticipated at the JHF. We have just succeeded the measurement of the γ ray from ${}^{7}_{\Lambda}Li$ hypernuclei with the Ge-ball detector for the first time [3]. With the improved intensity and larger scale Ge-ball detector with more than 10% peak efficiency such as Gammasphere or Euroball, one can measure the $\gamma - \gamma$ correlation of the hypernuclei. It enables not only the energy measurement but the spin-parity assignment of the excitated levels of hypernuclei. With the Ge-ball detector, the transition matrix elements, B(E1,E2) and B(M1), can be obtained by the lifetime measured using the doppler shifts of the γ rays. They provide the information on the electro-magnetic property and structure of hypernuclei. Thanks to the high intensity, the γ spectroscopy of heavier hypernuclei (than p-shell nuclei) becomes possible at JHF.

2.2 Spectroscopy of double strangeness nuclei

Since the production of double strangeness nuclei is very limited by the kaon intensity available today. Only a few double Λ hypernuclei was identified only in the emulsion. The counter experiments are being performed to search for the double Λ hypernuclei by (K^-, K^+) reactions from Be and C targets at BNL [4]. The statistics is the key ponit in these experiments. As for the Ξ hypernuclei, (K^-, K^+) spectroscopy has been tried to obtain the information on Ξ-nuclear poteintial on $^{1}2_{\Xi}Be$ [5]. Although the depth of the potential was indicated by their data, both the energy resolution and statistics are not sufficient to identify the expected narrow states of Ξ hypernuclear state. The H-search has been extensicely made also by using (K^-, K^+) reactions. Only possible evidence observed is a resonance state near the $\Lambda\Lambda$ threshold [6]. Again one needs more statistics to establish it.

The study of double starngeness nuclei, thus, desparately needs higher intensity K^- beams. At JHF we can obtain 2.0 GeV K^- beam of 10 times higher intensity which will enable the experiments above with much more statistics and better energy resolution. We plan to build a 2.0 GeV K beam line specifically for these physics. The SKS will be used for the spectroscopy of double strangeness nuclei with the energy resolution of about 2 MeV which is comparable to the expected width of Ξ hypernuclei.

The γ spectroscopy of double Λ hypernuclei with the Ge-ball is also discussed as a possible experimental program at JHF.

2.3 Hyperon-nucleon scattering

One of the most important goal of the strangeness nuclear physics is to understand the baryon-baryon interaction in the $SU_f(3)$ scheme. The hyperon-nucleon scattering is most direct way to study the hyperon-nucleon interaction. There are many questions on the hyperon-nucleon interaction such as spin-dependence, existence of hard-core or not, existence of resonance state etc. So far only a very limited experiments has been made. Difficulty of the experiment is mostly due to short flight lengths (a few cm) of hyperons. M.Ieiri et al., recently did the Σ^+p and Σ^-p scattering experiments with use of the scintillating fiber (SCIFI) active target [7]. The data taking rate for hyperon production events becomes much higher than those of old bubble chamber experiments. However, the rate is limited by both analysis and data taking of image data which has a huge data size.

There are several studies for the experiments at JHF. One is to develop the new technology of scintillating fiber detector. To increase the rate capability of the detector itself, it is important to develop the image delay tube which store the SCIFI image (photo-electron) in the vacuum tube for the trigger. This improvement will make the SCIFI rate capability from present 100kHz to 10 MHz. Another important item for SCIFI is to develop the faster readout system of the CCD (readout system of the image data). This development is going on because the CCD is now widely used in the various fields.

If we use a liquid hydrogen as a scatterer which is much cleaner than SCIFI (CH_n), we have to detect both an incoming or scattered hyperon and a recoil proton. In this case especially, the size of the pion or kaon beam should be as small as possible, so that the tracking detectors for short-flight-length hyperons can be placed out of the high intensity beam. The pencil or thin beam is desirable for the hyperon-nucleon scattering experiments. These high quality beams are possible

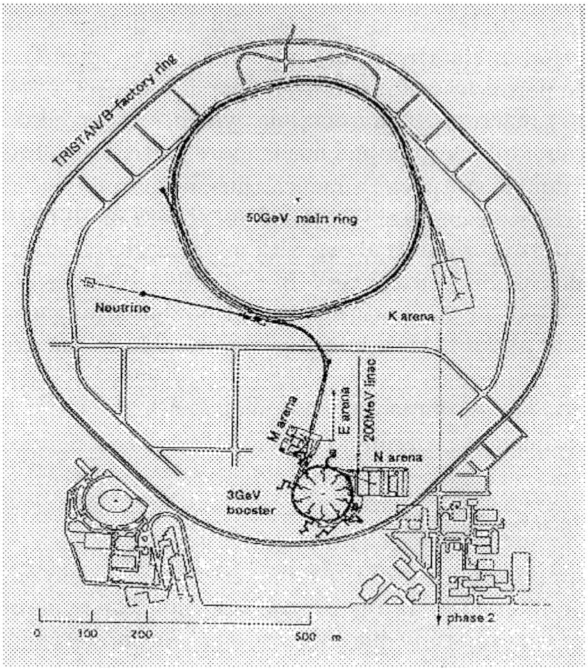

Figure 1: JHF Hadron Accelerator complex in the KEK–Tsukuba site.

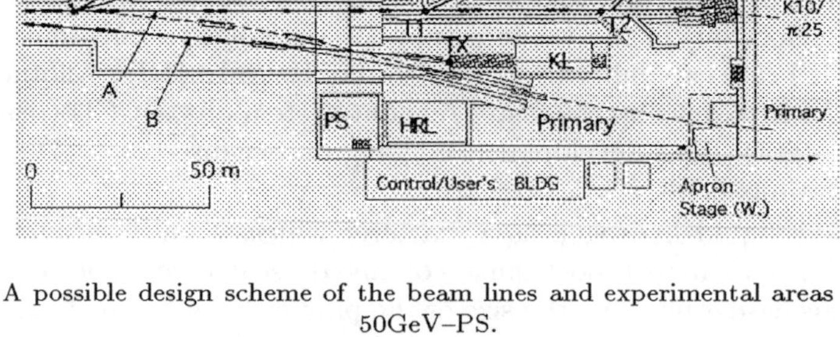

Figure 2: A possible design scheme of the beam lines and experimental areas of the JHF 50GeV–PS.

from the high intensity primary beam provided at JHF. The Ξ-nucleon scattering experiment is proposed as a typical experiment at JHF which needs high intensity K^- beams [8].

2.4 New physics

The high intensity K^- beam may open new physics. One of the possiblity is the production of nuclei with S=-3. The (K^-, K^+K^0) reaction can produce S=-3 system as like the (K^-, K^+) reaction for the study of double strangeness systems. The Ω^- can be produced abundantly. However, the momentum of Ω^- is quite large in the quasi-free reaction to be slowed down and stopped in material. Another possibility is the production of charmed nuclei. There are discussed a few ideas how to produce the charmed nuclei [9]. The p-A reaction and ν-A reaction have been discussed as candidate methods to produce the charmed nuclei. The imporatnt point is that the (D, π) reaction has a magic momentum of 0.6 GeV/c just like the (K^-, π^-) reaction for the Λ hypernuclei.

3 Present status of the project

The JHF including 50 GeV PS was proposed in 1995. In order to realize the plan, in 1997, the former INS and KEK were merged into the new organization called again KEK (the same name in short but different in full name). Under the new KEK, a team was formed to design the facility and do some R&D works. The proposal to build the JHF inside the TRITAN ring at KEK was submitted in 1998. Last year, it was suggested to combine the JHF and the high power pulsed neutron source proposed by JAERI (Japanese Atomic Energy Research Institute) into one big accelerator facility. After an extensive discussion between the two laboratories and among all scientists involved in both projects, the agreement was recently made to make a joint proposal of the combined project. The new proposal is still being worked out between the two laboratories. The main points of the new proposal are: 1) The accelerator is constructed on the site of JAERI's Tokai laboratory which is located about 60km north-east of KEK. 2) The design concept of the accelerator is unchanged. Especially the parameters of 50 GeV PS and its experimental facility are the same as before. 3) The linac and the 3GeV booster synchrotron will be upgraded to provide 1MW proton beam to the pulsed neutron source. 4) The facility to study the transmutation of the nuclear waste is newly added to the original JHF.

The new proposal will be reviewed by the international review committee and submitted to the government for FY2000 budget. If approved, we plan to start the construction of the facility in 2000 and get the first beam for physics in 2005. The JHF accelerator complex (previous design) is shown in Fig.1 The present plan of experimental area at 50 GeV PS is also shown in Fig.2. The beam lines are not fixed yet. We like to encourage everyone to come up with new ideas and physics. The different beam lines from those in Fig.2 will be made if there is good physics.

References

[1] Y.Mori et al., Proc. of Euro. Part. Accel. conf., Stockholm, 1998.

[2] H.Noumi, Proc.of Sendai International Workshop on the Spectroscopy of Hypernuclei, (1998) p.65.
[3] H.Tamura, in this Proceedings.
[4] M.May and G.Franklin, BNL-E885 and T.Fukuda, BNL-E906, T.Fukuda, in this Proceedings and K.Yamamoto, in this Proceedings.
[5] T.Fukuda et al., Phys. Rev. C58 (1998) 1306 and P. Khaustov et al., in preparation.
[6] J.K.Ahn et al., Phys. Lett. B444 (1998) 267.
[7] M.Ieiri, KEK-PS E251 and E289.
[8] M.Ieiri, private communication.
[9] J.C.Peng, Proceedings of the International Workshop on Physics with 50 GeV PS (1995) p93 and K.Imai, to be published.

CONCLUDING REMARKS

K. YAZAKI

Department of Physics, University of Tokyo, Bunkyo-ku, Tokyo 113-0033, Japan
E-mail: yazaki@phys.s.u-tokyo.ac.jp
present address: Physics Lab., College of Arts and Sciences, Tokyo Woman's Christian University, Suginami-ku, Tokyo 167-8585, Japan

A personal impression on this workshop is given. All the talks are first classified into several subjects to have an overview of the workshop. Topics from each subject are then briefly discussed.

1 An Overview

The present status of the strangeness nuclear physics reminds me of the ordinary nuclear physics about three decades ago. The nucleon-nucleon interaction had been studied experimentally and the so-called realistic nuclear forces were getting available at that time. The in-beam gamma ray spectroscopy was revealing various features of nuclear excitations and theoretical models for them became elaborate and quantitative. The theory of effective interaction had been "established" and the microscopic calculations based on the realistic nuclear forces were considered to be the ultimate goal of the nuclear structure theory.

The strangeness nuclear physics at present seems to be in a sense approaching a similar stage, though it has certainly many distinct features. "Realistic" hyperon-nucleon and hyperon-hyperon interactions are getting available though much more experimental efforts are still required to establish them. For the moment, the microscopic calculations based on the realistic interactions seem to be an important source of information on the interactions themselves. The high-resolution gamma ray spectroscopy is already producing crucial data on the excited states of hypernuclei and will become available for the study of higher excitations and heavier elements. Theoretical tools for hypernuclear spectroscopy are much more advanced than they were for nuclear spectroscopy three decades ago.

Let me now try to remind you of what have been discussed in this workshop.

From the most fundamemtal side, the properties of a single baryon were theoretically discussed by Kim, Jeong and Cheon.

The experimental status of the baryon-baryon interactions was given by Kondo and Ahn, while the theoretical discussions were given by Rijken, Suzuki and Kohno. Based on these "bare" baryon-baryon interactions, Hiyama performed exact calculations for few-body systems.

The baryon-baryon effective interactions to be used for microscopic calculations were discussed by Kuo, Yamamoto, Fujii and Akaishi. In connection with the effective interaction, Kamimura reported cluster model calculations while Fetisov and Millener did shell model analyses.

The hypernuclear spectroscopy was experimantally the main subject of this workshop. The experimental studies of Λ hypernuclei were discussed by Youn,

Tanida, Nagae, Hashimoto, Sakaguchi and Tamura, those of Σ hypernuclei by Chrien and those of strangeness -2 hypernuclei by Fukuda, Yamamoto, Ichikawa, Nakazawa and Ra. The theoretical analyses were reported by Hong (Λ), Dabrowski and Harada (Σ) and Afnan and Yamada ($S = -2$).

A recent experiment on the weak decay of hypernuclei was reported by Bhang and theoretical studies were discussed by Inoue, Garbarino, Itonaga, Gibson. Oka and Jun.

Hypernucleus production processes were theoretically studied by Ohnishi, Nakajima, Kim, Cheoun, Han, Koike and Paryev.

The strange matter and the related problems were discussed by Tatsumi, Takatsuka and Lanskoy.

The on-going and future experimental projects were explained by Fujiwara, Hungerford, Bertini, Bressani, Tang, Grzonka/Kilian and Imai.

After this very brief overview of the workshop in which I have hopefully named all the speakers, I will now talk on each subject with my personal prejudice.

2 Single Baryon and Baryon-Baryon Interactions

The strangeness contents of the nucleon was studied by H.-C.Kim by the chiral quark soliton model producing reasonable results. The chiral bag model was used by M.T.Jeong to calculate KNY coupling constants and the cloudy bag model was employed by I.-T.Cheon to study the Λ decay. Many models have been proposed for describing the internal structure of a baryon and we are now in the stage of selecting those which give natural explanations of all the observed single-baryon properties.

The experiment on $\Sigma^- p$ elastic scattering was reported by Y. Kondo, who showed preliminary data on the cross sections.

Experimental studies on two-baryon systems in $S = -2$ sector were discussed by J.K.Ahn, T.Fukuda, K.Yamamoto, K.Nakazawa and A.Ichikawa/ Y.S.Ra. The obtained $\Xi^- p$ elastic scattering data seem to exclude some of baryon-baryon interaction models. The observed enhancement of $\Lambda\Lambda$ invariant mass spectrum near the threshold was an issue of great excitement. A.Ohnishi tried to extract $a_{\Lambda\Lambda}$ from the spectrum.

On the theoretical side of baryon-baryon interactions, the meson exchange models of the Nijmegen group are probably the most "realistic" and T.Rijken talked on the most recent soft-core versions, NSC97a-f. They are still being improved by comparing "microscopic" calculations with data on hypernuclei but the two-baryon data are crucial in discriminating the many versions. The quark models discussed by Y.Suzuki and M.Kohno are more "microscopic" in the sub-nucleon level but suffer from lower quality of fit to the existing data. A new feature in this direction was an attempt to construct the G-matrix from the intrinsically non-local interactions due to the quark exchange.

In constructing theoretical models of baryon-baryon interactions, I would like to draw your attention to a work by Kaplan and Manohar[1], in which they gave a $1/N_c$ classification of the various components of nuclear force. If the nucleon-nucleon interaction is expressed in the form

$$V_{NN} = V^{(0)} + (\tau_1\tau_2)V^{(1)}, \tag{1}$$

where

$$V^{(i)} = V_0^{(i)} + V_\sigma^{(i)}\sigma_1\sigma_2 + V_{LS}^{(i)}LS + V_T^{(i)}S_{12} + V_Q^{(i)}Q_{12}, \tag{2}$$

with

$$S_{12} = 3\sigma_1\hat{r}\sigma_2\hat{r} - \sigma_1\sigma_2, \tag{3}$$

and

$$Q_{12} = \frac{1}{2}((\sigma_1 L),(\sigma_2 L)), \tag{4}$$

they argue that $V_0^{(0)}$, $V_\sigma^{(1)}$ and $V_T^{(1)}$ are $O(N_c)$, $V_0^{(1)}$, $V_\sigma^{(0)}$, $V_{LS}^{(1,2)}$, $V_T^{(0)}$ and $V_Q^{(1)}$ are $O(1/N_c)$ and $V_Q^{(0)}$ is $O((1/N_c)^3)$. These features are actually realized in most of the realistic nuclear forces. Unfortunately, it seems non-trivial to extend their discussion from $SU(2)_f$ to $SU(3)_f$ but the extended version would be extremely useful as a guide for constructing models.

3 Hypernuclear Spectroscopy and Effective Interactions

Y.Akaishi and E.Hiyama studied the ΛN spin-spin interaction in connection with the problem of $0^+ - 1^+$ splitting in ${}^4_\Lambda H$ and ${}^4_\Lambda He$. The $\Lambda - \Sigma$ conversion is known to play an important role but the details seem to depend sensitively on the underlying baryon-baryon interactions.

K.Tanida, H.Tamura and O.Hashimoto reported gamma-ray spectroscopy experiments with Ge detectors on p-shell hypernuclei. The E2 and M1 transitions in ${}^7_\Lambda Li$ were observed for the first time and the suppression of the E2 transition was established, which was consistent with the cluster model calculation by M.Kamimura indicating the glue-like role of Λ in this hypernucleus.

A.Sakaguchi reported a ${}^{13}C(K^-,\pi^-\gamma){}^{13}_\Lambda C$ experiment suggesting a possible spin-orbit splitting.

T.Nagae and O Hashimoto stressed the importance of high-resolution experiments based on their study of ${}^{12}_\Lambda C$ by (π^+,K^+) reaction with the SKS spectrometer. They showed that the improvement of the resolution by a factor of 1.5 gave a qualitative difference in data. The data were analyzed by S.W.Hong with the continuum TDA(Tamm-Dancoff approximation).

M.Youn reported the recent experimental study of the charge exchange reaction ${}^{12}C(K^-_{stopped},\pi^0){}^{12}_\Lambda B$ with the NMS(neutral meson spectrometer).

In addition to the above mentioned calculation on ${}^7_\Lambda Li$, M.Kamimura performed elaborate cluster model calculations on ${}^9_\Lambda Be$ and ${}^{13}_\Lambda C$.

V.Fetisov and J.Millener made shell-model analyses of p-shell hypernuclei with spin-dependent effective interactions suggesting a strong anti-symmetric spin-orbit term. A general discussion on the relation between the bare baryon-baryon interaction and the effective interaction was given by Y.Yamamoto.

Λ single particle orbits were experimentally studied through (π^+,K^+) reaction with SKS spectrometer and the results were reported by T.Nagae and O.Hashimoto.

The spectrum observed in $^{89}Y(\pi^+, K^+)^{89}_{\Lambda}Y$ showed some structures in a peak corresponding to the f (l=3) single particle orbit and the possibility of a relatively large spin-orbit splitting was discussed, though there were many other mechanisms of producing such structures.

Calculations of the Λ single particle potential with the effective interactions derived from the meson exchange and the quark model $\Lambda - N$ interactions were reported by S.Fujii and M.Kohno, respectively. The procedures of deriving effective interactions from bare hyperon-nucleon interactions seem to have been fairly well established but the consistency among different methods should still be examined and T.T.S.Kuo pointed out the importance of properly treating the Pauli operator in the standard G-matrix theory.

A historical survey of Σ hypernuclei was given by R.Chrien, who told us the story of eliminating many possible "bound" states. The only state which has survived is $^4_{\Sigma}He$ and was theoretically analyzed by T.Harada, who argued that not only the peak corresponding to the bound state but also the continuum spectrum would carry useful information on Σ-nucleus interactions. The strong iso-spin dependence is one of the explanations for the existence of the bound state and the Lane term in the Σ single particle potential was discussed by J.Dabrowski.

The experimental studies of S = -2 hypernuclei looking for double Λ hypernuclei and cascade (Ξ) hypernuclei were discussed by T.Fukuda, K.Yamamoto, A.Ichikawa, Y.S.Ra and K.Nakazawa. 4 events possibly due to pionic decays of these hypernuclei were reported.

The $\Lambda\Lambda$ - ΞN coupling in $^6_{\Lambda\Lambda}He$ was studied by I.Afnan who concluded that the coupling effect was small. The same coupling was also discussed by Y.Akaishi for $^5_{\Lambda\Lambda}H$, where the large binding energy of ^4He helped to bring down $^5_{\Xi}H$ states enhancing the coupling effects. T.Yamada reported an ambitious calculation of $^{13}_{\Lambda\Lambda}B$ as a system of ^{11}B and 6 quarks examining the possibility of H-nuclear states.

There were several theoretical studies of hyperon production reactions reported in this workshop. Subthreshold $pA \to K^+X$ reactions were discussed by E.Paryev to clarify important mechanisms of hyperon production, while the mechanisms of hyperon electro- and photo productions on a nucleon were discussed by W.Kim, M.K.Cheoun and B.S.Han. T.Koike reported detailed analyses of K^-, $\bar{p}$ and Ξ^- captures on ^4He. The polarization of hyperons produced in high energy proton-nucleus collisions was discused by N.Nakajima.

4 Weak Decay of Hypernuclei

Improved data on the lifetime of hypernuclei with A = 11 to 56 were reported by H.C.Bhang. The data indicate almost A-independent lifetime which is shorter than that of the free Λ by a factor of 1.2.

The non-mesonic decay of hypernuclei was theoreticaly studied by G.Garbarino and K.Itonaga with the meson exchage mechanisim, by T.Inoue and M.Oka including the direct quark mechanisim and by J.H.Jun with a phenomenological model. The results of lifetime calculations were now closer to the observed one but the long- standing puzzle on the ratio, Γ_n/Γ_p, remained. The admixture of $\Delta I = 3/2$ interaction has been known to be crucial and the direct quark mechanism gives a substantial contribution to this interaction but the contribution is not enough to explain the observed large ratio.

The π^+ decay of ${}^4_\Lambda He$ was a new interesting topic and B.Gibson stressed the importance of virtual Σ^+ in the hypernucleus while M.Oka used the soft-pion theorem to relate the relevant amplitude to the $\Delta I = 3/2$ interaction discussed above. Unfortunately, it seems almost impossible to discriminate the two mechanisms experimentally.

It is well known that the weak hamiltonian on the quark level has substantial $\Delta I = 3/2$ components and the mechanism of their suppression has not yet been fully understood. Weak decays of hypernuclei may shed light on this interesting problem of interplay between strong and electro-weak interactions.

5 Strange Matter and Strangeness Astrophysics

Strangeness might be playing important roles in astrophysical phenomena. T.Tatsumi discussed the possibility of kaon condensation in neutron matter which could be crucial in understanding the inner part of a neutron star and its cooling mechanism. The hyperonic nuclear matter is another possibility in the interior of a neutron star and was discussed by T.Takatsuka, who argued that it would be in a superfluidity phase. A similar problem concerning multi-strangeness hypernuclei was also discussed by D.Lanskoy.

6 On-going and Future Projects

The electron facility at CEBAF (TJNAL) is active also in hypernuclear physics and E.Hungerford talked on electroproduction while L.Tang talked on lifetime measurement. The photon beam from SPring 8 will become available in near future and the project was discussed by M.Fujiwara.

A specific experiment on $\vec{p}p \to pK^+\vec{Y}$ at Saturne was explained by R.Bertini and strangeness experiments at COSY were discussed by D.Grzonka.

DAΦNE at Frascati will be the nearest future facility for strangeness nuclear physics and T.Bressani told us about the FINUDA experiment.

The high-resolution gamma-spectroscopy has already been proved to be the most fruitful tool for hypernuclear physics and H.Tamura discussed future plans for it.

Finally, K.Imai talked on the prospects of Japan Hadron Facility (JHF), which would become a part of a joint project of the Japan Atomic Energy Research Institute (JAERI) and the High Energy Accelerator Research Organization (KEK). JHF is now in a critical stage and world-wide supports both scientific and spiritual are indispensable for its approval.

The strangeness nuclear physics is now at the entrance to its most flourishing stage and is attracting many young researchers as we have seen in this workshop. Experiments are essential in this field and I sincerely hope that all the proposed facilities will be realized in near future.

Acknowledgments

I would like to thank the organizers, the chairmen I.T.Cheon and T.Motoba in particular, for giving me the opportunity of attending this exciting workshop. I would also thank the scientific secretaries, S.W.Hong and Y.D.Kim, for their excellent arrangements. Special thanks are due to the host institute, the Asia Pacific Center for Theoretical Physics, and its representative, C.N.Yang and Y.M.Cho , for their hospitality.

Let's meet again at HYP2000.

References

1. D.B.Kaplan and A.V.Manohar, *Phys. Rev.* C **56**, 76 (1997).

CLOSING

H. Bhang
Department of Physics, Seoul National University

Before closing the workshop, I would like to take a few minutes to say several things on behalf of the organizing committee. It has been a real pleasure for me to have the SNP99 Workshop at our University.

First of all, we would like to thank all of you for making this workshop so exciting and warm in spite of the freezing temperature here in Seoul. We are very grateful to all the speakers who gave very high quality presentations, and to all the participants for their helpful comments and questions, especially those hot and spicy ones. We would like to express sincere thanks to those speakers coming from the distant continents: from the United States, Europe, and Australia.

Last year when I met Prof. Imai, he proposed this workshop and it proved to be much better workshop than I imagined at that time. We express sincere thanks to all the colleagues from Japan. Only with their help, we could make this workshop so successful.

We admire the heroic effort of the chairman, Prof. Motoba, and the hard and careful work of Prof. Hong in the course of preparing this workshop. Please let's thank them once again. I also would like to take this opportunity to introduce our students who greatly helped us. They belong to the nuclear physics groups of Prof. Hong and my laboratory.

Finally, we would like to express our deep appreciation for the support from the Asia Pacific Center for Theoretical Physics, Seoul, and for the Grant-in-Aid from the Ministry of Education, Science, Sports and Culture of Japan. Special thanks are also due to the secretary Ms. Ahran Joo for her efficient care of all the aspects of the workshop.

Before closing this enjoyable meeting, I wish you to have nice days for the rest of your stay here in Korea and a pleasant trip back home.

SCIENTIFIC PROGRAM

February 19 – 22, 1999, Seoul

February 19 (Friday) – Morning

Session 19A

Chairman : **B. T. Kim** (Sung Kyun Kwan) *Secretary* : S. Fujii (Kyushu)

Y. M. Cho (APCTP) (5) Welcome address

Il-T. Cheon (Yonsei) (5) Opening address

T. Motoba (Osaka E-C) (5) Opening address

T. Rijken (Nijmegen) (25+5)
The Nijmegen hyperon-nucleon and hyperon-hyperon interactions

Y. Kondo (Kyoto) (15+5)
Hyperon-nucleon scattering experiment (KEK-PS E289)

J. K. Ahn (RCNP) (15+5)
$\Xi^- p$ scattering and stopped-Ξ^- ^{12}C reactions

Session 19B

Chairman : **B. Gibson** (Los Alamos) *Secretary* : T. Yamada (Kanto Gakuin)

Y. Akaishi (KEK-Tanashi) (25+5)
Few-body systems with strangeness -1 and -2

E. Hiyama (RIKEN) (20+5)
Four-body calculation of ${}^4_\Lambda$H and ${}^4_\Lambda$He with realistic YN and NN Interactions

Y. Suzuki (Niigata) (25+5)
Quark Pauli principle in Λ-hypernuclear systems

February 19 (Friday) – Afternoon

Session 19C

Chairman : **J. Dabrowski** (Soltan, Warsaw) *Secretary* : J. K. Ahn (RCNP)

T.T.S. Kuo (SUNY) (25+5)
Two-frequency shell model for hypernuclei

Y. Yamamoto (Tsuru) (20+5)
Hypernuclear properties derived from G-matrix interactions

V. N. Fetisov (Lebedev) (20+5)
Estimate of influence of the ΛNN forces and the nuclear core size on the ${}^{7}_{\Lambda}$Li level spectrum

M. Youn (Houston) (15+5)
Hypernuclear spectroscopy for ${}^{12}_{\Lambda}$B through (stopped K^-, π^0) reaction using NMS

Session 19D

Chairman : **R. E. Chrien** (BNL) *Secretary* : E. Hiyama (RIKEN)

D. J. Millener (BNL) (25+5)
Structure of p-shell hypernuclei

M. Kamimura (Kyushu) (20+5)
Three-body model study of γ transition in A=7 hypernuclei

K. Tanida (Tokyo) (15+5)
Hypernuclear γ-ray spectroscopy experiments with germanium detectors

S. W. Hong (Sung Kyun Kwan) (15+5)
Continuum Tamm-Dancoff approximation calculation of Λ-hypernucleus

T. Nagae (KEK) (20+5)
Structure of ${}^{89}_{\Lambda}$Y measured with the SKS spectrometer

February 20 (Saturday) – Morning

Session 20A

Chairman : **E. V. Hungerford** (Houston) *Secretary* : T. Tamagawa (Tokyo)

O. Hashimoto (Tohoku) (25+5)
Reaction spectroscopy of Λ hypernuclei

R. E. Chrien (BNL) (25+5)
The story of Σ hypernuclei — a modern fable

J. Dabrowski (Warsaw) (20+5)
The isotopic spin dependent (Lane) part of the single-particle potential of the Σ hyperon in nuclear matter

Session 20B

Chairman : **O. Hashimoto** (Tohoku) *Secretary* : H. -C. Kim (Pusan)

T. Harada (Sapporo) (15+5)
Σ excitation in nuclei and its decay

I. Afnan (Flinders) (25+5)
The $\Lambda\Lambda - \Xi N$ coupling and the binding energy of ${}^{6}_{\Lambda\Lambda}$He

T. Yamada (Yokohama) (20+5)
Hyperon mixing in light $S{=}{-}2$ nuclei

T. Fukuda (KEK) (20+5)
Production of double Λ hypernuclei : BNL-AGS E906

February 20 (Saturday) – Afternoon

Session 20C (*in parallel with* Session 20D)

Chairman : **I. R. Afnan** (Flinders) *Secretary* : T. Tamagawa (Tokyo)

A. Ohnishi (Hokkaido) (20+5)
$\Lambda\Lambda$ interferometry in (K^-, K^+) and AA reactions

K. Yamamoto (Kyoto) (15+5)
Search for the H-dibaryon and $S=-2$ nuclei via the (K^-, K^+) reaction (BNL-AGS E885)

A. Ichikawa (Kyoto) (15+5)
Study of double-strangeness nuclei with hybrid-emulsion method (KEK-PS E373)

K. Nakazawa (Gifu) (15+5)
An experiment to study $S=-2$ nuclear system (KEK-PS E373)
— for fully automated scanning of Ξ^- tracks —

Session 20D (*in parallel with* Session 20C)

Chairman : **K. S. Sim** (Korea) *Secretary* : H. Takahashi (Kyoto)

T. Tatsumi (Kyoto) (20+5)
KN interaction in medium, chiral symmetry and kaon condensation

H.-C. Kim (Pusan) (15+5)
Strangeness in the nucleon in the SU(3) chiral quark-soliton model

N. Nakajima (Osaka) (15+5)
Quark recombination model for polarizations in inclusive hyperon productions at high energy

M. T. Jeong (Donshin) (15+5)
KNY coupling constants and form factors from the chiral bag model

T. Inoue (Tokyo) (15+5)
Weak production of Λ in the proton-neutron scattering

February 20 (Saturday) – Afternoon continued

Session 20E (*in parallel with* Session 20F)

Chairman : **T. T. S. Kuo** (SUNY, Stony Brook) *Secretary* : T. Inoue (Tokyo)

A. Sakaguchi (Osaka) (15+5)
Measurement of spin-orbit splitting in ${}^{13}_{\Lambda}C$ hypernucleus

S. Fujii (Kyushu) (15+5)
Energy-independent and hermitian ΛN effective interaction and Properties of Λ in hypernuclei

M. Kohno (Kyushu) (15+5)
Implication of the SU_6 quark model YN potential for Λ and Σ hypernuclear states

G. Garbarino (Torino) (20+5)
Decay rates of medium-heavy Λ-hypernuclei within the propagator method

K. Itonaga (Miyazaki) (20+5)
Hypernuclear nonmesonic decay and $\Delta I = 1/2$ and $\Delta I = 3/2$ interactions in the meson-exchange model

Session 20F (*in parallel with* Session 20E)

Chairman : **V. Fetisov** (Lebedev, Moscow) *Secretary* : M. T. Jeong (Dongshin)

M. Fujiwara (RCNP/JAERI) (20+5)
Laser-electron-photon project at SPring-8; where do we stand and where do we go?

W. Kim (Kyungpook) (15+5)
Longitudinal and transverse cross sections in ${}^1H(e,e'K^+)\Lambda$ reaction

M. K. Cheoun (Yonsei) (15+5)
The Kaon photo-production near threshold

B. S. Han (Yonsei) (15+5)
Kaon photoproduction on the nucleon up to $E_\gamma^{\rm Lab} = 2.1\text{GeV}$

T. Koike (KEK-Tanashi) (15+5)
Cascade calculation of exotic helium atoms — s-orbit vs. p-orbit absorption rates —

February 21 (Sunday) – Morning

Session 21A

Chairman : **K. Ikeda** (Niigata) *Secretary* : M. K. Cheoun (Yonsei)

B. Gibson, (Los Alamos) (25+5)
The π^+ decay of light hypernuclei

H. Bhang (Seoul) (20+5)
Mass dependence of lifetimes of Λ-hypernuclei

M. Oka (Tokyo) (25+5)
Weak decay of Λ in nuclei: Direct quark mechanism vs meson exchanges

Session 21B

Chairman : **K. S. Lee** (Chonnam) *Secretary* : T. Koike (KEK-Tanashi)

J. H. Jun (Chung-Ang) (20+5)
Four-baryon point interactions for non-mesonic weak decay of hypernuclei

T. Takatsuka (Iwate) (25+5)
Λ-hyperon superfluidity in neutron stars and in hyperonic nuclear matter

D. E. Lanskoy (Moscow) (20+5)
Structure of multi-strange nuclear matter

February 22 (Monday) – Morning

Session 22A

Chairman : **T. Fukuda** (KEK) *Secretary* : B. S. Han (Yonsei)

E. Hungerford (Houston) (25+5)
The electroproduction of light hypernuclei

E. Ya. Paryev (Moscow) (20+5)
Subthreshold and near threshold K^+ meson production on light nuclei by protons

R. Bertini (Torino) (25+5)
Spin observables in hyperon production

Session 22B

Chairman : **K. S. Lee** (Chonnam) *Secretary* : K. S. Kim (Yonsei)

Il-T. Cheon (Yonsei) (25+5)
Quark description of interaction for pionic decay of Λ particle

T. Bressani (Torino) (25+5)
The FINUDA experiment at DAΦNE

L. Tang (Hampton) (20+5)
Direct measurement of the lifetime of heavy hypernuclei at CEBAF using low pressure LPMWPC Technique

February 22 (Monday) – Afternoon

Session 22C

Chairman : **T. Bressani** (INFN Torino)   *Secretary* : K. Tanida (Tokyo)

D. Grzonka/K. Kilian, (Forschungszentrum) (25+5)
Strangeness experiments at COSY

H. Tamura (Tohoku) (20+5)
Future plans of high-resolution hypernuclear gamma-spectroscopy

K. Imai (Kyoto) (25+5)
Prospects of Japan Hadron Facility

K. Yazaki (Tokyo) (25+5)
Concluding remarks

B. Bhang (Seoul)
Closing

List of Participants

Iraj R. Afnan
(iraj.afnan@flinders.edu.au)
Flinders University
Adelaide 5001, Australia

Jung Keun Ahn
(jkahn@ne.scphy.kyoto-u.ac.jp)
RCNP, Osaka University
Osaka 567-0047, Japan

Y. Akaishi
(akaishi@tanashi.kek.jp)
KEK-Tanashi
Tanashi, Tokyo 188, Japan

K. M. Baek
Korea University
Seoul 136-701, Korea

Raimondo Bertini
(bertini@to.infn.it)
Inst. A. Avogadro and INFN
10121 Torino, Italy

Hyoung C. Bhang
(bhang@phya.snu.ac.kr)
Seoul National University
Seoul 151-742, Korea

Tullio Bressani
(bressani@to.infn.it)
INFN
10125 Torino, Italy

Daniela Calvo
(calvo@to.infn.it)
INFN
10125 Torino, Italy

Il-Tong Cheon
(itcheon@phya.snu.ac.kr)
Yonsei University
Seoul 120-749, Korea

Myung Ki Cheoun
(cheoun@phya.yonsei.ac.kr)
Yonsei University
Seoul 120-749, Korea

C. H. Chin
National Chiao Tung University
Taiwan

Y. M. Cho
(ymcho@apctp.kaist.ac.kr)
APCTP/Seoul National University
Seoul 151-742, Korea

Robert E. Chrien
(chrien@bnl.gov)
Brookhaven National Lab.
Upton, New York 11973 USA.

Janusz Dabrowski
(dabrnucl@fuw.edu.pl)
Soltan Inst. for Nuclear Problems
00-681 Warsaw, Poland

Vladimir Fetisov
(fetisov@sci.lebedev.ru)
Lebedev Physical Inst.
Moscow 117924, Russia

Shinichiro Fujii
(fuji2scp@mbox.nc.kyushu-u.ac.jp)
Kyushu University
Fukuoka 812-8581, Japan

Hisako Fujimura
(fujimura@rcnp.osaka-u.ac.jp)
RCNP, Osaka University
Osaka 567-0047, Japan

Mamoru Fujiwara
(fujiwara@rcnp.osaka-u.ac.jp)
RCNP, Osaka University / JAERI
Osaka 567-0047, Japan

Tomokazu Fukuda
(fukuda@jhfpc1.kek.jp)
KEK
Tsukuba, Ibaraki 305-0801, Japan

Gianni Garbarino
(garbarino@to.infn.it)
University di Torino
Torino, Italy

Benjamin Gibson
(gibson@t5.lanl.gov)
Los Alamos National Lab.
Los Alamos, New Mexico 87545, USA

D. Grzonka
(d.grzonka@fz-juelich.de)
IKP1, Forschungszentrum
D-52425 Juelich, Germany

Bong Soo Han
(bshan@phya.yonsei.ac.kr)
Yonsei Uni.
Seoul 120-749, Korea

I. S. Han
Ewha Women's University
Seoul 120-750, Korea

Toru Harada
(harada@earth.sgu.ac.jp)
Sapporo Gakuin University
Hokkaido 069-8555, Japan

Osamu Hashimoto
(hashimoto@lambda.phys.tohoku.ac.jp)
Tohoku University
Sendai 980-8578, Japan

Emiko Hiyama
(hiyama@postman.riken.go.jp)
RIKEN
Wako, Saitama 351-01, Japan

Byungsik Hong
(bhong@kuccnx.korea.ac.kr)
Korea University
Seoul 136-701, Korea

Seung-Woo Hong
(swhong@skku.ac.kr)
Sung Kyun Kwan University
Suwon 440-746, Korea

Hideaki Hotchi
(hotchi@kekvax.kek.jp)
University of Tokyo
Tokyo 113, Japan

Ed V. Hungerford
(Hunger@UH.EDU)
University of Houston
Houston, Texas 77204

Felice Iazzi
(iazzi@iazzi.polito.it)
Politecnico di Torino
10129 Torino, Italy

Atsuko Ichikawa
(ichikawa@ne.scphys.kyoto-u.ac.jp)
Kyoto University
Kyoto 606-8502, Japan

Kiyomi Ikeda *(ikeda@nt.sc.niigata-u.ac.jp)*	Niigata University Niigata 950-2181, Japan
Kenichi Imai *(imai@ne.scphys.kyoto-u.ac.jp)*	Kyoto University Kyoto 606-8502, Japan
Takashi Inoue *(inoue@nt.phys.s.u-tokyo.ac.jp)*	University of Tokyo Bunkyo, Tokyo 113, Japan
Kazunori Itonaga *(itonaga@physics.miyazaki-med.ac.jp)*	Miyazaki Medical College Miyazaki 889-1692, Japan
Moon Taeg Jeong *(mjeong@dsweb.dongshinu.ac.kr)*	Dongshin University Naju 520-714, Korea
Jung-Hwan Jun *(jhjun@cau.ac.kr)*	Chung-Ang University Seoul 156-756, Korea
Masayasu Kamimura *(kami2scp@mbox.nc.kyushu-u.ac.jp)*	Kyusu University Fukuoka 812, Japan
Kurt Kilian *(k.kilian@fz-juelich.de)*	Forschungszentrum D 52425 Juelich, Germany
Byung-taik Kim *(btkim@skku.ac.kr)*	Sung Kyun Kwan University Suwon 440-746, Korea
Hungchong Kim *(hckim@th.phys.titech.ac.jp)*	Tokyo Inst. of Technology Tokyo, Japan
Hyun-Chul Kim *(hchkim@hyowon.pusan.ac.kr)*	Pusan National University Pusan 609-735, Korea
J. C. Kim *(jckim@phya.snu.ac.kr)*	Seoul National University Seoul 151-742, Korea
J. Y. Kim	Seoul National University Seoul 151-742, Korea
S. H. Kim	Kyungsang University Jinju, Korea
Y. D. Kim *(ydkim@kunja.sejong.ac.kr)*	Sejong University Seoul 143-747, Korea
Wooyoung Kim *(wooyoung@bh.kyungpook.ac.kr)*	Kyungpook National University Taegu 702-701, Korea
Michio Kohno *(kohno@kyu-dent.ac.jp)*	Kyushu Dental College Kitakyushu, Japan

Takahisa Koike
(tkoike@tanashi.kek.jp)
KEK-Tanashi
Tanashi, Tokyo 188, Japan

Yasuhiro Kondo
(kondo@ne.scphys.kyoto-u.ac.jp)
Kyoto University
Kyoto 606-8502, Japan

Thomas T. S. Kuo
(kuo@nuclear.physics.sunysb.edu)
State University of New York
Stony Brook, New York 11794-3800

Dmitry Lanskoy
(lanskoy@npi.msu.su)
Moscow State University
Moscow 119899, Russia

Do-Hee Lee
(dhlee@nature.skku.ac.kr)
Sung Kyun Kwan University
Suwon 440-746, Korea

Jeonghan Lee
(citadel@hanmail.net)
Pusan National University
Pusan 609-735, Korea

K. S. Lee
(kslee@chonnam.chonnam.ac.kr)
Chonnam National University
Kwangju 500-757, Korea

S. E. Lee
(seungen@color.skku.ac.kr)
Sung Kyun Kwan University
Suwon 440-746, Korea

John Millener
(millener@bnl.gov)
Brookhaven National Lab.
Upton, New York 11973

Dong-Pil Min
(dpmin@phya.snu.ac.kr)
Seoul National University
Seoul 151-742, Korea

Toshio Motoba
(motoba@isc.osakac.ac.jp)
Osaka Electro-Communication University
Osaka 572-8530, Japan

Tomofumi Nagae
(nagae@mail.kek.jp)
KEK
Tsukuba, Ibaraki 305-0801, Japan

N. Nakajima
(nakaji@miho.rcnp.osaka-u.ac.jp)
RCNP, Osaka University
Osaka 567-0047, Japan

Kazuma Nakazawa
(nakazawa@kekps.kek.jp)
Gifu University
Gifu 501-1193, Japan

Akira Ohnishi
(ohnishi@nucl.sci.hokudai.ac.jp)
Hokkaido University
Sapporo 060-0810, Japan

Makoto Oka
(oka@th.phys.titech.ac.jp)
Tokyo Inst. of Technology
Tokyo 152-8551, Japan

Haruhiko Outa
(outa@kekvax.kek.jp)
KEK
Tsukuba Ibaraki 305-0801, Japan

Eduard Paryev *(paryev@al20.inr.troitsk.ru)*	Inst. for Nuclear Research Moscow 117312, Russia
H. C. Ra	Jung-Ang University Seoul 156-756, Korea
H. S. Roh *(hroh@nature.skku.ac.kr)*	Sung Kyun Kwan University Suwon 440-746, Korea
Thomas A. Rijken *(rijken@tanashi.kek.jp)*	University of Nijmegen/KEK-Tanashi Nijmegen, Netherland
Chaiho Rim *(rim@apctp.kaist.ac.kr)*	APCTP Seoul , Korea
Chung-yeol Ryu *(cyryu@color.skku.ac.kr)*	Sung Kyun Kwan University Suwon 440-746, Korea
Atsushi Sakaguchi *(sakaguch@phys.sci.osaka-u.ac.jp)*	Osaka University Osaka 560-0043, Japan
Michiko Sekimoto *(sekimoto@post.kek.jp)*	KEK Tsukuba, Ibaraki 305-0801, Japan
Kwang-Souk Sim *(kssim@kucc08.korea.ac.kr)*	Korea University Seoul 136-701, Korea
S. I. Shim *(shim@knu.kongju.ac.kr)*	Kongju University Kongju 314-701, Korea
Chaejun Song *(chaejun@fire.snu.ac.kr)*	Seoul National University Seoul 151-742, Korea
W. Y. So *(wyso@color.skku.ac.kr)*	Sejong University Seoul 143-747, Korea
Yasuyuki Suzuki *(suzuki@nt.sc.niigata-u.ac.jp)*	Niigata University Niigata 950-2181, Japan
Hitoshi Takahashi *(thitoshi@ne.scphys.kyoto-u.ac.jp)*	Kyoto University Kyoto 606-8502, Japan
Miho Takahashi *(miho@econ.niigata-u.ac.jp)*	Niigata University Niigata 950-2181, Japan
Tatsuyuki Takatsuka *(takatuka@iwate-u.ac.jp)*	Iwate University Morioka 020-0066, Japan
Toru Tamagawa *(tamagawa@jhfpc1.kek.jp)*	University of Tokyo Tokyo 113, Japan

H. Tamura
(tamura@lambda.phys.tohoku.ac.jp)
Tohuko University
Sendai 980-8578, Japan

Liguang Tang
(tangl@jlab.org)
Hampton University
Hampton, VA 23668, USA

Kiyoshi Tanida
(tanida@nucl.phys.s.u-tokyo.ac.jp)
University of Tokyo
Tokyo 113, Japan

T. Tatsumi
(tatsumi@ruby.scphys.kyoto-u.ac.jp)
Kyoto University
Kyoto 606-8502, Japan

Noriyuki Ushida
(nushida@auecc.aichi-edu.ac.jp)
Aichi University of Education
Aichi 448-8542, Japan

Taiichi Yamada
(yamada@kanto-gakuin.ac.jp)
Kanto Gakuin University
Yokohama 236-8501, Japan

Kazuhiro Yamamoto
(yamamoto@ne.scphys.kyoto-u.ac.jp)
Kyoto University
Kyoto 606-8502, Japan

Yasuo Yamamoto
(yamamoto@tsuru.ac.jp)
Tsuru University
Yamanashi 402-0054, Japan

Ghil-Seok Yang
(ghsyang@hanmail.net)
Pusan National University
Pusan 609-735, Korea

K. Yazaki
(yazaki@phys.s.u-tokyo.ac.jp)
University of Tokyo
Tokyo 113, Japan

J. H. Yoon
(jinyoon@dragon.inha.ac.kr)
Inha University
Incheon 402-751, Korea

Minyoung Youn
(myoun@bnl.gov)
University of Houton
BNL Upton NY 11973

Byung Geel Yu
(gyu@mail.hangkong.ac.kr)
Hankuk Aviation University
Koyang 412-791, Korea